T0073747

SYSTEMS DEVELOPMENT METHODS FOR DATABASES, ENTERPRISE MODELING, AND WORKFLOW MANAGEMENT

SYSTEMS DEVELOPMENT METHODS FOR DATABASES, ENTERPRISE MODELING, AND WORKFLOW MANAGEMENT

Edited by

Wita Wojtkowski
W. Gregory Wojtkowski
Boise State University
Boise, Idaho

Staniław Wrycza
University of Gdansk
Gdansk, Poland

and

Jože Zupančič
University of Marbor
Kranj, Slovenia

Kluwer Academic / Plenum Publishers
New York, Boston, Dordrecht, London, Moscow

Library of Congress Cataloging-in-Publication Data

Systems development methods for databases, enterprise modeling, and workflow
management/edited by Wita Wojtkowski ... [et al.].
  p. cm.
Result of the ISD'99, Eighth International Conference on Information Systems
Development: Methods and Tools, Theory and Practice held August 11–13, 1999.
Includes bibliographical references.
ISBN 0-306-46299-0
   I. System design—Congresses.   2. Computer software—Development—Congresses.   I.
Wojtkowski, Wita, 1944–   II. International Conference on Information Systems and
Development: Methods and Tools, Theory, and Practice (8th: 1999; Boise, Idaho)

QA76.9.s88 s977 1999
004.2'1—dc21
                                                                 99-049588

ISBN 0-306-46299-0

Proceedings of the Eighth International Conference on Information System Development: Methods and
Tools, Theory and Practice, held August 11–13, 1999, in Boise, Idaho

©1999 Kluwer Academic / Plenum Publishers, New York
233 Spring Street, New York, N.Y. 10013

http://www.wkap.com

Printed in the United States of America

PREFACE

This book is a result of the ISD'99, Eight International Conference on Information Systems Development-Methods and Tools, Theory, and Practice held August 11-13, 1999 in Boise, Idaho, USA. The purpose of this conference was to address the issues facing academia and industry when specifying, developing, managing, and improving information systems.
ISD'99 consisted not only of the technical program represented in these Proceedings, but also of plenary sessions on product support and content management systems for the Internet environment, workshop on a new paradigm for successful acquisition of information systems, and a panel discussion on current pedagogical issues in systems analysis and design.

The selection of papers for ISD'99 was carried out by the International Program Committee. Papers presented during the conference and printed in this volume have been selected from submissions after formal double-blind reviewing process and have been revised by their authors based on the recommendations of reviewers. Papers were judged according to their originality, relevance, and presentation quality. All papers were judged purely on their own merits, independently of other submissions.

We would like to thank the authors of papers accepted for ISD'99 who all made gallant efforts to provide us with electronic copies of their manuscripts conforming to common guidelines. We thank them for thoughtfully responding to reviewers comments and carefully preparing their final contributions.

We thank Daryl Jones, provost of Boise State University and William Lathen, dean, College of Business and Economics, for their support and encouragement.

ISD Conference was brought into existence over ten years ago and it continues fine tradition of the first Polish-Scandinavian Seminar on Current Trends in Information Systems Development Methodologies, held in Gdansk, Poland in 1988. ISD'97 and ISD'99 met in USA, in Boise, Idaho. ISD'98 was held in Slovenia, in Lake Bled. ISD'2000 Conference will return to Scandinavia and we will gather in Kristiansand, Norway.

August 1999, Boise, Idaho

Wita Wojtkowski
W. Gregory Wojtkowski
Stanisław Wrycza
Jože Zupančič

ORGANIZING COMMITTEE--CO-CHAIRMEN

Wita Wojtkowski	Boise State University	(USA)
Gregory Wojtkowski	Boise State University	(USA)
Stanisław Wrycza	University of Gdansk	(Poland)
Jože Zupančič	University of Maribor	(Slovenia)

The co-chairmen express their thanks to the International Program Committee. They provided authors of submitted manuscripts with many useful recommendations and detailed comments on how to make improvements to the manuscripts. Their contribution was essential.

INTERNATIONAL PROGRAM COMMITTEE

Witold Abramowicz	Economic University Poznan	(Poland)
Gary Allen	University of Huddersfield	(UK)
Evelyn Andreewsky	INSERM-TLNP	(France)
Josie Arnold	Swinburn University of Technology	(Australia)
Hans Aus	University of Wurzburg	(Germany)
Andrzej Baborsk	Wroclaw University of Economics	(Poland)
Susan Balint	University of Paisley	(UK)
Sjaak Brinkemper	Baan Company	(Netherlands)
Daniele Bourcier	University of Paris	(France)
Yves Dennebouy	Ecole Polytechnique Federale de Luasanne	(Switzerland)
Oscar Diaz	Universitad del Pais Vasco	(Spain)
Edwin Gray	Glasgow Caledonian University	(UK)
Igor Hawryszkiewycz	University of Technology, Sydney	(Australia)
Alfred Helmerich	Research Institute for Applied Technology	(Germany)
Lech Janczewski	University of Auckland	(New Zealand
Marius Janson	University of Missouri, St. Louis	(USA)
Jens Kaasbol	University of Oslo	(Norway)
Marite Kirikova	Riga Technical University	(Latvia)
Robert Leskovar	University of Maribor	(Slovenia)
Henry Linger	Monash University	(Australia)
Leszek Maciaszek	Macquarie University	(Australia)
Heinrich Mayr	University of Klagenfurt	(Germany)
Emerson Maxson	Boise State University	(USA)
Elisabeth Metais	Universite de Versailles	(France)
Murli Nagasundaram	Boise State University	(USA)

Anders Nilsson	Stockholm School of Economics	(Sweden)
Eugene Ovsyannikov	The Academy of Sciences	(Russia)
Tore Ørvik	Agder University College	(Norway)
Jaroslav Pokorny	Charles University, Prague	(Czech Republic)
Jan Pour	Prague University of Economics	(Czech Republic)
Stephen Probert	University of London	(UK)
Vaclav Repa	Prague University of Economics	(Czech Republic)
Moung K. Sein	Agder University College	(Norway)
Roland Stamper	University Twente	(Netherlands)
Eberhard Stickel	Europa-Universitat, Frankfurt	(Germany)
Sharon Tabor	Boise State University	(USA)
Jacek Unold	Wroclaw University of Economics	(Poland)
Douglas Vogel	City University of Hong Kong	(Hong Kong)
Alexander Zak	Jet Propulsion Laboratory	(USA)
Josef Zurada	University of Louisville	(USA)

CONTENTS

STATE DEPENDENT BEHAVIORAL ASPECTS OF OBJECTS: 1
AUTOMATIC SYNTHESIS
      Rubina Polovina

A CORBA COMPONENT MODELING METHOD BASED ON   31
SOFTWARE ARCHITECTURE
      Chang-Joo Moon, Sun-jung Lee, and Doo-Kwon Baik

DYNAMIC DATABASE OBJECT HORIZONTAL FRAGMENTATION   51
      C. I. Ezeife and Jian Zheng

AN ELECTRONIC EKD METHOD GUIDE OVER THE WEB   61
      Judith Barrios, Christophe Gnaho, and Georges Grosz

CASE-BASED REASONING, GENETIC ALGORITHMS, AND THE PILE   77
FOUNDATION INFORMATION SYSTEM
      Otakar Babka, Celestino Lei, and Laurinda A. G. Garanito

JAVA, WEB, AND DISTRIBUTED OBJECTS   87
      Yin-Wah Chiou

A DATA REGISTRY-BASED ENVIRONMENT FOR SHARING   97
XML DOCUMENTS
      Hong-Seok Na, Jin-Seok Chae, and Doo-Kwon Baik

EVALUATION OF A METHODOLOGY FOR CONSTRUCTION   109
OF TMN AGENTS IN THE ASPECT OF COST AND PERFORMANCE
      Soo-Hyun Park and Doo-Kwon Baik

A TAXONOMY OF STRATEGIES FOR DEVELOPING SPATIAL   139
DECISION SUPPORT SYSTEM
      Silvio Luis Rafaeli Neto and Marcos Rodrigues

UNDERSTANDING IS STRATEGIES IMPLEMENTATION: 157
A PROCESS THEORY BASED FRAMEWORK
      Carl Erik Moe, Hallgeir Nilsen, and Tore U. Ørvik

ON METHOD SUPPORT FOR DEVELOPING PRE-USAGE 169
EVALUATION FRAMEWORKS FOR CASE-TOOLS
    Björn Lundell and Brian Lings

ASSISTING ASYNCHRONOUS SOFTWARE INSPECTION BY 183
AUTO-COLLATION OF DEFECT LISTS
    Fraser Macdonald, James Miller, and John D. Ferguson

META-MODELING: 199
THEORY AND PRACTICAL IMPLICATIONS
    Milan Drbohlav

CRITICAL SYSTEMS THINKING AND THE DEVELOPMENT 209
OF INFORMATION TECHNOLOGY IN ORGANISATIONS:
ANOTHER PERSPECTIVE
    Diego Ricardo Torres Martinez

INFORMATION SYSTEMS AND PROCESS ORIENTATION: 233
EVALUATION AND CHANGE USING BUSINESS ACTIONS THEORY
    Ulf Melin and Goran Goldkuhl

TOWARDS THE SHIFT-FREE INTEGRATION OF 245
HARD AND SOFT IS METHODS
    Stephen K. Probert and Athena Rogers

IMPROVED SOFTWARE QUALITY THROUGH REVIEWS 257
AND DEFECT PREVENTION:
AN INCREMENTAL IMPROVEMENT PATH
    Edwin M. Gary, Oddur Benediktsson, and Warren Smith

IDENTIFYING SUCCESS IN INFORMATION SYSTEMS: 263
IMPLICATIONS FOR DEVELOPMENT METHODOLOGIES
    Steve Page

RESEARCHING ORGANISATIONAL MEMORY 279
    Frada Burstein and Henry Linger

DATA MINING USING NEURAL NETWORKS 299
AND STATISTICAL TECHNIQUES:
A COMPARISON
    Jozef Zurada and Al F. Salam

THE YEAR 2000 PROBLEM IN SMALL COMPANIES IN SLOVENIA 313
    Borut Verber, Uroš Jere, and Jože Zupančič

USING DISCOURSE ANALYSIS TO UNDERSTAND IS PLANNING 327
AND DEVELOPMENT IN ORGANIZATION:
A CASE STUDY
    John A. A. Sillince and G. Harindranath

INFORMATION SYSTEM DEVELOPMENT METHODOLOGY: 337
THE BPR CHALLENGE
    Vaclav Repa

PERVASIVE IT SYSTEM: 349
ENSURING A POSITIVE CONTRIBUTION TO THE BOTTOM LINE
    Robert Moreton

THE CHALLENGE: 357
PROVIDING AN INFORMATION SYSTEM FOR A CLINICAL LABORATORY
    H. M. Aus, M. Haucke, and A. Steimer

THE CONCEPT OF IMPROVING AND DEVELOPING AN INFORMATION 367
SYSTEM FOR THE NEEDS OF MANAGING A POLISH ENTERPRISE
    Adam Nowicki

THE IMPLICATIONS OF CHANGING ORGANISATIONAL STRUCTURES 375
FOR THE FUTURE OF THE INFORMATION SYSTEM (IS) FUNCTION
    Robert Moreton and Myrvin Chester

MIDDLEWARE ORIENTATION: 385
INVERSE SOFTWARE DEVELOPMENT STRATEGY
    Jaroslav Kral

SELECTED DETERMINANTS OF APPLYING MULTIMEDIA 397
TECHNIQUES FOR DESIGNING DECISION SUPPORT SYSTEMS
    Celina M. Olszak

TOWARD EFFECTIVE RISK MANAGEMENT IN IS DEVELOPMENT: 403
INTRODUCTION TO RISK BASED SCORECARD
    Piotr Krawczyk

HYPERMEDIA BASED DISTANCE LEARNING SYSTEMS 417
AND ITS APPLICATIONS
    Christian Andreas Schumann, Karsten Wagner, Andrey Oleg Luntovskiy,
    Eugene K. Ovsyannikov, and Elena M. Osipova

MULTI-MEDIA INSTRUCTIONAL SUPPORT: 431
STATUS, ISSUES, AND TRENDS
    Doug Vogel, Johanna Klassen, and Derrick Stone

ELECTRONIC DELIVERY OF CURRICULUM: 441
PREPRODUCTION OF CYBERSCRIPTS
    Josie Arnold and Kitty Vigo

A NEW PARADIGM FOR SUCCESSFUL ACQUISITION 455
OF INFORMATION SYSTEMS
    Michael Gorman

TOTAL QUALITY MEASUREMENT: 465
ISSUES AND METHODS FOR MEASURING PRODUCTIVITY FOR
INFORMATION SYSTEMS
          Richard McCarthy

MUSING ON WIENER'S CYBERNETICS AND ITS APPLICATIONS IN   479
THE INFORMATION PROCESSES ANALYSIS
          Jacek Unold

"MAN IS A CREATURE MADE AT THE END OF THE WEEK...   487
WHEN GOD WAS TIRED":
SOME REFLECTIONS ON THE IMPACT OF HUMAN ERROR UPON
INFORMATION SYSTEMS
          George J. Bakehouse

AUTHOR INDEX   495

STATE DEPENDENT BEHAVIORAL ASPECTS OF OBJECTS:
AUTOMATIC SYNTHESIS

Rubina Polovina
SOLECT, 55 University Avenue, Suite 200
Toronto, Ontario M5J 2L2 CANADA

INTRODUCTION

This paper shows the benefits of the formal approach in the OO analysis. A use case is a piece of functionality in the system that provides a certain result to the user of the system. Event flows of use cases are specified by a set of scenarios. Each scenario represents a sequence of events in the use case. Many sequences are possible. If scenarios are formalized, then they may be used for automatic synthesis (in the limited way) of state dependent behavioral aspects of objects that participate in the use cases. The technique is reduced to inference of Finite Automata from the given (positive) sample of strings. The given Sequence Diagrams are used to extract the strings of interaction items among the participants. An interaction item consists of an event noticed by the object, and a string of events that are invoked by that object in a response. The obtained strings of the interaction items of an object constitute the sample. This sample serves as input in the Muggleton's KC-algorithm (1990) that infers the finite automaton. Input in the KC algorithm is slightly modified to synthesize a finite translation automaton as the result. The described technique is from the domain of Inductive Logic Programming.

SYSTEM EVENTS

The objects of the system operate in synergy to fulfill the functionality required from the system. An *event* has been defined as something that is happening at a point in time (Rumbaugh, 1991). We have found relevant events of receiving and sending messages, expiration of conditions, and events caused by the system clock.

Events exist only within the time they occur and have no duration. We know that nothing is really instantaneous. Therefore, we may say that events actually consist of micro-events or actions. For example, in the event of interchanging a message, there are at least two micro-events of sending the message by one object, and receiving the same message by another object. Time difference between those micro-events is not considered.

Dynamic relationships (communication) of the objects in the system will be expressed in terms of *events* and *event sequences*.

Systems Development Methods for Databases, Enterprise Modeling, and Workflow Management,
Edited by W. Wojtkowski, *et al.* Kluwer Academic/Plenum Publishing, New York, 1999.

1

Let O_S denotes all objects of the system S. Let E_S be a finite set of all *events* that may happen in the system S. The system S has two special events – *default* event $\alpha \in E_S$ considered to occur at any time and *null* event $\varepsilon \in E_S$ that never occurs.

Let seq_S be a finite *event sequence* drawn from the set of all event E_S of the system S:

$$seq_S = e_1\, e_2 ... e_n \qquad\qquad e_1,\, e_2,\, ...,\, e_n \in E_S$$

Let *EventSeq_S* denote a finite set of all event sequences that may happen in the system S:

$$EventSeq_S = \{seq_S^1,\, seq_S^2, ...,\, seq_S^i\}$$

If an event $e \in E_S$ is an atomic unit of communication in the system, then an event sequence $seq_S \in EventSeq_S$ determines a "chunk" of intercommunication in which some functionality required from the system will be executed. In execution time, the objects contribute as a whole to the proper execution of the functionality expected from the system.

SEQUENCE DIAGRAMS

There are many ways to specify an interaction between objects (Muller, 1997), but one of the most popular techniques between modelers is to use sequence diagrams. They are particularly well suited for the representation of complex interaction. Because of their self-explanatory qualities and user-friendliness, sequence diagrams are widely used to model object dynamics. Most of the modelers would rather escape finite automata, and base dynamic models on sequence diagrams (Koskimies, Makinen, 1994).

A sequence diagram consists of vertical and horizontal bars. Vertical bars correspond to objects collaborating. In our example (figures 1to 11), the objects are the *Automatic Teller Machine* (*ATM*), *Account, Bank* and *Client*. Horizontal bars represent events. The arrows are used in a manner that expresses causality. Obviously, an arrow pointing to the vertical bar represents registration of the event by the corresponding object. This is an input event that might invoke the reaction of that object. An arrow pointing from the vertical bar denotes that the corresponding object invokes the event.

We believe that object dynamics are not recognized as an inherent object quality , and generally, little is done in the investigation and improvement of dynamic models. We will mention four positive examples in OO methodologies that use a kind of sequence diagrams. *Interaction diagrams, trace diagrams* and *sequence diagrams* are very much the same. The difference is in their content and the way they are used, as described further in the text.

A *use case* consists of event sequences called scenarios. They determine interaction with the system from the perspective of the actors. The object that has a special role in the interaction with the system is called an *actor*. A use case corresponds with the event sequences that provide completion of certain functionality required from the system. Observing and specifying actor by actor determines use cases, the event sequences as observed from the application domain. The *application domain* is immediate surroundings of the system, or the context in which the target system is to be used. An application domain includes users, and the work they carry out and to which the target system is requested.

Use cases in the analysis phase of Jacobson's OOSE (1992) are only distinguished as abstractions, assuming that scenarios will be determined later. Even

2

interaction diagrams that are used later to take further into the system, assumes that an object design exists, so that the interaction diagrams may specify what goes on in each scenario. Also inter-user perspectives are not encouraged and supported.

Rumbaugh used event *trace diagrams* in his OMT (1991), to do something of the same sort. The main difference is that, specific objects were in the center of the attention, rather then the actors or usage as in Jacobson's OOSE (1992). However, OMT's (1991) activity of transforming information from trace diagrams into a state model could be of interest as shifting a view from the proper cooperation between objects to an intra-object issue.

Unfortunately, these authors used the described diagrams only as descriptive tools. Similarly is in Unified Modeling Process and UML. Although they are not considered for mandatory aspect of the analysis model (Jacobson, Booch, Rumbaugh, 1999), sequence diagrams may be used in the analysis phase (Muller, 1997). They are introduced descriptively within the modeling (analysis) stage and prior to the arrows corresponding to a message being sent, the difference between control flows and data flows are not generally established.

Fusion (Coleman, *et al.*)] uses a kind of sequence diagram to specify some examples of the possible interaction with the system. Unfortunately, the modeler does not have formal support to induce state dependent behavior of the system. Instead, the modeler must generalize scenarios himself, and transform them into regular expressions. Those regular expressions specify sequences of events that may be received or responded to by the system. These are called life-cycle models. Later in the implementation phase, the determined regular expressions will be automatically transformed into finite automata. Transformation of regular expressions to finite automata is described by many authors such as Aho and Ullman (1987). Automata results from the technique described in the scope of Fusion (Coleman, *et al.*) This methodology does not consider events generated by the system, which means that they miss causality that links noticed events and those that the system invoked in response.

Stephen Muggleton (1990) investigated techniques for constructing finite automata from a sample of a language. His work is based upon grammatical induction where the grammar is discovered from sample sentences. The author presented an excellent survey covering several algorithms that synthesize finite automata from positive samples. The general algorithm is defined so it may use different heuristics. The outputs differ depending on used heuristics. Such an approach (of using heuristics) required neither an *ad hoc* numerical measure, nor the need for negative data.

Muggleton (1990) also proposed new heuristics and appropriate algorithms. One of them, which is denoted as the KC algorithm, infers languages that the author determined as k-contextual languages. It has been shown as the most appropriate when applied to our problem domain. This algorithm and the associated formal concepts are described in full detail in the further text.

Our technique uses sequence diagrams to extract sequences of *interaction items*. Interaction items are precisely defined pairs of noticed events, and an event sequence generated in response. Semantically, the noticed event causes the reaction. A reaction is (among the other internal changes that are happening in the object) the generation of an event sequence.

For example, the interaction between a client and an Automatic Teller Machine (ATM) starts when the client inserts his cash card. The card slot of the ATM's interface raises an event - *Card Inserted*. That is the event that invokes the *ATM* object, which controls the ATM. The *ATM* object reacts by asking the client to enter his password. The *ATM* object sends a message to the display device and a short message that reads *"Enter your password!"* will appear on the display. Thus, this interaction item consists of the event *Card Inserted*, which invokes the *ATM* object to respond by asking for client's password:

Card Inserted / Display *(*Enter your password!*)*

A slash "/" marks the end of the noticed event, and an asterisk "*" denotes comments.

In the interaction item that follows, the client has typed his password, and the *ATM* interface has sent the typed password to the *ATM* object. In turn, the *Password* event caused the *ATM* object to perform two actions. First, it informed the client that account verification has started. Second, it sent the necessary data to account verification to another object that is responsible for verification of the clients' accounts. Thus, this interaction item consists of the event that invokes the *ATM* object – the *Password* event, and a sequence of the two events generated in response – *Display; Verify Account.*

Password / Display *(*Account verification*)*; Verify Account

A semicolon ";" separates events in the generated event sequence.

Sequences of interaction items of an object are called *interaction information*. *Interaction information* enables the modeler to infer the following aspects of the finite translation automaton:

- unlabeled states (or at least states whose labels have limited or no semantic meaning),
- sets of events that the object notices and invokes (relevant for the interaction in which the object participates).

Sequence diagrams represent the order in which the events occur and the links between them. That is unfortunately, insufficient for the inference of complete object specifications including attributes and rules for their evaluation. On the other hand, it is enough to determine in the limit state dependent behavior of the object.

Observed internally, events cause an object to change states. The environment notices those changes in behavior. Sequence diagrams contain information about events but not about the intra-object states and transitions. For example, a transition is associated with attribute changes, semantic rules for evaluation of attribute values, dependencies between noticed and generated events, etc. Sequence diagrams do not contain that information but only the information about the chronology and causality relevant for the interaction in the object group. Since intra-object changes (i.e., those concerned attribute values) are not specified in sequence diagrams, states are too abstract to be synthesized automatically. Their labels have semantic meaning for the modeler that cannot be synthesized from sequence diagrams only.

Interaction Information

Semantically, each sequence diagram represents a finite sequence of events happening in chronological order. A vertical bar represents each object. Time elapsed appears from top to bottom and horizontal bars represent events.

$$SD = e_1\, e_2 \ldots e_n \quad \text{where} \quad e_1,\, e_2, \ldots,\, e_n \in E^{SD}$$

E^{SD} denotes the finite set of all events that appear in the sequence diagram *SD*. Events that are represented in the sequence diagram *SD* are a subset of the system events $E^{SD} \in E_S$. The event $e_1 \in E^{SD}$ initializes the interaction between objects.

Functions on Sequence Diagrams

The additional functions on sequence diagrams are defined as follows:

GetEvents is the function that maps the event sequence from the sequence diagram *SD* into the finite set of events E^{SD} that appears in that diagram

$$GetEvents : SD \to E^{SD}$$

GetObjects is the function that maps the event sequence from the sequence diagram *SD* into the finite set of objects that participate in the interaction

$$GetObjects : SD \rightarrow O^{SD}$$

If the sequence diagram *SD* determines the event sequence of the system *S*, then the objects that appear in the sequence diagram *SD* must belong to the system *S*. That means O^{SD} is a subset of objects of the system *S* $(O^{SD} \subseteq O_S)$, and E^{SD} is a subset of events of the system *S* $(E^{SD} \subseteq E_S)$.

The four categories of objects may be distinguished by observing their behavior:

- passive objects
- active objects
- agents
- generators

A *passive object* reacts to a message by modifying its state and emitting a response, which may invoke a sequence of further events or actions. Passive objects are also called *servers* because they never initiate interaction but provide services for the others. They wait for another object to require their actions.

The behavior of a passive object $obj \in O^{SD}$ may be specified as a sequence of *interaction items*. *Interaction items* are pairs of the form *(t, d)*. $t \in T_{obj}$ denotes an event from the set of system events. $T_{obj} \subseteq E_S$ invokes the reaction of the object $obj \in O^{SD}$. $d \in D_{obj}^*$ is a sequence of events from the set of system events $D_{obj} \subseteq E_S$ that the object $obj \in O^{SD}$ generates as the response. We may say that D_{obj} are events by which the object $obj \in O^{SD}$ influences its environment.

An *interaction item* determines a unit of interaction by specifying an event $t \in T_{obj}$ that causes the generation of the response $d \in D_{obj}^*$. A sequence of interaction items of a passive object is of the form:

$$((t_1, d_1), \ (t_2, d_2),..., \ (t_n, d_n)), \text{ where } \ t_1, t_2, \ ... \ , \ t_n \in T_{obj} \cup \{\varepsilon\}, \ d_1, d_2, \ ... \ , d_n \in D_{obj}^*$$

First event t_1 initiates object $obj \in O^{SD}$. Sequences of interaction items for the corresponding objects are represented on the *vertical bars* of sequence diagrams.

For example, the *ATM* object is a passive object. In the example that follows, the *ATM* behavior is determined in the sequence of interaction items. For example, one sequence of interaction items that determines the basic interaction between a client, *ATM* object, and other objects in the model, is given below:

(*Sequence of interaction items from the basic scenario*)

Card Inserted	/ Display *(*Enter your password!*)*
Password	/ Display *(*Account verification*)*; Verify Account
Account OK	/ Display *(*Enter amount!*)*
Enter *(*Amount*)*	/ Withdrawal; Display *(*Transaction processing*)*
Success	/ Display *(*Take cash!*)*; Dispense *(*Cash*)*; Print Receipt
Cash Slot Emptied	/ Display *(*Take your card!*)*; Eject Card
Card Slot Emptied	/ Display *(*Insert your card!*)*

GetInteractionItems is a function that extracts *interaction item sequences* from the sequence diagram *SD* for the specified object $obj \in O^{SD}$.

$$GetInteractionItems : SD \rightarrow InteractionItems_{obj}^{SD}, \quad obj \in O^{SD}$$

Note that a sequence of interaction items $IntreactionItems_{obj}^{SD}$ from the vertical bar corresponding to the object $obj \in O^{SD}$ is relevant to the further synthesis of state dependent behavioral aspects for this object. The technique of the synthesis is described in the further text.

Let $SequenceDiagrams_S$ be the finite set of all sequence diagrams of the system S that we have on disposition. Note that it is a training set in the terminology of machine learning.

$$SequenceDiagrams_S = \{SD_1, SD_2, ..., SD_n\}$$

The function $GetInteractionInfo$ provides interaction items sequences for the object $obj \in O_S$. This function is applied on the set of all sequence diagrams $SequenceDiagrams_S$ that we have on disposition for the system S.

$$GetInteractionInfo : SequenceDiagrams_S \rightarrow InteractionInfo_{obj}, obj \in O_S$$

$InteractionInfo_{obj}$ is the finite set of interaction items sequences. Note that $InteractionInfo_{obj}$ will be later used as input into the algorithm that synthesizes state dependent behavioral aspects for the object $obj \in O_S$.

An *active object* initiates an interaction. Active objects have a control on the system, and responsibility for other objects. They are also called *clients* because they require services from the others. An active object invokes the reaction of the passive object and passes the control to the object service of which it required. The control is passed back to the active object after the execution of the service.

An active object with a special role in the interaction with the system is called an *actor*. For example, a client at the keypad of an Automatic Teller Machine (ATM) is an actor. The client controls the ATM object, which is passive. The client generates events that he finds convenient. A vertical bar on a sequence diagram corresponding to an actor may contain any event sequence.

In general, actor objects

- may be persons, other systems, etc. that belong to the application domain rather then to the system itself,
- have intentions that are usually unpredictable – events may be in any order, so we cannot determine event sequences coming from them,
- can, intuitively, behave according two modes: "correctly" when they do not cause excessive events that may affect the consistency of the system, and "incorrectly" when they do cause excessive events.

Because of this, we cannot determine their behavior from interaction information, but we can register the sets of events that the actor objects notice or invoke (T_{act} and D_{act}, respectively).

An *agent* combines characteristics from both passive and active objects. Agents may either invoke interaction with other objects, or response to external information.

Some systems may need components, which generate (pseudo) random events. An example may be a simulation system. Thus, objects that generate (pseudo) random events are called *generators*. Intuitively, the sequence of events invoked by those objects cannot be predicted in any way.

6

INFERENCE OF STATE DEPENDENT BEHAVIORAL ASPECTS OF AN OBJECT FROM SEQUENCE DIAGRAMS

It appears that an attempt to infer object behavior from interaction information endeavors to unify contradicting principles.

Reusability of objects (welcomed benefit of the OO paradigm) requires the generalization of objects. For example, good design principles recommend that the address of the client, to whom an invoice is sent should not be "hard coded" in the object. The object should be designed to permit sending invoices to different clients. This should be natural, because a company provides goods or services for many clients. However, even invoices sent to the same clients do not necessarily contain the same data.

Knowing this, no restrictions are specified to the address of the receiver. What if the object will belong to the system where it will be allowed to send an invoice only to the particular class of clients? Without the possibility to control its own behavior internally (and generation of the client's address), the object may not be incorporated in the system because of the possibility of violating security principles.

The security level guaranteed by the system design may sometimes require restrictions on generalization, and the object behaves only as it was specified in the sequence diagrams. From this point of view, generalization may cause behavior that violates security principles, on the other hand, does support reusability of objects. Faced with this choice, we have chosen the algorithm that infers more general objects.

The technique for inference of state dependent object behavior from interaction information is simple. It starts by the application of the function *GetInteracctionInfo* on all sequence diagrams that we have on disposition. Note that the set *SequenceDiagrams$_S$* usually contains only a sample of all possible event sequences that may happen in the system S. Specifying the interaction between objects in the system S by writing all possible event sequences may require exponential time because of combinatorial explosion.

$$GetInteractionInfo : SequenceDiagrams_S \rightarrow Sample_{obj}^{+}, \qquad obj \in O_S$$

The result is the positive sample $Sample_{obj}^{+}$. It contains sequences of interaction items for the object $obj \in O_S$.

In the second step, we let run Muggleton's KC algorithm (1990):

> **Input:** a non-empty positive sample $Sample_{obj}^{+}$
> **Output:** a finite translation automaton FTA$_{obj}$
> **Enter KC Algorithm with k=1**
>   Let FTA_{obj} be KC($Sample_{obj}^{+}$, *1*);
> **Termination**
>   Output (*FTA$_{obj}$*);
> **End;**

The KC algorithm (Muggleton, 1990) accepts sequences of interaction items, and infers the finite translation automaton of the specified object, which determines its state dependent behavior. The KC algorithm (Muggleton, 1990) and related concepts are described in detail at the end of this chapter. We can do without the restriction that the resulting automaton has to be deterministic. Non determinism of the underlying automaton may be resolved later by adding a certain condition to each transition that causes non determinism. The synthesized finite translation automaton FTA_{ATM} has one more states than distinct interaction items. This is a consequence of letting the KC algorithm (Muggleton, 1990) run for the value $k=1$, as it is described further. Additionally, all the transitions associated to the same interaction item are directed to the same state.

REALISTIC EXAMPLE OF INTERACTION

We will illustrate the synthesis of object's state dependent behavioral aspects on the example of interaction with an Automatic Teller Machine (ATM). Assume a bank planned to have an ATM network. Each ATM in the network is supposed to communicate with the central bank computer and update clients' accounts. Security aspects must be included too.

An ATM has a client interface that consisting of a simple display, a keypad with digits and three control keys – Cancel, Enter and Clear. It also has a card slot for the clients' cards, cash slot and unit for printing receipts. The *ATM* object is supposed to control client interface. It accepts the client's messages coming through the interface, and transfers it to the bank computer. In order to keep our example simple, let us consider that this *ATM* is used for withdrawals only, although an ATM may be programmed for other retail banking transactions.

The clients' accounts are on the bank computer. The objects that model accounts may differ depending on the other bank's systems, but generally they can process messages coming from the network of ATMs. They approve or reject withdrawals providing some explanations, when necessary. Each client's account has the number given based on the bank's business rules.

An ATM accepts a client's card. The bank issues all cards. Therefore, each card has a code unique to the bank. To maintain simplicity, we will assume that one card may be used for one bank account only. Because of security reasons, each card has a unique password too.

The client starts the transaction by inserting the card. Generally, a withdrawal transaction consists of the three phases. The first is authorization of the client that starts when the *ATM* object checks if the password entered by the client corresponds with that one on the card. If it corresponds, the *ATM* will send a message to the *Bank* object requiring further verification of the account. If the *Bank* object authorizes the client and if the account is verified, the *Bank* object will also send the client's account number to the *ATM* object, which now has data necessary to require the settlement.

The second phase of the transaction starts when the client enters the amount for a withdrawal, and the *ATM* sends data for the settlement directly to the client's *Account* object. If the withdrawal is legitimate, the *Account* object will confirm the settlement. When the *ATM* receives this confirmation, it enters the last phase and finishes the withdrawal transaction through the dispensation of cash to the client, the printing the receipt, and, finally, the ejection of the card. In the case of failure, the *ATM* informs the client of the reason, and ejects the card. If the reason for failure is not at the level of the *ATM*, the object will proceed with the messages coming from the responsible objects.

Thus, we must assume that the *ATM* is programmed for only one use case – *Cash Withdrawal*. The eleven sequence diagrams on figures 1 to 11 represent some of the possible scenarios. The positive sample is extracted in the form of sequences of interaction items. Comments accompany events where necessary to ensure better understanding.

The first sequence diagram (Figure 1) specifies *basic* cash withdrawal when the actor or *Client* sends correct messages, which is recognized and processed without errors. The client inserts the card, and the *ATM* asked him to enter the password. Since the password corresponds to the password on the client's card, the *ATM* continues authorization by sending data (the ATM code and the card number) to the *Bank* object. The *Bank* object provides the account number, and verifies the account. When the *Bank* object receives the confirmation that the withdrawal is allowed (*Client's Account OK*), it sends the account number to the *ATM* object (*Account OK*), which then asks the client to enter the amount for the withdrawal. The client types in the amount and the *ATM* object sends the withdrawal data directly to the client's *Account* object. It also informs the client that the transaction is in processing. The *Account* object will settle the withdrawal, and send the

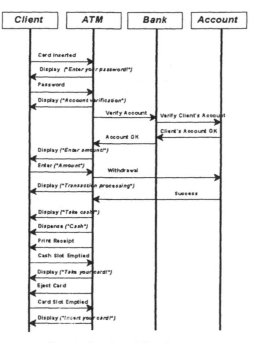

Figure 1 Scenario No. 1 *(Basic Scenario)*

9

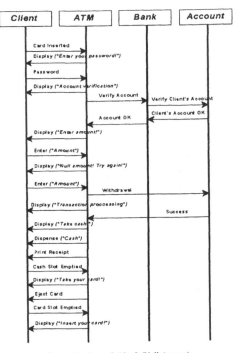

Figure 2   Scenario No. 2 *(Null Amount)*

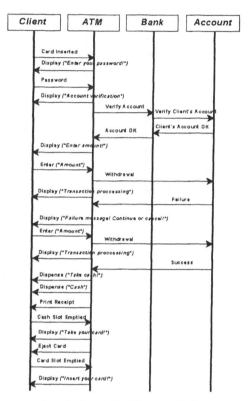

Figure 3   Scenario No. 3 *(Amount too High)*

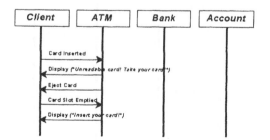

Figure 4   Scenario No. 4 *(Illegal State of the Account)*

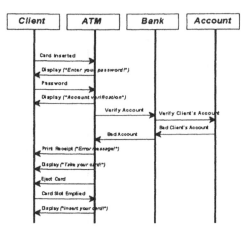

Figure 5 Scenario No. 5 *(Unreadable Card)*

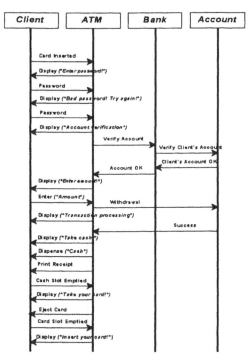

Figure 6 Scenario No. 6 *(Bad Password)*

12

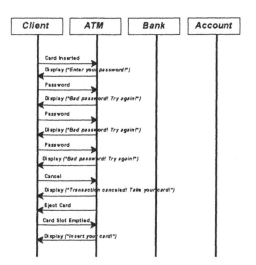

Figure 7   Scenario No. 7 *(Unauthorized Client)*

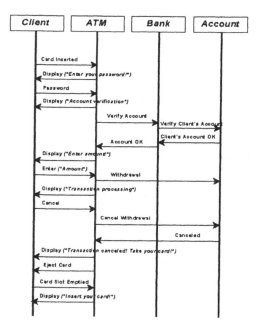

Figure 8   Scenario No. 8 *(Cancel on Transaction Processing)*

confirmation to the *ATM* (*Success*). When the *ATM* receives this message, it continues by controlling the cash dispensed and by printing the receipt. The client takes the cash and the card. The *ATM* controls both events. At the end of this scenario, the *ATM* goes back to the initial state and displays the message that is ready to accept a new client's card.

Figure 2 represents a trivial case where the *ATM* object rejects the withdrawal because the client entered *a null amount*. The *ATM* asked the client to try to enter the amount again, and when the client provides it, the transaction is completed as in the previous scenario.

The two following figures (3 and 4) describe what is happening when the withdrawal transaction is rejected by the object *Account*. In the first scenario (Figure 3), the reason is trivial – the required *amount is too high* (higher than allowed). The *Account* object sends the error-message, which the *ATM* forwards to the client. In such a case, a description of the error is modeled as an attribute of the *Bad Account* event outside the sequence diagram. The client then, corrects the mistake by entering lower amount. With the correct amount, the transaction may be continued the same as in the first two scenarios (figures 1 and 2).

The situation is more serious in the scenario in Figure 4. This time the reason for the transaction failure cannot be corrected due to *illegal state of the account* (i.e., the amount on the account is already at the lowest permitted level, account is canceled, etc.). The *Account* sends the error-message. The *ATM* forwards the error-message to the client. The error-message is also printed, and client's card is ejected.

The sequence diagram in Figure 5 specifies the behavior of the *ATM* object in a case when client's *card is unreadable* (i.e., the card is mechanically damaged). The *ATM* will inform the client, and control the ejection of the card.

Similarly, as in figures 1, 2 and 3, in the scenario in Figure 6, the client manages to complete the withdrawal. The only difference is that in this scenario, the client has entered a *bad password*. The *ATM* noticed the bad password, and asked the client to enter the password again. When the client enters the valid password (which was controlled by the *ATM* again), the transaction continues.

The scenario in Figure 7 represents the situation in which an *unauthorized client*, unable to enter the valid password, cancels the transaction after several attempts.

In the remaining sequence diagrams, we represent scenarios in which the client decides to cancel the transaction.

Figure 8 represents a scenario in which the client has decided to *cancel the transaction although processing has already started*. The *Client* has already been authorized and has entered the requested amount. As you may see in the second part of the diagram, the *ATM* accepts the canceled message, and sends it to the *Account* (*Cancel Withdrawal*). The *Account* object confirms the cancellation (*Canceled*). The *ATM* informs the client that the transaction is canceled and ejects the client's card.

A similar scenario occurs in Figure 9, with the exception that the client has decided *to cancel the transaction before he has entered the amount*. Since such a situation has no consequence on the *Account* object, the *ATM* finishes the scenario by ejecting the client's card.

Even more simple scenarios are represented in figures 10 and 11. In Figure 10, the client *cancels the withdrawal while the ATM authorizing him*. In the last scenario (Figure 11), the client has canceled the transaction immediately as he has inserted the card, even before he *enters the password*.

Thus, we have determined the interaction between an ATM (from the network of ATMs), and the client by ten sequence diagrams

$$SequenceDiagrams_{ATMNet} = \{SD_1, SD_2, ..., SD_{11}\}$$

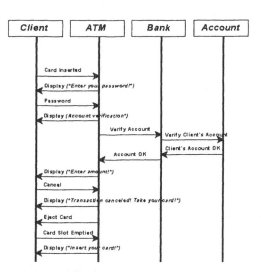

Figure 9   Scenario No. 9 *(Cancel on Amount Entry)*

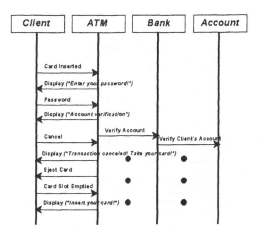

Figure 10   Scenario No. 10 *(Cancel on Account Verification)*

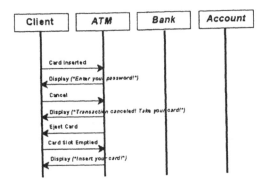

Figure 11   Scenario No. 11 *(Cancel on Password Entering)*

The function *GetInteractionInfo* extracts the positive sample of sequences of interaction items for the object *ATM*

$$GetInteractionInfo : SequenceDiagrams_{ATMNet} \rightarrow Sample_{ATM}^{+}$$

The positive sample $Sample_{ATM}^{+}$ that consists of sequences of interaction items, is represented below. Note that the interaction items *(t, d)* are written here in the form of *t /d*. Recall that $t \in T_{obj}$ is an event that invokes the object *obj* to react by generating event sequence $d \in D_{obj}^{*}$. T_{obj} is the set of event that the object *obj* may notice, and D_{obj} is the set of event that object *obj* may generate.

Note that asterisks "*" denote comments. A slash "/" separates a noticed event from the sequence of generated events, and semicolons ";" separated events in sequences generated in response.

Sample$_{ATM}^{+}$

(*Sequence of interaction items from the scenario no. 1 - Basic Scenario*)

Card Inserted	/ Display *(*Enter your password!*)*
Password	/ Display *(*Account verification*)*; Verify Account
Account OK	/ Display *(*Enter amount!*)*
Enter *(*Amount*)*	/ Withdrawal; Display *(*Transaction processing*)*
Success	/ Display *(*Take cash!*)*; Dispense *(*Cash*)*; Print Receipt
Cash Slot Emptied	/ Display *(*Take your card!*)*; Eject Card
Card Slot Emptied	/ Display *(*Insert your card!*)*

(*Sequence of interaction items from the scenario no. 2 - Null Amount*)

Card Inserted	/ Display *(*Enter your password!*)*
Password	/ Display *(*Account verification*)*; Verify Account
Account OK	/ Display *(*Enter amount!*)*
Enter *(*Amount*)*	/ Display *(*Null amount! Try again!*)*
Enter *(*Amount*)*	/ Withdrawal; Display *(*Transaction processing*)*
Success	/ Display *(*Take cash!*)*; Dispense *(*Cash*)*; Print Receipt
Cash Slot Emptied	/ Display *(*Take your card!*)*; Eject Card
Card Slot Emptied	/ Display *(*Insert your card!*)*

16

(*Sequence of interaction items from the scenario no. 3 - Amount too High*)

Card Inserted	/ Display *(*Enter your password!*)*
Password	/ Display *(*Account verification*)*; Verify Account
Account OK	/ Display *(*Enter amount!*)*
Enter *(*Amount*)*	/ Withdrawal; Display *(*Transaction processing*)*
Failure	/ Display *(*Failure message! Continue or cancel!*)*
Enter *(*Amount*)*	/ Withdrawal; Display *(*Transaction processing*)*
Success	/ Display *(*Take cash!*)*; Dispense *(*Cash*)*; Print Receipt
Cash Slot Emptied	/ Display *(*Take your card!*)*; Eject Card;
Card Slot Emptied	/ Display *(*Insert your card!*)*

(*Sequence of interaction items from the scenario no. 4 - Illegal State of the Account *)

Card Inserted	/ Display *(*Enter your password!*)*
Password	/ Display *(*Account verification*)*; Verify Account
Bad Account	/ Print Receipt *(*Error message!*)*; Display *(*Take your card!*)*; Eject Card
Card Slot Emptied	/ Display *(*Insert your card!*)*

(*Sequence of interaction items from the scenario no. 5 - Unreadable Card*)

Card Unreadable	/ Display *(*Card unreadable! Take your card*)*; Eject Card
Card Slot Emptied	/ Display *(*Insert your card!*)*

(*Sequence of interaction items from the scenario no. 6 - Bad Password*)

Card Inserted	/ Display *(*Enter your password!*)*
Password	/ Display *(*Bad password! Try again!*)*
Password	/ Display *(*Account verification*)*; Verify Account
Account OK	/ Display *(*Enter amount!*)*
Enter *(*Amount*)*	/ Withdrawal; Display *(*Transaction processing*)*
Success	/ Display *(*Take cash!*)*; Dispense *(*Cash*)*; Print Receipt
Cash Slot Emptied	/ Display *(*Take your card!*)*; Eject Card;
Card Slot Emptied	/ Display *(*Insert your card!*)*

(*Sequence of interaction items from the scenario no. 7 - Unauthorized Client*)

Card Inserted	/ Display *(*Enter your password!*)*
Password	/ Display *(*Bad password! Try again!*)*
Password	/ Display *(*Bad password! Try again!*)*
Password	/ Display *(*Bad password! Try again!*)*
Cancel	/ Display *(*Transaction canceled! Take your card!*)*; Eject Card
Card Slot Emptied	/ Display *(*Insert your card!*)*

(*Sequence of interaction items from the scenario no. 8 - Cancel on Transaction Processing*)

Card Inserted	/ Display *(*Enter your password!*)*
Password	/ Display *(*Account verification*)*; Verify Account
Account OK	/ Display *(*Enter amount!*)*
Enter *(*Amount*)*	/ Withdrawal; Display *(*Transaction processing*)*
Cancel	/ Cancel Withdrawal
Canceled	/ Display *(*Transaction canceled! Take your card!*)*; Eject Card
Card Slot Emptied	/ Display *(*Insert your card!*)*

(*Sequence of interaction items from the scenario no. 9 - Cancel on Amount Entry*)

Card Inserted	/ Display *(*Enter your password!*)*
Password	/ Display *(*Account verification*)*; Verify Account
Account OK	/ Display *(*Enter amount!*)*
Cancel	/ Display *(*Transaction canceled! Take your card!*)*; Eject Card
Card Slot Emptied	/ Display *(*Insert your card!*)*

(*Sequence of interaction items from the scenario no. 10 - Cancel on Account Verification*)

Card Inserted	/ Display *(*Enter your password!*)*
Password	/ Display *(*Account verification*)*; Verify Account
Cancel	/ Display *(*Transaction canceled! Take your card!*)*; Eject Card
Card Slot Emptied	/ Display *(*Insert your card!*)*

(*Sequence of interaction items from the scenario no. 11 - Cancel on Password Entering*)

Card Inserted	/ Display *(*Enter your password!*)*
Cancel	/ Display *(*Transaction canceled! Take your card!*)*; Eject Card
Card Slot Emptied	/ Display *(*Insert your card!*)*

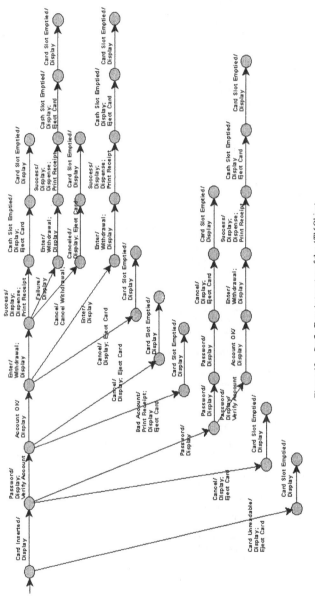

Figure 12 Prefix Tree Automaton of the *ATM* Object

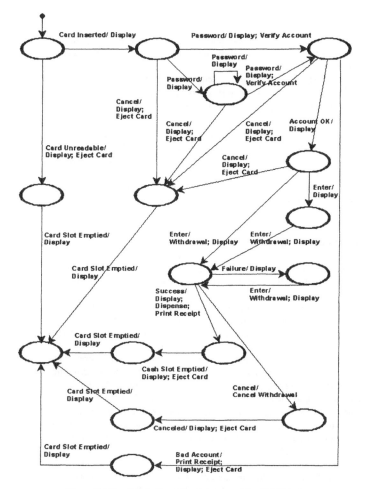

Figure 13 Inferred Finite Translation Automaton of the *ATM* Object

These eleven sequences of interaction items are the input into the KC algorithm (Muggleton, 1990). Note that the KC algorithm also requires a k parameter on input. The semantic meaning of this parameter is described in detail at the end of this chapter. The KC algorithm is looking for substrings of the strings from the positive sample, whose length is exactly k. When the same substrings are found, the states to which these substrings "lead", will be merged. The consequence is that for $k=1$, all the states to which the transitions are labeled by the same interaction items will be merged, and only one state will be formed. For example, after each unit of interaction:

Cancel　　/ Display *(*Transaction canceled! Take your card!*)*; Eject Card

that starts by the cancellation of the transaction, and continues by informing the client that the transaction is canceled, and continues with the ejection of the cash card, the *ATM* object finishes in the same state.

In one of its first steps, the KC algorithm constructs the prefix tree automaton $PT(Sample_{ATM}^+)$ from Figure 12. It continues by merging $PT(Sample_{ATM}^+)$ states through iterations and at the end, infers the finite translation automaton FTA_{ATM} in Figure 13.

Note that the inferred automaton (Figure 13) has only 16 states (or 17 if the instantiation state is included). That is one more then the number of different interaction items in the positive sample. This is the consequence of letting the KC run the algorithm for $k=1$, which is described in more detail at the end of the chapter. Smaller number of states makes the inferred automaton much more compact and more natural then the prefix tree automaton with 42 states (Figure 12).

AUTOMATA INFERENCE FROM POSITIVE SAMPLE

In the following pages, we will describe the algorithm that we used for the inference of finite translation automata from the interaction information. We also describe related concepts in more detail.

As a short introduction, we will present several of the main facts from the Muggleton survey (Muggleton, 1990) on finite automaton inference from the given sample of language. His survey was limited to the inference of automata that accepts regular languages only.

Gold has proved (1967) that no algorithm uniquely determines the entire set of regular languages from a positive sample only. This can easily be seen by assuming C to be the set of regular languages over the alphabet Σ, and showing that for any positive sample set $Sample^+$, there are at least two languages, which can be postulated. The first one is the universal language Σ^* and the second, the finite language containing only the members of $Sample^+$. Logical constraints contained in a positive sample only are insufficient for the determination of a unique regular language, which fits the sample.

The solution requires some form of additional constraint that must guarantee identification within the limit. Therefore, the authors who are active in this domain proposed several algorithms for language identification that use additional information sufficient for limited identification of the language, which the positive sample belongs to. According to Muggleton's survey (1990), the additional information may be of the form:

- Negative samples when finite automata are inferred from combination of positive and negative samples.
- Limit on a total number of states.
- Heuristic related to the compactness of the output automata.
- Presentation of semantic information in the positive samples.

Muggleton (1990) studied algorithms that used parameterized constraint predicates, which allowed identification of languages in the limit. All these algorithms have the same quality that the proposed language L does at least contain the sample set Sample[+]. The heuristic described by Muggleton (1990), provides a set of possible additional constraints which are used to make such a unique determination.

Definitions

Let *Sample*[+] be a subset of a language *L*. *Sample*[+] represents a *positive sample* of the language *L*.

Let *Pr(L)* be a set of the *prefixes* of elements of *L*, $Pr(L) = \{u \mid \text{for some } v \in \Sigma^*, uv \in L\}$.

Let $Tail_L(u)$ be the *left-quotient* of *u* in *L*, $Tail_L(u) = \{v \mid uv \in L\}$. $Tail_L^k(u)$ is the set of *k-tails* of *u* in *L*, $Tail_L^k(u) = \{v \mid v \in Tail_L(u), |v| \leq k\}$.

A *(non-deterministic) finite automaton A* is a five-tuple $A = (Q, \Sigma, \delta, I, F)$ where *Q* is non-empty finite set of the *automaton states*, Σ is an *input alphabet*, $\delta : Q \times (\Sigma \cup \varepsilon) \rightarrow 2^Q$ is a *transition function*. $I \subseteq Q$ is the set of *initial states* and $F \subseteq Q$ is the set of *final states*.

For a *deterministic automaton* where $q_0 \in Q$, the initial state is $I = \{q_0\}$, and the transition function is $\delta : Q \times \Sigma \rightarrow Q$.

The *prefix tree automaton* of the positive sample *Sample*[+] is a finite automaton, which accepts all strings from the positive sample *Sample*[+]. The *prefix tree automaton* is defined as following:

$$PT(Sample^+) = (Q, \Sigma, \delta, I, F)$$

where:

$Q = Pr(Sample^+)$,
$I = \{\varepsilon\}$ if *Sample*[+] $\neq \varnothing$, otherwise $I = \varnothing$,
$F = Sample^+$,
$\delta(u, a) = ua$ whenever $u, ua \in Pr(Sample^+)$

Note that strings from the input alphabet Σ label states.

For example, for the given sample

$$Sample^+ = \{ab, aba, abba, abbab\}$$

where the set of prefixes is

$$Pr(Sample^+) = \{\varepsilon, a, ab, aba, abb, abba, abbab\}$$

The corresponding prefix tree automaton is represented on Figure 14. The prefix tree automaton of the sample *Sample*[+] will accept no more and no less than the strings from the sample.

An automaton *A* is *isomorphic* to the automaton *A'* if and only if exists a bijective mapping $h : Q \rightarrow Q'$ such that $h(I) = I'$, $h(F) = F'$, and for every $q \in Q$ and $b \in \Sigma$, $h(\delta(q, b)) = \delta'(h(q), b)$. In short, two automata are *isomorphic* if renaming of their states makes them identical.

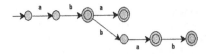

Figure 14   Prefix Tree Automaton $PT(Sample^+)$

A *canonical automaton* or a *minimal automaton* for the language L that accepts only the sample $Sample^+$ is defined as:

$$A(L) = (Q, \Sigma, \delta, I, F)$$

where:

$Q = \{Tail_L(u) \mid u \in Pr(L)\}$
$I = \{Tail_L(\varepsilon)\}$ if $L \neq \varnothing$, otherwise $I = \varnothing$
$F = \{Tail_L(u) \mid u \in L\}$
$\delta(Tail_L(u), a) = Tail_L(ua)$ if $u, ua \in Pr(L)$

The canonical automaton $A(L)$ has the *minimum* number of states possible for a finite automaton of L. Any automaton A' that is isomorphic to $A(L)$ is called canonical.

A finite automaton A' is a *subautomaton* of A if and only if $Q' \subseteq Q$, $I' \subseteq I$, $F' \subseteq F$ and for every $q' \in Q'$ and $b \in \Sigma$, $\delta'(q',b) \subseteq \delta(q',b)$. Alternatively, A' is a *subautomaton* of A if and only if $L(A') \subseteq L(A)$. Diagrammatically a *subautomaton* is formed from a finite automaton by removing some states and transitions from the transition diagram of the original finite automaton.

π_S is a *partition* of the set S, that is, π_S is a set of pairwise disjoint non-empty subsets of S such that the union of all sets in π_S is equal to S.

$B(s, \pi_S)$ is the unique *block* of π_S containing the element $s \in S$.

The quotient A/π_Q of the state set of the automaton A and some partition π_Q of Q is defined as follows. Let Q' be the set of blocks of π_Q, and let I' be the set of blocks of π_Q that contain at least one element of I. Similarly, F' is the set of all blocks of π_Q that contains at least one element of F. Block B_2 is a member of $\delta'(B_1, a)$ if and only if exists $q_1 \in B_1$ and $q_2 \in B_2$ such that $q_2 \in \delta(q_1, a)$.

Inference

Muggleton (1990) proposed a general algorithm in which the characteristic *predicate* χ is a heuristic with value that depends on matching certain local properties of compared states. By including different χ, the general algorithm infers finite automata that vary in numbers of states and transitions. We concentrate on the one of Muggleton's algorithms (1990), which exhibits particular good semantic qualities when applied to our domain.

Input into the algorithm is a non-empty positive sample $Sample^+$. The output is the automaton $A_0/\pi_{Pr(Sample+)}$ that denotes the quotient of state set of the prefix tree automaton A_0 and a partition of the prefix tree $\pi_{Pr(Sample+)}$ compressed in the iterations of the algorithm.

The inference algorithm starts by producing the *prefix tree automaton* *PT(Sample⁺)* that accepts no more and no less than the strings of the given sample *Sample⁺*. At first glance, it may seem that the solution may be the application of any algorithm that reduces the number of states of the given prefix tree automaton, and forms the *canonical automaton* by merging some of its states. Recall that the canonical automaton is formed to accept only the sample *Sample⁺* but with the least number of states.

Another issue that must be considered. Usually, the modeler does not specify all event sequences that may happen in the system. Specification of all possible interactions between objects in the system may require exponential time because of a combinatorial explosion. Instead, the modeler determines only a sample of event sequences. Therefore, we have to be aware that the resulting finite automaton is supposed to accept a regular language from which the specified sample is only a subset.

Due to the reduction of the number of states and allowance for the automaton to accept new event sequences, we continue merging the prefix tree automaton *PT(Sample⁺)* in a way that the resulting finite automaton accepts more strings. In other words, the algorithm infers regular languages that are generalizations of the sample *Sample⁺*, and of which the sample *Sample⁺* is a proper subset.

If we continue inferring even further, the algorithm will produce a *universal automaton* that accepts any strings that consist of symbols from the original sample *Sample⁺*. Since such an automaton is usually an *over-generalization*, that is the *characteristic predicate* χ that serves as a criterion for merging and a restraining factor. During the inference process, the algorithm performs testing of every possible pair of states of the prefix tree automaton to decide whether they should be merged.

KC Algorithm

We represent the KC algorithm in the same manner as Muggleton (1990):

Algorithm KC

> **Input:** a non-empty positive sample *Sample⁺* and k parameter
> **Output:** a k-contextual automaton A.
> *Initialization*
> Let $A_0 = (Q_0, \Sigma, \delta_0, I_0, F_0)$ be *PT(Sample⁺)*;
> Let π_0 be $\{\{u\} \mid u \in Q_0, \ |u| < k\}$;
> Let Q_0' be $Q_0 - \cup \pi_0$;
> Let $i = 0$;
> *Merging*
> for each state $u_1v \in Q_0'$ where $|v| = k$ do
> > begin
> > > if there exists some block B_1 such that $B_1 = B(u_2v, \ \pi_i)$ then
> > > > $B_2 = B_1 \cup \{u_1v\}$;
> > > > Let $\pi_{i+1} = \pi_i$ with B_2 replacing B_1;
> > > else
> > > > $B_2 = \{u_1v\}$;
> > > > Let $\pi_{i+1} = \pi_i \cup B_2$;
> > > endif;
> > > Increase i by 1;
> > end;

Termination
Let $f=i$;
Output A_0/π_f;
End;

In the KC algorithm, the predicate χ is of the form:

$$(\exists B_1 \in \pi_i)(\exists u_2)(u_2 v \in B_1) \quad \text{where } (\exists v)(|v|=k \ \& \ u_1 v \in Q_0')$$

k-parameter represents the length of substrings that will be compared. The KC algorithm recognizes same substrings whose length is exactly k. The algorithm merges states to which those substrings "lead." The KC algorithm does not merge states of the substrings. For example, the KC algorithm recognizes substring u whose length $|u|=k$. At the end, all state to which substring u "leads" will be merged.

To illustrate we will use the positive sample:

$$Sample^+ = \{ab, aba, abba, abbab\}$$

The KC algorithm merges states of the prefix tree automaton and for $k=1$ from Figure 14 and based on the similarity of the last symbol in the prefix forms the partition:

$$\pi_f = \{\varepsilon, \{a, aba, abba\}, \{ab, abb, abbab\}\}$$

The finite automaton generated by the KC algorithm for $k=1$ is represented on Figure 15. It accepts the language $L = (ab^*)^+$.

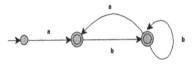

Figure 15   Inferred Automaton

The KC (Muggleton, 1990) algorithm exhibits time-complexity $O(n)$ only (n is one more than the sum of the lengths of strings in $Sample^+$). The KC algorithm starts by constructing the prefix tree automaton $PT(Sample^+)$ that contains no more than n states. That the first step is completed in time $O(n)$. Since both π_0 and Q_0' can be created in a single pass over all strings in $Sample^+$, it also takes time $O(n)$. Because Q_0' contains at most n strings and each pass through the iteration can be completed in constant time if given a hashing mechanism for finding the appropriate block B_1, merging also takes $O(n)$ time. At the end, the output A_0/π_f may also be created in time $O(n)$. To conclude, no operation takes more than time $O(n)$ which implies that the KC algorithm completes within $O(n)$ time.

The KC algorithm (Muggleton, 1990) has an advantage that it infers automata from the positive sample that contains only one string. Muggleton (1990) proved that other algorithms described in his survey cannot infer further the prefix tree automaton $PT(Sample^+)$ if the $|Sample^+|=1$. Since Muggleton (1990) described the issue in detail (including the proofs of the appropriate theorems), we will present only this ability of the KC algorithm.

Let us suppose that the positive sample $Sample^+$ consists of one string only:

$Sample^+ = \{aaaabbbbb\}$

A human being no trouble concluding that L is a^*b^* and that the associated finite automaton may have two states only. The KC algorithm first produces the prefix tree automaton $PT(Sample^+)$ based on the set of prefixes:

$Pr(Sample^+) = \{\varepsilon, a, aa, aaa, aaaa, aaaab, aaaabb, aaaabbb, aaaabbbb, aaaabbbbb\}$

The prefix tree automaton the $PT(Sample^+)$ has ten states because each prefix from $Pr(Sample^+)$ corresponds to one state. The $PT(Sample^+)$ is represented in Figure 16. The KC algorithm infers from $Pr(Sample^+)$ and for $k=1$ the automaton based on the partition that has three states only (Figure 17). The inferred automaton accepts the language a^+b^+.

$\pi_0 = \{\varepsilon, \{a, aa, aaa, aaaa\}, \{aaaab, aaaabb, aaaabbb, aaaabbbb, aaaabbbbb\}\}$

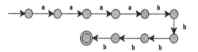

Figure 16   Prefix Tree Automaton

Figure 17   Inferred Automaton

Generally, the KC algorithm recognizes substrings of the length $|k|$, and merges states to which transitions are caused by the same symbols. For example:

$Sample^+ = \{cacb, bacc\}$

$Pr(Sample^+) = \{\varepsilon, b, c, ba, ca, bac, cac, bacc, cacb\}$

The KC algorithm with $k=1$ infers the partition:

$\pi_f = \{\{\varepsilon\}, \{c, bac, cac, bacc\}, \{b, cacb\}, \{ca, ba\}\}$

The resulting finite automaton has only four states. Thus, from the prefix tree automaton $PT(Sample^+)$ in Figure 18, the KC algorithm infers the finite automaton in Figure 19.

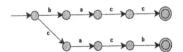

Figure 18   Prefix Tree Automaton

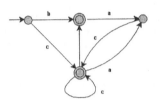

Figure 19   Inferred Automaton

This is an obvious advantage because it is important for an analyst to have a tool that is able to recognize any similarity and regularity in object behavior.

K-contextuality

Muggleton (1990) gave the following definitions and proved their consequences concerned *k*-contextuality and related entities.

Let L be a regular language. Then L is k-contextual if and only if whenever u_1vw_1 and u_2vw_2 are in L and $|v|=k$, $Tail_L(u_1v)=Tail_L(u_2v)$.

A finite automaton A is k-contextual if and only if L(A) is a k-contextual language.

$A=(Q, \Sigma, \delta, \{q_0\}, F)$ is k-contextual if and only if for all strings u_1vw_1 and u_2vw_2, accepted by A, where $|v|=k$, there is a unique state q such that $\delta(q_0, u_1v)=q=\delta(q_0, u_2v)$.

If a language L is k-contextual and contains two not necessarily distinct strings u_1vw_1 and u_2vw_2, where $|v|=k$, then L also contains u_1vw_2 and u_2vw_1.
Any 0-contextual language L containing two not necessarily distinct strings u_1w_1 and u_2w_2 also contains u_1w_2 and u_2w_1.

Any 0-contextual non-empty language L is equal to Σ^* (the universal language) where $b \in \Sigma$ if and only if there is some $ubv \in L$.

Muggleton (1990) gave the following theorem, which proves that his KC algorithm may identify in the limit, any language *L*:

Let L be a non-empty k-contextual language for some natural number k. Let w_1, w_2, w_3,... be a positive sample of L, and A_1, A_2, A_3,... be the output of KC_∞ on this input. Then $L(A_1)$, $L(A_2)$, $L(A_3)$,... converges to L after a finite number of steps.

It is particularly important to our application to note that the set of constraints on the resulting automaton is more natural. This implies the finite automaton that may be accepted as a satisfactory result of the inference looks more natural. The KC algorithm (Muggleton, 1990) recognizes substrings of the length $|k|$, and merge states to which they "lead".

Lemma: *Assume that $Sample^+$ is a non-empty positive sample, k a natural number, and A_0/π_f is the automaton output by the KC algorithm on input $Sample^+$ and k. Then π_f is the finest partition of the states of A_0 such that A_0/π_f is k-contextual.*

Let $A_0=(Q_0, \Sigma_0, \delta_0, I_0, F_0)$. Note that the initialization and merging sections of KC guarantee that every state of Q_0 will be placed into exactly one block of π_f. Thus π_f is a partition of Q_0, and A_0/π_f is a legal finite automaton. In addition, a trivial inductive argument can be used to show that every block B of π_f contains either a single state $u \in Q_0$ for which $|u| < k$, or all states $uv \in Q_0$ for which uv has a particular suffix v of length k.

Assume u_1vw_1 and u_2vw_2 to be two strings of a language L, where $|v|=k$. From the definition of k-contextual languages, L is k-contextual if and only if $T_L(u_1v)=T_L(u_2v)$, i.e. u_1v and u_2v lead to the same state in $A(L)$. Since all states $uv \in Q_0$ for which uv has a particular suffix v of length k are contained within the same block of π_f, it follows that u_1v and u_2v lead to the same state in A_0/π_f for any u_1vw_1, $u_2vw_2 \in Sample^+$. Thus A_0/π_f is k-contextual.

Increment of the Positive Sample

Sequential diagrams may be (and usually are) determined in an incremental fashion. Therefore, the inference algorithm should produce a gradually changing output given progressive increment of the sample. In the case of using any inference algorithm that cannot perform like this, there is no way to predict the effect that any particular new string added to the sample is likely to have on the resulting finite automaton.

For this reason, Muggleton (1990) proposed the definition of incremental modification concerned to finite automata inference algorithms. We present it in the similar manner as Mugleton (1990).

Let A be the finite automaton that is the output of some inference algorithm InfAlg on the input of positive $Sample^+$. Let the finite automaton A' be the output of InfAlg on input $Sample^+ \cup \{w\}$. We say that InfAlg is incremental if and only if A is a subautomaton of A'.

Muggleton (1990) proved that the KC algorithm is incremental on input k and positive $Sample^+$ of some k-contextual language L (for a given fixed number k).

Negative Sample

The KC algorithm (Muggleton, 1990) does not exclude using a negative sample (i.e. sequences of interaction items that must not happen in the system). If negative sample is available then the KC algorithm (Muggleton, 1990) may be used in the following way.

Let us assume that we have on disposition the positive and negative samples *(Sample⁺, Sample⁻)*. These samples are disjoint sets. First, we let the KC algorithm run (Muggleton, 1990) on the positive sample *Sample⁺* and infer k-contextual languages for $k = 0, 1, 2, ...$, until the algorithm finds some k for which the computed language does not contain any of the strings from the negative sample *Sample⁻*.

Future Work

We have proposed that each OO CASE tool has an editor of sequence diagrams. Here are some examples how these editors may be extended and improved.

The KC algorithm is used in the same form as it is originally introduced by Muggleton (1990). The algorithm is looking for the same substrings u of the length $|u|=k$. It merges the states to which substrings u leads, but does not merge states along the substring u. A possible improvement, in the sense that exhibits better semantical properties when applied to our domain, may be to extend the KC algorithm to merge those states.

The KC algorithm addressed automata as formalism only, without any semantics. Future work may be directed to extensions of the algorithm, in the sense that some information about the object is included. The first candidates are information about an instantiation state, destruction states, initial states, and final states. We believe that by incorporating this information into the KC algorithm, the resulting automaton may look even more natural. Merging some initial and final states (when appropriate by semantics), may decrease number of states.

We will also experiment with attributed events. It is obvious that for each event in a sequence diagram, two attributes may be extracted – an attribute that points to the object that invokes the event, and the other object that notices the event. Thus, let us assume an event $e \in E^{SD}$ from the set of events E^{SD} on the sequence diagram SD. Let O^{SD} be the set of objects that appear in the sequence diagram SD. We will also assume also the finite set of attributes Att associated with the event $e \in E^{SD}$, which is denoted as $Att(e)$. There is a subset of $Address(e) \subseteq Att(e)$, which contain the information about the objects interested in the event (they may invoke or be invoked by this event). Automatic determination of these attributes from sequence information may be useful extension in the scope of an editor of a sequence diagram.

When a modeler works in the editor of sequence diagrams, he may want to search for a given sequence of interaction items. For example, assume that we have sequences of interaction items that determine interaction between objects in the system. The modeler may decide to change a certain pattern of interaction, which consists of several interaction items only. Instead of looking for each occurrence of the interaction pattern, the modeler may automatically search for the pattern. The problem may be reduced to the problem of pattern matching (pattern of interaction items) in a text (a sequence diagram).

ACKNOWLEDGMENTS

The work was commenced in cooperation with the members of the AI Group, Department of Control, Faculty of Electrical Engineering, Czech Technical University in Prague. I would like to thank Prof. Olga Stepankova for a critical reading of the manuscript.

REFERENCES

Adelberger, H.H., Lazansky, J., Marik, V. 1995. Information Management in Computer Integrated Manufacturing. Lecture Notes in Computer Science, Springer Verlag Berlin Heidelberg.

Aho, A.V., Sethi, R., Ullman, J.D. Compilers. Addison-Wesley, 1987.

Banerji, B.R. The Logic of Learning: A Basis for Pattern Recognition and for Improvement of Performance. Advances in Computers, Vol. 24.

Batory, D., O'Mallay, S. The Design and Implementation of Hierarchical Software Systems with Reusable Components. ACM Transactions on Software Engineering and Methodology, Oct. 1992, Vol. 1, No. 4, pp. 355-398.

Blumel, E., et. al. Managing and Controlling Growing Harbor Terminals. The Society for Computer Simulation International (Europe), 1997. ISBN 1-56555-113-3.

Booch, G. Object Oriented Analysis and Design with Applications. Benjamin Cummings, 1994.

Booch, G., Rumbaugh, J., Jacobson, I. The Unified Modeling Language User Guide. Addison-Wesley, 1999. ISBN 0-201-57168-4.

Camurati, P., Corno, F., Prinetto, P. An efficient tool for system-level verification of behaviors and temporal properties. IEEE 1993.

Coad, P., Yourdan, E. Object-Oriented Analysis. Yourdan Press, 1991.

Coad, P., Yourdan, E. Object-Oriented Design. Yourdan Press, 1991. ISBN 0-13-630070-7.

Coleman, D., Arnold, P., Bodoff, H., Gilchrist, F., Jeremaes, P. Object-Oriented Development - The Fusion Method. Prentice Hall, 1994.

de Champeaux, D., Faure, P. A Comparative Study of Object-Oriented Analysis Methods. Journal of Object-Oriented Programming, Mar./Apr. 1992, pp. 21-33.Feldman, Y.A., Scheider, H. Simulating Reactive Systems by Deduction. ACM Transactions on Software Engineering and Methodology, Apr. 1992, Vol. 2., No. 2, pp. 128-175.

Gibson, E. Objects - Born and Bred. BYTE Oct. 1990, pp. 245-254.

Graham, I. Object-Oriented Methods. Addison-Wesley, 1994. ISBN 0-201-59371-8.

Hachtel, D. Gary., Somezi, Fabio. Logic Synthesis and Verification Algorithms, Kluwer Academic Publishers (USA), 1996.

Jacobson, I. Object-Oriented Software Engineering. ACM Press, 1992.

Jacobson, I., Booch, G., Rumbaugh, J. The Unified Software Development Process. Addison-Wesley, 1999. ISBN 0-201-57169-2.

Klir, G.J. Facets of Systems Science. Plenum Press, New York (USA), 1991.

Koskimies, K., Makinen, E. Automatic Syntheses of State Machines from Trace Diagrams. Software - Practice and Experience, Jul. 1994, Vol. 24(7), pp. 643-658.

Kunzig, R. A Head for Numbers. Discover, July 1997.

Lifschitz, V., Rabinov, A. Things That Change by Themselves.

Liskov, B.H., Wing, J.M. A Behavioral Notion of Subtyping. ACM Transactions on Programming Languages and Systems, Nov. 1994, Vol. 16, No. 6, pp. 1811-1841.

Marzik, V., Stepankova, O., Lazansky, J., et all. Artificial Intelligence I, in Czech. Prague Academy, 1993.

Marzik, V., Stepankova, O., Lazansky, J., et all. Artificial Intelligence II, in Czech. Prague Academy, 1997.

Melewski, D. Ready for Prime Time? Application Development Trends, Nov. 1997, Vol. 4, No. II, pp. 30-44.

Meyer, B. Object-Oriented Software Construction, Prentice Hall International (UK), 1988.

Muggleton, S. Inductive Acquisition of Expert Knowledge. Addison-Wesley Publishers, Ltd., 1990.

Muller, P. Instant UML. Wrox Press, 1997. ISBN 1-861000-87-1.

Osborn, L.S., Yu, L. Unifying Data, Behaviors, and Messages in Object-Oriented Databases. October, 1994.

Parsaye, K., Chignell, M., Knoshafian, S., Wong, H. Intelligent Databases. John Wiley & Sons, 1989.

Polovina, R. Formal Object Specifications in OO Modeling and their Consequences. Ph.D. Thesis. Department of Computer Science and Engineering, Faculty of Electrical Engineering, Czech Technical University in Prague, 1999.

Polovina, R., Stepankova, O., Petrus, M. Automatic Synthesis of K-contextual automaton by the KC-algorithm, in Czech. The Mobile Robot Group, The Gerstner Laboratory for Intelligent Decision and Control, Faculty of Electrical Engineering, Czech Technical University in Prague, Czech Republic, 1998.

Rumbaugh, J., Blaha, M., Premerlani, W., Eddy, F., Lorensen, W. Object-Oriented Modeling and Design. Prentice-Hall International, 1991. ISBN 0-13-630054-5.

Rumbaugh, J., Jacobson, I., Booch, G. The Unified Modeling Language Reference Manual. Addison-Wesley, 1999. ISBN 0-201-30998-X.

Valmari, A., Savola, R. Verification of Behaviors of Reactive Software with CFFD - Semantics and ARA Tools. International Symposium on On-Board Real-Time Software, Nov. 1995, Noordwijk, The Netherlands.

Wallace, M. Software Reliability - What are the Contributory Factors? Application Development Trends, Nov. 1997, Vol. 4, No. II, pp. 26-28

Wegner, P. Dimensions of Object-Oriented Modeling, IEEE Computer, Oct. 1992, pp. 12-20.

Wojtkowski, W.G., Wojtkovwki, W., Wrycza, S., Zupancic J. Systems Development Methods for the Next Century. International Conference of Information Systems Development - Methods and Tools, Theory and Practice, Aug. 1997, Boise, Idaho, USA.

Wooldrige, M., Jennings, R.N. Intelligent Agents: Theory and Practice. Lecture Notes in Artificial Intelligence, Springer Verlag Heidelberg, 1995.

A CORBA COMPONENT MODELING METHOD BASED ON SOFTWARE ARCHITECTURE

Chang-Joo Moon, Sun-jung Lee, and Doo-Kwon Baik
Software System Laboratory, Dept. of Computer Science & Engineering
Korea University 1, 5-ka, Anam-dong, Sungbuk-gu, 136-701, Korea
Ph.: +82-2-925-3706, Fax: +82-2-953-0771
E-mail: {mcj,silee.baik}@swsys2.korea.ac.kr

Key words: CORBA, Software Architecture, IDL, Component Modeling, CTL, Connector

Abstract: To develop a conventional CORBA(Common Object Request Broker Architecture)-based distributed object system, first the system's requirements must be analyzed in requirements analysis step and system's overall interface and details must be specified and built using IDL(Interface Definition Language) in design step. However, there are not too many models available that can be used as a benchmark in specifying the requirements and details of the CDO(CORBA-based distributed object) system. Conversion to IDL from requirements analysis step or preliminary design stage is a difficult task and an effective method for verification of IDL is not currently available. In the light of unique features of distributed object system, IDL errors that occur in the beginning of software development become the main causes of errors for the system as a whole. This paper suggests SACC (Software Architecture-based CORBA Component) model as a solution to the problems mentioned above in specifying details and verifying CDO system. SACC model uses SCI (SACC Converted IDL) algorithms to be converted to IDL. SACC model utilizes software architecture approach. SACC modeling method uses CTL (Computational Tree Logic) to verify CORBA system that is specified by SACC model. In conclusion, SACC modeling method is proven effective in specifying details and verifying the CDO system. Furthermore, it helps programmers to effectively convert to IDL, which is highly reliable.

1. INTRODUCTION

Software architecture is a focus on reasoning about the structural issues of system. These structural issues are design-related . software architecture is, after all, a form of software design that occurs earliest in a system's creation . but at a more abstract level than algorithms and data structures. Structural issues include gross organization and global control structure; protocols for communication, synchronization, and data access; assignment of functionality to design elements; physical distribution; composition of design elements; scaling and performance; and selection among design alternatives [4], [11], [16], [9], [8]. Similar to the software architecture, distributed object systems like CORBA and DCOM (Distributed Component Object Model) utilize the

Systems Development Methods for Databases, Enterprise Modeling, and Workflow Management,
Edited by W. Wojtkowski, *et al.* Kluwer Academic/Plenum Publishing, New York, 1999.

31

components. Unlike traditional objects, Components as distributed object, it can interoperate across language, tools, operating system, and networks. But components are also object-like in the sense that they support inheritance, polymorphism, and encapsulation [18]. Through utilization of standardization and IDL(Interface Definition Language), CORBA components significantly improved and solved the incompatibility problems that exist between software architecture components.

The approach used in developing components in CORBA environment is a top-down approach by creating IDL. Before developing the system itself, IDL that specifies details of the system and defines interfaces should be created. To develop a conventional CDO(CORBA-based distributed object) system, system's requirements must be analyzed in requirements analysis step and system's overall interface and details must be specified and built using IDL in design step. In implementation step, the language which is supported by ORB(Object Request Broker) is chosen and then developer implements CDO system. However, it is a difficult task to specify the details of the CDO system due to a lack of benchmark model. Analysis of system requirements and the conversion to IDL are also difficult task and verification of IDL in specifying the system is very difficult task. Especially, the creation of IDL in CDO system development of larger scale is a much more difficult process. A number of system developer complaint about the additional creating IDL tasks and difficulty in creating IDL despite of many advantages that CORBA system offers. However, the creation of IDL in the beginning of the development is important given the unique characteristics of the distributed object system. This is because stub formed when pre-compilation is done using created IDL. Because the components are created on this stub, error which is occurred during the creation of IDL coincides with error in the system. Therefore, the creation process of IDL in the beginning of CORBA system development affects the successful development of the system and errors that occur during that period also become the main cause of devastating errors occurring later in the system.

Since CORBA components possess characteristics of conventional objects such as inheritance and encapsulation, it is quite difficult to support object oriented concept with graphic notations, which is available in conventional ADL (Architecture Description Language). In contrast to conventional objects, CORBA components should be treated as an independent unit which has the characteristics of autonomy, automation, and cooperation of an application. For this reason, the expression of the components with conventional object-oriented modeling methods, such as OMT (Object Modeling Technique) and UML (Unified Modeling Language), becomes a difficult task at hand. Therefore, a new model dedicated to modeling of the distributed object system is required.

This paper suggests SACC(Software Architecture-based CORBA Component) model as a solution to the problems mentioned above in specifying details and verifying CDO system. To specify the CDO system, SACC model utilizes software architecture approach. This approach specializes general software architecture approach to express the CORBA component interaction. Therefore, structural elements of the SACC model express the behaviors of CORBA components with high accuracy and apply graphic codes that are based on the underlying concepts of software architecture. SACC modeling method uses CTL(Computational Tree Logic) to verify CORBA system that is specified by SACC model. A CORBA system created from a SACC model can be easily converted to CTL. SACC model expresses the computation during the components' interaction through a computation that changes state along each path of CTL. By establishing a corresponding criteria between the structural elements of SACC model and IDL, SCI algorithm that changes from SACC model to IDL is suggested. If the verified SACC model converts into IDL through the conversion algorithm, the IDL created is also verified, which in turn, reduces the burden on developers of the system. The corresponding criteria between the structural elements of SACC model and IDL and between the elements of IDL and actual implemented codes means that there is a direct relationship between the elements of SACC model and actual implemented codes. Therefore, unlike the conventional software architecture methods, which is limited to expression of interface in a modeling stage, SACC models can be used as a blueprint of the system development through the implementation step. With a SACC modeling method, it is now possible for system developers to create a reliable low-cost CORBA system.

This paper is divided into 7 sections as follows: Related studies are examined in Section 2 and discussion on structural elements of SACC modeling and SCI algorithm is conducted in Section 3. In Section 4, an example of SACC model is verified using CTL. Section 5 compares and analyzes the SACC model with other modeling methods. An example of successful implementation of SACC model is suggested in Section 6. Lastly, Section 7 concludes this paper and talks about future plans.

2. RELATED STUDIES

2.1 Architectural Representation

1) ADL
Architecture Description Languages(ADLs) are language for architecture models. ADLs use graphics and text to express architecture information. There are a variety of ADL emerging from various industrial and academic research groups. ADL that is suitable all situation is not yet. Therefore you select ADL to fit situation or modify ADL[5].

2) Architecture Representation Notation
The expression of software structure in a highly abstract environment becomes the basis of architecture. However, this does not provide an adequate amount of information needed to develop the system. Therefore, it is important to bring together the different perspectives in expressing software architecture. Recently, a few attempts were made to optimally express the architectural structure through bringing together the formal notation and the informal prose [8].

3) Formal method for architecture representation
After a system is designed and created, there must be a way to verify it's correctness in satisfying the requirements of the system and any errors evident in the system. Formal method is used for this purpose. This approach is included in ADL and is experimented as a new method to overcome the limits of ADL. The existing theories of formal methods are being applied to software architecture[10], [6], [1].

2.2   CORBA Component and IDL

CORBA components are not confined by language, tools, operating system, network, address domain, vendor and compiler and similar to objects, they support inheritance, polymorphism and encapsulation. IDL assigns the scope of components and specifies the interface between components. It is purely declarative. This means that it provides no implementation details. IDL provides operating system and programming language independent interfaces with all the services and components that reside on a CORBA bus [Robert98].

2.3   The Limits of Components

1) Components of Software Architecture
Components of software architecture can be defined as a computation at a level equal to that of an application. Currently available component models have big gap when implementation of model due to their lack of adequate expression. Components are reusable blocks, which are important element in increasing the efficiency of software production. However, the incompatibility problem in components still remains to be solved. Usually, the incompatibility problems arise from lower level components such as programming language, operating system and database structure. Connectors available currently are not adequate for seamless interactions between components.

2) CORBA Component
A component is a unit of a job and distribution. Incompatibility problems in the lower level components are solved by defining standards and using IDL as the interface language. However, there are still weaknesses in system modeling and analysis due to the lack of benchmark models and methodology in system development. Since all component interfaces are defined in IDL, there is a burden on developers in creating IDL. Basically, components interact through method calls on a client/server base, thus a connector is rather simple in construction. Therefore, it does not support interactions in a variety of ways.

3) JavaBean Components
JavaSoft adopted CORBA as a model for an distributed object system and Enterprise JavaBean and RMI API are executed on CORBA/IIOP. Therefore, even both Enterprise JavaBean and RMI can not exceed the scope of CORBA significantly.

3.   SACC MODELING METHOD

3.1   Basic Concepts

SACC is a model that connects software architecture and CORBA system. It is a customized model that derived from a software architecture approach to effectively support a CORBA system. Therefore, the structural elements of SACC effectively express CORBA components. In comparison with graphic expression, such as text boxes and arrows, of conventional software architecture, SACC models use much more precise expressions to specify details of the system, serve as a blueprint for implementation and as a supporting materials for system verification. SACC modeling method utilizes SCI algorithm to convert from SACC model to IDL, which reduces burden on developers. Furthermore, it serves as a verification method using CTL. This verification method can create verified IDL and CDO system.

SACC modeling method is shown in figure 1 (b). The three boxes of bold lines in (b) are the areas where the SACC modeling method is applied. In conventional CORBA system, the designer or system developer must follow the circulation through ①,②,③ in figure 1 (a) to correct development errors. Errors can occur in any of the three sections and when error occure, it confuses the system developer because the source of error is unknown or unclear. Especially, errors occurring in section② are most frequent and require a tremendous amount of code modification. Most of system developers complain about difficulties they encounter in section②. However, if the SACC modeling method in figure 1 (b) is used, errors occur only in section④, which in turn limits the developers search for errors to that section only. The most problematic section② has been excluded from the source of errors, if the SACC method is applied, thus allowing the developers to concentrate on implementation of the system.

3.2   Structural Elements of SACC Model

Structural elements of SACC model are as shown in figure 2.

1) Process Component

A process component is a unit of a job and distribution. The component processes the requested job by processing the data in the data component and saving the result. Inheritance is supported in this component, which allows inheritance of features of previous components to create a new expanded component. It is expressed through interface on IDL.

2) Data Component

Data component refers to data that is created, processed, saved and shared in the process component.

3) Port

Port is an interface of the process component and programmer names each port to use the naming service of CORBA system. Through these names, each component detects and connects to one another. A call-back function is available to use two-way port. When there is a call-back port, two separate components have two types of interactions against a single port. Therefore, a single component computes as if there are two separate components.

4) Connector

A connector includes the interaction between components and a connect protocol. In software architecture, there are numerous connectors available. However, in SACC model, the connectors are limited to CORBA IIOP(Internet Inter-ORB Protocol) and the method call, which remote controlled between components. CORBA IIOP can be used as a protocol for remote controlling of components. Interaction is expressed remote controlled methods and returned values. Again, if the call-back function is available, two-way remote control is possible.

5) Inheritance

IDL is a subset of C++ that possesses the characteristics of object-oriented language. The inheritance of components are possible only on IDL. Therefore, a complete inheritance of previously created components is not

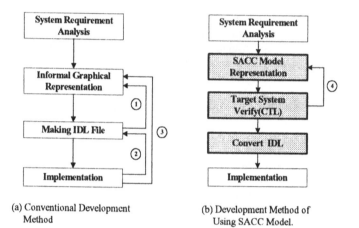

(a) Conventional Development Method

(b) Development Method of Using SACC Model.

Figure 1. SACC Modeling Method

possible. However, the interaction between components on IDL can be inherited to be used again.

6) Matchmaker

Matchmaker displays a data component that is matched with a process component.

7) Architecture Style

Architecture style is set of structural elements. By sharing the structure and contents of a system that is frequently used, system designer can get the re-usage of previous designs. This also allows system developers to easily understand the system structure and increases the level of code re-usage.

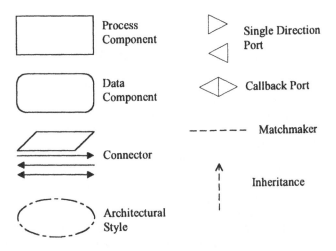

Process
Component

Single Direction
Port

Data
Component

Callback Port

Connector

— — — — — — Matchmaker

Architectural
Style

Inheritance

Figure 2. SACC Model Structural Elements

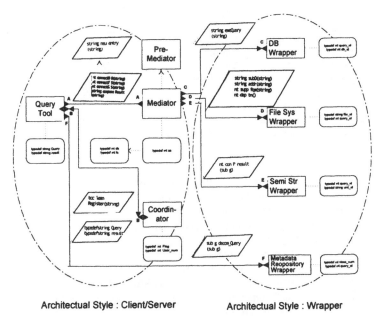

Architectual Style : Client/Server   Architectual Style : Wrapper

Figure 3. Example of SACC Model

Figure 3 is example of SACC model. This example is query system for relational database, file system and object database in reference to meta-data repository in a CORBA environment. Meta-data repository is served as a reference by query processors to process complex and intelligent queries and to resolved semantic conflicts in the processing.

3.3 Conversion to CORBA IDL in SACC Model

The structural elements of SACC model are matched with IDL element as shown in Table 1. Through this matching, SACC model provides guidelines from the beginning of design step through implementation step. Therefore, SACC models can be applied not only to the implementation step but to used as an understanding tool in the higher-level designing. SACC model expresses the CDO system and through an adequate analysis and verification, the SACC model converts into IDL through SCI(SACC Converted IDL) algorithm. The converted IDL is a proven IDL, which accurately expresses the intended system structure and solves any problem arising from erroneous IDL created in the implementation stage. Generally, conventional software architecture expresses only the interface of the system. However, the converted IDL from a SACC model provides a direct connection between the software architecture

Table 1. SACC Model, IDL, Implementation Comparison

SACC Model Structural Elements	IDL	Implementation
Target System	Module	Package
Architectural Style	Interface Set	Class Set
Process Component	Interface	Class
Data Component	Type, Constant	Field
	Attribute	Field, Get Method, Set Method
Port		Naming Service
Connector	Method	Method, IIOP

```
/* TS : Target System    PC : Process Component */
/* DC : Data Component    C : Connector      */
search TS from SACC Model
2.    write TS name as Module item in IDL file
3.    while(PCj is not null)
3.1.  write PCj name as Interface item in IDL file
3.2.  if (exist super interface of PCj)
3.2.1     search super interfaces PCj from SACC Model
3.2.2     write " : super interfaces name" next of Interface name
3.3.  write DCj in IDL File
3.4.  write Cj in IDL File
3.5   increment j
4.    end
```

Figure 4. SCI Algorithm

and the compiled codes from which the software architecture derived. This is useful in actual implementation of the system. Figure 4 shows SCI algorithm.

4. VERIFICATION OF SACC MODEL THROUGH CTL

4.1 Application of CTL on SACC Model

The distributed object system becomes a single system through interactions between several components. In the light of unique features of the CDO system, components interact with each other through method calls on ORB. Therefore, the overall status of the CDO system depends on what method of component currently being executed, what method of component will be executed in the future ,what values are returned to the component. If the order in which the methods are executed and the location where the return value will ultimately reach is tracked down, system developers can verify whether the system characteristics are satisfactory to the system. Figure 5 shows that the SACC model in figure 3 is converted to CTL construction. Each S_i $(i \geq 0)$ status in figure 5 is matched with components in SACC model. In another mean, each status in CTL is viewed as a single component.

The problem in tracking the computational status in CTL is similar to the problem that is finding what component being executed along time in the CDO system. Therefore, viewpoint that is to verify SACC model by CTL, is a quite convincing viewpoint. The transition of the computations follows the

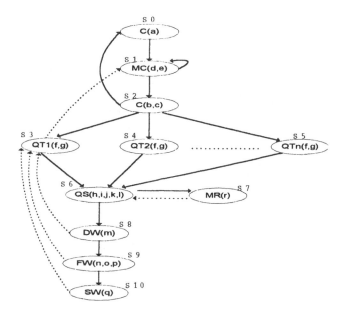

Figure 5. CTL Structure of Query System

arrow in CTL. Each Si(i ≥ 0) status in CTL possesses a propositions. These propositions have TRUE value in themselves belonged status. These propositions are equal to connectors in SACC model that expresses the interactions between components. Methods used in the interactions have true value and methods that are not used in the interactions possess false value. When the methods are called, it becomes evident that the computation transferred from a component to another and those methods can be called by others component. Therefore, those methods can be represented as a TRUE proposition value in Si status. C(b,c) in figure 5 is an abbreviation for C.b, C.c, which are methods appearing on IDL. For example, C of S2 is a component name and b, c of S2 is a method name that can change the state in coordinator component. In reality, each component possesses much more methods than as shown in figure 5. However, methods not shown in figure 5 do not affect the state change, which are not considered in CTL. Because components with a call-back function service two purposes, although one component carries out two computations. Therefore, it is shown as two status in CTL. In figure 5, unlike the conventional CTL, the status transition arising from a method call is shown with a straight-line arrow. The status transition from a return value is shown with a dotted-line arrow. Status transitions from S8, S9, S10 to S4, S5 and from S4, S5 to S1 are the same with dotted-line arrow transition in the left side of S6, thus they are omitted in figure 5.

Figure 6 shows the computational tree in figure 5. Computational trees are used to create algorithms for verification process. Branches of the tree represent all possible paths of activities in figure 5. Figure 7 lists the abbreviations of component methods used in figure 5

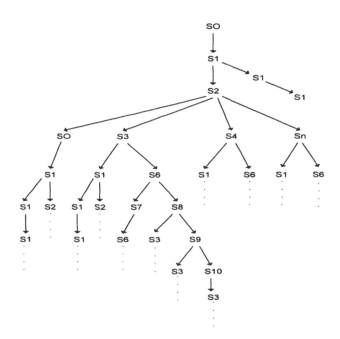

Figure 6. Computation Tree of Query System

Coordinator (register, startQuery1, sendstatus_re) :   C(a,b,c)
    ? Receive enrollment of users and control users.
    MultiConsole(startQuery2, displaystatusresult1): MC(d,e)
    ? Display of user's query result and query processing status.
    QueryTool(startQuery3, displaystatusresult2) :   QT(f,g)
    ? Tool of query
    QueryServer(QueryDB, QueryFS, QuerySS, discompo_Query,
    compo_Query_result) : QS(h,i,j,k,l)
    ? Receive users query and query processing then return query
        processing result.
    Meta_Repo(exeRef) : MR(m)
    ? Provide information that is required query processing.
    DBWrapper(exeQuery) : DW(m)
    FSWrapper(Monthly, dept_no, division) : FW(n,o,p)
    SSWrapper(exeQuery) : SW(q)
    ? Legacy system links up with distributed system through wrappers

Figure 7. Method Abbreviation of Query System and Role of Component

4.2   Property Verification through CTL

1) Verification of Response Properties
This property is interpreted one proposition as a guaranteed response to the
other property [13] . The example in figure 5 is a query system, which means
that a user must receive a query processing service at least once if the user is
registered and the command for the query is started. Considering response
property in figure 5, we can see that if users are enrolled(C.a) coordinator and
query tool(QT) receives command for query start(MC.d), users must get a
query processing service(QS.1) in all path of computation. If this property is
expressed as CTL formula, it will be as follows: For details of CTL , refer to
[3], [21], [14], [12].

(M,S0) ? AG ((C.a ? MC.d) ? AF QS.i)   ->
(M,S0) ? AG ~ C.a ? AG ~ MC.d ? AG (AF QS.i))

That is, if each status of CTL in all path of computation satisfies at least
one out of three proposition, ~ C.a, ~ MC.d or AF QS.i, it possesses a response
property. States in figure 5 satisfy AG (AF QS.i), thus the query system in
figure 5 possesses the response property.

2) Verification of Persistence Property
Persistence property is used to describe the eventual stability of some state
of system [13]. This property is interpreted that if computation reach one state,
this state is continuing. In figure 5, after the results of user's query are
displayed, if multi-console (MC.d) not receives query request from users, the
system remains in S1 state. In a path of computation where the status changes
to S1 after S3, the stable status must be sustained. The property can be
expressed using CTL formula as follows:

(M,S0) ? AG (QT1.g ? EFG MC.e) ->
(M,S0) ? AG ~ QT1.g ? AG(EFG MC.e)

Table 2. Comparison between OMT and SACC Model

	OMT	SACC Model
Main Expression	Object Relationship: Links among objects are used to express object relation (Inheritance, Link, Association, Multiplicity, Role Name, Ordering, Qualification, Aggregation, Generalization)	Connections and Interactions Components: Links objects are used for method calls (Port name, Method calls by links, Return value)
Data Flow	Function Model: Use DFD (Date Flow Diagram). Expressed through actor, process, arrows, etc. Difficult to follow actual data flow objects.	Data flow and interactions between components can be identified through connectors.
Dynamic Model Distributed Model Concurrency Model	Dynamics and concurrency can be expressed through a scenario model and a status diagram. However, expression of distributed model is inadequate.	Dynamics and distributed model can be expressed through expression of interactions components. Expression of concurrency model is inadequate.
Inheritance	Direct inheritance objects.	Only indirect inheritance through IDL.
Activity Range	Single Machine	Global Network
Object Independence	Included in the entire system, which disallows it to be mentioned when entire system activity is discussed	Provides application-level independence

Table 3. Characteristic Comparison of SACC Model and ADL

	SACC	Rapide	Wright	Darwin	Aesop
Specification	Graphical Notation	Graphical Notation	Text	Graphical Notation	Text
Compositionality	IDL	Map	-	-	-
Scalability	Variable number of Component	Variable number of Component	Exact number of Component	-	-
Real-time	-	Support	-	Support	-
Distribute System	Exclusive	-	Support	-	-
Verification	Enable	Disable	Enable	Disable	Disable

43

Table 4. Component Comparison of SACC Model and ADL

	SACC	Rapide	Wright	Darwin	Aesop
Interface	Port	Constituent	Port	-	Port
Types	Distributed Object	Parameteri-zation	-	Parameteri-zation	
Semantic	CTL	Partially ordered event sets	-	p-calculus	
Constraint	CORBA environment	Algebric Language	Protocol of Interaction	-	
Evolution	Inheritance in IDL	Inheritance	-	-	

44

Table 5. Connect Comparison of SACC Model and ADL

	SACC	Rapide	Wright	Darwin	Aesop
Interface	Port	-	Connector Interface as Role	-	Role
Types	Protocol	First class entity	Protocol	First Class Entity	Protocol
Semantic	-	Poset	Glue with CSP	-	-
Constraint	Method call interaction	-	Specifying protocol for each role	-	-
Evolution	Inheritance in IDL	-	-	-	Behavior preserving subtyping

The system in figure 5 possesses the persistence property if any of two above, ~ QT1.g or FG MC.e, is satisfied in some path of computation.

If each status of CTL in all path of computation satisfies at least one out of two proposition, QT1.g, EFG MC.e, it possesses a persistence property. States in figure 5 satisfy AG (AF QS.i), thus the query system in figure 5 possesses the persistence property.

5. COMPARISONS WITH OTHER METHODS

5.1 Comparison between SACC Model and OMT (Object Modeling Technique)

OMT [20] is a representative object-oriented modeling technique, which refers to conventional objects. Distributed objects that are the subjects of SACC model are represented as classes similar to class categorization used in OMT. However, because of distinct difference between conventional and distributed object, it is not appropriate to model distributed objects with conventional objects modeling method such as OMT. By comparing the two models in the following Table 2, the characteristics and advantages of SACC model are clearly visible.

5.2 Comparison between SACC Model and ADL

In Table 3, Table 4, Table 5 [15], we have compared conventional ADL with SACC model, it's components, connectors and characteristics in this paper. It is difficult job to compare SACC model with ADL since SACC model is limited to a distributed object model. However, the comparison is useful taking into consideration that there are no ADL that are widely used and ADL is selected by each purpose.

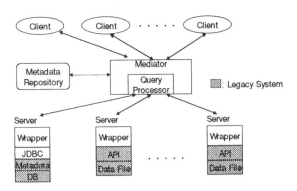

Figure 8. Layout of Target System

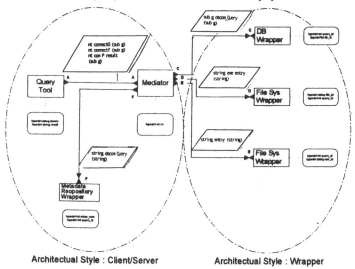

Target System : 3-tier Multi-data Source Query System

Architectual Style : Client/Server   **Architectual Style : Wrapper**

Figure 9. Application Example of SACC

6.   APPLICATION EXAMPLE OF SACC MODEL

SACC model suggested in this paper has been applied to design and implement 3-tier multi-data source query system[17]. If the 3-tier multi-data source query system in figure 8 is expressed using a SACC model, it will look like figure 9. SACC model in figure 9 can be verified with CTL and converted IDL as shown in figure 10 by the SCI algorithm. Therefore, since the final IDL produced is verified by SACC mdeling method, it is an ideal IDL to be used for system implementation in figure 8.

7.   CONCLUSION

Because development model that specifies and verifies CORBA distributed object system is lack, an effective CORBA-based distributed object system development was difficult. Conversion to IDL from system requirement analysis and verification of converted IDL is a difficult job for developers. To solve these problems, this paper suggests SACC model. SACC model is specialized form of conventional software architecture approach to effectively develop CORBA components. It is possible to specify and verify CORBA distributed object system on high abstract level with SACC model. Also, in SACC modeling method, structural elements of the SACC model and the implemented codes can be matched, therefore SACC model is used as a blueprint for the entire development of the system. SACC modeling method

```
Model
module  TQT
  {
     interface QP
       {
          typedef string query;
          attribute long count;
          typedef string stringResult;
          typedef string fsResult;
          int connectDB(string DB_addr)
          int connectFS(string FS_addr)
          string composeResult (string query)
        };

     interface FS1
       {
          typedef string fsResult1;
          string filequery1(string query);
        };

     interface FS2
       {
          typedef string fsResult2;
          string filequery2(string query);
        };

     interface MD
       {
          typedef string mdResult;
          string filequery2(string query);
        };

  } ;
```

Figure 10. Produced IDL from SACC Model

applies CTL to verify the SACC models then SACC model that is verified by CTL is converted IDL using SCI algorithms. Consequently we can remove burden of making IDL and acquire reliable IDL.

Our future research will include developing expression methods that can support parallel and concurrency processing. Ultimately, automated tools that support SACC modeling method should also be developed.

REFERENCE

[1] Robert J. Allen, " A Formal Approach to Software Architecture" Ph.D. Thesis, Carnegie Mellon University, Technical Report Number: CMU-CS-97-144, May, 1997.

[2] Linda Brownsword, Paul C. Clements, and Ulf Olsson, " Successful Product Line Engineering: A Case Study,". Software Technology Conference, Salt Lake City, April 1996.

[3] E. M. Clarke, E. A. Emerson, and A. P. Sistla, " Automatic Verification of Finite-State Concurrent Systems Using T1919emporal Logic Specifications" , ACM Transaction on Programming Languages ana System, Vol. 8, No.2,pp 244-263 April 1986.

[4] Paul C. Clements and Linda M. Northrop, " Software Architecture : An Executive Overview" http://www.sei.cmu.edu/architecture/projects.html 1996.

[5] Paul C. Clements, " A Survey of Architecture Description Languages " Eighth Intl. Workshop in Software Specification and Design, Paderborn, Germany, March 1996.

[6] Paul C. Clements, " Formal Methods in Describing Architectures" Monterey Workshop on Formal Methods and Architecture, Monterey CA, September 1995.

[7] Paul Clements, "From Subroutines to Subsystems: Component-Based Software Development," The American Programmer, vol. 8. no. 11, November 1995.

[8] Cristina Gacek, Ahmed Abd-Allah, and Bradford Clark, " On the Definition of Software System Architecture " USC Technical Report: USC-CSE-95-500, Nov 14, 1994.

[9] David Garlan, "An Introduction to the Aesop System" http://www.cs. cmu.edu/afs/cs/project/able/www/aesop/acsop_home.html 1995.

[10] David Garlan, " Introduction to the Special Issue on Software Architecture " IEEE Transactions on Software Engineering, April 1995.

[11] David Garlan and M. Shaw, " An Introduction to Software Architecture. " Advances in Software Engineering and Knowledge Engineering. Vol1. River Edge, NJ; World Scientific Publishing Company, 1993.

[12] Hiroaki Iwashita, Tsuneo Nakata, and Fumiyasu Hirose, "CTL Model Checking Based on Forward State Traversal" Proceedings of the 1996 IEEE/ACM international conference on Computer-aided design , 1996, Page 82.

[13] Zohar Manna and Amir Pnueli, " The Temporal Logic of Reactive and Concurrent Systems" Springer- Verlag pp 179- 221, 275-296.

[14] K. L. McMillan, " Symbolic Model Checking" , Kluwer Academic Publishers, pp 11-24.

[15] Nenad Medvidovic and Richard N. Taylor, "A Framework for Classifying and Comparing Architecture Description Language", Software Engineering Notes, V.22 N.6.

[16] Robert T.Monroe, " Stylized Architecture, Design Patterns, and Objects " IEEE Software January, 1997. pp. 43-52.

[17] Chang-Joo Moon, Sun-Jung Lee, Sung-Kong Park, and Doo-Kwon Baik, "An Interoperability Solving Method using Metadata at CORBA Environment between RDBMS and File System," KISS conference, spring, 1998.

[18] Robert Orfali, Harkey, and Edwards, " The Essential Distributed Objects " John Wiley & Sons, 1996.

[19] Robert Orfali, " Client/Server Programming with JAVA and CORBA " John Wiley & Sons, 1998.

[20] James Rumbaugh, "Object-Oriented Modeling and Design", 1991 , Prentice Hall

[21] Yves-Marie Quemener and Thierry Jeron, " Model-Checking of CTL on infinite Kripke Structures Defined by Simple Graph Grammer ".

DYNAMIC DATABASE OBJECT HORIZONTAL FRAGMENTATION

C. I. Ezeife* and Jian Zheng
School of Computer Science, University of Windsor
Windsor, Ontario N9B 3P4, CANADA
E-mail: Cezeife@cs.uwindsor.ca
http://www.cs.uwindsor.ca/user/c/cezeife

Abstract

Applying an object based horizontal fragmentation scheme to an object oriented database system, creates subsets of class extents which are allocated to sites where they are most needed. As new class instances are created or database schema evolves or application queries access patterns change at distributed sites, system performance may drop. Restoring the system performance requires re-running the object horizontal fragmentation scheme after conducting a full static system requirements analysis. This paper proposes an object horizontal distributed design architecture which determines system performance threshold, monitors changes in distributed design input which may affect performance and dynamically triggers a re-fragmentation of the system if performance drops below the system performance threshold.

Keywords: Object-oriented databases, Horizontal fragmentation, Distribution, Performance analysis

1 Introduction

Horizontally fragmenting database classes and placing class fragments at distributed sites has the advantage of reducing the amount of irrelevant data accessed locally by applications as well as reducing the amount of remote access made to data by applications running at other sites. Other advantages of data fragmentation include increased throughput since more than one application can run on fragments of a class placed at distributed sites concurrently. Fragmentation also reduces data transmission costs by transmitting only part of a class and not the whole class. Recent works on fragmentation of database relations and classes include [3, 4, 8]. All of these earlier distributed database design works are static designs because input to the design process, which include queries accessing database and their frequencies as well as the database schema, are obtained from an earlier full requirements analysis. This implies that when the system undergoes sufficient changes, a new requirements study is carried out to re-run the fragmentation and allocation schemes. In order to make these systems more acceptable by users, a mechanism for determining what change is considered sufficient to trigger a re-fragmentation is needed.

This paper adopts the technique of setting a distributed system performance threshold as a factor of the measured system performance immediately after fragmentation and allocation and a system determined variable which measures the average change in system input over time.

---

*This research was supported by the Natural Science and Engineering Research Council (NSERC) of Canada under an operating grant (OGP-0194134) and a University of Windsor grant.

Systems Development Methods for Databases, Enterprise Modeling, and Workflow Management.
Edited by W. Wojtkowski, *et al.* Kluwer Academic/Plenum Publishing, New York, 1999.

51

The paper proposes an architecture that includes a monitor component which keeps track of changes in the system and periodically re-computes the system performance values using the new input states. A re-fragmentation is triggered if the performance values fall below the threshold. The component used to evaluate the performance of the distributed system is the object horizontal partition evaluator presented in [5], which extends the initial work presented in [2] for vertical fragmentation of relations. Chakravarthy et al. in [2] presents an objective function for measuring the performance of vertical fragments of relations. [5] extends this work to measure object horizontal fragmentation including features of encapsulation, inheritance, class composition and method nesting hierarchies.

1.1 Data Model

Object-oriented databases represent data entities as objects supporting features of encapsulation. Thus database objects have a set of attributes ($\mathcal{A}$) and a set of methods ($\mathcal{M}$) for accessing these attributes. Objects with common attributes and methods belong to the same class C with a class identifier K. Every object has an object identifier corresponding to its class identifier prefixed to a unique object number. All objects of a class form its instance objects ($\mathcal{I}$) or class extents. Thus, we can represent a class as an ordered relation C = ($K,\mathcal{A},\mathcal{M},\mathcal{I}$). For example, a class *Person* has class identifier Person. Its attributes are social security number, name, age. Its methods are getname, getage etc. The instances of the class have names, John, Mary etc. Object-oriented database systems also support inheritance whereby a class B called subclass of class A can be defined as a specialization of class A. A class A is called the superclass of class B and class B can inherit all attributes and methods of its superclass A but not all of its extents. Thus if class B is a subclass of class A, A is represented as $C^A = (KA, A^A, M^A, I^A)$ while B can be represented as $C^B = (KB, A^A \cup A^B, M^A \cup M^B, I^B \odot I^A)$ where the symbol $\odot$ stands for object pointer, meaning that every $I \in I^B$ is connected to some $I \in I^A$. For example, in a database with class *Person* and *Student*, *Student* is a subclass of class *Person*. Object-oriented database systems also support class composition and method nesting hierarchies. Each horizontal fragment (C_h) of a class contains all attributes and methods of the class but only some instance objects ($I' \subseteq I$) [4]. Vertical fragments of a class contain subsets of attributes and methods but all instance objects of the class, while hybrid fragments contain subsets of both attributes, methods and instances of the class.

1.2 Related Work

In horizontally fragmenting relations, primary fragmentation is performed on all owner relations while derived fragmentation is performed on all member relations of links according to a selection operation specified on its owner relation. Ozsu and Valduriez [8] give a formal definition of simple predicates as $P_j : A_i \ \theta \ Value$ (e.g. Studentgpa ≥ 4.0) and that given a set of simple predicates $P_{ri} = \{P_{i1}, P_{i2}, ..., p_{im}\}$, the set of minterm predicates $M_i = \{m_{i1}, m_{i2},...,m_{iz}\}$ is defined as $M_i = \{m_{ij}|m_{ij} = \wedge P_{ik}^*\}$, $P_{ik} \in P_i$, $1 \leq k \leq m$, $1 \leq j \leq z$ where $P_{ik}^* = P_{ik}$ or $P_{ik}^* = \neg P_{ik}$. For example, with P_1 as studentgpa ≥ 4.0 and P_2 as major $=$"Math", the defined minterms are $m_1 = P_1 \wedge P_2$, $m_2 = P_1 \wedge \neg P_2$, $m_3 = \neg P_1 \wedge P_2$, $m_4 = \neg P_1 \wedge \neg P_2$. They also propose an iterative algorithm (COM_MIN) that basically eliminates minterm predicates which are either redundant or meaningless.

Object-oriented database systems that support some form of distribution or based on client/server model include GOBLIN [6], THOR [7] and for improved performance of such systems, techniques for proper fragmentation of objects are needed. Thus, Ezeife and Barker in [4] present a set of horizontal fragmentation algorithms (called object oriented horizontal fragmentation scheme OOHF), for four types of class models including those with both simple attributes and methods, simple attributes and complex methods, complex attributes and simple methods, and complex attributes and methods. Since the model with complex attributes and methods is the most encompassing, this algorithm is summarized next. The algorithm defines primary horizontal fragments of the class using predicates from queries accessing this class directly. Then, derived fragments of this class are defined using all primary fragments of all its owner classes from all link graphs to accommodate inheritance, class composition hierarchies and possibly method nesting hierarchies. Finally, the sets of primary and derived fragments of the class are combined following the rule that every derived fragment should be merged with a primary fragment it has the highest affinity count with. However, if a class has no primary fragments, its derived fragments become its final fragments ensuring all instance objects are placed in some fragment. Other object horizontal fragmentation schemes include the class dependence graph techniques in [9] and an approach by Bellatreche, Karlapalem and Simonet in [1] which defines primary horizontal fragmentation on simple queries while derived horizontal fragmentation is defined on component queries. Chakravarthy *et al.* in [2] presents an objective function called Partition Evaluator (PE) for evaluating the goodness of any vertical partitioning algorithm. The PE is used to measure the total costs of processing all applications at all distributed sites if each access to a data fragment amounts to a unit cost. It is assumed that there are no data redundancies and following fragmentation, the fragments have been allocated to sites where they are most needed. The two components of PE are irrelevant local method access cost which minimizes the square error for a fixed number of fragments and assigns a penalty factor whenever irrelevant attributes are accessed in a particular fragment. Ezeife and Barker extend the relational vertical Partition Evaluator (PE) to handle measuring the costs of local and remote accesses incurred by the object vertical fragments. Ezeife and Zheng in [5] presents an object horizontal partition evaluator algorithm which computes the processing cost of horizontally distributed objects by first computing the application instance object sets. Then, the performance of each class of the object base is measured by its partition evaluator value (PE). The PE of a class is the sum of its local irrelevant access cost (E_M^2) and its remote relevant access cost (E_R^2). To accommodate features of encapsulation, inheritance, class composition and method nesting hierarchies, access to a class, includes all queries accessing the class instances directly, those accessing it indirectly through its descendant classes, containing classes and complex method classes.

1.3 Contributions

This paper contributes by proposing an architecture for dynamic object horizontal fragmentation. Components of the dynamic object horizontal distribution design include: (1) a mechanism for determining the performance threshold of an object horizontal distributed system which is based on the partition evaluator measure. (2) a monitor algorithm for keeping track of changes in the system. This component eliminates the need for a full static requirements analysis and provides input for measuring system performance in order to determine when a re-fragmentation should be initiated. Dynamic fragmentation saves processing cost by eliminating the cost of full requirements study before every re-fragmentation. It also improves on overall system performance because it is able to detect early when the system performance drops below acceptable threshold. Furthermore, user confidence in the system is increased since the user is more reassured to know there is a mechanism in place for detecting and correcting low system performance.

1.4 Outline of the Paper

Section 2 presents the architecture of the dynamic object horizontal distributed database design and the algorithm for determining performance threshold (OOThreshold algorithm). Section 3 presents the distributed object-oriented monitor algorithm (DOBS-Monitor algorithm). Section 4 presents an example while section 5 presents conclusions.

2 Dynamic Distributed Objectbase Horizontal Design Architecture

This section presents the architecture of the dynamic object horizontal distributed database design (section 2.1), including the algorithm for determining the system performance threshold (section 2.2).

2.1 Dynamic Distributed Objectbase Architecture and Components

Figure 1 shows the architecture of the dynamic distributed object horizontal design. At the onset of the design, the input of application queries (AQ) and the sites where they run, query access frequencies (AF), class inheritance hierarchy (CH), class composition hierarchy (CCH), method nesting hierarchies (MNH) and the objectbase class instances (IS) are fed to an object horizontal fragmentation scheme like OOHF in [4] or Bell's in [1].

The effect is that horizontal fragments of objects is defined and allocated to distributed sites. The instance object fragments allocated to distributed sites, and other inputs to distribution design (AQ, AF, CH, CCH and MNH) now constitute input to the object horizontal partition evaluator algorithm [5]. The object horizontal partition evaluator algorithm immediately following a fragmentation and allocation, computes the best system partition evaluator using the input data. Next, the dynamic architecture system computes the system performance threshold using the computed best partition evaluator and system change variable (S). The system change variable measures or represents how frequently the system's input set that may affect performance change. The S is computed as the inverse of the average percentage change in system input, multiplied by 10 to allow a reasonable margin of degrading performance. The inverse is used because a frequently changing system would require more frequent re-fragmentation and thus a lower S value since a lower S value means a threshold that is closer to the best PE value. The lower the threshold, the more likely it is that the system computed PE value at any other check time is higher than threshold and would call for a re-fragmentation. The implication is that if a system's set of input change frequently, then its average change is very high making its inverse very low and close to zero. Then, the system threshold which is computed as the addition of the best PE following a fragmentation and its product with the system change variable S (that is (threshold = best PE + (best PE * S))) is close to best PE value. The margin of lower performance allowed by the system before calling for a re-fragmentation is the fraction of the best PE determined by S. This margin is closer to best PE if S is closer to 0 and S is closer to 0 as the system average input change increases. The more frequently a system changes, the lower its S value; while a high S value represents a slowly changing system with a low average percentage change in its input. Note that S has a positive value and a higher PE value represents a lower performance by the system.

The dynamic architecture also includes a Monitor algorithm which keeps track of all changes in the input data and runs periodically to determine if the change is big enough to call for a re-evaluation of the system performance. A re-evaluation of the system performance is accomplished by computing the system's current PE value using the object horizontal partition evaluator algorithm. It is only when the current PE value of the system is computed that the current PE value is compared with performance threshold value. A current PE value bigger than the performance threshold value calls for the system to be re-fragmented using the new input data from the monitor. The monitor keeps track of the system average change for each input type over time and compares the new change in this input type with the average change. Its output is a value (check-pe) to recommend whether to run the object horizontal partition evaluator for evaluating system performance or not. It also outputs the new system average change for each input data as the average of the old and new changes. In Figure 1, the algorithms are shown in rectangles while the solid input and output data are shown in left-corner trimmed rectangles. In addition to the overall system architecture, the algorithms, OOMonitor and Compute Threshold are contributions of this paper.

2.2   The Algorithm Compute Threshold

Determining the system performance threshold is necessary if changes in the input that would drop system performance below acceptable limits are to be detected. Thus, the system change variable S is used to determine the performance threshold value of the system. We propose a method for computing the threshold value as the sum of the best system PE obtained immediately after a re-fragmentation and a fraction of this best PE value determined by S. This means that threshold is computed as best PE + (best PE * S). The value of S is determined as the inverse of the average change in the system input data. Since it is not easy deciding how to measure changes in such system inputs as class inheritance, class composition hierarchies, we define a technique for obtaining the changes in the input data in section 3. This computes the change in an input data like access frequency as a percentage of the original access frequency, while creation, destruction or re-location of a query each counts as a change measured against the total number of queries in the system. More detailed and formal presentations of these definitions are discussed later since it is the monitor program that is responsible for computing those changes.

The average changes are gathered by the monitor algorithm and passed in as input to the compute threshold algorithm. Once the input changes are computed, the S variable is simply computed as the 10 divided by the average of the percentage changes in all of these input data. For example, in system with a 25% change in class inheritance hierarchy, 35% change in class composition hierarchies, 10% change in application queries, 20% change in application frequencies, no change in method nesting hierarchies and 10% change in object instance fragments, the S value is computed as $10/((25 + 35 + 10 + 20 + 10)/5) = 0.5$. On the other hand, if the average percentage change is 50%, the S is computed as 10/50 or 0.2. This shows that a higher change produces lower S while lower percentage changes in system input produces higher S value. The threshold value is now determined as best PE value + (best PE

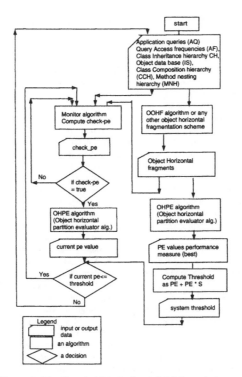

Figure 1: Dynamic Distributed object horizontal database design architecture

value * S). This means that any input change less or equal to the average input change would not trigger a re-fragmentation. The formal presentation of the Compute Threshold algorithm is given as Figure 2.

Algorithm 2.1 *(Algorithm (OOThreshold- computes the system performance threshold))*

Algorithm OOThreshold
Input: (input changes: $\triangle AF$, $\triangle AQ$, $\triangle CH$, $\triangle CCH$, $\triangle MNH$, $\triangle IS$ fragments),
            best PE
Output: Performance threshold
Begin
          $S = 10/((\triangle AF + \triangle AQ + \triangle CH + \triangle CCH + \triangle MNH + \triangle IS)/(\text{number of inputs with change} > 0))$
          Threshold = best PE + (best PE * S)
end // of oothreshold //

Figure 2: The Algorithm Compute Threshold

3   The Distributed Object-oriented Monitor Algorithm (OOMonitor)

The object-oriented monitor algorithm serves to compute the input changes used to determine the system change variable S by the Compute Threshold algorithm (OOThreshold). The monitor also determines whether the system performance should be checked or not. By determining the times it is necessary to re-compute the system current PE value, the overhead incurred doing the checks is reduced. The monitor would recommend that the system PE values be re-computed only if during any system input change check, it encounters an input change that is higher than the average system input change. It re-computes the average system input change only after a re-fragmentation. The other question is determining how frequently the monitor itself should run. We recommend that the monitor be run periodically, once a day for some applications but may be set to suit application requirements.

The input to the monitor include the previous AQ, AF, CH, CCH, MNH and IS fragments and their sites. Other input to the algorithm are current AQ, AF, CH, CCH, MNH and IS fragments. An additional input is periodic wait time which is used to determine how often the monitor should run. In order to compute the average changes in the input data, the monitor needs to remember the previous change in each input data so that it can compute the average change by finding the sum of the previous change and the current change. The monitor output includes a boolean variable which is either true or false where true means that the system PE value should be re-computed and false means that it should not. The other output data from the monitor are the average changes in the input data of AQ, AF, CH, CCH, MNH and IS fragments. These average input changes are used to compute the system change variable S by algorithm OOThreshold. A change in any input data of AQ, AF, CH, CCH, MNH and IS fragments is computed as the percentage of the original data. This means that in a class hierarchy with 8 classes, if a class gets added, is dropped or moves to another location in the hierarchy, the change is defined as $1/8*100 = 12.5\%$. If a total of two classes is either created, destroyed or relocates, the change is defined as $2/8 * 100 = 25\%$ change. This definition applies to all hierarchies. In the case of object instance fragments, if a new object gets created in any site, it stays in that site until a re-fragmentation. To compute the change in object instance fragments, we count any new object that is created at any site as 1, any destruction or relocation of an object also counts as 1. The total change in the object instance fragments of a class is the sum of all changes at all sites divided by the number of instance objects in the class * 100. The change in application queries is defined as the percentage of new queries that access a class, stop accessing a class or change the site of access to a class. Thus, the change in application queries accessing a class is obtained as the sum of the number newly accessing it, or no longer accessing it divided by the total number of queries originally accessing the class multiplied by 100. The change in access frequency of queries is defined as the maximum of the percentage change in any one query at all sites. Thus, the change in access frequency of queries is computed as maximum of the total change at all sites for each query divided by total access at all sites for that query multiplied by 100. The formal presentation of the Monitor algorithm is given as Figure 3. The monitor computes change in each input data during each run as the average of the previous change and the new change.

4 An Example

This section shows an example object database system with its class inheritance hierarchy similar to the example in [4], which is first fragmented horizontally using the object horizontal fragmentation scheme OOHF in [4]. The sample object database represents a university database with classes *Person, Prof, Student, Grad, Undergrad and Dept*. Classes Prof and Students are subclasses of Person while classes Grad and Undergrad are subclasses of class Student. Both classes Person and Dept are subclasses of the Root class. This section shows how the performance threshold of the distributed system is measured and how a change in the access frequencies of queries accessing these classes at distributed sites is used by the monitor system proposed in this paper to trigger a re-fragmentation of the classes whose re-computed PE values are higher than the threshold PE values. Using the given query access frequency and other input data, the object horizontal fragmentation scheme (OOHF) produces the following fragments for the classes in this database, where the I's represent the classes' instance objects in each fragment.

Class Person

$$F_1^h = \{I_1, I_7\}, \ F_2^h = \{I_3, I_4\}, \ F_3^h = \{I_2, I_8\}, \ F_4^h = \{I_{10}\}, \ F_5^h = \{I_6, I_9\}, \ F_6^h = \{I_5\},$$

Class Student

$$F_1^h = \{I_1, I_5\}, \ F_2^h = \{I_2, I_4\}, \ F_3^h = \{I_3, I_6\},$$

Class Prof

$$F_1^h = \{I_2, I_3, I_4\}, \ F_2^h = \{I_1\},$$

Class Dept

$$F_1^h = \{I_2, I_3\}, \ F_2^h = \{I_1\}, \ F_3^h = \{I_4\},$$

Class Grad

$$F_1^h = \{I_1\}, \ F_2^h = \{I_2, I_3\},$$

Algorithm 3.1 *(Algorithm (OOMonitor - computes when to re-evaluate the system))*

Algorithm OOMonitor
Input: Previous AQ, AF, CH, CCH, MNH, IS
    Current AQ, AF, CH, CCH, MNH, IS
    monitor_wait_time(length of time in seconds)
    Previous system change averages ($\triangle AQ, \triangle AF, \triangle CH, \triangle CCH, \triangle MNH, \triangle IS$)
Output: check_current_PE (boolean)
    check_current_PE (boolean)
    new system change averages ($\triangle AQ, \triangle AF, \triangle CH, \triangle CCH, \triangle MNH, \triangle IS$)
begin
//Initialize variables //
check_current_pe = false
time = systemtime - lastruntime
Read all input data
While not(check_current_pe) and (time >= monitor_wait_time)
Begin
  Current change AQ = $|currentAQ - previousAQ|$
  Current change AF = $|currentAF - previousAF|$
  Current change CH = $|currentCH - previousCH|$
  Current change CCH = $|currentCCH - previousCCH|$
  Current change MNH = $|currentMNH - previousMNH|$
  Current change IS = $|currentIS = previousIS|$
  //Monitor first determines if the system pe should be recomputed//
  //before re-computing the new average system changes. //
  if current change AQ > previous change AQ then check_current_pe =true

else if current change AF > previous change AF then check_current_pe =true
else if current change CH > previous change CH then check_current_pe =true
else if current change CCH>previous change CCH then check_current_pe =true
else if current change MNH>previous change MNH then check_current_pe =true
else if current change IS>previous change AQ then check_current_pe =true
average $\triangle AQ$ = (current $\triangle AQ$ + previous $\triangle AQ$)/2
average $\triangle AF$ = (current $\triangle AF$ + previous $\triangle AF$)/2
average $\triangle CH$ = (current $\triangle CH$ + previous $\triangle CH$)/2
average $\triangle CCH$ = (current $\triangle CCH$ + previous $\triangle CCH$)/2
average $\triangle MNH$ = (current $\triangle MNH$ + previous $\triangle MNH$)/2
average $\triangle IS$ = (current $\triangle IS$ + previous $\triangle IS$)/2
lastruntime = systemtimenow
end // of while //
end //of monitor //

Figure 3: The Distributed Object-oriented Monitor Algorithm

Class	Site 1	Site 2	Site3
Person	F1={I1, I7} F5={I6, I9} F4={I10}	F6={I5} F2={I3, I4}	F3={I2, I8}
Student	F1={I1, I5}	F2={I2, I4}	F3={I3, I6}
Prof	F1={I2, I3, I4}	F2={I1}	
Dept	F1={I2, I3}	F2={I1}	F3={I4}
Grad	F1={I1}	F2={I2, I3}	

Figure 4: Allocation of Fragments to Distributed Sites

	S1	S2	S3	Running directly on
q1	10	20	0	Grad
q2	20	30	15	Prof
q3	25	10	0	Prof
q4	15	25	10	Student
q5	20	20	20	Dept

Figure 5: Access Frequencies of queries at distributed sites

The fragments above are allocated to three distributed sites using the simple allocation scheme that places fragments at sites where they are most needed by queries, while Figure 4 shows the placement of above fragments to sites and Figure 5 shows the access frequencies of queries at the three sites.

The local irrelevant access costs, remote relevant access costs and the PE values obtained after executing the OHPE algorithm are shown in the table below:

Class	Local irrelevant	Remote relevant	PE value
Person	250	0	250
Student	250	0	250
Prof	0	0	0
Grad	0	0	0
Dept	0	0	0

Assume that overtime the change in the system change variable is determined by the OOThreshold algorithm to be 0.3, then the OOthreshold computes the system performance threshold as best PE of the class plus best PE of the class multiplied by S. This gives for classes *Person* and *Student* the threshold values of 333 and for the rest of the classes threshold values of 0. This is because the algorithm computes S as 10/30, and threshold as best PE + (S * best PE), which gives 325 + (10/30 * 325). Thus, if every other input to the system remains constant but there is a change in the access frequencies of the five queries to give the new access frequencies shown in Figure 6, then, the OOMonitor algorithm will compute the new change

	S1	S2	S3	Running directly on
q1	20	30	10	Grad
q2	25	35	20	Prof
q3	30	15	5	Prof
q4	25	35	20	Student
q5	25	25	25	Dept

Figure 6: The new Access Frequencies of queries at distributed sites

in query 1 as ((20-10)+(30-20)+(10-0))/30 * 100 which gives a percentage change in the access frequency of 100% or 1.0. Since this change is greater than the S value of 0.3, the Monitor algorithm will recommend that the system PE-value be re-computed. The system PE values are again re-computed with the OHPE algorithm of [5] using the new access frequencies and the results obtained are given below:

Class	Local irrelevant	Remote relevant	PE value
Person	650	0	650
Student	650	0	650
Prof	0	0	0
Grad	0	0	0
Dept	0	0	0

5 Conclusions

The importance of dynamic object fragmentation had been argued by many earlier works. The value of such a technique increases when it is horizontal fragmentation which is more commonly used in the industry. A dynamic horizontal fragmentation approach provides the means to detect when the distributed system performance degrades below acceptable limit.

The advantage of such a design includes reduced processing cost by eliminating the need for full systems requirements analysis, user confidence because system is participating in alerting the user when there is low performance, and higher productivity.

A re-design of the distribution technique may be needed when due to changes in application access pattern or object database schema, system performance has dropped below an acceptable threshold. This paper contributes by first presenting a dynamic distribution design architecture and then, a set of algorithms that would determine the system performance threshold based on the computed partition evaluator value and the average rate of change of system input. The paper also uses the average system change variable and the threshold to determine when a re-design is necessary. A re-design is necessary when the computed current partition evaluator value of the class is higher than the performance threshold value. A re-design would entail running the object horizontal fragmentation scheme with the most current input data before allocation to distributed sites. Future work should investigate reducing the cost of re-design of the system by performing incremental re-fragmentation whereby only the changes in the input data are used to re-fragment the system.

References

[1] L. Bellatreche, K. Karlapalem, and A. Simonet. Horizontal Class Partitioning Design in Object-Oriented databases. In *Lecture Notes in Computer Science volume 1308 DEXA*, pages 58–67, Toulouse-France, September 1997.

[2] S. Chakravarthy, J. Muthuraj, R. Varadarajan, and S. B. Navathe. An Objective Function for Vertically Partitioning Relations in Distributed Databases and its Analysis. *Distributed and Parallel Databases*, 2(1):183–207, 1993.

[3] C. I. Ezeife and Ken Barker. Distributed Object Based Design: Vertical Fragmentation of Classes. *Journal of Distributed and Parallel Database Systems*, 6(4):327 – 360, 1998. Kluwer academic Publishers.

[4] C.I. Ezeife and Ken Barker. A Comprehensive Approach to Horizontal Class Fragmentation in a Distributed Object Based System. *International Journal of Distributed and Parallel Databases*, 1:247–273, 1995. Kluwer Academic Publishers.

[5] C.I. Ezeife and Jian Zheng. Measuring the Performance of Database Object Horizontal Fragmentation Schemes. submitted to the 3rd international database engineering and Applications Symposium (IDEAS99), 1999.

[6] M.L. Kersten, S. Plomp, and C.A Van Den Berg. Object Storage Management in Goblin. In M. Tamer Ozsu, U. Dayal, and P. Valduriez, editors, *Distributed Object Management*. Morgan Kaufmann Publishers, 1994.

[7] Barbara Liskov, A. Adya, M. Day, M. Castro, S. Ghemawat, R. Gruber, L. Shrira, A.C. Myers, and U. Masheshwari. Safe and efficient sharing of persistent objects in THOR. In *proceedings of ACM SIGMOD International Conference on Management of Data*. Association of Computing Machinery, 1996.

[8] M.T. Ozsu and P. Valduriez. *Principles of Distributed Database Systems*. Prentice Hall, 1991.

[9] M. Savonnet, M. Terrasse, and K. Yetongnon. Fragtique: A Methodology for Distributing Object Oriented Databases. In *Proceedings of the 9th International Conference of Computing and Information ICCI'98 Winnipeg*, pages 148–159, 1998.

AN ELECTRONIC EKD METHOD GUIDE OVER THE WEB[1]

Judith Barrios[2], Christophe Gnaho, and Georges Grosz
Université Paris 1 Panthéon-Sorbonne
Centre de Recherché en Informatique
90, rue de Tolbiac75013 PARIS, France
E-mail: {barrios, gnaho, grosz}@univ-paris1.fr.
Ph.: 01 40 77 46 04, Fax: 01 40 77 19 54

Key words: method engineering, process guidance, world wide web, EKD method,
information system development

Abstract: In this paper we propose a web based tool called « The EKD Electronic Guide
Book » which forms the base line of the guidance component of the EKD
method. The EKD method proposes a set of models for describing enterprise
knowledge for the purpose of handling situations of change such as business
process transformation or improvement, information system re-engineering,
information system requirements elicitation, etc.. The core of the « EKD
Electronic Guide Book » is a set of process guidelines which constitute the
expression of EKD method knowledge and heuristics. These guidelines help in
both the understanding of the EKD models concepts and the use of these
concepts in the construction of the different EKD models. Following the
process modelling approach developed in the ESPRIT project
NATURE(Novel Approaches to Theories Underlying Requirements
Engineering), they are expressed as the contexts formed of a situation and a
decision. The paper describes the structure of the process guidelines
repository, the structure of the tool along with its main functionalities. All
these are exemplified.

INTRODUCTION

The Enterprise Knowledge Development (EKD) method proposes a set of
models for describing enterprise knowledge for the purpose of handling situations of
change such as business process transformation or improvement, information system
re-engineering, information system requirements elicitation, etc.. The expression of

---

[1] This work is partly funded by the European Commission under the ESPRIT project
ELEKTRA (N° 22927).
[2] The author is supported by the University of Los Andes. Venezuela

the enterprise knowledge takes the form of different inter-related models, namely object, goal, actor/role, role/activity and rule models ([LOUC97, BUBE97]). These models are referred to as the product component of the EKD method.

Complementary to the product component, guidelines have been defined. These guidelines are the expression of knowledge and heuristics that help in both the understanding of the EKD concepts and the use of these concepts in the construction of the different EKD models.

These guidelines are accessible through a WEB based tool, called the « The EKD Electronic Guide Book » which forms the base line of the guidance component of the EKD method. Thanks to the WEB based approach, the tool can be either used in a standalone manner (i.e. without the help of any other tool) or in connection with a CASE tool dedicated to the EKD method. In the latter case, the EKD Electronic Guide Book can be seen as an on-line powerful intelligent help that can be activated by an EKD engineer at any time while using the different editors of the CASE tool. In the former case, the EKD Electronic Guide Book, that indeed can be used through the Internet, acts as an hypertext media like encyclopaedia that allows any user to gain knowledge on the use of the EKD concepts.

We view an EKD process, and more generally any system engineering process, as a decision based process [ROLL94], i.e. an EKD engineer progresses in models construction by making decisions on what to represent and how to represent it. Following this line, the description of the guidelines is based on a decision oriented approach that is presented in the section 4. EKD guidelines are stored in a repository that constitute the core component of the EKD Electronic Guide Book. The rest of the paper is organized as follows. Section 2 present some related works. Section 3 presents an overview of the EKD method. Section 4 presents the decision oriented approach. The global architecture and the main functionalities of the « EKD Electronic Guide Book » are described in section 5. In section 6, the guidelines that have been developed so far are listed. Conclusions and future work are drawn in section 7.

RELATED WORKS

Both software and method engineering communities are investigating the issue of using Web based Technology for browsing and presented method related information to method participants. We refer in the following to the contributions more relevant to our proposal.

The V-Model Browser [V-Model-Browser, 1993] provides access to a German national software development standard in widespread use within governmental and industrial software development projects. The browser uses hyper-linked, electronic version of the standard to provide access to process elements. However, only one web page is visible at a time, so it is not possible to simultaneously see an overview of that page, navigation information and related information. It also offers no diagrams or other graphics.

In recent years, several firms have developed Process Asset Libraries (PALs) [Kellner93]. PALs are an instance of the more general notion of a corporate " memory ". It is usually a repository of historical data useful for decision-making. Intranets are sometimes used to assist in organizing the data. In comparison to our work, PALs are focused on decision making needs of project managers rather than the method understanding needs of Information Systems Engineers.

The Electronic Process Guide (EPG) [Kellner98] is closer to our vision. However, the EPG process guide pages are not generated dynamically from a database. Moreover, the tool does not provide an effective guidance mechanism as the searching mechanism is based on simple keyword matching.

OVERVIEW OF THE EKD METHOD

The Enterprise Knowledge Development (EKD) approach represents the confluence of a number of interrelated methodological components whose main objective is the development of an enterprise knowledge model according to a current business situation. This model identifies the enterprise entities their attributes and explicit relationships between them.

Taking into account the different areas of business concern, EKD model is partitioned into coherent sub-models :
- the *enterprise goal* sub-model which expresses the concepts involved in modelling enterprise objectives,
- the *enterprise actor- role* sub-model which expresses the concepts involved in modelling the enterprise actors and their roles, together with their interactions,
- the *enterprise role - activity* sub-model which expresses the concepts involved in the execution of any enterprise process and/or activity as well as individual roles played in the fulfilling of enterprise functions.
- the *enterprise object* sub-model which expresses the concepts involved in modelling physical and informational objects.
- the *enterprise rules* sub-model which expresses the concepts involved in modelling the policy, constraints, procedures etc. setting the bases for the execution and control of the way that processes, activities, roles and resources may behave in the enterprise.

All these sub-models are linked by the interaction sub-model, which brings together into an integrated view, the enterprise partitioned knowledge represented by each of EKD sub-models, its interrelations and dependencies. This kind of models permits design appropriate discussions about how things interact and influence each other, leading to more complete representations of business knowledge perceptions.

As we can see, the EKD method is complex because the EKD engineer needs to manage several interrelated models. Therefore, the EKD engineer must be guided effectively and efficiently while using the method assuring that the final business model represents reality as nearly as possible.

The guidance proposed by the « EKD Electronic Guide Book » is based on a decision oriented approach. This approach allows to represent design decisions and trace them. It also facilitates the understanding of the design process, its decisions and arguments providing EKD engineers with a more effective means to use the EKD method.

NOTION OF METHOD CHUNK

In this section, we describe the formalism used to express EKD guidelines. Indeed, the method knowledge base is structured according to this formalism.

Generally speaking, and following [ROLL95], we consider a method as being composed of a set of method chunks. In the context of the EKD method, a method

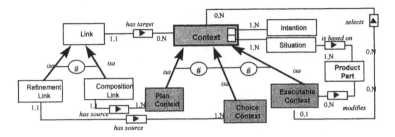

Figure 1. The structure of a method chunk

chunk describes guidelines related to the use of EKD models, it suggests how to progress at a given point of the EKD process, how to fulfil a design goal that an EKD engineer may have. A method chunk might be looked upon as a structured module of knowledge for supporting decision making in EKD processes.

Following the contextual approach developed within the ESPRIT project NATURE [NATU96], we propose to describe a method chunk using the concept of *context*. As depicted in Figure 1, a *context* is defined as a pair *<situation ; intention>*. An *intention* represents a design goal an EKD engineer wants to fulfil at a given point in time during the development process. A *situation* is a part of the product it makes sense to make a decision on. What we mean here by product refers to the different EKD models. A context has an associated process guidelines that describes how to achieve the intention with regard to the context's situation.

Contexts can be linked repeatedly in a hierarchical manner. A hierarchy of contexts is composed of *contexts* and *links*. As shown in Figure 1, links are of two kinds : *refinement links* which allow the refinement of a large-grained context into finer ones and *composition links* for the decomposition of a context into component contexts.

Finally, contexts are of three types, namely executable, choice or plan. Each type of context plays a specific role. We now consider each of these types of context in turn.

1. Plan Context

In order to represent situations requiring a set of decisions to be made for fulfilling a certain intention the process modelling formalism includes a first type of context called the plan contexts. A plan context can be looked upon as a macro issue which is decomposed into sub-issues, each of which corresponds to a sub decision associated to a component situation of the macro one. Components of a plan context are also contexts but related through *composition links*. In Figure 2, the context *<(Goal) ; Use an actor ability based strategy>* is a plan context composed of three component contexts, namely *<(Goal) ; Identify role>*, *<(Role identified) ; Identify sub-goal>*, *<(Sub-goals identified) ; Perform an AND reduction between sub-goals>*. This means that, when using the actor ability based strategy, the EKD

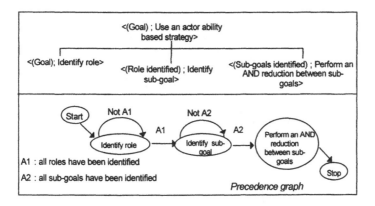

Figure 2. Example of a plan context

engineer has first to identify the different roles involved in the fulfilment of the goal, then identify a sub-goal for each role and finally perform an AND reduction between the identified sub-goals and the goal of the context' situation. Figure 2 provides the graphical description of this example of plan context.

Component contexts of a plan context can be organised in a sequential, iterative and/or parallel manner. This is graphically represented as a *precedence graph* (see the example in Figure 2). There is one graph per plan context. The nodes of the precedence graph are the component contexts of the plan while the links, called *precedence links,* define either the possible *ordered transitions* between contexts or their possible *parallel execution.* Arguments can be provided on the precedence links defining when the transition can be performed (e.g. A1 and A2 in Figure 2).

2. Choice Context

When building an EKD product, an EKD engineer may have several alternative ways to solve an issue. Therefore, he/she has to select the most appropriate one among the set of possible choices. In order to model such a piece of process knowledge, we introduce the second type of context, namely the *choice context.* The execution of such a context consists of choosing one of its alternatives, i.e. selecting a context representing a particular strategy for the resolution of the issue raised by the context.

For example in Figure 3, the context *<(Goal) ; Reduce goal>* is a choice context introducing three alternatives to goal reduction.

Arguments are defined to support or object to the various alternatives of a choice context. For example, the *Use an Actor driven strategy* is the right decision to make when the satisfaction of the goal requires the contributions of one or several actors that can be isolated. Description arguments play an important role. They help in capturing heuristics followed by the EKD engineer in choosing the appropriate

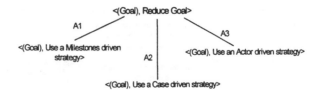

A1 : Achieving the goal requires some intermediate states to be reached in which some property about the product must hold.
A2 : One can identify an exhaustive list of cases in which the goal has to hold and, the work to be achieved is different for each case.
A3 : The contribution of one or several actors in the goal satisfaction can be isolated.

Figure 3. An example of a Choice Context

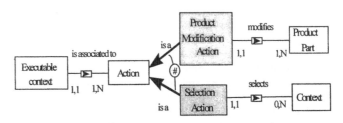

Figure 4. The detailed structure of an executable context

Let us exemplify the two types of actions.

problem solving strategy. Figure 3 presents, through an example, the graphical notations of a choice context.

Finally, it is important to notice that the alternatives of a choice context are contexts too. In the example, all alternatives contexts are plan contexts. But they could have been choice contexts introducing what is referred to as a refinement based hierarchies of contexts.

Similarly, components of a plan context are contexts too. As mentioned earlier, this leads to define method chunks as *hierarchies of contexts* having executable contexts as leaves of those hierarchies (Figure 1).

3. Executable Context

An *executable context* corresponds to an intention which is directly applicable through an *action*. This action leads either to a modification of the EKD specification under development or to the selection of a context. Figure 4 introduces the details of the relationship between an executable context, the different actions it can trigger, and what is impacted by these actions.

Let us consider again the context: $<(role\ identified),\ identify\ sub\mbox{-}goal >$ introduced in Figure 2. This context is an executable context, it is associated to an action that requires the EKD engineer to identify a sub-goal that is associated to the role of the situation.

Let us now consider the following context: $<(goal,\ goal.state\ =\ acquired)\ ;$ $progress\ by\ reducing\ goal>$. This context is also an executable context but its associated action is a "selection action". It suggests the EKD engineer to progress in the goal modelling activity, once a business goal have been acquired, by reducing the goal. Indeed, reducing a goal is handled by a specific method chunk. Therefore, the action of this context selects the context <(Goal) ; Reduce Goal> presented in Figure 3.

In fact this distinction between actions leads to distinguish two types of method chunks.

The first type concerns method chunks that lead to product modifications, these provide what is called « Point Guidance » in [SISA97]. The leaves of such a type of method chunk comprise executable context whose action are all of the type "Product Modification Action". Such method chunks offer guidelines in the satisfaction of an intention.

The second type concerns method chunks that lead to context selection, (their executable contexts are associated with "Selection Action"), these provide « Flow Guidance » [SISA97]. Such method chunks help in selecting the next intention to be fulfilled to make the process proceeds.

THE EKD ELECTRONIC GUIDE : ARCHITECTURE AND FUNCTIONALITIES

In this section, we present the global architecture of the tool and its main functionalities. For the sake of clarity, the functionalities are presented through the description of snapshots taken from the tool. All these snapshots partially describe the execution of a context.

Our approach emphasises the integration of Web based and hypertext based technologies with methodological guidance tailored to the EKD method. The hypertext technology allows users to easily understand and take advantage of the myriad of inter-relationships in any knowledge base. It helps by streamlining access to and providing rich navigational features around related information, thereby increasing user comprehension of information and its context [BIEB95]. Used in the context of the EKD method, hypertext facilities permit access to EKD guidelines in a simple manner, providing different levels of details, allowing to combine textual descriptions along with graphical descriptions. It also defines links between pieces of knowledge and leads to a smooth navigation within the method knowledge base. The Web based technology offers accessibility, portability and eases the distribution.

There are many potential approaches for integrating Web servers with knowledge bases, such as Common Gateway Interface (CGI), server API (e.g., Netscape's NSAPI and Microsoft's ISAPI), Applets, ActiveX, etc ; but none of them is systematic [CHIU97] nor complete or fully standardised. Our approach uses conjointly traditional CGI, applets and client scripting written as Visual Basic Scripts (VBScript) and Java Scripts (JScript).

1. Global Architecture

As shown in Figure 5, the architecture of the EKD Electronic Guide Book is based on the Web intrinsic client-server environment. Therefore, the tool is composed by two inter-related modules: the *User Guidelines Browser* (which runs at the client site) and *Guidelines Search Engine* (which runs at the server site).

The *User Guidelines Browser* is the interface provided to the EKD engineer, it consists of a set of HTML forms which allows him/her to interact with the tool and to access the method knowledge base (arrows labelled 1 in Figure 5) through either Internet, Intranet or locally. These HTML forms can be displayed by any traditional web navigator (form-compatible Web browser), such as Microsoft Internet Explorer ® or Netscape ®, among others.

We embed, directly into User Guidelines Browser HTML forms, VBScript ® code represented as standard ASCII text. VBscript is a simple subset of Microsoft Visual Basic for Applications (VBA) [BROP96]. This code runs entirely within the User Guidelines Browser at the client site and aims at helping CGI applications by pre-processing user's actions and information entered through the HTML forms. Some client-side applications, for instance a graph browser written in the Java language, are downloaded through the network from the Web server. They are also called Java Applets.

The *Guidelines Search Engine* is composed of a Web server, a set of CGI Applications, and the method knowledge base. Indeed, it runs at the server site. The Web server is responsible for all communication exchanges between the client(s) and the CGI Applications (arrows labelled 1 and 2 in Figure 5) that run on the server. It receives requests from the client(s) and sends the resulting Web pages that are generated « on the fly » from the method knowledge base (arrows labelled 1 and 6 in Figure 5). We chose to use the software called « Website ®» for the development of our Web server. This software provides many features which bring power and flexibility in the development. For instance, Website provides a fully graphical «property sheet » administration tool that makes web server construction and configuration very easy to use, it also offers an integrated toolkit for developing CGI applications using Visual Basic. The Web Browser and Server communicate each other by using the Hypertext Transfer Protocol (HTTP).

The interactions between the method knowledge base and the Web server are handled by a set of CGI applications. The CGI applications are organised in two modules: the Dynamic Form Processing module and the Query module. The first one transforms method knowledge base output data, embeds them into HTML forms (arrows labelled 4 and 5 in Figure 5) and returns them to the WWW Server. The second module manages all queries that emanate from the EKD engineer. This module interacts directly with the method base by transforming users queries into SQL queries (see arrows 2 and 3 in Figure 5).

The method knowledge base is implemented with Microsoft Access ® RDBMS. It is structured according to the formalism presented in section 2. Thank to the multi media ability of Microsoft Access, images are directly stored into the database. Note that the tuples of the database do not contain any HTML statement, as already mentioned, these are always generated on the fly.

2. Tool Functionalities

The first contact with the EKD Electronic Guide Book is through the home page shown at the top of Figure 6 (window 1). The « Starting the EKD Electronic Guide

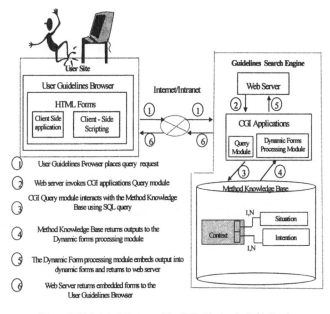

Figure 5. Global Architecture of the EKD Electronic Guide Book

Book » window presents to user the three alternative ways of starting the EKD Electronic Guide Book :

- using a context based selection,
- using a guidance based selection and
- using a form based selection.

Each of these alternatives aims at retrieving from the methods knowledge base a set of candidate contexts for which the tool is able to provide guidance to the EKD engineer. We now describe each of these alternative in turn.

The first alternative, called « Using a Context Based Selection », permits the EKD engineer to directly choose the situation(s) he/she wants to work on and/or the intention(s) he/she has in mind. This alternative requires the EKD engineer to be familiar with the formalism used to represent EKD guidelines (to understand what a situation and an intention are) and the vocabulary used in the EKD method (goal, role, actor, etc.).

The second alternative, called « Using a Context Based Guidance » is useful for beginners (e.g. people who use the tool for the first time). It is a fully guided mode. The selection of this alternative leads to guide the EKD engineer step by step, starting with a very high level intention : « Progress using the EKD method ». Guidance is provided using successive refinement and/or decomposition of this high level intention. The starting point of this alternative is always the same whereas it can be different in the two other alternatives.

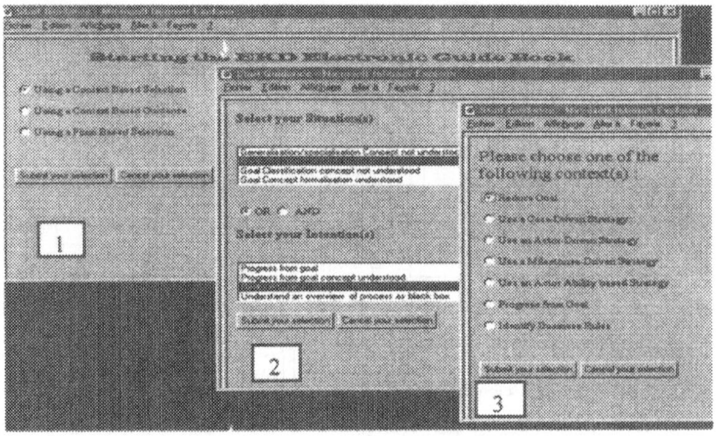

Figure 6. Starting with the "EKD Electronic Guide Book"

The last alternative is called « Using a Form Based Selection ». It offers the EKD engineer a structured form where different fields can be filled. Each field represents a facet of a context such as : the situation, the intention (similarly to the first alternative) but also the target (the result the EKD engineer expects) and the approach (e.g. top-down, bottom-up, etc.). After typed in the values of different fields he/she is aware of, the User Guideline Browser sends a query to the Guideline Search Engine with the elements provided. All contexts that match these elements (even partially) are retrieved from the method knowledge base and sent back to the EKD engineer for selection. This alternative is an intermediate one with regards to the two previous ones. It is more flexible than the second one because the starting point is not pre-defined but depends on the inputs. It is less complex than the first one because it does not require the EKD engineer to know exactly the structure of a context and the semantics of its components.

Assume that the EKD engineer has chosen the first alternative. As shown in window number 2 of Figure 6, the engineer is asked :
• to select one or several situations in the list,
• to select the « AND /OR » option, specifying that the resulting contexts will match the situation(s) and/or the intention(s) respectively,
• to select one or several intentions in the list.

Note that both the list of situations and intentions are constructed dynamically, based on the current content of the method knowledge base.

Based on the selection, a query is sent to the server and all contexts that match the EKD engineer requirements are retrieved from the method base and sent back to him/her.

At this point, window 3 of Figure 6 is displayed on the EKD engineer' screen. He/she selects in this window the context he/she wants to execute. After having

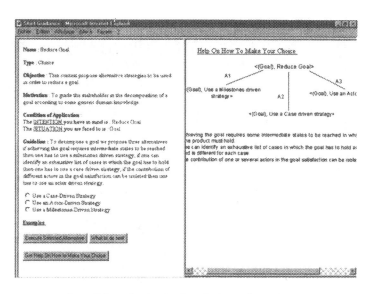

Figure 7. Executing the <(goal) ; reduction goal> context

clicked on « submit your selection» button , guidance effectively starts.

Assume the EKD engineer selects the context <(Goal) ; reduce goal>, a query is sent to the server asking for the execution of the context. The resulting new window is the one of Figure 7. Both the textual and graphical definitions of the context are prompted to the engineer in the left and the right frame respectively.

The textual representation concerns general information such as context name, type, objective, motivation, situation, intention and guidelines (see the context's description in section 5.1). It also contains some dynamic hypertext links. For example when the user clicks on the word « situation » (respectively « intention »), in this frame, a detail and formal description of the situation (respectively the intention) appears in the right frame.

The EKD engineer can also get some examples by clicking on the word « example » at the bottom of the left frame. The example appears in the right frame. It aims at helping the engineer in understanding what can be achieved when executing this context through a concrete example. Concerning the context displayed in Figure 7, the example is shown in Figure 8.

Because the context <(Goal) ; reduce goal> is a choice context, the EKD engineer has to select the appropriate alternative in order to proceed in the EKD process. He/she can get some assistance for making his/her decision by studying the arguments associated to each alternative. These arguments are accessible when clicking on the button « Get Help On How to Make Your Choice » shown in Figure 7. After selecting the appropriate alternative, the EKD engineer could have browsed

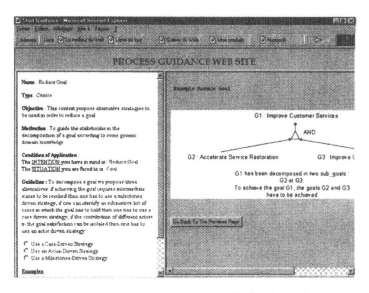

Figure 8. The example associated to the context <(Goal) ; reduce goal>

through the details (guidelines, example, intention, situation, etc.) of this context as much as he/she likes.

The process proceeds by a succession of context executions until the top level intention (the intention of the starting context) is fulfilled. The EKD engineer is then proposed to select a new starting context (through the home page shown in Figure 6); the execution of the new context follows exactly the same procedure as the one we just described.

DESCRIPTION OF AVAILABLE CHUNKS

According to section 2, two types of method chunks are provided. The first one suggests EKD engineers to make product transformations, and the second one leads EKD engineer to *progress* in the change process. It makes the change process to proceed in accordance with process steps already executed. However, a second classification can be made according to the content of the method chunks instead of their type. Therefore, the method chunks can be classified in two separated groups. The first group, called « *understanding EKD concepts*» chunks, aims at guiding EKD engineers in learning EKD models concepts, it can be seen as a tutorial facility that can be access either during a learning session or any time during the construction of any EKD models. The second group, called « *working with EKD*» chunks, provides guidance to EKD engineers in the construction of the different

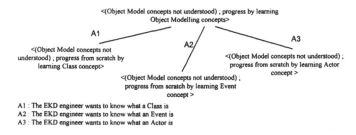

A1 : The EKD engineer wants to know what a Class is
A2 : The EKD engineer wants to know what an Event is
A3 : The EKD engineer wants to know what an Actor is

Figure 9. Understanding EKD Object Model concepts

EKD models, and more generally, in progressing in the change process. It is within this group of method chunks that the knowledge and heuristics related to the EKD method is embedded.

1. Method Chunks for « understanding EKD concepts »

The «*understanding EKD concepts* » chunks are conceived for giving EKD engineer a support for understanding any concepts of any EKD models as well as all concepts underlying the definition of method chunk (context, situation, intention, etc.). Therefore, this set of chunks allows to guide, step by step, the learning process of both the product and the guidance component of EKD.

The situation associated to these chunks aims at characterising the EKD engineer knowledge about EKD concepts. For instance, the chunk <(Object model concepts not understood) ; Progress by learning object modelling concepts> (see Figure 10), aims at guiding an EKD engineer in learning the concepts for object modelling, assuming that he/she has not understood yet those concepts. The intention of this chunk describes the EKD engineer goal : he/she wants to progress by learning object modelling concepts.

2. Method Chunks for « working with EKD»

The « *working with EKD*» method chunks allow any EKD engineers to be guided, step by step, while constructing any EKD models : object, actor/role, process and goal model and more generally for progressing in the process of change. The situations of the associated context deal with EKD products. The intentions of these contexts are mainly concerned with transformations of EKD models. All examples provided in section 4 belong to this type of method chunks.

For the time being, the « working with EKD » method chunks that we have defined are those related with object model construction, and with some goal modelling activities : acquire goals and reduce goals.

CONCLUSION AND FUTURE WORK

In this paper, we have described the « The EKD Electronic Guide Book », a web based tool that aims at supporting an EKD engineer with a set of process guidelines in the construction of the EKD models.

The description of the guidelines is based on the *ESPRIT project NATURE* decision oriented approach. Guidelines are described as method chunks. According to this approach, an EKD engineer progresses in models construction by making decisions on what to represent and how to represent it. A method chunk encapsulates a prescriptive description of the possible paths of decision making.

Therefore, the core of « The EKD Electronic Guide Book »is a method knowledge base where we have stored EKD process guidelines as method chunks.

As the tool has been built to be used over the web. It can be either used in a standalone manner (i.e. without the help of any other tool) or in connection with others software tools ; and it provides a uniform interface across platforms that allows EKD engineers working in incompatible environment to share for instance a wide variety formats of methods knowledge .

Examples are included along the paper to illustrate both how the tool support EKD modelling processes and the tool functionalities.

By now, the tool implementation includes all functionalities described in this paper. The repository includes guidelines dealing with EKD concepts understanding, EKD object and goal models construction.

Future work will includes :

(1) the completion of the EKD models knowledge with the method chunks corresponding to actor/role, role/activity and rule models ;

(2) the extension of the tool knowledge base and new functionalities to other information system methods such as OMT and UML among others. The ultimate objective is to obtain a method independent « Electronic Guide Book » ;

(3) addition of a graphical trace functionality, so that the user can have a global view of current process sequences ;

(4) the completion of the tool with functionalities dedicated to a « method » engineer allowing him to describe method specific guidelines (for a new method or to complete an existing method) which can be directly used within the « The Electronic Guide Book ».

REFERENCES

Bieber, M. and Kacmar, C. « Designing Hypertext Support for Computational Applications » , Communication of the ACM, 38(8), 1995, 99 - 107.

Keith Brophy, Timothy Koets, « Le programmeur VBScript » S & SM, France 1996.

Bubenko J, Brash D, Stirna J. EKD user guide. ELEKTRA Project Internal Document. February 1997.

Chao Min Chiu , Michael Bieber « Integrating Information Systems into the WWW », Proceedings of the Americas

Conference on Information systems , Indianapolis, Indiana USA, August 15 - 17 1997.

Kellner M. phillips R. Pratical Technology for Process Assets. 107-112. Proc. 8th International Software Process Workshop : State of the Practice in Process technology. Wadern, Germany, March 2-5, 1993. IEEE Computer Soc. Press, 1993.

Kellner et al. Process guides : Effective Guidance for Process Participants, Proc. of the 5th International Conference on the Software Process, 14-17 June 1998.

Nature team. « Defining Visions in Context : models, processes and tools for requirements engineering ». Information Systems. Vol. 21, No.6, pp. 515-547, 1996.

C. Rolland. « A contextual approach to modelling the requirements engineering process ». SEKE'94. 6th. International conference on software engineering. Vilnius, Lithuania, 1994.

S.Si-Said, C Rolland. « Guidance for requirements engineering processes ». Proceedings of CAISE'97 conference, Barcelona, Spain, June 1997.

CASE-BASED REASONING, GENETIC ALGORITHMS, AND THE PILE FOUNDATION INFORMATION SYSTEM

Otakar Babka, Celestino Lei, and Laurinda A. G. Garanito
University of Macau, Faculty of Science and Technology
P. O. Box 3001, Macau (via Hong Kong)
Ph.: +853 3974 471, Fax: +853 838 314

Key words: Case-based Reasoning, Genetic Algorithms, Pile Foundation

Abstract:
Case-based support of decision making process relating to pile foundation is one of the most important parts of the *Pile Foundation Information System*. Case-based reasoning (CBR) can draw a lesson from a previously realized pile foundation of the similar attributes (geotechnical situation, as well as the geometric characteristics of the piles). Two case libraries were created based on previously realized sites. Representativeness of the case libraries and intensification of the search process is facilitated by *genetic algorithm* technology.

1.   INTRODUCTION

Case-based reasoning facilitates effective reuse of previously accepted results. The similar old solution is retrieved from recorded, previously used successful solutions. This retrieved case can also be adapted to fit closer the new conditions. This paradigm is applied to model pile foundation as one of the principal parts of the *Pile Foundation Information System.*

Main objectives of *Case-based Reasoning* are described in the next paragraph. Application of this paradigm to *Pile Foundation* is studied then, along with an intensification of the retrieve using *Genetic Algorithms.*

Systems Development Methods for Databases, Enterprise Modeling, and Workflow Management,
Edited by W. Wojtkowski, *et al.* Kluwer Academic/Plenum Publishing, New York, 1999.

77

2.   CASE-BASED REASONING

Differing from more traditional methods, case-based reasoning (Aamodt and Plaza 1994, Kolodner 1993) is one of the most effective paradigms of knowledge-based systems. Relying on case history is its principal feature. For a new problem, case-based reasoning strives for a similar old solution. This old solution is chosen based on a correspondence of the new problem to some old problem (which was successfully solved by this solution).

Representation and indexing are important issues of this methodology. The description of the case should comprise the problem, solution of the problem, and any other information describing the context for which the solution can be reused. Most important source of information is the case library. The case library serves as a knowledge base of the case-based reasoning system. The system can learn by acquiring knowledge from these old cases. Learning is basically achieved in two ways: (i) accumulating new cases, and (ii) through the assignment of indexes. When solving a new case, the most similar old case is retrieved from the case library. The suggested solution of the new case is generated in conformity with this retrieved old case. The search for a similar old case from the case library represents an important operation of the case-based reasoning paradigm. Retrieval relies basically on two methods: nearest neighbor, and induction. Complexity of the first method is linear, while the complexity of a search in a decision tree developed by the latter method is only logarithmic. This reasonable complexity enables the effective maintenance of large case libraries.

Rather the verified ways are preferred using case-based reasoning, in accordance with the anthropomorphic and pragmatic approach. Why to start from scratch again if we can learn from the past? Once a suitable old solution, tied to the new context, can be retrieved by the case-based reasoning part of the framework, such a tested solution should be preferred.

Because situations are often repeating during the life cycle of an applied system, case-based reasoning relies on the previous cases. Further, after a short period of time the most frequent situations should be identified (and documented in the case library). So, the case library can usually cover common situations after a short time.

3.   SUPPORTING PILE FOUNDATION DECISIONS

Case-based reasoning paradigm was applied to Pile Foundation in the presented research. The cases are organised to define a situation – geotechnical information associated with pile test results as well as pile geometric characterists. The main purpose is to retrieve information related to piles driven and tested in different sites in Macau, where the geotechnical data is also available, and use that information as a reference in a new site, whenever the new situation can be compared to a known old one, i.e., when the soil characteristics for a new site are known and similar to an experienced case. Using the system here described, it is possible to evaluate the bearing resistance of a pile, from a similar stored case where piles were tested static or dynamically and the geotechnical soil information is known. Moreover, the parameters used in the Hiley formula previously calibrated by pile load tests can be reused in a similar situation.

In agreement with the purpose of our system, features describing the geographical and geotechnical soil information mean *input* features and are known for all stored points. The inquired data about these points are called output features. In the presented application, features describing the pile characteristics and the pile test results associated with the geotechnical information are the *output* features.

To govern the design of the geotechnical structures, a site investigation must be done making use of some boreholes and performing Standard Penetration Tests (SPT), in order to know the geotechnical properties of the soil on the top of which the structure will be built.

As soon as the geotechnical properties are known, if piles are chosen, the pile foundation is calculated, thus defining the pile characteristics and pile design load based on the SPT results, on the structure service load and on the specifications given by the pile factory.

For civil construction purposes, the number of boreholes are normally less than the number of piles. Normally, the relation borehole-pile is *1:N*, i.e., the relation one borehole to several piles. This is the reason why it was decided to maintain two case libraries for the application: data related to boreholes stored in one library while data related to piles in a different one. Boreholes library defines the problem space and piles library defines the solution space.

For previously examined cases, both input features and output features are known and recorded. So, these old cases can serve as a source of knowledge of case-based reasoning system. Learning process of case-based reasoning system passes through the accumulation of the old cases. Based on this knowledge, output features can be consequently found for similar newly examined cases.

For the presented application, recorded data of earlier completed pile foundation areas can serve for this purpose. Features of selected piles are collected in a case library. Features describing the correspondent geographical and geotechnical soil information are collected in a second case library. Once collected, case-based reasoning can partially behavior of pile foundation if their properties, along with the geotechnical situation, are similar.

Examining a point of the new area, the case corresponding to a previously examined point with the most similar geotechnical features is retrieved from the case library. To fit closer to the new situation, the retrieved old case can also be adapted. Taking into account the differences between features of the new and the old retrieved point, output features can be recalculated before suggesting new results.

4.   PILE FOUNDATION OBJECTIVES

Different types of foundations are used for civil engineer construction purposes, depending on the geological composition of the site and the type of construction to be built on it.

Pile foundations are used when the upper soil layers can not sustain the load of the structure to be built on it. Piles transfer the load of that structure to the lower and more resistant layers than the upper ones, those being able to sustain the upper load, without causing any detrimental settlement to the structure that they support.

The system here referred considers purely pre-fabricated concrete piles (PHC - pile high-strength concrete), driven in different sites in Macau under the action of a hammer.

Figure 1 shows the different steps usually followed up during a civil construction work in Macau, in terms of geotechnical foundations.

The site investigation allows the collection of the necessary soil information, thus gsiving an idea of the appropriate foundation type to be chosen. Normally, site investigation includes drilling boreholes in the construction area, from the ground surface to the bedrock or strong soil layer, and performing standard penetration tests (SPT) at 1.5m to 2m intervals.

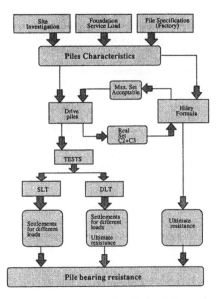

Figure 1. Schedule used during pile foundation works

After this description of some of the steps performed while developing decisions related to pile foundations, the next paragraph focuses on the issues involved when applying the case-based reasoning paradigm in this application area.

5.   DATA PROCESSING

Borehole library – The features stored in the borehole CBR library are: soil penetration resistance values, borehole total depth and soil layers.

For a better analysis, it was consider more convenient to have a sequence of blows that increases with the depth, instead of the simple N-values. Therefore, instead of considering the number of blows itself, it was calculated the accumulated number of blows related to each depth, for each borehole. Based on these values and by means of a regression analysis, a trend line chosen as a six-degree polynomial allows predicting the accumulated number of blows from the depth values, for each borehole. The R^2 value (coefficient of determination between the real accumulated N-values and the correspondent ones obtained from the regression analysis) was also calculated and helps to determine the line of the best fit. The R^2 values referred to the boreholes stored in the library are greater than 0.98 and most of them around 0.99.

Basically, the standard penetration test consists of driving a standard sampler by dropping a 63.5kg hammer onto a drive head from a height of 76cm. The number of blows required for a penetration of 300mm after an initial setting drive of 150mm is called standard penetration test value (N-value). The SPT values yield information on engineering characteristics of the soil, and on the nature and sequence of soil layers.

Considering the type of construction, pile service load, and site investigation results, the designer chooses appropriate geotechnical structures. If piles are chosen as the structure, related calculations need to be performed to define pile characteristics, its bearing resistance, the total number of piles needed and their distribution on the site.

The design of a pile foundation depends on the pile working load and on the depth of the strong layer given by the standard penetration test. The diameter of a pre-cast driven pile should be chosen in accordance with the structural design of the pile, which is recommended by the manufacturer.

The construction phase generally starts with pile driving. After this phase, some piles are statically or dynamically tested in order to estimate the piles' behavior.

The Hiley formula (Ferreira et al. 1997, Garanito 1997), has been widely used for the design of driven piles. The formula defines the maximum pile penetration (average penetration under the last few blows) accepted during pile driving operations. The calculation is based on the designed pile load, the geometry of the pile, hammer's characteristics, and the soil parameters that can be obtained by the pile load test results.

The borehole library has the following structure:

a) borehole number;
b) borehole depth;
c) different soil layers related with respective depth;
d) borehole coordinates (X,Y);
e) depths where the N-values were taken;
f) average of the N-values along the pile;
g) N-values at the pile base;
h) piles number driven nearby the borehole;
i) coefficient of distribution;
j) the coefficient of x (a_1);
k) the coefficient of x^2 (a_2);
l) the coefficient of x^3 (a_3);
m) the coefficient of x^4 (a_4);
n) the coefficient of x^5 (a_5);
o) the coefficient of x^6 (a_6).

Pile library - In terms of piles, our aim is to find their geometric characteristics as well as the load test results, besides the parameters used in the Hiley Formula. Consequently, features included in the pile library are grouped: (i) *Pile characteristics*, that are predominantly the geometric characteristics of the pile; (ii) *DLT* data, which refers to *Dynamic Load Test* results; (ii) *SLT* data, which refers to *Static Load Test* results; (iv) *Hiley Formula* data, with fields that are used in the formula.

Pile characteristics, the following fields were considered:

a) pile number;
b) pile nearest borehole;
c) pile load service;
d) pile embedded length;
e) pile total length;
f) pile external diameter;
g) pile internal diameter;
h) pile thickness;
i) pile type;
j) pile weight per volume.

The *DLT* stored data is:

a) settlement at half of service load;
b) settlement at service load;
c) settlement at 1.5 times the service load;
d) settlement at 2 times the service load;
e) skin friction and thickness for the 1st soil layer;
f) skin friction and thickness for the 2nd soil layer;
g) skin friction and thickness for the 3rd soil layer;
h) skin friction and thickness for the 4th soil layer;
i) skin friction and thickness for the 5th soil layer;
j) total skin friction;
k) toe resistance;
l) ultimate load.

The *SLT* stored data is:

a) settlement at half of service load;
b) settlement at service load;
c) settlement at 1.5 times the service load;
d) settlement at 2 times the service load;
e) settlement at 2 times the service load + 24H.

Values used by the *Hiley formula* are stored in the following fields:

a) temporary compression for pile head and cap;
b) temporary compression for pile and quake of ground;
c) hammer efficiency;
d) final set - pile penetration for last blow;
e) hammer drop;
f) hammer type;
g) hammer weight;
h) coefficient of restitution;
i) Hiley ultimate load.

In this stage of the presented research the sources for the case libraries was limited. Several obstacles also hampered a faster and wider progress of the application:

– SPT, SLT and DLT are made by and/or for different organisations, some of them no longer existing in Macao. Actually it is quite difficult to retrieve the necessary information, from a reasonable number of sites.
– For most sites, the data is not complete. Either it is missing borehole data or pile tests information, so turning it unfeasible the combination boreholes - pile tests.

Nevertheless, CBR facilitates incremental learning. CBR system can be deployed with a limited set of "seed cases" to be augmented progressively. This research allowed defining which information each case must contain and deliberate which data should be retained by one single organisation in order to update and increase the reliability of this system. According to experts in geotechnical area, this application can be very useful for future foundation designs, helping them on predicting the behaviour of piles without the need of so many piles tests. However, more and varied cases must be added to the libraries in order to increase the representativeness of the libraries and the accuracy of the decision.

This system can be also used for foundation calculations, since different fields are defined as Formulas. For instance, the ultimate load (PHU) given by Hiley Formula can be calculated directly by this model.

6.   CASE LIBRARY REDUCTION

Case-based reasoning relies on past case history. For a new problem, case-based reasoning strives to locate a similar previous solution.

When solving a new problem, the most similar old case is retrieved from the case library. Retrieval methods (Kolodner 1993) are based mainly on some modification of the nearest neighbor algorithm or induction As the case library gets larger in size, the retrieval process becomes more time-consuming, especially for nearest neighbor methods, where time complexity is generally linear.

A case-based reasoning system can only be as good as its case library (Kolodner 1993), and the quality of case libraries can be judged in two aspects:

- *Representativeness of the library* – The quality of the decision, and especially its accuracy, can be improved by employing a more appropriate representation of the case and with a cautious selection of cases.
- *Effectiveness of the retrieval* – Effectiveness is mainly based on (i) the complexity of retrieval algorithms, and (ii) the size and organization of the library.

With the progress of case-based reasoning, more complex case libraries have been constructed. Researches have begun addressing more practical problems assisting case authors (Aha and Breslow 1997, Heider et al. 1997, McKenna E., Smyth B., 1998). The similarity function plays a crucial role in nearest neighbor retrieval. This function is sensitive to imperfect features, some of them being *noisy*, *irrelevant*, *redundant*, or *interacting* (Wettschereck and Aha 1995). A cautious selection of features can reduce dimensionality, improving efficiency and accuracy.

A group of *weight-setting methods* faces this problem by parameterizing the similarity function with feature weights (Wettschereck and Aha 1995). Binary weights of features were used in the presented research. This means that either a feature is used in the similarity function with a weight of 1 or it is ignored completely (i.e. with a weight of 0). The problem with possible imperfect features was solved by the reduction of the features. Motivated by Skalak (1993, 1994) we decided to use *genetic algorithms* as a core of our reduction method. This method is flexible, not too complex for realization, and no domain-specific knowledge is needed. One of reason for this intensification of the retrieval process was our positive experience with a similar reduction used for the handwritten digit recognition system (HWDR) (Babka et al. 1998). Due to the size of the case library, the recognition process was time consuming. Therefore, the case library was reduced using genetic algorithms and the original case library with 4086 cases was reduced to 300 cases. After this reduction, the recognition accuracy of the system was changed from 95.96% to 94.63%. Although with a slight decrease in recognition accuracy, the time required for case retrieval was reduced dramatically, to less than 1% of the original time required.

Similar experiments were done for reducing the number of features of the case library, with very positive results. The number of features was reduced from 342 to 50, with a slight decrease in recognition accuracy from 95.96% to 95.47%. The method, based on genetic algorithms, showed capability to select representative features from that wide range of features, still maintaining accuracy after this massive reduction for both features and cases. Motivated by this good result, we decided to test the ability of this methodology also on the pile foundation application.

The presented research is focusing on features reduction, since the particular case libraries used for testing possibly contain redundant or irrelevant features. On the other hand, the number of cases is rather limited in this stage of the research. The reduction of cases will be reconsidered after extension of the case libraries.

7.   REDUCTION RESULTS

As stated previously, the experiments were conducted in order to reduce the original borehole library to a different number of features. The graph below (Fig. 2) shows the evolution of generations when reducing the original borehole library with 20 features to 10 features.

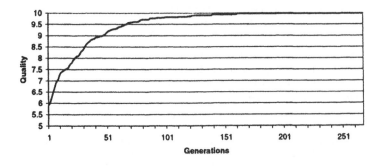

Fig. 2. Graph showing the evolution of the individuals' quality when reducing to 10 features

The values shown on the graph are averages for 72 experiments. In the worst case of the experiments, 267 generations were needed in order to find a subset of features that gives the same level of classification as using all features. On average, 59.75 generations are needed to achieve that level of accuracy.

As one might expect, when reducing to 8 features (Table 1), the number of generations required for producing a reduced library which could have the same accuracy as the original library is greater than that for reducing to 10 features. In the worst case, 606 generations were needed and 205.37 generations are needed on average.

There is a significant difference between the results for reduction from to 8 and 7 features (Table 1). For the case of reducing to 7 features, in the worst case, 3527 generations were needed to obtain the same level of classification as using all the 20 features. On average 951.25 generations are needed for that level of accuracy.

The Table 1 summarizes the experimental results for successful experiments.

From Table 1, it can be found that the less the number of reduced features, the more the number of generations required and the standard deviation related to the number of generations gets higher as well. This table further shows that the number of generations required to reduce the number of features varies greatly between experiments.

From these experiments, one can notice that there is a trade-off between the degree of reduction of the case library and the time required for that reduction.

Table 1. Statistical information regarding each set of experiments

	Number of reduced features					
	7	8	9	10	13	15
No. of experiments	28	27	21	72	85	13
Generations in the worst situation	3527	606	802	267	37	3
Generations in the average situation	951.25	205.37	114.05	59.75	10.96	1.31
Standard deviation	933.28	151.05	163.89	45.81	10.15	0.75

After conducting the many tests of reducing the number of features to different values, it was found that the features that are finally discovered by the system do not differ very much. This shows that, although the initial situation of each experiment is different (since the initial population is randomly generated), the final collection do converge to some similar subset of features. This is one of the very positive results found.

Trying to find limits, experiments were also conducted for reducing to 6, 5, and 3 features. However, as expected, this is obviously an excessive reduction. For instance, when reducing to 3 features, after 5461 generations, only 60% of the test cases were classified according to the reference classification done with all features considered.

From the experiments, it was found that features could be substantially reduced in this application. On the other hand, with the help of this method, limits of the reduction for the given date are recognizable. For the presented library, the number of features should not be less then eight (see Fig. 6, 7, 8).

8.   CONCLUSION

The presented application was developed on a research level. Generally, borehole and pile library were provided with limited number of cases in present time. Data is not complete for all sites. However, despite this limited source of data, the general results are promising. CBR facilitates incremental learning. The environment can be deployed with a limited set of "seed cases" to be augmented progressively. According to experts in the geotechnical area, this application can be very useful for future foundation designs, helping them predict the behavior of piles without the need of so many piles tests.

Selection of representative features of the borehole library was studied in this research. Imperfect features were reduced with the help of genetic algorithms. Adopting this methodology, that we have originally developed for pattern recognition application, feature reduction of the borehole library was also positive, although the nature of the application, and the source, characteristics and size of data differ significantly. Although a possible generalization of this conclusion is limited by two experiments, results suggest a positive evaluation for the presented approach.

The feature reduction of the pile library is logically the next step of the research. After these libraries are supplied with more cases, we will again employ case reduction, using the same approach.

As another direction for further research, during the application of the genetic algorithm for case library reduction, the method itself could decide for the appropriate number of features. This might be accomplished by giving higher quality values to those individuals with a lesser number of features. In this way, the genetic algorithm might be influenced towards generating individuals with less number of features.

Furthermore, the genetic algorithm could also decide on the suitable population size according to needs.

REFERENCES

Aamodt A. and Plaza E., 1994, *Case-Based Reasoning: Foundational Issues, Methodological, and System Approaches*, IOS Press.

Aha D. W. and Breslow L. A.,1997, Refining conversational case libraries, in *Proceedings of the Second International Conference on Case-Based Reasoning*, pp. 267-278, Providence, RI, Springer-Verlag.

Babka O.,S. I. Leong, C. Lei, M. W. Pang, 1998, Reusing Data Mining Methods for Handwritten Digit Recognition, in *Proceedings of World Multiconference on Systems, Cybernetics and Informatics (SCI'98)*.

Babka O. and Garanito L. A. G., 1998, Case-based Reasoning for Pile Foundation, in *Proceedings of Symposium on Science & Technology and Development of Macau*, Dec. 1-4, 1998, Macau, pp. 295-300.Smyth B., 1996, Case Adaptation & Reuse in Déjà Vu, in *Proceedings of ECAI-96 Workshop on Adaptation in Case-Based Reasoning*, ftp://ftpagr.informatik.uni-kl.de/pub/ ECAI96-ADAPT-WS/smyth.ps

Berchtold S. and Ertl B., Keim D. A., Kriegel H-P. and Seidl T., 1998, *Fast Nearest Neighbor Search in High-dimensional Space*, in ICDE'98.

Ferreira H. N., Lamas L. N., L. Hong Sai, S. Qiang, 1997,*Guia de Fundações LECM*, Macau.

Garanito L. A. G., 1997, *Case-based Reasoning and Piles Foundations*, Master's Thesis, University of Macau.

Heider R., AuriolE., Tartarin E., Manago M., 1997, Improving quality of case bases for building of better decision support system. in R. Bergmann and W. Wilke (Eds.) *Fifth German Workshop on CBR: Foundation, Systems, and Application (Technical Report LSA-97-01E)*, University of Kaiserlautern, Department of Computer Science, 1997.

Kolodner J., 1993, *Case-based Reasoning*, Morgan Kaufmann Publ., U.S.A.

Leong S. I., C. Lei, M. W. Pang, 1998, *HWDR – Handwritten Digit Recognition – Final Report*, Bachelor degree graduation project report, University of Macau.

McKenna E., Smyth B., 1998, A Competence Model for Case-Based Reasoning, in *9th Irish Conference on Artificial Intelligence and Cognitive Science*, Ireland, 1998, http://ww.cs.ucd.ie/staff/bsmyth/papers/Cascade%20Submit.doc

Skalak D. B., 1993, Using a genetic algorithm to learn prototypes for case retrieval and classification, in *Proceedings of the AAAI-93 Case-based Reasoning Workshop*, 1993.

Skalak D. B., 1994, Prototype of feature selection by sampling and random mutation hill algorithms, in *Proceedings of the Eleventh International Machine Learning Conference*, Morgan Kaufmann Publ., U.S.A., pp. 293-301, 1994.Wettschereck D.and Dietrich T. G., 1995, An experimental comparison of the nearest neighbor and nearest hyperectangle algorithms, in *Machine Learning*, 19, 5-28.

Wettschereck D. and Aha D. W., 1995, Weighting features, in *Proceedings of the First International Conference on Case-Based Reasoning*, Lisbon, Portugal, Springer-Verlag.

Wilke W. and Bergmann R., 1996, Adaptation with the INRECA - System, in *Proceedings of ECAI-96 Workshop on Adaptation in Case-Based Reasoning*, ftp://ftpagr.informatik.uni-kl. de/pub/ECAI96-ADAPT-WS/wilke.ps

JAVA, WEB, AND DISTRIBUTED OBJECTS

Yin-Wah Chiou
Dept. of Information Management
Chinese Culture University
87 Yen-Hou Street, Hsinchu
Taipei, Taiwan, R. O. C.

ABSTRACT

Java, Web, and distributed objects are playing critical roles in the Internet computing. In this paper, we present a survey of some of the most important features for Java, Web, and distributed object technologies. We begin with a description of two key issues in Java technology, including Java component models (i.e., *JavaBeans* and *Enterprise JavaBeans*), and *Java Database Connectivity* (JDBC). We then describe the relevant techniques on advanced Web documents: *Dynamic HTML* (DHTML), *Document Object Model* (DOM), and *Extensible Markup Language* (XML). Further, distributed object standards are reviewed, including OMG's *CORBA* and Microsoft's *COM/DCOM/COM+*. Finally, we explore *Java distributed objects*, which represents an integration of Java and distributed object technologies to provide a new and promising paradigm for software applications and development.

Keywords: JavaBeans, JDBC, DHTML, DOM, XML, CORBA, DCOM

1. INTRODUCTION

Java is a simple and powerful object-oriented programming language with syntax similar to C++. There are two types of Java programs: applications and applets. The Java application is a standalone application, whereas Java applet is a mini-application distributed over the Internet. In Section 2 of this paper, we will focus on the two key issues of Java: *Java component models* and *Java Database Connectivity* (JDBC). *JavaBeans* component model represents a combination of Java and component technologies. The *Enterprise JavaBeans* (EJB) is an extension of JavaBeans component model for supporting server components. The *JDBC* is a set of Java classes to provide a Java object interface to SQL databases.

Hypertext Markup Language (HTML) is designed to specify the logical organization of text documents and to allow for hyperlinks between documents. The *Dynamic HTML* (DHTML), *Document Object Model* (DOM), and *Extensible Markup Language* (XML) represent the integration of Web and object-oriented technologies to enhance HTML capabilities. In Section 3, we will describe the

Systems Development Methods for Databases, Enterprise Modeling, and Workflow Management,
Edited by W. Wojtkowski, *et al.* Kluwer Academic/Plenum Publishing, New York, 1999.

DHTML, DOM, and XML for advanced Web documents.

Distributed Object Computing (DOC) has shown a good solution in providing software components to plug-and-play, work across diverse network (i.e., interoperability), run on different platforms (i.e., portability), integrate (or coexist) with legacy applications via object wrapping techniques, and allow them to be reused. The most important issue with DOC is standards. In Section 4, we present an overview of DOC standards, including OMG's (Object Management Group's) *CORBA* (Common Object Request Broker Architecture) and Microsoft's Component Object Model (*COM*), Distributed COM (*DCOM*), and *COM+*.

Java Distributed Objects (JDO) represents a combination of Java and distributed object technologies to provide powerful new software applications and development methods. In Section 5, we describe two major techniques of JDO: *CORBA for Java* (i.e., CORBA/Java merger or Java ORB) and *DCOM for Java* (i.e., DCOM/Java merger). There are three prominent *Java ORB* products: Inprise's (Visigenic's or Borland's) *VisiBroker for Java*, Iona's *OrbixWeb*, and Sun's *Java IDL*. The example product of DCOM/Java merger is *Visual J++*. Finally, the conclusion is given in Section 6.

2. JAVA TECHNOLOGY

Java is a simple and powerful object-oriented programming language from SunSoft. This section describes two popular issues in Java technology, including Java component models and JDBC. The Java component models cover JavaBeans and Enterprise JavaBeans (EJB). The JDBC is employed to use Java with databases.

2.1 Java Component Models

In Java, a *component* is a set of related Java classes. *JavaBeans* component capabilities are implemented as a set of language extensions to the standard Java class library; thus, JavaBeans is a set of specialized Java programming language interfaces [6]. The JavaBean components (or Java classes) are called *Beans* (reusable software components). To develop Beans, it is necessary to have *Beans Development Kit* (BDK) 1.0 and the *Java Development Kit* (JDK) 1.1. A *Java builder tool* can be used to build Java classes, Java applets, Java applications, or JavaBeans. The *Beans* can be visually manipulated using Java builder tool.

The following is a presentation of the key concepts for Beans [2, 12]:

- *Properties*: Attributes of the component that can be read (and, optionally, written). Properties are a Bean's appearance and behavior attributes that can be changed at design time.
- *Events*: Notifications of state changes that the component is capable of publishing to other interested components. Beans use events to communicate with other Beans.
- *Persistence*: Persistence enables Beans to save their state, and restore that state later. JavaBeans uses *Java Object Serialization* to support persistence.
- *Methods*: Services the component provides that can be invoked by others. A Bean's methods can be called from other Beans or a scripting environment.
- *Introspection*: Builder tools discover a Bean's properties, methods, and events by introspection.

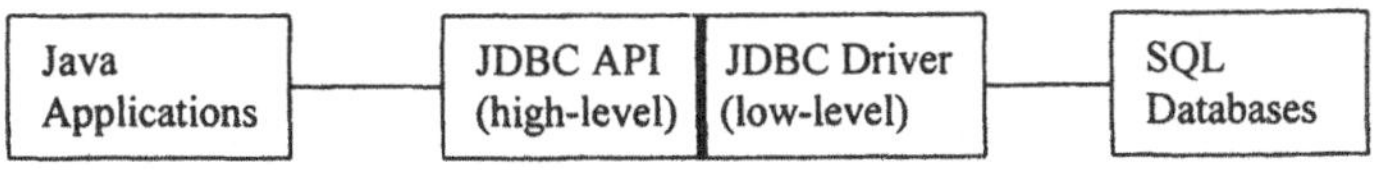

Java Applications	JDBC API (high-level)	JDBC Driver (low-level)	SQL Databases

Figure 1. Relationship of JDBC API and Driver

The *Enterprise JavaBeans* (*EJB*) extends the JavaBeans component model to support *server components* (i.e., application components that run on a server) [15]. Potential EJB environments include TP monitors, Component Transaction Server (CTS), CORBA platforms, DBMSs, and Web servers. The JavaBeans component model and EJB provide component portability, interoperability, reusability, ease of use, highly customizable, and so on.

2.2 Java Database Connectivity

To use Java with databases, JDK1.1 already includes *Java Database Connectivity* (JDBC) module to allow a Java programmer connecting and querying remote databases. That is, JDBC is a set of Java classes to provide a Java object interface to SQL databases. JDBC can be viewed from a *high-level* abstract view or from a *low-level* database specific view [1, 14]:

- *JDBC API* (high-level view): The JDBC *application programming interface* (API) defines Java classes to represent database connections, SQL statements, result sets, database metadata, and so on. It allows a Java programmer to issue SQL statements and process the results. JDBC is the primary API for database access in Java. In short, the JDBC API provides methods allowing an application to connect, query, and manipulate databases.
- *JDBC Driver* (low-level view): A database specific implementation of the JDBC abstract classes, called a JDBC driver, must be provided in order for the Java database programmer to access the database. JDBC drivers can either be entirely written in Java so that they can be downloaded as part of an applet, or they can be implemented using native methods to bridge to existing database access libraries. So, a database application obtaining database access through the JDBC API will work with any data source providing a JDBC driver.

Figure 1 shows the relationship of JDBC API and driver. In brevity, the JDBC API is implemented via a JDBC driver.

3. ADVANCED WEB DOCUMENTS

The *Dynamic HTML* (DHTML), *Document Object Model* (DOM), and *Extensible Markup Language* (XML) represent the integration of Web and object-oriented technologies to enhance dynamic capabilities of HTML documents. This section describes the salient features of DHTML, DOM, and XML for advanced Web documents.

3.1 DHTML and DOM

The *DHTML* is developed by Microsoft and Netscape. The WWW Consortium (W3C) is currently working on a *DOM*. The following presents some of the important features for DHTML and DOM [8, 16, 19, 20]:

- DHTML is a term used by some vendors to describe the combination of HTML, style sheets and scripts that allows documents to be animated. DHTML is becoming the norm on the Web. Personalized Web pages can now be served to individual users on a customized basis, because DHTML lets authors control every element of a Web page and change the page's content and style at any time, without having to refresh the page from the server. This can significantly improve network performance by minimizing roundtrip traffic to the Web server.
- DHTML is like "*programmable*" HTML. That is, the contents of the HTML document can be manipulated dynamically by script code embedded in the page. Web developers will be able to create interactive and visually enhanced DHTML Web pages easily and direct them to specific audiences. These technologies will radically improve the gathering and communicating of information to a targeted audience.
- DOM is a generalization of DHTML facilities. The DOM is a platform- and language-neutral interface that will allow programs and scripts to dynamically access and update the content, structure and style of documents. The document can be further processed and the results of that processing can be incorporated into the presented page.
- DOM defines an object-oriented API for HTML or XML documents which a Web client can present to programs that need to process the documents. Through this API, scripts or programs access and manipulate a document's content as a collection of objects. In short, DOM makes all Web page elements programmable objects.

3.2 XML

The *XML* is a subset of *SGML* (Standard Generalized Markup Language) to enable generic SGML to be easily used on the Web. Since the HTML document standard lacks expressiveness, the Web's data representation is shifting from *structural HTML markup* to *semantic XML markup*. The *structural* markup distinguishes the elements of a page with tags and declaring the physical relationships among the various document elements, while *semantic* markup identifies what each particular elements means on its own [5]. In addition, XML has been designed for ease of implementation and for interoperability with both SGML and HTML [21]. Obviously, XML is more powerful than HTML.

The important differences between HTML and XML [6, 8] are listed here:

- HTML defines a fixed set of tags for specifying content and format. In contrast, XML allows the definition of customized markup languages with *application-specific* tags for exchanging information in particular application domains. For example, XML tags could be defined for classifying component applets according to company- or industry-specific classifications.
- The full hypertext linking capabilities of XML are much more powerful than those of HTML, providing support for both bidirectional and multi-way links, as well as links to span of text (i.e., a subset of the document) within the same or other documents.

4. DISTRIBUTED OBJECT COMPUTING

Distributed Object Computing (DOC) represents a combination of client-server computing and object-oriented technology to achieve interoperability and portability of heterogeneous distributed components. The most important issue with DOC is *standards*. This section provides an overview of OMG's CORBA and Microsoft's COM/DCOM/COM+ standards.

4.1 OMG's CORBA

CORBA is a global object bus for distributed components. It provides communication and cooperation facilities for distributed object applications. A client object sends the request to a server object. Clients use the *object reference* (object identity) to connect to a particular object.

An object's *interface* is a collection of related operations (or methods) to serve as a binding contract between clients and servers. *Interface Definition Language* (*IDL*) is used to define interfaces to client and server objects. These objects can be written in different programming languages to interoperate across networks and operating systems. That is, *OMG IDL* is used as a way of defining interfaces neutral to any particular programming language. CORBA supports *language mappings* (or *bindings*) for C, C++, Smalltalk, Java, Cobol, and Ada.

Figure 2 shows an overview of CORBA 2.0 elements. The *ORB core* is an intermediary between client and object implementation. It functions as a communication infrastructure to transparently relay object requests. The *object implementation* is the codes to implement the operations in the interface. The other CORBA elements [10, 18] are briefly described as follows:

- *Static IDL Stubs* (or *Client Stubs*) define how clients invoke corresponding services on the servers.
- *Static IDL Skeletons* (or *Server Stubs*) provide static interfaces to each service exported by the server.
- *Static Invocation Interface* (SII) provides compile-time checking (type-safe) to improve developing productivity and results in more reliable code.
- *Dynamic Invocation Interface* (DII) provides a mechanism for discovering the method to be invoked at run time.
- *Interface Repository* (IFR) is a run time database that stores on-line descriptions of known IDL-defined interfaces.
- *Dynamic Skeleton Interface* (DSI) is the server equivalent of a client's DII.
- *Object Adapter* comprises the interface between the ORB and the object implementation.
- *Implementation Repository* (IMR) provides a runtime repository of information about classes a server supports, the objects that are instantiated, and their IDs.
- *ORB Interface* is located on both client and server sides to provide access to the IFR and IMR, and some operations on object references that the ORB can perform.

The *CORBA object services* (*CORBAservices*) define the system-level object frameworks that extend the CORBA object bus to augment and complement the functionality of the ORB. The object services include naming, event, life cycle,

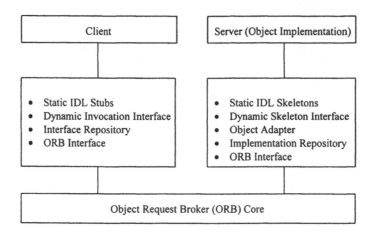

Client	Server (Object Implementation)
• Static IDL Stubs • Dynamic Invocation Interface • Interface Repository • ORB Interface	• Static IDL Skeletons • Dynamic Skeleton Interface • Object Adapter • Implementation Repository • ORB Interface

Object Request Broker (ORB) Core

Figure 2.   Overview of CORBA 2.0 Elements

object transaction, concurrency control, persistence, query, relationship (or association), externalization, licensing, property, trader, time, collection, security, and change management. The high level application-oriented services are known as *CORBAfacilities*.

CORBA is an evolutionary and fast-changing technology. The new release CORBA 3.0 will include many new features [10, 17]: messaging, multiple interfaces, objects-by-value, IIOP proxy for firewall support, Portable Object Adapter, new CORBA persistence service for supporting automatic persistence, a Common Facility for Mobile Agents, a JavaBeans-based Business Object Framework, a Workflow Facility, and Domain-level Frameworks.

4.2 Microsoft's COM/DCOM/COM+

COM does not support remote method invocations or distributed objects. *DCOM* is an extension of COM implementation to provide a distribution model. A *component* may have one or more public interfaces. An *interface* defines a set of properties, methods, and events through which external entities can connect to, and communicate with, the components [6]. At the run time, each interface is identified by a *globally unique interface identifier* (GUID).

We provide a brief comparison between DCOM and CORBA as follows:

• Like CORBA, DCOM separates the object interface, declared in IDL, from the implementation. However, CORBA only uses a single IDL. In DCOM, there are two IDL-like mechanisms: *Microsoft IDL* (*MIDL*) and *Object Description Language* (*ODL*).
• The *Type Library* (*TL*) is a run-time repository to store ODL-defined object interfaces and parameters. TL is the DCOM version of CORBA Interface Repository.

- A DCOM server contains one or more classes. A unique *class identifier* (CLSID) is used to identify each DCOM class. A *class factory* in the server is used to create a new object. The *Registry* is the DCOM version of CORBA Implementation Repository, which stores information about classes that a server supports.
- Like CORBA, DCOM also provides both static and dynamic method invocations. The static *VTBL* (virtual table or *vtable*) *binding* makes use of tables containing addresses of code. The dynamic *IDispatch binding* includes some operations to tell the component what method is being invoked.

COM+ is an evolution of COM/DCOM. The salient features of COM+ [19, 22] are briefly summarized as follows. COM+ has been designed to more closely model some of the features of CORBA and Java. Therefore, COM+ is more object-oriented than COM/DCOM. Because of the close similarities of COM+ and CORBA and Java, a cleaner abstraction can be derived. For example, COM+ introduces user-defined exceptions to Microsoft's model. This is already in CORBA. COM+ will more fully integrate and automate middle-tier application services, including dynamic load balancing, queued components (asynchronous deferred invocation), distributed event notification, in-memory database, and persistence.

5. JAVA DISTRIBUTED OBJECTS

Java Distributed Objects (JDO) represents an integration of Java and distributed object technologies to provide new software applications and development methods. This section describes two major techniques of JDO, including *CORBA for Java* and *DCOM for Java*.

5.1 CORBA for Java

The CORBA and Java converge when a mapping is defined from OMG IDL to Java; when combined' with a run-time system which supports this language mapping, the result is a *Java ORB* [18]. All Java ORBs are compliant with *CORBA/IIOP* and *OMG IDL/Java Mapping* standards. The *IIOP* (Internet Inter-ORB Protocol) is a messaging protocol for the ORBs from different vendors to interoperate over the Internet. Let's take a brief look at three prominent *Java ORBs* as follows:

- *VisiBroker for Java* (VBJ) [3, 11]: Inprise's (Visigenic's or Borland's) VisiBroker for Java ORB implements the CORBA 2.0 and IIOP standards. VisiBroker 3.2 for Java can run on any platform with a JDK 1.1-compatible Java VM (Virtual Machine).
- *OrbixWeb* [4]: Iona's OrbixWeb extends the functionality of *Orbix* to Java developers, and provides a complete solution for the creation of Java-based middleware applications. The OrbixWeb requires that JDK 1.1 or above be pre-installed on the development machine.
- *Java IDL* [7, 13]: Sun's Java IDL is compliant with the CORBA/IIOP 2.0 specification and the IDL-to-Java language mapping. Java IDL connects Java-based Web applications to servers provided by an IIOP-compliant ORB such as OrbixWeb or VisiBroker. To use Java IDL, you'll need both *idltojava* compiler and the current release of the JDK 1.2 software.

5.2 DCOM for Java

In DCOM/Java combination, Microsoft's *Visual J++* provides Java language bindings for DCOM. With Visual J++ 6.0, Java clients can use DCOM to invoke remote Java objects. The following provides a good summary of the salient features for *Visual J++ 6.0* [9]:

- Visual J++ 6.0 is the fastest way to harness the productivity of the Java language and power of Windows to build and deploy high-performance, data-driven client/server solutions.
- Visual J++ 6.0 introduces a powerful new language feature: *delegates*. Unlike function pointers, delegates are object-oriented, type-safe, and secure. Delegates and Interfaces are similar in that they enable the separation of specification and implementation.
- Visual J++ 6.0 contains many new features designed to help developers build commercial-quality Windows applications in the Java language. Visual J++ enables developers to build and reuse software components including Microsoft *ActiveX* and *COM* objects.
- Visual J++ gives developers an easy way to design, code, and deploy their client- and server-side solutions, including *Active Server Page* (ASP) components, *Microsoft Transaction Server* (MTS) components, and *DCOM* components.
- Visual J++ 6.0 supports the creation of the next generation of Web-based applications using *DHTML* classes. With Visual J++ 6.0's DHTML library, developers have the ability to design and deploy truly integrated Web and Windows-based applications that can be executed on multiple platforms.

6. CONCLUSIONS

We have explored some of the important features for Java, Web, and distributed objects. The Java component model represents a combination of Java and component technologies. The JDBC module allows a Java programmer connecting and querying remote databases. The Web's data representation is shifting from structural HTML markup to semantic XML markup. DHTML, DOM, and XML constitute important enhancements to the Web documents. The prominent distributed object computing standards include OMG's CORBA and Microsoft's COM/DCOM/COM+. Java Distributed Objects (JDO) represents an integration of Java and distributed objects. The CORBA/Java ORBs include Inprise's VisBroker for Java, Iona's OrbixWeb, and Sun's Java IDL. Microsoft's Visual J++ is a DCOM/Java merger.

The goal of *WebComputing* model is to extend the current Web application model such that the benefits of distributed object computing systems such as the OMG's CORBA and Microsoft's COM+ can be realized in a Web native fashion [22]. In addition, Orfali and Harkey [10] claim: "The next big wave of client/server computing will be catalyzed by the marriage of distributed objects and the Web; we call this marriage the *Object Web*." Obviously, the *Object Web Computing* will be the mainstream of Internet technology.

REFERENCES

1. Burton, B. F. and Marek, V. W., "Applications of Java Programming Language to Database Management," *ACM SIGMOD Record*, March 1998, pp.27-34.
2. D'Souza, D., "JavaBeans: Coding and Design," *Journal of Object-Oriented Programming*, January 1998, pp.14-16.
3. Inprise, Inc., "*VisiBroker for Java*," http://www.inprise.com/prod/vbjpd.html, 1998.
4. Iona Technologies, Inc., "*OrbixWeb*," http://www.iona.com/products/internet/orbixweb, 1998.
5. Khare, R. and Rifkin, A., "XML: A Door to Automated Web Applications," *IEEE Internet Computing*, July/August 1997, pp.78-87.
6. Krieger, D. and Adler, R. M., "The Emergence of Distributed Component Platforms," *IEEE Computer*, March 1998, pp.43-53.
7. Lewis, G., Barber, S., and Siegel, S., "*Programming with Java IDL: Developing Web Applications with Java and CORBA*," John Wiley & Sons, New York, 1998.
8. Manola, F., "Towards a Richer Web Object Model," *ACM SIGMOD Record*, March 1998, pp.76-80.
9. Microsoft, Inc., "*Visual J++*," http://www.microsoft.com/visualj, 1998.
10. Orfali, R. and Harkey, D., "*Client/Server Programming with Java and CORBA*," 2nd ed., John Wiley & Sons, 1998.
11. Pedrick, D. et al., "*Programming with VisiBroker: A Developer's Guide to VisiBroker for Java*," John Wiley & Sons, New York, 1998.
12. Sun Microsystems, Inc., "*JavaBeans*," http://java.sun.com/beans/, 1998.
13. Sun Microsystems, Inc., "*Java IDL*," http://java.sun.com/products/jdk/idl/, 1998.
14. Sun Microsystems, Inc., "*JDBC API*," http://java.sun.com/products/jdbc, 1998.
15. Tomas, A., "*Enterprise JavaBeans Server Component Model for Java*," Prepared for Sun Microsystems, Inc., http://java.sun.com/products/ejb/white-paper.html, December 1997.
16. Vetter, R. and Kroeker, K. L., "The Internet in the Year Ahead," *IEEE Computer*, January 1998, pp.143-144.
17. Vinoski, S., "New Features for CORBA 3.0," *Communications of the ACM*, October 1998, pp.44-52.
18. Vogel, A. and Duddy, K., "*Java Programming with CORBA: Advanced Techniques for Building Distributed Applications*," 2nd ed., John Wiley & Sons, New York, 1998.
19. Voth, G. R., Kindel, C., and Fujioka, J., "Distributed Application Development for Three-Tier Architectures: Microsoft on Windows DNA," *IEEE Internet Computing*, March/April 1998, pp. 41-45.
20. W3C, "*Document Object Model (DOM)*," http://www.w3.org/dom, 1998.
21. W3C, "*Extensible Markup Language (XML)* 1.0," http://www.w3.org/TR/REC-xml, 1998.
22. W3C, "*WebBroker: Distributed Object Communication on the Web*," http://www.w3.org/TR/1998/Note-webbroker/, 1998.

A DATA REGISTRY-BASED ENVIRONMENT FOR SHARING XML DOCUMENTS

Hong-Seok Na, Jin-Seok Chae, and Doo-Kwon Baik
Software System Laboratory, Dept. of Computer Science & Engineering
Korea University 1, 5-ka, Anam-dong, Sungbuk-gu, 136-701, Korea

Key words: XML(Extensible Markup Language), XML document, XML DTD(Document Type Definition), Data Registry, Data Element, Document Sharing Environment

Abstract: In the web-based distributed computing environment, XML may be a solution to the data-sharing problem between applications. However, data exchange and sharing using XML supposes the use of a common and standardized XML DTD. Applications, using their own DTD, cannot understand the meaning of XML documents of others, therefore need human intervention to exchange their XML documents. In this paper, we proposed a method to represent the semantics of tags in XML documents using data registry and designed a document sharing environment supporting automatic exchange of XML documents. The key idea of the method is to attach a data element identifier in data registry to the corresponding tags in XML documents. By representing the exact meaning of tags using the data element identifier, applications can understand the meaning of tagsets in XML documents even if the documents are made on the basis of different DTDs. As a result, it is possible to exchange XML documents of different DTDs without human intervention

1. INTRODUCTION

The biggest inhibitor to electronic commerce today is neither security nor reliability. Instead, it's the sharing of data between applications[1]. Moving data between applications, whether on an intranet or an extranet, requires human intervention. That is, physically mapping each and every application's data to all your applications is time consuming, expensive and unreliable work.

A solution to this problem is XML(Extensible Markup Language). Developed by the World Wide Web Consortium(W3C), XML is a simplified(but strict) subset of the ISO's SGML(Standard Generalized Markup Language) that maintains SGML's features of extensibility, structure, and validation[2]. With XML, application designers can create sets of data element tags and structures that define and describe the information contained in a

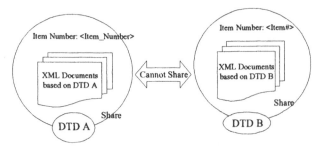

Figure 1. Current Environment of Data Exchange using XML

document, database, object, catalog or general application, all in the name of facilitating data interchange[1]. For example, in an XML free-form text document, users can easily retrieve and find the "Item Number" of a product they want to buy by looking for an XML construct such as <ItemNumber>xxxxx</ItemNumber>.

However, data interchange between applications using XML supposes sharing semantics of the tagsets defined in DTD(Document Type Definition). That is, two applications in an XML-based data interchange environment must know the semantics that the tag "Item_Number" means the number of a product.

Figure 1 shows the current state of data exchange using XML. If one application uses "Item_Number" as the tag of a product number but the other application uses "Item#", it is impossible for two applications to exchange the product's data automatically. XML provides facilities to represent the syntactic and structural elements defined in DTDs, but doesn't support semantic representation of the tag elements. So, if an application doesn't know the semantic meaning of tag elements in DTD of others, it can not understand XML documents but only browse it to users as current web browsers do.

In this paper, we propose an environment that make it possible for applications using different DTDs to share, interchange and search XML documents without human intervention. The main idea of this paper is to attach semantic meaning defined in a data registry to tag elements in DTD.

2.   RELATED WORKS

In order to understand other system's data automatically, there must be common agreements or rules between the two systems. Currently, there are two methods for sharing data between applications of different domain. One is data dictionary based approach which is used for business data interchange of EDI applications. And the other is an approach using standardized DTDs for XML documents.

Work has been concentrated on the development of specialized data dictionary to interchange business data electronically in EDI parts. The two representative transmission standards(dictionaries) are X12 and EDIFACT[1].

The data dictionary used in EDI requires that all applications must transmit their data according to specification defined in it. And applications can understand the received data of standard format. In addition, since these standards were made for electronic commerce and force each system to follow their data format, they have many constraints on supporting general-purpose applications in world wide web environment.

On the contrary, XML is a standardized text format designed specifically for transmitting structured data to web applications. XML documents, the unit of data exchange, are based on DTD(Document Type Definition). DTD is a document defining the structure of tagsets, used in XML documents, and XML standard specification defines the syntax of DTDs.

One important point is that XML documents based on same DTD have same structure and tagsets. So, if a system understands the meaning of DTD, it also understands the structure and meaning of XML documents. That is, it is possible to share data between applications if they use standardized DTDs. Such standardized DTDs include CML DTD[8] used in chemistry domain, Math DTD[7] in mathematics. Using standardized DTD makes it possible to share data elements and attributes defined in the DTD, and understand semantics of the XML documents automatically.

However, if a system cannot understand tags and attributes in the DTD, it cannot understand XML document based on the DTD, either. Therefore, web environment in which many different applications interact each other, requires human intervention including mapping processes.

There have also been some resource-description methods using metadata. Efforts such as the Platform for Internet Content Selection(PICS)[9] and the Resource Description Framework(RDF)[6] provide mechanisms for transferring machine-readable metadata describing resources among communities. PICS attaches labels to Web resources, using a URL to identify the rating service and rating scheme. RDF combines the PICS extensions with the metadata model in Netscape's Meta Content Framework(MCF), yielding both a metadata representation model and XML-based syntax for metadata capture and transfer. An RDF schema, named using a URL, gives a human- and machine-readable set of assertions of attribute-value pairs. The application of technologies such as PICS and RDF in community ontologies helps determine commonly understandable meanings for those tags within any given community that will be using them.

Though both the two approaches concentrate on self-descriptive and machine-readable mechanism for web resources, they have a limit that the owner himself/herself should describe his/her web resource. This can be some burden to the owner who makes the web resources.

In our approach, the users, preparing XML documents, have no burden to describe their documents. They only write XML documents according to DTDs. It is the responsibility of the person who makes the DTD. General users who use the DTD to make an XML document need not know the existence of data registry in our method.

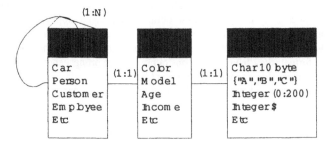

Figure 2. The Structure of a Data Element

3. A SEMANTIC REPRESENTATION METHOD BASED ON DATA REGISTRY

3.1 Data Registry

To share a data between applications, both user and owner of the data must clearly understand semantics, representation and identification of the data. Also, in order to understand the meaning of data, the description of the data should be provided to users, and users should have an easy method for accessing the descriptions. A data registry plays a role of unique identification, registration and service about data.

The describing unit of data registered to a data registry is data element[3]. A data element is a basic unit of identification, description and value representation unit of a data. Figure 2 shows the structure of a data element.

A data element is composed of three parts as follows[3]:

• The **object class** is a set of ideas, abstractions, or things in the real world that can be identified with explicit boundaries and meaning and whose properties and behavior follow the same rules.

• The **property** is a peculiarity common to all members of an object class, and

• The **representation** describes how the data are represented, i.e. Combination of a value domain, datatype, and, if necessary, a unit of measure or a character set.

100

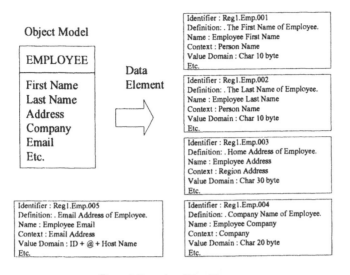

Figure 3. Examples of Data Elements

Object classes are the things about which we wish to collect and store data. Examples of object classes are cars, persons, households, employees, orders, etc. Properties are what humans use to distinguish or describe objects. Examples of properties are color, model, sex, age, income, address, price, etc. The most important aspect of the representation part of a data element is the value domain. A value domain is a set of permissible(or valid) values for a data element. For example, the data element representing annual household income may have the set of non-negative integers(with the unit of dollars) as a set of valid values.

Figure 3 shows an object "Employee" and its properties. The combination of an object class and a property is a data element concept(DEC). A DEC is a concept that can be represented in the form of a data element, described independently of any particular representation. The combination of the data element concept and a value domain makes a data element. A data registry supports registration, authentication and service of such data elements.

Each data element has an unique identifier, all users and programs using the data registry identify a data element by the identifier. A data element also contains its definition, name, context, representation format, and so on.

3.2   Representing Semantics using Data Element

XML documents are based on DTD(Document Type Definition). XML users can make their own document structure and tagsets by freely defining

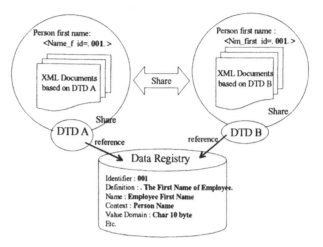

Figure 4. Data Registry-based XML Document Sharing

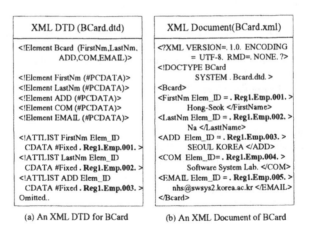

XML DTD (BCard.dtd)	XML Document(BCard.xml)
<!Element Bcard (FirstNm,LastNm, ADD,COM,EMAIL)>	<?XML VERSION=. 1.0. ENCODING = UTF-8. RMD=. NONE. ?> <!DOCTYPE BCard SYSTEM . Bcard.dtd. >
<!Element FirstNm (#PCDATA)> <!Element LastNm (#PCDATA)> <!Element ADD (#PCDATA)> <!Element COM (#PCDATA)> <!Element EMAIL (#PCDATA)>	<Bcard> <FirstNm Elem_ID = . Reg1.Emp.001. > Hong-Seok </FirstName> <LastNm Elem_ID = . Reg1.Emp.002. > Na </LasttName> <ADD Elem_ID = . Reg1.Emp.003. > SEOUL KOREA </ADD>
<!ATTLIST FirstNm Elem_ID CDATA #Fixed . Reg1.Emp.001. > <!ATTLIST LastNm Elem_ID CDATA #Fixed . Reg1.Emp.002. > <!ATTLIST ADD Elem_ID CDATA #Fixed . Reg1.Emp.003. > Omitted..	<COM Elem_ID= . Reg1.Emp.004. > Software System Lab. </COM> <EMAIL Elem_ID = . Reg1.Emp.005. > nhs@swsys2.korea.ac.kr </EMAIL> </Bcard>
(a) An XML DTD for BCard	(b) An XML Document of BCard

Figure 5. Examples of a DTD and an XML Document

102

the DTD with their intention. This freedom of defining tagsets and the structures, however, prevent the interchange of an XML document between two applications without knowing the other's DTD.

As shown in figure 4, XML documents using same DTD can be interchanged between applications (because they use same document structure and tagsets), but documents using different DTDs cannot share the meaning of the documents even if they have similar structures or contents.

Figure 5 shows a DTD, which defines the tagsets used to represent business card, and an XML document based on the DTD. The DTD is made using the data elements in figure 3. Figure 5(a) shows the contents of a DTD file named "BCard.dtd". The words of "!Element" defines tags used in XML documents. The first line of figure 5(a) shows that there is a tag named "Bcard" and the tag consists of five sub tags - "firstNm", "LastNm", "Add", "Com" and "Email". The word "!Attlist" defines attributes of each tags.

Figure 5(b) is an XML document written by the definition in DTD of figure 5(a). Each tags have an attribute named 'Elem_ID' with the value of a data element identifier in data registry. In XML, attributes of an element can have default values[2]. We use this characteristic of XML.

In figure 5(a), the part "<!Element Bcard (FirstNm, .)>" means that users of the DTD will use tag "Bcard" and tag "Bcard" is composed of five tags "FirstNm", "LastNm", "ADD", "COM" and "EMAIL" tags. The part "<!Element FirstNm (#PCDATA)>" means that the tag "FirstNm" will contain a value in XML documents. The most important part is "<!ATTLIST FirstNm Elem_ID CDATA #Fixed "Reg1.Emp.001">" that means the tag "FirstNm" has an attribute "Elem_ID" and this attribute has the fixed value "Reg1.Emp.001". Thus, all XML document based on Bcard.dtd use tag "FirstNm", and the attribute "Elem_ID" of the tag contains the fixed value "Reg1.Emp.001". This attribute-value pair is inserted automatically by general XML editors.

3.3   XML Document Interchange using Data Element

Let's suppose a situation that company A uses DTD A for personnel management and company B uses DTD B for the same object. If an application of company A would sent a person's personnel records to company B using XML document, it is necessary to make the document understood by the receiving application of the company B.

In our approach that attaches a data element identifier representing the concept of tags in XML document, the converting process can be done automatically if XML documents are made by the DTDs referencing a common data registry.

The converting process consists of three steps:

• **Step 1**: Reading two different DTDs, and one XML document

• **Step 2** : Constructing a mapping table

• **Step 3** : Replace the tags in the XML document with the corresponding tags in mapping table

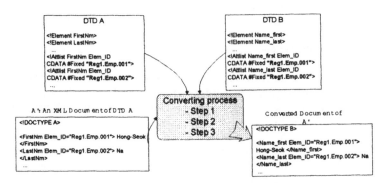

Figure 6. Input and Output of the Converting Process

Table 1. An Example of Mapping Table

Elem_ID	DTD A	DTD B
Reg1.Emp.001	FirstNm	Name_first
Reg1.Emp.001	LastNm	Name_first
Reg1.Emp.001	ADD	Address
Reg1.Emp.001	COM	Company

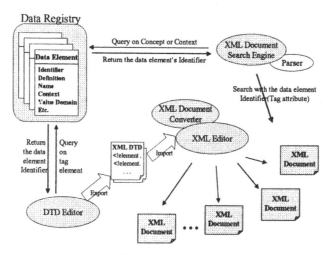

Figure 7. An XML Document Sharing Environment

In the first step, the process takes two DTDs, the document sender's DTD and the receiver's DTD, and one XML document to be converted for sending. After parsing the two DTD, it constructs a mapping table as `the second step. An example of mapping table is shown in table 1. The table consists of three columns - "Elem_ID", "DTD A" and "DTD B". The column "Elem_ID" contains the data element identifier in the data registry as the unique concept. The column "DTD A" has the tag names defined in the DTD of sender's corresponding to the identifier, and the column "DTD B" has the tag names defined in DTD of receiver's.

Then, we can automatically replace the tags in the XML document with the corresponding tags in mapping table, generating a document that the receiver can understand. In figure 6, the tag <FirstNm> in document A' is replaced with the tag <Name_first>, since they have the same data element identifier "Reg.Emp.001". Like this, the tag <LastNm> is replaced with the tag <Name_last> and so on.

As a result the receivers can get XML documents converted as their DTDs, understand and use the documents directly.

4.   AN ENVIRONMENT FOR SHARING XML DOCUMENTS

The XML document sharing environment suggested in this paper makes it possible to interchange the meaning of tags and attributes between XML applications, and to understand the contents as well as the structures of the documents. That is, the environment can support automatic document interchange and data interchange.

The structure of the document sharing environment is shown in figure 7. Centering around a data registry, this environment consists of four major independent tools . the DTD editor, the XML editor, the XML converter, and the document search engine.

The DTD editor supports a person to make out his DTD by defining tagsets and attributes referring to the data registry. Thus, it has one additional function to query the registry on tag element and get the data element identifier from the registry than existing DTD editors..

The XML editor supports editing of valid XML documents by reading a DTD and structuring the XML document. Since there are already data element identifiers in a DTD referencing the data registry, whatever XML editors we use, the data element identifiers in DTD will be located also in XML documents automatically.

The XML converter supports that one application with it's own DTD can understand the other's XML documents by converting the tags of the other's XML document to the application's tags using the data element identifiers in tags.

In the last, the XML document search engine accomplishes searching documents produced in the suggested environment. In addition to the keyword and full text search, the engine can understand the meaning of tags and attributes in XML documents by referencing the data registry, and thus we can get accurate results more closed to the concept and context of user's intention.

5. CONCLUSION

XML provides a dynamic and structural mechanism for data sharing in web based distributed computing environment, but it supposes the use of standardized DTDs.

In this paper, we proposed a method and environment that make it possible for applications using different DTDs to share, interchange and search XML documents without human intervention. By identifying the exact meaning of tag in XML documents with the unique data element identifier in data registry, applications can automatically interchange XML documents even if the documents use different tagsets.

We believe that the proposed method and environment will increase the interoperability of XML documents between applications in different domains.

We have implemented a prototype for the data registry and are presently working on the implementation of the DTD editor and the XML converter. Our future research will include developing a semantic representation scheme in case of multiple data registries and implementing the XML document search engine.

REFERENCES

[1] Rik Drummond and Kay Spearman, "XML Set to Change the Face of E-Commerce", Network Computing, V.9, N.8, pp.140-144, May 1998.

[2] Charles F. Goldfarb and Paul Prescod, "The XML Handbook", Prentice Hall, 1998.

[3] "Information technology - Specification and standardi- zation of data element", ISO/IEC 11179-1 Final Committee Draft, June 1998.

[4] Rohit Khare and Adam Rifkin, "The origin of (document) species", Computer Networks & Isdn Systems ,V.30, N.1-7, pp.389-397, April 1998.

[5] Rohit Khare and Adam Rifkin, " XML: A Door to Automated Web Applications", IEEE Internet Computing, pp.78-87, July & August 1997.

[6] O. Lassila and R. Swick, "Resource Description Framework(RDF) model and syntax", World Wide Web Consortium Working Draft(Work in Progress), October 1997. available at http://www.w3.org/Metadata /RDF.

[7] "Mathematical Markup Language (MathML) 1.0 Specification, W3C, July 1998. available at http://www.w3c.org/TR/1998/REC-MathML-19980407.

[8] P. Murray-Rust, " Chemical Markup Language",Version 1.0, Jan.1997. available at http://www.venus.co.uk/omf/cml/.

[9] Paul Resnick and Jim Miller, "PICS: Internet Access Controls Without Censorship", Communications of the ACM, vol. 39(10), pp. 87-93. 1996. available at http://www.w3.org/PICS/iacwcv2.htm.

EVALUATION OF A METHODOLOGY FOR CONSTRUCTION OF TMN AGENTS IN THE ASPECT OF COST AND PERFORMANCE

Soo-Hyun Park* and Doo-Kwon Baik**
*Dept. of Computer Engineering, Dongeui University
San 4, Kaya-dong, Pusan Jin-ku, Pusan, 614-714, Korea
Ph.: +82 51 890 1725, Fax: +82 51 890 1704
E-mail: Shp@hymin.dongeui.ac.kr

**Software System Laboratory, Dept. of Computer Science & Engineering
Korea University 1, 5-ka, Anam-dong, Sungbuk-gu, 136-701, Korea
Ph.: +82-2-925-3706, Fax: +82-2-921 9317
E-mail: Baik@swsys2.korea.ac.kr

Key words: TMN Agent, Farmer Model, Farming Methodology

Abstract: *TMN agent performs the function of the agent since the elements such as the manager that uses the agent service, the real sources that the agent takes full charge, and Guideline for Definition of Managed Object(GDMO) compiler that creates the code of the agent interact mutually. In this book, the result that is received when the Farming methodology is applied to the actual implementation of TMN agent is evaluated by comparing with TMN system that is developed by not applying the concept of the Farming (non-Farming methodology). As a result, we can find that if the network that composes Data Communication Network(DCN) is developed as a network with the speed higher or equal to ATM, the performance does not show much difference between the Farming methodology and the non-Farming methodology. However, in the aspect of the development cost, the Farming methodology is relatively better. Therefore, if the network with the speed higher or equal to 100Mbps is composed of DCN, it is desirable to develop the system by the Farming methodology*

1. Introduction

1.1 Background and Purpose of the Research

By managing various communication networks collectively, Telecommunication Management Network(TMN) has appeared as a concept to aim for the unified and effective communication network operation and

Systems Development Methods for Databases, Enterprise Modeling, and Workflow Management,
Edited by W. Wojtkowski, *et al.* Kluwer Academic/Plenum Publishing, New York, 1999.

maintenance. Since TMN has been developed by different operating systems and different versions of hardware platforms in the implementation process, several problems have been found in the step of developing and maintaining the class of TMN system agent[Park1 97 – Park4 98]. First, since the classes in the agent are implemented and maintained by having the dependency on the platform such as the hardware or the operating system, various number of agents and managers must maintain all different classes that perform the identical function in duplication. Because of this, it is not easy to maintain the class version that performs the identical function in the whole network management point of view. Second, the support of the multi-platform becomes impossible, and eventually, it is hard to provide the consistent interface in case of developing Q3 interface between Data Communication Network(DCN), Operation System(OS), Mediation Device(MD), and Network Element(NE). Third, since the different networks are maintaining the different network management system, the compatibility between the operation and maintenance systems is not guaranteed in case of implementing TMN that is the concentrated network management system [Park1 97 – Park4 98][KUTMN 98][ITU-1 92][ITU-2 92][Aidar 94][DSET 95].

In order to solve these problems, this paper suggests the Farming methodology[Park1 97][Park3 97]. The software components in the distributed object such as the manager and the agent are created in the componentware type, and they are maintained in the Platform Independent Class Repository(PICR)[Park1 97 - Park4 98]. The Farming methodology is the methodology that executes these componentwares by downloading to the distributed object dynamically or statically if necessary. In case of composing the agent in TMN system, this methodology defines the function blocks of the distributed object as the framework list in order to perform the given role, composes the distributed object framework by transplanting the software components that composes this framework list from PICR, and composes the whole agent by transplanting the data component needed for the execution of the agent. Especially, OLB[Park4 98] is executed by downloading it from PICR to the agent on demand dynamically.

The Farming methodology is based on the Farmer model[Park4 98] that the knowledge representation model. The Farmer model is the frame structure model that introduces the concept of the system entity structure, and it is based on Entity-Aspect(EA) model[Chkim 89][Dycho 91][Zeigl 84] that is the knowledge representation model. The main purpose of the Farmer model is to separate and extract the component elements that compose the agent by analyzing the agent to be actually designed by the aspect. As a result of this, the component elements are located on the leaf node of the Farmer model tree, and these component elements are finally stored in PICR through the network. Furthermore, the Farmer model is the formal model that adds the concept of the Farming which downloads the component elements from PICR to the agent. The core concept of the Farming methodology is the Farming and the Aspect-Object oriented programming is introduced as the basic programming paradigm of the Farming methodology. The Farming model becomes the basic model of the programming method that uses the Aspect- Object[Park2 97][Park4 98].

The Farmer model is composed of the components such a the entity node type that describes the objective and abstract object of the real world, the aspect node type that describes this entity, the link type that identifies the relationship between the entity and the aspect, the attribute type that identifies the characteristic of the entity, the uniformity entity node type and the uniformity aspect node type that are defined by the axiom of the uniformity, that is, the two nodes with the same name have the same attributes and the same subtrees, IM-component type node (ILB multiplicity component type node) that has the attribute corresponding to ILB component in case of Farming, and OM-component type node (OLB Multiplicity component type node) that has the attribute of OLB component. The Farmer model is also composed of three abstract concepts such as, generalization, aggregation, and multiplicity. Among these, IM-component type node and OM-component type node are related to the concept of the multiplicity, and they become the elements of multiplicity set. IM-component type nodes and OM-component type nodes are connected to the upper entity node by the Multiplicity Instance Link.

In this paper, the result that is received when the Farming methodology is applied to the actual implementation of TMN agent is evaluated by comparing with TMN system that is developed by not applying the concept of the Farming (non-Farming methodology).

1.2 Related Research

1.2.1 Way to Develop the Design and the Configuration of the System through the Network

There is a phrase "Network is computer". This phrase explains the importance of the network in the configuration of the computer. The way to develop the design and the configuration of the system through the network can be divided into the following 3 steps.

Step 1 : Unification of the Hardware through the Internet[Bapat 94]

Step 2 : Unification of Software by using Platform Independent Language and the Package

The research such as software architecture[Gacek 97][Garla 93], ComponentWare Consortium(CWC) [CWC1 95][CWC2 95][CWC3 95][CWC4 95], Active-X[Chen 97], and OLE[Chen 97] are on progress. Java is often used for plug and play concept and platform independent language, and CORBA is often used for the unification of the heterogeneous systems.

Step 3 : Unification of the Data

In step 2, the unification of the software by using the component concept is the major interested area, but it is not meaningful for the software unification methodology without the unification of the data. In this step, it suggests the unification of the data by using the concept of the meta data.

Since step 2 and 3 should be thought as mutual relation rather than considering them separately, the methodology to unify the hardware, software, and data through the network generically is needed. In this paper, the Farming methodology is suggested for this unifying methodology. The

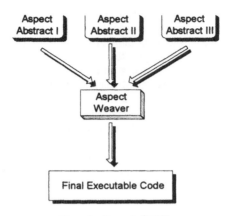

Figure 1. Concept of AOP

Farmer model is the model that becomes the basis for this Farming methodology.

1.2.2 Comparison with AOP

For the programming method to design a system, the most commonly used method since mid 1980's is Object-Oriented Analysis & Modeling (OOAM) [Coad 94][Kempe 94][Lee 97][Marti 92][Winb 92]. However, since this OOAM can describe the real world entity in only one fixed aspect, it ignores many different aspects that is needed and helpful to understand the entity in case of designing and modeling the entity.

The basic method to construct the software system is the method to divide the system to be constructed into the unit of modules such as the subroutine, the procedure, the object, the client, and the server. However, in order to construct a very complicated formless system such as MEMS[Berli 95], a very careful analysis is needed for the function executed by each module in MEMS. The system aspects like this affects between the aspects and between the final executable code related to this aspect. For example, the communication between MEMS units becomes the type that is tangled complicatedly by the distribution aspect, the power consumption aspect, the communication processing failure processing aspect, and the synchronization aspect, and affects each other. The various aspects of the system can be considered by separating them conceptually, but they should be unified in the implementation level. However, the module that executes the function should match with the executable code block that executes the corresponding function directly, and the executable code block is affected by other related aspects. So the modules become the tangled mess of aspects. This TANGLED-MESS-OF-ASPECTS phenomenon is a problem that becomes the core of the currently existing complicated software system. This cannot be solved by using the most commonly used modularization method.

112

Table 1. Comparison between the Farming methodology and AOP methodology

Items to Compare	Farming Methodology	AOP Methodology
Model to support	Farmer Model (Component)	Object-Oriented Model (Open Implementation)
Component sourcing	Farming from PICR (Outsourcing)	Not Support
Description of the Relationship Between The entities	Horizontal / phased description	Horizontal description (Multi-step model is needed for the hierarchical description.)
Way to analyze and access the system	Classify the relationship between the entity and the aspect	Classify only the aspect of the entity.
Method for describing the entity	Aspect-Object	Separated aspect description using special language
Way to unify the entities	No need for integrator	Aspect Weaver
Time to unify the entities	When the aspect object is defined	After completing the aspect description

In order to solve this, Xerox Parc suggests Aspect Oriented Programming(AOP)[Grego 94][Grego 95] for the method to separating a certain system by considering the aspect and to unify them in the implementation level. That is, after the system analyzer describes and specifies the system to be developed into a natural type by separating each aspect, the specifications that are separated by the tool called aspect weaver are unified into a single executable code. The aspectual decomposition that is the first step is a concept that is clearly distinguished from the modularity.

AOP uses the separated aspect description for the aspects created in the aspectual decomposition step by using the special language, and at this time, the description is made in the abstraction type, and it is unified to a executable code in the unification step. Figure 1 illustrates the concept of AOP. However, the case of AOP oriented methodology includes many problems. It cannot describe the relationship by hierarchies between the entities, the system analysis method can identify only the aspect of the entity, and it cannot identify the relation between the entity and the aspect that entity has. Furthermore, the entities that are identified by the aspect use the integrator called aspect weaver, and the time to unify the entities is not when the aspect is defined but after the separated aspect description that uses the special language is completed. Finally, AOP does not support the concept of the Farming that outsources the componentware from the outside.

Table 1 compares the Farming methodology and AOP methodology by the major items.

2. Design and Prototype Implementation of PCN TMN Agents

2.1 Composition of TMN Agents

2.1.1 Introduction and Structure

TMN agent[Kutmn 98] performs the function of the agent since the elements such as the manager that uses the agent service, the real sources that the agent takes full charge, and Guideline for Definition of Managed Object(GDMO) compiler [DSET 95] that creates the code of the agent interact mutually. In fact, the TMN platform that creates TMN agent supports this mutual interaction to be done effectively and comfortably. Figure 2 illustrates the mutual interaction between the elements of TMN system.

In case of creating the TMN agent, the elements that should be considered are the inheritance tree related module that maintains the information of the managed object class which is managed by the agent, the containment tree related module that includes the information about the managed object instance which is created by the agent, the communication module that supports the communication with the manager, the conversion module that manages the conversion between the Protocol Definition Unit(PDU) related data and the internal processing data, and the object framework that manages the mutual interaction with the real resource.

In case of creating the new managed object class through GDMO compiler, GDMO compiler informs this fact to the agent, and the agent that

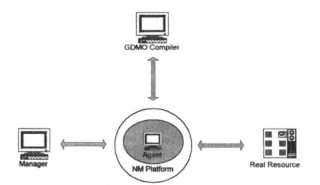

Figure 2. Mutual Interaction between the Elements of TMN System

receives this message does not accept any more connection request and performs only the currently requested service. When the currently requested service terminates, the agent reads the information about the new managed object class and registers on the inheritance tree.

When the manager requests the management function, the agent receives CMIP PDU from the communication module, interprets it, and converts this to the internal data structure through the conversion module. After the conversion, CMIP PDU is sent to the containment tree management module, so that the corresponding managed objects can be found. The management function of the next managed object is executed and the result is sent to the communication module though the conversion module, so that CMIP PDU can be made and returned. At this time, if the mutual interaction with the real resource is needed, the support of the object framework can be received.

In the object framework, the function that monitors the real resource periodically and the function that processes the message which comes from the real resource and informs it to the corresponding managed object. Figure 3 illustrates this module composition of the TMN agent.

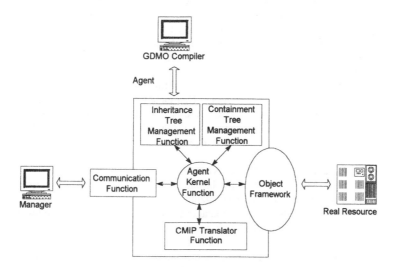

Figure 3. Module Composition of the TMN Agent

2.1.2 Module Composition

l) Agent Kernel

Agent kernel manages all the operations of the actual agent. First of all, in case of starting the agent, the inheritance tree related module is called, the inheritance tree is composed, the object framework is initialized, the managed object instance is created by reading the information of the continuity managed object instance from the file, and this instance is registered on the containment tree. When all the initialization operations are terminated, the agent waits for the management request message to come from the manager. This request message is performed when it is received. In addition, when the creation message of the new managed object class is arrived from GDMO compiler, the inheritance tree related data are reinitialized after terminating the currently processing service.

2) Communication Module

The communication module interprets the inputted CMIP PDU, and creates CMIP PDU that will be sent. This is supported as Application Programming Interface(API) type, so that the function provided by the communication module can be used. In addition to CMIP/S, XOM/XMP can also be supported.

3) Conversion Module

The data structure that is proper for PDU creation and the data structure that is proper for the agent internal processing can be different. Therefore, the conversion module manages the conversion function between these. For example, when the managed objects or the attributes are expressed as the object identifier in PDU, the function to convert this to the integer type that makes the internal processing easy or the function to convert the filter part of CMIP PDU logically belongs to this module.

4) Inheritance Tree Related Module

The inheritance tree includes the informations for the managed object class that is supported by the agent. This includes the informations such as the basic Abstract Syntax Notaion 1(ASN.1) syntax, the attribute type, the attribute group, the notification, the action, the name binding, and the creation function of each managed object.

5) Containment Tree Related Module

When the managed object instance is created in inside of the agent, this should be registered on the containment tree. In fact, all the management function is made through the containment tree. Therefore, the function to register on the containment tree in case of creating the new object, the function to delete from the containment tree in case of deleting the managed object, and the function to process the category to find the corresponding managed object in case of requesting the management function should be provided.

6) Object Framework Module

The one to help the mutual interaction with the real resource is the object framework. There are two functions that are supported here. The scheduler function that supports the operation started in the agent and the monitor function that supports the operation started from the real resource are provided. In the scheduler, the function to register a specific management function in order for each managed object to execute it periodically, the function to delete this, and the function to execute the management function

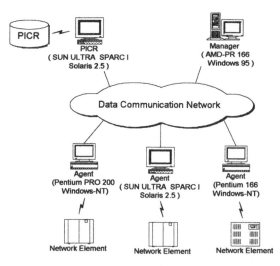

Figure 4. Implementation Environment of PCN TMN Agent Prototype

on the time requested by the managed object are provided. The monitor includes the function to register the function that interprets the message in order to interpret the message from the real resource and informs to the corresponding managed object, the function to delete this, and the function to execute the function of the managed object.

2.2 Creation of the Agent by the Framework

The method to implement by the Farming methodology has been explained by the description of each step from 1 to 6 in [Park4 98] and by the construction example of PCN TMN agent. This section describes the items that should be considered in case of designing PCN TMN system by each element, the construction method according to the framework, and the down loading method of the componentware in case of executing the agent in detail. The Farming methodology is based on the Farmer model, and therefore, the agent is considered as the managed entity.

2.2.1 Implementation of PICR

PICR used in the Farming methodology is designed and implemented by considering the following.

1) First of all, design the agent system by using the Farmer model diagram.

2) Convert the diagram of the previous step to Aspect-Object Definition Language(ADL).

3) The function blocks that are separated from the 2nd step "Analysis of the Distributed Object System and the Extraction of the Componentware Element" of the Farming methodology are divided into ILB/OLB.

4) ILB/OLB's are implemented by using JavaBeans[Youn97] that is the platform independent language by using J++ Builder and KAWA[Jryul 97]. After this , move them to PICR.

5) For the implementation environment, this research uses UniSQL that is executed on SUN ULTRA SPARC I / Solaris 2.5. Figure 4 illustrates this implementation environment.

6) PICR server daemon process that executes the function to download the necessary JAVA class by accepting the request message from the agent. The algorithm of this program is as follows.

[Algorithm 2.1] PICR server deamon

1 **While true do**
2 Receive the component down loading request message from the agent
3 Check if the requested component exists in PICR.
4 **IF Find OK THEN**
4.1 *Send the component to the agent*
5 **ELSE**
5.1 *Exception handling*
6 **END IF;**
7 **End While**

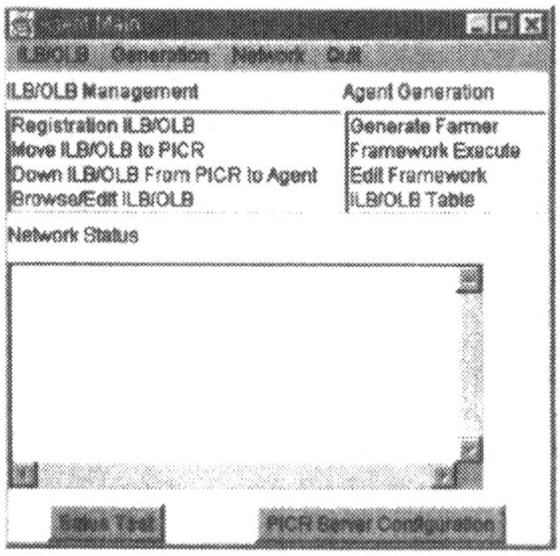

Figure 5. Agent Constructor

118

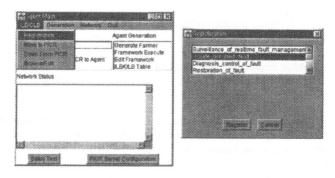

Figure 6. Registration ILB/OLB

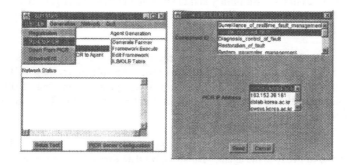

Figure 7. Move ILB/OLB to PICR

2.2.2 Implementation of the Agent

The process of design and implementation of the agent is connected to the design and the step 1 and the steps after step 2 of the implementation process of PICR. ADL that is created in the process of constructing PICR is used as a framework to implement the agent.

1) Implement the agent constructor that reads and executes the framework written in ADL code.

2) The agent constructor performs the function like the following. Figure 5 illustrates the main window of this agent constructor.

2-1) After creating the new necessary agent by using J++ Builder in the agent, execute the function to store the created ILB/OLB component in the local library in the agent platform before moving it to PICR.

"Registration ILB/OLB" item in ILB/OLB management window of Figure 5 belongs to this, and this is connected to Figure 6. Figure 6 illustrates the browse function to find ILB/OLB component in the local library.

2-2) After discussing with TMN system administrator, execute the function to transmit ILB/OLB components that are permitted to be stored in PICR to PICR server. Execute this function by using **"Move ILB/OLB to PICR"** item in ILB/OLB management window in Figure 5. In order to execute this, the function to browse the component that is needed to registered in the local library and the function browse the target PICR server are needed for the specific function. Figure 7 illustrates the window that executes this function.

In the Farming methodology, the function of 2-1) and 2-2) is called seeding. Therefore, since the user performs the role of creating ILB/OLB components and storing in PICR when an agent cannot find ILB/OLB component that has the function needed by itself only in PICR, this is called seeding.

2-3) Execute the function to download ILB/OLB component that is stored in PICR to the agent. **"Down ILB/OLB From PICR to Agent"** item in ILB/OLB management window in Figure 5 belongs to this, and this is connected to Figure 8. Unlike the function to download the component dynamically or statically by the framework, the goal of this function is the component management in the agent level. This is the function to download the component defined as ILB from PICR and to store in the local library. Figure 8 illustrates the window that executes this function. The component in PICR that is browsed by using the first window shows the profile of the component through the second window. If the agent administrator decides that this ILB component is needed in his or her agent after reviewing the profile, it is stored in his or her local library by clicking **"save to local lib"** button.

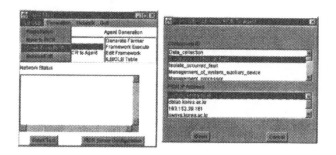

Figure 8. Down ILB/OLB from PICR to Agent

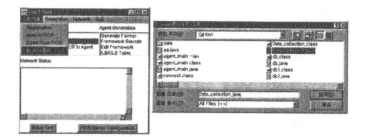

Figure 9. Browse/Edit ILB/OLB

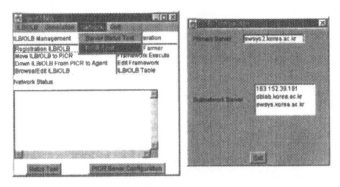

Figure 10. PICR Server Configuration

2-4) **"Browse/Edit ILB/OLB"** item in ILB/OLB management window in Figure 5 is the function used when it is necessary to modify the existing component. After browsing ILB/OLB component that needs to modify the function in PICR by the window shown in Figure 9, if **"Edit"** button is clicked, the connection is made to J++ Builder. After the modification is completed, it can be registered in PICR by using **"Move ILB/OLB to PICR"** item in ILB/OLB management window of Figure 5. This step also needs to discuss with the administrator in order to get the permission for the move.

2-5) **"Generate Farmer"** item in the main window of Figure 5 creates the Farmer class. The Farmer class is brought from PICR to agent as a component type to be executed. The creation result of the Farmer is displayed on the status message window of the agent constructor main window shown in Figure 5.

2-6) **"Framework Execution"** item in the agent constructor main window of Figure 5 corresponds to the step 6 of the Farming methodology. The componentwares defined as OLB in the framework are examined if they are marked in OLB table that has been already constructed. The function like the following is executed through the mutual interaction with CRBP. By the component type(ILB or OLB) requested by the Farmer,

■ In case of ILB, the corresponding component is brought from the local library.
■ In case of OLB, IP address of the server that has PICR is resolved. After sending downloading request message of the component, the requested component is downloaded. Then, the component is retransmitted to the Farmer.

122

2-7) **"ILB/OLB Table"** item in the agent construct window of Figure 5 is the function that displays the list of ILB/OLB component which exists in the current agent platform.

2-8) **"Status Test"** item in the network status window executes the function to confirm the status of the communication channel between the current agent and PICR server.

2-9) **"PICR Server Configuration"** item in the network status window executes the function that displays the information for all the PICR servers that exists in current TMN system. Figure 10 illustrates this function

2.2.3 System Configuration

In this research, TMN PICR, the agent, and the prototype of the sample manager are implemented under the following environment(Figure 4).

1) PICR : SUN Ultra SPARC I, Solaris 2.5

2) Manager : AMD-PR 166, Windows 95

3) Agent 1 : Pentium 166, Windows NT
Agent 2 : Pentium PRO 200, Windows NT
Agent 3 : SUN Ultra SPARC I, Solaris 2.5

4) Communication protocol between the manager, the agent, and PICR (DCN)
TCP/IP under LAN environment is used. The communication between these can be considered though the exchange of the class by using Load_class. In this case, the communication between the manager and the agent is carried out by JAVA classes. A Java class is a bytecode format which is transported across the network to run on a machine which contains a Java interpreter. The BER (Basic Encoding Rules) used in the current CMIP standard is eliminated in this methodology since the interoperability is supported by the bytecode in the Java Language Environment (JLE).

3. Evaluation of the Farming Methodology

In this chapter, the result that is received when the Farming methodology is applied to the actual implementation of TMN agent is evaluated by comparing with TMN system that is developed by not applying the concept of the Farming (non-Farming methodology). Until now, TMN agent has been developed by using the agent creating toolkit such as DSET[DSET 95] and RETIX[RETX1 96 - RETX4 96], and these methods are based on the object-oriented programming method. However, if these methods are used, the agent that is dependent on the operating system and the hardware that composes the agent must be developed. The developing method is also called the non-Farming methodology in this paper. The criteria for the evaluation considers the following 2 aspects.

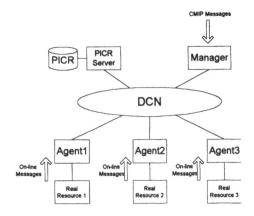

Figure 11. Simulation Environment

■ Performance aspect

Response time for the transaction processing such as CMIP message and notification message(on-line message) that occurs on the network

■ Cost aspect

The cost needed in case of developing TMN agent by the non-Farming methodology and the cost needed in case of developing by the Farming methodology are compared with each other. The costs needed in developing by the Farming methodology for each proportion of ILB/OLB that is stored and maintained in PICR are evaluated by comparing with each other.

3.1 Analysis of the Performance by the Transaction Response Time

The simulation evaluation model that is related to the transaction t that handles CMIP message and the notification message which occurs in the agent is defined. For the performance factor, this paper considers CPU processing time, repository access time, and the communication time between the network elements in each distributed object are considered, but the network topology is not considered. First of all, the terminology's used in this evaluation model are defined as follows.

s : transaction t (CMIP message, notification message processing) occurrence node
d : node that actually handles transaction t (agent, manager)
r : real resource that is managed by the agent
p : PICR server site
N : set of sites (s, d, p $\in$ N)
T : set of transactions (t $\in$ T)

The model of this simulation becomes the distributed environment like Figure 11.

The definition for the completion time CT_{total} of the transaction is as follows. In order to define CT_{total} , the following terminologies are defined in addition.

$CPU_T(x)$: CPU processing time of the node x
$COM_T(s,d)$: communication time between the node s and the node d
$COM_T(y,r)$: communication time between the node y (agent) and the real resource
$ACC(p)$: processing time of PICR access transaction
$T_m(r)$: processing time of r
$T_m(p)$: processing time of PICR server ($= CPU_T (p) + ACC(p)$)
CT_{t_CMIP} : processing time needed for the completion of the execution of transaction t (CMIP message)
$CT_{t\_notification}$: processing time needed for the completion of the execution of transaction t (notification message)
CT_{total} : processing time needed for the completion of the execution of transaction t ($= CT_{t\_cmip} + CT_{t\_notification}$)

3.1.1 Definition of CTt_CMIP

In Figure 11, it is defined in the time needed for the handling and the completion when the manager transmits CMIP message to a certain agent in TMN. CTt_CMIP can be descried as the sum of the CPU time of each node such as the manager, the agent, and the PICR server, the transmission time between the nodes though the network, the input and the output time of PICR disk, the communication time between the agent and the real resource, and the processing time of the real resource.

$$CT_{t\_CMIP} = CPU_T(s) + COM_T(s, d) + CPU_T(d) + COM_T(d, p) + CPU_T(p) + ACC(p) + CPU_T(p) + COM_T (p, d) + CPU_T(d) + COM_T(d, r) + Tm(r) + COM_T(d, r) + CPU_T(d) + COM_T(s, d) + CPU_T(s)$$

$$= \sum_\alpha CPU_T(s) + \sum_\beta COM_T(s, d) + \sum_\gamma CPU_T(d) + \sum_\delta COM_T(d, p) + \sum_\varepsilon CPU_T(p) + \sum_\zeta ACC(p) + \sum_\eta Tm(r) + \sum_\theta COM_T(d, r)$$

(3.1)

The definitions from α to θ are as follows.
   α : number of CPU access on the transaction t occurrence node
   β : number of communication though the unit communication path between s and d
   γ : number of CPU access on the transaction t processing node
   δ : number of communication between transaction t processing node and PICR server (*)
   ε : number of CPU access in PICR server (*)
   ζ : number of input and output in PICR disk (*)
   η : processing time in real resource
   θ : number of communication for the real resource and the agent

Among the variables defined above, the values for δ , ε , and ζ that are indicated by * are determined by the proportion of ILB and OLB that are stored in PICR. So, as the proportion of ILB becomes higher and the proportion of the OLB becomes lower in the framework of the agent, the

Table 2 System Parameter Values in simulation

Expr. (3.1)	Expr. (3.2)	Meaning of Variables
$COM_T(d, r)$	$COM_T (r, s)$	Communication time between the real resource and the agent
$CPU_T(d)$	$CPU_T(s)$	CPU processing time of agent
$CPU_T(s)$	$CPU_T (d)$	CPU processing time of manager
$COM_T (d, p)$	$COM_T (s, p)$, $COM_T (p, s)$	Communication time between agent and PICR Server

values for these variables become smaller, and the values for these variables become bigger in the opposite case. Therefore, these variables manages a important role in the simulation.

3.1.2 Definition of $CT_{t\_notification}$

TMN agents make the on-line notification message such as the occurrence of the fault in the real resource or the periodic status report, and they report it to the manager. $CT_{t\_notification}$ is defined as a time needed to complete the processing of this on-line message. $CT_{t\_notification}$ can be described as a sum of the CPU time of each node such as manager, agent, and PICR server, the communication time between the nodes through the network, the input and the output time of PICR disk, and the communication time between the agent and the real source.

$$CT_{t\_notification} = COM_T(r, s) + CPU_T(s) + COM_T(s, p) + CPU_T(p) + ACC(p)$$
$$+ CPU_T(p) + COM_T(p, s) + CPU_T(s) + COM_T(s, d)$$
$$+ CPU_T(d) \qquad (3.2)$$

Some variables used in expression (3.1) and expression (3.2) has a same meaning, and this is shown in Table 2.

3.1.3 Definition of CT_{total}

CT_{total} signifies the total processing time needed for completing the execution of the transaction t, and the equation like the following is formed.

$$CT_{total} = CT_{t\text{-}cmip} + CT_{t\_notification} \qquad (3.3)$$

According to Table 2, all the variables except Tm(r) are shared by (3.1) and (3.2). When CMIP message is processed n time and the notification message is processed m times in actual TMN system, the change for the values of α, β, ... , θ are shown in Table 3.

126

Table 3 Values of Each Variable in case of CMIP Message n Times / Notification Message m Times

Variables	(3.1)	(3.2)	(3.3)
α	2n	m	2n + m
β	2n	m	2n + m
γ	2n	2m	2(n + m)
δ	n	2m	n + 2m
ε	2n	2m	2(n + m)
ζ	N	m	n + m
η	N	0	n
θ	2n	2m	2(n + m)
Total	13n	11m	13n + 11m

Table 4 Rc & Ro

Message Type	Frequency Information					
	Place of Origin	Ro Per Place of Origin	Rc per Agents		Number of Message
CMIP	Manager	0.0	Agent 1	0.5	10^3
			Agent 2	0.3	
			Agent 3	0.2	
On-line	Agent 1	0.46	0.0		10^5
	Agent 2	0.29			
	Agent 3	0.25			

Table 5 System Parameter Values in Simulation

Parameters	Manager	agent1	agent2	agent3	PICR server
$CPU_T(x)$	2.0	2.5	2.0	3.0	2.0
ACC(p)	0.0	0.0	0.0	0.0	1.0
Tm(r)	0.0	0.0	0.0	0.0	0.23

Table 6 Values of Parameter $COM_T(s,d)$

DCN	Tx Speed
Frame Relay (I)	2,590
Frame Relay (II)	1, 953
FDDI FDDI-II	40
ATM	25.0
Giga-Bit Net	2.0

3.1.4 Experiment and Analysis of the Result

For the simulation of TMN system that is implemented by the Farming methodology suggested in this paper, SLAM II and FORTRAN 77 that are executed in Sun-Sparc 3 are used. TMN system that is considered to be applied is consisted of the communication link that connects 1 manager, 3 agents, and 1 PICR. For this applying objects, there are CMIP message that occurs in the manager and the on-line message that occurs in the real resource 1, 2, and 3. The frequency of on-line message occurrence Ro for each agent, the frequecy of processing for each agent of CMIP message that occurs in the manager, and the values of the system parameter variables are shown in Table 4 and 5. These numbers are gotten by the measurement of ATM TMN system, The communication time between the nodes like $COM_T(s, d)$ is a network dependent value, and this experiment compares the total performance of the system by adjusting DCN as Frame Relay(1.544 Mbps or 2.048 Mbps), FDDI or FDDI-II (100 Mbps), ATM(155 Mbps), and Giga-Bit Net(2G). If the average size of the message that is used in the network is 500K, the value of $COM_T(s, d)$ for each protocol is shown in Table 6. The value of $COM_T(p, s)$ that is the communication time between PICR and PICR server is defined to be 1 ms. The Figure 12 is a graph that

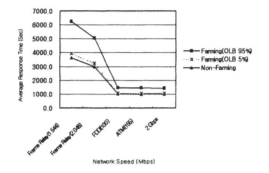

Figure 12. Average Response Time for Total Transactions (CT_{total})

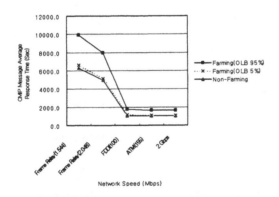

Figure 13. Average Response Time for the CMIP Message($CT_{t\text{-}cmip}$)

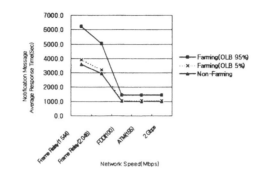

Figure 14. Average Response Time for the On-line message($CT_{t\,notification}$)

compares the average response time for the whole transaction that processes CMIP message that is received as the result of the experiment by changing the values for the variables of $COM_T(s, d)$ and processes on-line message that occurs in the agent as shown in Table 6. As a result of the experiment, in case of using the Farming methodology, the low speed network shows a lot of difference in the response time with the non-Farming methodology, but as the speed of the network gets higher, the difference becomes smaller. Figure 13 and 14 shows the average response time for processing CMIP message in the agent, and the average response time for processing the on-line message. These Figures show that as the speed of the network gets higher, the difference between the Farming methodology and the non-Farming methodology becomes smaller like Figure 12.

3.2 Analysis for the Software Development Cost

In this section, the cost for developing TMN agent by the non-Farming methodology that does not use the concept of the Farming and the cost for developing by the Farming methodology are compared with each other, and the cost for developing by the Farming methodology according to the proportion of ILB/OLB that are stored and managed in PICR are evaluated by comparing with each other.

3.2.1 COCOMO Model

For the analysis of this development cost, the basic principle of COCOMO (Constructive Cost Model) [Boehm 95] suggested by B.Boehm is used. Among the assumption used in case of using the basic expressions of COCOMO, this section uses the following assumption and condition.

1) For the size of the software, only KDSI (1,000 Delivered Source Instruction), that is the number of lines of the execution code is

acknowledged. comments is not acknowledged.

2) The principle of the software engineering is applied to the whole development process systematically.

For the analysis of this development cost, the organic, that is, a general application program is used for the type of the software. Therefore, the calculation for the total number of people(MM) and the development period(M) needed for the development uses the following basic equation.

$$MM = 2.4 \times [\text{ KDSI }]^{1.05} \qquad (3.4)$$

$$M = 2.5 \times [MM]^{0.38} \qquad (3.5)$$

where MM : total number of people
   M : development period (unit of month)
   KDSI : 1,000 Delivered Source Instruction

For the next step, COCOMO must consider the characteristic of the project that gets out of the basic standard and the base(characteristic). This characteristic is indicated by the effort multiplier(Ψ) that correspondents to the effort adjustment factor, and the effort multiplier has a value that depends on the product attribute(s), the computer attribute(c), the characteristic of the developing man(d) and the project attribute(p), and Ψ is indicated by the multiple of these attributes.

$$\Psi = scdp \qquad (3.6)$$
$$C_{total} = MM \times M \times \Psi \qquad (3.7)$$

The scope of each effort multiplier is as follows. This scope is gotten by the min/max scope of the sub-attribute that each attribute has.

$0.75 < s < 1.65$
$0.87 < c < 1.66$
$0.70 < d < 1.14$
$0.82 < p < 1.24$

For analyzing this development cost, the values for s, c, d, p are all assumed to be 1 identically by either non-Farming methodology or the Farming methodology. Also the value for **KDSI** is assumed to be 10.

Finally, the analysis for this development cost applies (3.7) for calculating MM and M.

3.2.2 Analysis for the Development Cost

For analyzing the development cost, this paper considers 4 cases as shown in Table 7.

By increasing the number of agents to be set up in the unit of 10, C_{total} for each case is compared and evaluated. Figure 15 illustrates the result of this development cost analysis. Figure 16 illustrates the case 2, 3, and 4 of Figure 15 in more detail. C_{total} for the case of implementing the agent by using the non-Farming methodology on the heterogeneous platform

Table 7 4 Cases for the Analysis of Development Cost

Case	Methodology	Agent Development Environment
Case 1	Non-Farming	Heterogeneous Platform
Case 2	Non-Farming	Homogeneous Platform
Case 3	Farming	Heterogeneous Platform
Case 4	Farming	Homogeneous Platform

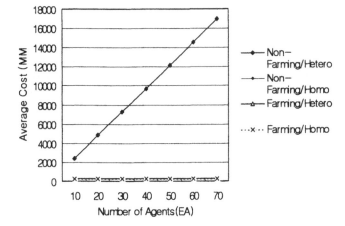

Figure 15. Cost Analysis by the Farming Methodology and non-Farming Methodology

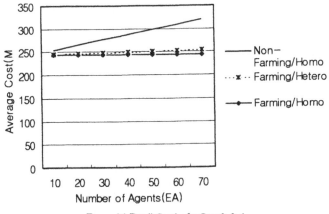

Figure 16. Detail Graphs for Case 2, 3, 4

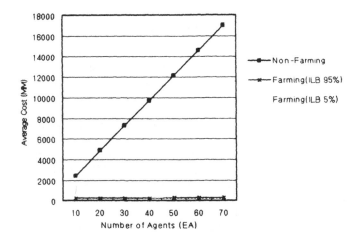

Figure 17. Cost Analysis for non-Farming / Farming(ILB 95%, ILB 5%)

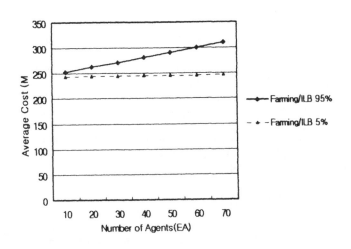

Figure 18. Cost Analysis for Farming(ILB 95% / ILB 5%)

increases proportional to the number of the agents. However, for the case of using the Farming methodology, there is a little difference in the development cost by the platform as shown in Figure 16, but the difference is almost fixed.

The reason why C_{total} for the case 3 is larger than case 4 is that the platform dependent component such as the protocol access by the platform exists although the Farming methodology itself is platform independent. C_{total} for case 2 is smaller than case 1, but as the number of the agents increases, C_{total} increases more than the case 3 and 4. In case of case 2, the reason why C_{total} increases as the number of the agents increases is that the software package cannot be installed automatically whenever each agent is grown or the software package is changed by using the non-Farming methodology. However, in case of the Farming methodology, since the changed or the new components such as ILB or OLB are installed only in PICR, this cost becomes small.

Next, in case of using the Farming methodology, C_{total} is analyzed by changing the composition proportion of ILB and OLB to be 95% to 5%, and 5% to 95%. Figure 17 illustrates this analyzed result by comparing with the non-Farming methodology, and Figure 18 illustrates the case of Farming/ ILB 95% and Farming/ILB 5% by separating from it.

In Figure 18, the reason why the case of ILB 95% needs more development cost than ILB 5% is that the case of ILB 95% needs more communication cost between PICR and the agent in case of the actual system implementation since the proportion of OLB is only 5%. So, after registering the new ILB in PICR, if PICR broadcasts this fact to each agent, each agent must register the new ILB in its local library by Farming it, but in case of OLB, it needs to be registered only in PICR server. Therefore, the case of ILB 95% and OLB 5% needs more C_{total}.

3.3 Relationship between the Performance and the Development Cost

As shown in Figure 12, 13, and 14, if the network that composes DCN is developed as a network with the speed higher or equal to ATM, the performance does not show much difference between the Farming methodology and the non-Farming methodology. However, in the aspect of the development cost, the Farming methodology is relatively better as shown in Figure 15 and 16. Therefore, if the network with the speed higher or equal to 100Mbps is composed of DCN, it is desirable to develop the system by the Farming methodology.

4. Conclusion

The TMN that has appeared to operate various communication networks in overall and effectively has been developed under different platform environment such as different hardware and operating system in the implementation process. So, there are several problems for developing and maintaining the class in the distributed object such as the manager and the agent. One of the main problems is that the agents in TMN system cannot make the standard for Q3 interface implementation and multi-platform cannot be supported. Furthermore, the compatibility between the maintenance system of different networks is not guaranteed. In order to solve these problems, this paper suggests the Farming methodology that is based on the Farmer model. The Farming methodology is the design and the implementation methodology that applies the concept of the client/server that is based on the high speed network and the framework and the software architecture in order to design and to implement the manager/agent system. The Farmer model is the frame structure model that introduces the concept of the system entity structure, and the main objective of the Farmer model is to model the concept of the farming by analyzing the agent that is to be actually designed and by separating and extracting the component elements that composes the agent by the aspect. This paper designs and implements the TMN agent of PCN by using the Farming methodology.

This paper can find its significance by the following aspect. First of all, the development methodology by using the platform independence concept is suggested for the first time. The reason why it is called the platform independence is that the component which is stored and farmed in PICR can be performed by itself or by the platform independence by defining it as a set of the component called the framework. So, the componentware that composes the framework and the framework of the Farming methodology is possible to be executed on basically any platform. The Farming methodology uses these platform independent componentwares by outsourcing from PICR. The Farmer model that is the basis for the Farming methodology is a model that has modeled the concept of the platform independent component and the concept of the Farming, and the Farming methodology and the Farmer model uses the concept called the platform independence. Second, the concept of the software architecture has been used. Third, the cheaper development methodology of the TMN agent is suggested. As a result of the simulation based on the implemented system, when the network that composes DCN is developed as a network with a speed higher or equal to ATM, there are not much difference in the performance between the Farming methodology and the non-farming methodology. However, the Farming methodology is better in the development aspect. Therefore, if the network with the speed higher or equal to 100Mbps is composed as DCN, it is desirable to implement the system by using the Farming methodology. Fourth, the Farming concept is modeled. It is significant since the concept of the component outsourcing through the high speed network is modeled. Fifth, the analysis for the complex system by the multiple aspect is attempted. The Farmer modeling method is suggested in order to analyze and design the system that has the aspects of the modules composed of the tangled mess of aspects such as the TMN agent. Finally, the arrangement and the concreteness of the multiplicity abstraction concept is included. The concept of the multiplicity suggested in the EA model is divided into the concept of the representative entity, IM-Component, OM-Component, and the multiplicity instance link, so that the mutual relationship can be defined.

The methodology suggested in this paper can expect the following result. First of all, the extension is possible by using the common development methodology that uses the high speed network and PICR. The Farming methodology is the methodology for designing and implementing the client/server that is based on the high speed network and the manager/agent system. It is a common methodology for designing and developing not only the communication network management agent, but also the different system that is based on the high speed network. Second, the development cost can be reduced in large amount. According to the computation of the cost by the COCOMO model, when the system is developed by the Farming methodology, its cost is much less than the development made by the non-farming methodology.

The Farming methodology suggested in this paper has the following advantages by comparing to the previous non-farming methodology. First of all, the development expense is much less. Second, it is stable since it uses the already verified component. Third, it can allow the access method that is based on CORBA architecture. The bottom-up access method is used in order to unify the legacy system by using CORBA architecture, and the top down access method for re-constructing is supported in case of designing and implementing the new system. In case of designing and implementing the new system by using CORBA, the top-down access is possible by using the Farmer model. Fourth, the multi-platform is supported. It does not belong to the platform that composes the system since it uses the concept called 'platform independence. Finally, the software version management is easy.

These advantages are direct advantages that the concept called the platform independence gives, and they become the reasons for using the concept called the platform independence in this paper. However, the Farming methodology has an disadvantage that the system performance for the Farming methodology is lower than the non-farming methodology as shown in the test result. In order to improve this research, the further research must be proceeded in the following area. First of all, the low system performance problem that can happen in case of using the platform independence interpreter language such as Javabeans must be solved in the first place. Second, the problems related to the topology of PICR in the network must be solved.

5. References

Salah Aidarous and Thomas Plevyak, *TMN into the 21st Century, Techniques, Standards, Technologies and Applications*, IEEE PRESS, 1994.

Alper Caglayan and Colin Harrison, *Agent Sourcebook*, John Wiley & Sons, Canada, 1997.

Subodh Bapat, *Object-oriented Networks : Models for Architecture, Operations, and Management*, Prentice Hall, 1994.

Berlin A., "Distributed Information Systems for MEMS, ISAT Study", *ISAT*, 1995.

Joseph P. Bigus and Jennifer Bigus, *Constructing Intelligent Agents with Java*, John Wiley & Sons, Canada, 1996.

Barry Boehm, Bradford Clark, *The COCOMO 2.0 Software Cost Estimation Model*, International Society of Parametric Analysts, 1995.

Weiying Chen, *ActiveX Programming UNLEASHED*, Sams net, 1997.

Chang-Hwa Kim and Doo-Kwon Baik, "Knowledge Representation Modeling Based on the EA Model", *Ph.d Thesis*, Korea University, 1989.

Peter Coad, *Object Models : Strategies, Pattern, and Applications*, Prentice Hall, 1994.

Douglas E. Comer, *Internetworking With TCP/IP, 3rd Edition*, Prentice Hall, Inc. 1995.

ComponentWare Consortium, "ComponentWare Architecture : A technical product description", *I Kinetics, Inc.*, 1995.

ComponentWare Consortium, "CWC Technical Plan Statement of Work", *I-Kinetics, Inc.*, 1995.

ComponentWare Consortium, "ComponentWare Database Component Version 3.0 : A technical product description", *I-Kinetics, Inc.*, 1995.

ComponentWare Consortium, "CWC Whitepaper : Realizing a Virtual Application Warehouse using ComponentWare", *I-Kinetics, Inc.*, 1995.

Dirk van Dalen, *Logic and Structure, Third, Augmented Edition*, Springer-Verlag, 1991.

DSET Corporation Version 1.2, "GDMO Agent Toolkit User Guide", 1995.

Dong-Young Cho and Chong-Sun Hwang, "The Object-Oriented Database Modeling with Multiple Aspects", *Ph.d Thesis*, Korea University, 1991.

David Flanagan, *Java in a Nutshell, A Desktop Quick Reference for Java Programmers,* O'Reilly & Associates, Inc. 1996.

Gacek C., Abd-Allah, B. K. Clark, and B. W. Boehm, "Focused Workshop on Software Architectures: Issue Paper", *In Proceedings of the USC-CSE Focused Workshop on Software Architectures*, 1994.

Michael D. Gallagher and Randall A. Snyder, *Mobile Telecommunications Networking With IS-41*, McGraw-Hill, 1997.

Garlan D. and M. Shaw, "An Introduction to Software Architecture", *In Advances in Software Engineering and Knowledge Engineering*, Vol. 1, World Scientific Publishing Company, 1993.

Gregor Kiczales and Andreas Paepcke, "Open Implementations and Metaobject Protocols, Tutorial Book", *The MIT Press*, 1994.

Gregor Kiczales, John Irwin, "Aspect-Oriented Programming, A Position Paper", *Xerox Parc*, 1995.

Eliotte Rusty Harold, *JAVA Network Programming*, O'Reilly, 1997.

Gilbert Held, *Lan Management with SNMP and RMON*, John Wiley & Sons, 1996.

ITU-T Recommendation M.3010, "Principles for a TMN", 1992.

ITU-T Recommendation M.3020, "TMN Interface Specification Methodology", 1992.

ITU-T Recommendation M.3100, "Generic Network Information Model", 1992.

ITU-T Recommendation M.3180, "Catalogue of TMN Management Information", 1992.

ITU-T Recommendation M.3200, "TMN Management Services : Overview", 1992.

ITU-T Recommendation M.3300, "TMN Management Facilities Presented at the F Interface", 1992.

ITU-T Recommendation M.3400, "TMN Management Function", 1992.

Ivar Jacobson, *Object-Oriented Software Engineering, A Use Case Driven Approach*, Addison-Wesley, 1992.

In-Kee Jung and Doo-Kweon Baik, "U-turn-Methodology : A Database Re-engineering Methodology Based on the Entity-Structure- Relationship Data Model", *In Proceedings 6th International Hong Kong Computer Society Database Reengineering Workshop,* Hong Kong Computer Society, pp.39 - 53, 1995.

In-Kee Jeong and Doo-Kwon Baik, "Data Reengineering Methodology using ESR Model", Ph.d Thesis, Korea University, 1996.

John Shirley and Word Rosenberry, *Microsoft RPC Programming Guide,* O'Reilly & Associates, Inc. 1995.

Jung-Ryul Park, *Borland's JBuiler, 21 days*, SAMGAKHYUNG Press, 1997.

Alfons Kemper and Guido Moerkotte, *Object-oriented Database Management : Applications in Engineering and Computer Science*, Prentice Hall, 1994.

Department of Electronics, Korea University, *The Final Report of TMN Agent Generation Platform,* 1998.

Richard C. Lee and William M. Tepfenhart, *UML and C++ - A Practical Guide to Object-Oriented Development*, Prentice Hall, 1997.

James Martin and James J. Odell, *Object Oriented Software Analysis & Design*, Prentice-Hall, 1992.

Thomas J. Mowbray and Ron Zahavi, *The Essential CORBA Systems Integration using Distributed Objects,* OMG, 1995.

Robert Orfali, Dan Harkey, and Jeri Edwards, *The Essential Distributed Objects, Survival Guide*, John Wiley & Sons, Canada, 1996.

Robert Orfali and Dan Harkey, *Client/Server Programming with JAVA and CORBA,* John Wiley & Sons, Canada, 1996.

Soo-Hyun Park, Sang-Hoon Park, and Doo-Kwon Baik, "15. Platform Independent TMN ComponentWare and Data Element Repository Based on Software Farming Methodology", *Systems Development Methods for the Next Century*, *Plenum Press*, pp.169 - 184, Edited by W. Gregory Wojtkowski, Wita Wojtkowski, Stanislaw Wrycza, and Joze Zupancic, 1997.

Soo-Hyun Park, Doo-Kwon Baik, and Sang-Hoon Park, "Farming Methodology for TMN Platform Independent Class Repository Design", *In Proceedings of The 21st Annual International Computer Software & Applications Conference (COMPSAC'97), IEEE Computer Society,* pp.352 - 355, Washington D.C, USA, 1997.

Soo-Hyun Park, Doo-Kwon Baik, and Sang-Hoon Park, "The TMN Agent in PCN based on Entity-Aspect Model", *In Proceeding of 1997 IEEE International Conference on Personal Wireless Communications (ICPWC'97),* pp. 394 - 398, Bombay, India, 1997.

Soo-Hyun Park, Sung-Gi Min, Doo-Kwon Baik, "Platform Independent TMN Agents Based on the Farming Methodology", *The IEICE Transactions on Fundamentals of Electronics, Communications and Computer Sciences,* The Institute of Electronics, Information and Communication Engineers (IEICE), VOL.E81-A, NO 6, pp.1152 - 1163, Japan, June, 1998

ISO CMIP Reference Guide, Retix, 1996

ISO CMIP Programmer Guide, Retix, 1996

Common Include and Common System SRM, Retix, 1996

OSI Management Toolkit Programmer Reference Guide, Retix, 1996

James Rumbaugh and Michael Blaha, *Object-Oriented Modeling and Design,* OMG, 1991

Jon Siegel, *CORBA, Fundamentals and Programming,* OMG, 1996

Bernard Van Haecke, *JDBC : Java Database Connectivity,* IDG Books Worldwide, 1997.

Ann L.Winblad, Samuel D. Edwards, and Davis R. King, *Object-Oriented Software,* Addison-Wesley, 1992.

Kyung-ku Youn, *JavaBeans Developer's Reference,* Daerim Publishing, 1997.

Zeigler B. P, *Multifaceted Modeling and Discrete Event Simulation,* Academic Press, 1984.

A TAXONOMY OF STRATEGIES FOR DEVELOPING SPATIAL DECISION SUPPORT SYSTEM

Silvio Luis Rafaeli Neto[1] and Marcos Rodrigues[2]

[1]*Deparatamento de Engenharia Rural, Universidade do Estado de Santa Catarina e doutorando do Curso de Engenharia de Transportes da Escola Politécnica da Universidade de São Paulo, Luis de Camões, Lages (SC) E-mail: a2srn@cav.udesc.br*

[2]*Escola Politécnica da Universidade de São Paulo, São Paulo (SP) E-mail: carta00@ruralsp.com.br*

Key words: Taxonomy, Spatial Decision Support System, SDSS, Geographic Information System, GIS, Scientific Modelling, Database, Modelbase, Coupling, Integration, Watershed.

Abstract: Due to their extension and complexity, problems concerned with geographic space are usually classified as ill or unstructured. For these reasons decision-making processes (DMP) that conclude with the selection of optimal or satisfactory solutions require effective and efficient means of support. Spatial Decision Support Systems (SDSS) are computer systems developed to support DMP in which the problems have geographic dimensions and whose structure is complex or impossible to delineate. These systems are functionally composed of data and scientific models managed with the aim of providing maximised support to DMP. The component represented by spatial data is one of the main obstacles that have to be overcome for SDSS to give effective support. Geographic Information Systems (GIS) have been a paradigm in SDSS development strategies, fundamentally due to their capacity to collect, store, and handle spatial data. The scientific modelling component, represented by mathematical models of natural physical processes, usually is implemented in SDSS through specific software subsystems. Especially in the last seven years there has been great scientific interest in SDSS accompanied by a proliferation of adjacent technologies. Some authors have attempted to classify SDSS without, however, reaching a generally accepted proposal. This lack of clarity has made detailed analysis of existing systems and development of new projects more difficult. The aims of this research were: a) identify and analyse the main variables that determine current application software development methodologies in the field of SDSS; b) provide a taxonomy of these methodologies, with the objectives of being generic, general, and practical; c) identify and analyse the strategy that presents the greatest flexibility to develop effective SDSS. The research was based on a bibliographic survey and is part of the doctoral thesis of first author. The paper presents and criticises relevant issues about different development strategies and the respective systems produced. Three main variables were identified and guided the development of the taxonomy of the strategies. The paper proposes five taxonomic classes of coupling GIS technology and scientific modelling subsystems. These classes are defined and the paper argues that they are sufficient to categorise the main current methodologies as well as suggest places to expect new technologies. Each class is also exemplified with a number of SDSS applied to the watershed DMP domain. Within these classes, the research identified and analysed the one most likely to show the greatest flexibility to develop effective SDSS

Systems Development Methods for Databases, Enterprise Modeling, and Workflow Management.
Edited by W. Wojtkowski, *et al.* Kluwer Academic/Plenum Publishing, New York, 1999.

139

1.   THE NEED FOR SPATIAL DECISION SUPPORT

Decision making processes on spatial problems comprehend a wide range of parameters and conflicting objectives. These characteristics obstruct the formulation of a structured decision making model. It means that a set of descriptors of solution-space, which may be geographically distributed, do not allow the logic or algorithmic formulation of the steps that lead to the solution-space.

Watershed spatial problems are related to water cycle in the soil, sub-soil and atmosphere. The situations that need formal decisions usually lead to a scientific modelling and analysis of physical processes, with views to effective forecasting, planning, management and operation. With respect to the hydrological processes, the scientific models follow two approaches: the lumped parameters models and distributed parameters models. In the lumped approach the model parameters consider the watershed an homogeneous single entity in that the input are represented by excess of rainfall and the output by hydrograph in the outlet, without taking into account the spatial variabilities within the watershed. According to DeVantier & Feldman (1993) when the model uses a smaller element than the scale size of the physical process it is named 'lumped-model'. In the distributed approach the objective is to capture the spatial variability as much as possible, dividing the watershed into a regular smaller areas or smaller watersheds (Mamillapalli *et al* 1996). The hydrological parameters are considered uniform within each subelement. The hydrological phenomenon is thus simulated inside each subarea and the result is routed to the outlet. The distributed models have raised the attention of scientists with the improvement on computer technologies. While lumped-models usually provide better computational performances distributed models present better results. The later require extensive handling of a vast amount of data. The AGNPS (*Agricultural Nonpoint Source Pollution*), as an example, may require up to 157 input parameters for each spatial cell (León et al 1998). The analysis of output on a tabular format is another difficult aspect of these models. These aspects have promoted the integration of Geographic Information Systems (GIS) with functionalities of scientific modelling subsystems (SMS)[1].

2.   SPATIAL DECISION SUPPORT SYSTEM

Spatial decision support system (SDSS) is a integrative and interactive computer system that has analytical tools and management capabilities designed to support decision makers to achieve solutions of relatively large and ill- or unstructured problems (Watkins & McKinney 1995). The basic architecture of SDSS is composed of a data management and a model management subsystems and interface subsystems among database-modelbase, database-user and modelbase-user (Sprague Jr. 1991, Pearson &

---

[1] At SDSS development approach will be used the term "scientific modelling subsystem" (SMS) to designate the class of application software which implement physical processes simulation within watershed, although the strategies presented are generics for other application areas.

140

Shin 1994). Functionalities of modelbase subsystem are similar to functionalities of database subsystem: to construct, store and activate the methods that implement scientific models of reality. According to Pearson & Shim (1994), the presence of modelbase management subsystem is the main difference among traditional computer-based systems and decision support systems. To these authors, the modelbase management subsystem has the follow capabilities:

    i.   allow the use of multiple scientific models to support diversified problems;
    ii. support the solution of ill- or unstructured problems;
    iii. easy and fast construction of scientific models;
    iv. track scientific models available in the modelbase (or model directory);
    v. allow integration of models with the SDSS;
    vi. create, store and retrieve models.

The current SDSS technologies have not implemented those concepts in an effective and efficient way. Some desired attributes of an effective SDSS include (Bennett 1997, Porto & Azevedo 1997):

    i.   provide effective and efficient support to decision making process and to scientific models development;
    ii. provide virtual environment where decision makers can explore the theory and evaluate management strategies;
    iii. provide iterative and didactic environment that develops user learning on space-problem and the possibilities and limitations of methods used;
    iv. they have to be practical devices that move to the formulated objectives in a simple, direct, clear and objective way;
    v. to consider administrative, socio-political and even psychological factors, such as level of decision, organisational environment, type of decision maker and hierarchical flow of the decision.

Moreover, effective SDSS must have specific attributes related to application domain (USDA 1997).

3.   SOFTWARE SUBSYSTEM

According to Pressman (1995) a computer-based system is a set of organised elements that execute a method, procedure or control to process information. Computationally a SDSS comprehend software, hardware, data and procedures subsystems. The software subsystem is composed of programs, data structure and related documentation that implement the method, process, or logical control of the system. The software subsystem may be classified as application software, which executes the information processing functions, and basic software, which executes the integration functions with other elements of the system (Pressman 1995), such as the hardware. The strategies for developing SDSS are methodologies to incorporate functions to application software. The study aimed at: a) identifying and analysing the main variables that determine current application software development methodologies in the field of SDSS; b) providing a taxonomy of these methodologies; c) identifying and analysing the strategy that allows greater flexibility for SDSS development. The research was based on a bibliographic survey and is part of the doctoral thesis of first author.

4.   THE ROLE OF GIS

It is a common practice in software engineering to arrange functions, in the software subsystem, into modules as a strategy to build large programs. The strategies for developing SDSS basically take advantage of existing modules or software. Most of these methodologies pursue the GIS concept either incorporating it in the scientific modelling software subsystem (SMS) or developing interconnections between GIS and SMS. GIS have been needed for its capacity to collect, store and handle a variety of spatial data, such as satellite imagery, raster and vector maps, graphics and digital terrain models. GIS has also been used due to its advanced display tools, that allow the user to visualise the spatial location, distribution and relationship of geographical phenomena modelled from real world. To Hydrology, important attributes such as area, flow extension, slope, roughness, soil type and land cover may be displayed from GIS spatial database. Attributes about soil, land use, land cover, water conditions, human systems, may be extracted from GIS aspatial database. SMS have been difficult to use because the tedious handling of inflexible data format (Handcock 1995) and its inability to handle the spatial variability of modelled process. One of the GIS roles in SDSS would be to give greatest quality parameters to SMS. Notwithstanding these attractive capabilities the GIS has not reached yet a broad acceptance in the hydrologic community. Although been used in Hydrology, from DeVantier & Feldman (1993) and practical experience, some issues remain unsolved:

i.   degree in that GIS can replace the activities that depend on engineering judgement;
ii. inclusion of temporal series on GIS database;
iii. suitable handling of time varying series;
iv. cost of implementing a GIS;
v.   lack of clear evidences whatever the superiority of GIS results to more traditional methods;
vi. lack of field studies of GIS technology applied to Hydrology supporting its real usefulness;
vii. disagreement on what GIS and SMS technologies should be applied;
viii.no agreement on spatial scale of application parameters and hydrological models;
ix. no definition of spatial resolution to the distributed models;
x.   the resolution of scientific models depends on the resolution provided by the GIS.

On the GIS and SMS integration, Dragosits *et al* (1996) points out:

i.   GIS assists reasoning, communication and experimentation;
ii. forecast of results where the direct measurement is impossible or expensive;
iii. execution of "what if" scenarios analysis;
iv. output of the best/worst scenarios;
v.   synergism of technologies.

Moreover, GIS technology has had a wide role in data pre-processing to SMS, allowing the user to focus on scientific modelling activities rather than on software details.

5. TAXONOMY OF STRATEGIES FOR SOFTWARE SUBSYSTEMS INTEGRATION

Software integration is activities which major objectives are enhance basic tools and increase the capabilities of the application (Turban 1995). The integration of software technologies to configure SDSS has been used a number of systems. Some authors have defined classes of integration methodologies mostly based on their own experiences with these systems (Goodchild *et al* 1992, Fedra 1993, Watkins *et al* 1996). In most cases, these classes are not clearly distinguishable because there are many attributes, which prevent comparative analysis. The taxonomy presented in this paper aims at:

    i. being applied to integration of any software subsystems that incorporate scientific modelling related to an application domain;

    ii. embracing potential future technological development;

    iii. guiding studies on different SDSS categories;

    iv. been useful to SDSS developers, for the different categories may be a starting point to a SDSS project.

5.1 Taxonomy of Criteria

Integration strategies may be grouped into two basic categories: partial integration and full integration. On partial integration each subsystem is an independent entity, which require determined data type, file format, file type, and execute specific functions under an application domain. The technologies are integrated through mechanisms that act on data files, which are, essentially, the joining link. The variations among the different approaches in partial integration are, fundamentally, in the implementation of the integration control mechanism and in the data source. On the full integration, one of technologies is configured to run tasks or functions of the other. One would, thus, have a single entity with extended functions. The variations within the full integration are in the method for that technology is extended.

These two categories are not enough to discriminate of the various methodologies of software subsystems integration for developing SDSS. This study grouped criteria into three basic categories: scientific modelling, data and integration control (Fig 1). These criteria are discussed.

5.1.1 Scientific modelling

This criterion distinguishes integration or coupling strategies in accordance with the degree of physical proximity[2] of involved subsystems. When the physical proximity is minimum or null the subsystems are *independent* (Fig 1). Each subsystem constitutes an independent entity, which perform specific functions under a distinct application domain. When scientific modelling is independent some authors have used the "black box" term to refer to the system which passes data out to another system and receives output back. In the SMS and GIS case, the contact among

---

[2] At this paper physical proximity refers to level of code incorporation of one subsystem into other one subsystem.

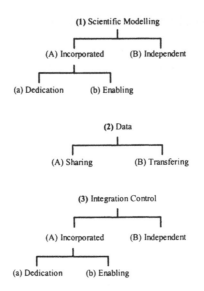

Figure 1. Taxonomy of criteria of strategies for software subsystems integration.

subsystems does not exist, one only knows about input and output into respective data structures (Handcock 1995). If the physical proximity is maximum the subsystems are *incorporated*, that is, one software subsystem absorbs the full code needed to run the tasks executed by the other subsystem. If incorporation is made by compiled programming languages such as C, Fortran, C++, Pascal, or by compiled programming language of the incorporator subsystem, the incorporation is made by *dedication*. If incorporation is made by high level language, interpreted at run time, and which is owned by one of the subsystems, the incorporation is made by *enabling*. The former software subsystem may be said to be "dedicated" to the tasks of the absorbed subsystem, besides its own. Thus, at the incorporation there is a single entity with extended functions.

The dedication level is more demanding with respect to software development and the involvement between the application and the incorporator subsystem is maximum. To implement the dedication means dealing with software engineering concepts. The development of software systems usually follows the concepts of life cycle systems development. The life cycle is chain of phases where the outputs of one phase are the inputs of the next phase. Different approaches were developed under life cycle concept and constitute a paradigm of software engineering, such as the cascade model, the prototyping model, the spiral model and fourth generation techniques. These paradigms may be combined so as to take advantage each ones best quality. Nevertheless, they are usually described as alternative approaches instead of complementary (Pressman 1995).

According to Bayas (1995) the structured techniques normally adopt the cascade model in which the systems are developed according to the Analysis

144

phase ("what" the system have to do, abstracting "how"), the Design phase (where is defined the "how") and Implementation phase (where the programs are coded, tested and concluded). There are distinct data and behaviour abstraction levels between analysis and implementation phases.

5.1.2   Data

This criterion distinguishes strategies according to devices and mechanisms used to accommodate data shared by subsystems. Among methodologies the study distinguishes when there is data *sharing* or data *transferring* between subsystems (Fig 1). Data sharing defines logical proximity between subsystems. The logical proximity is maximum when one subsystem access directly data stored according to data model and data structure of the other. Data accessed remains readily needing no data files translation. Logical proximity is minimum or null when, before being used by subsystems, data are extracted from storing structure (database or memory) to one or more intermediary files. The integrated subsystems may only access this data through transfer file. Transfer file acts as to speak as a translator whose language is its storing format.

5.1.3   Integration control

The analysis of this criterion is similar to the analysis of the scientific modelling criterion. Integration control discriminates the strategies according to proximity between integration mechanisms and the subsystems integrated by these mechanisms (Fig 1). The physical proximity is maximum when the code responsible for integration control is entirely incorporated into the code of one of the integrated subsystems. There is a situation in which such code may be segmented and distributed close to more than one subsystem. However, this situation is not common on usual SDSS developments. As with scientific modelling integration control may be made at *dedication* level or *enabling* level.

5.1.4   User support

This criterion discriminate strategy in accordance with the degree of user-computer interaction offered by the software tools. Because it relate to the interaction user and computer and not between subsystems to be integrated, the user support criterion will not be developed within the scope of this text.

5.2   Taxonomy of Criteria Applied to Integration of GIS-SMS Subsystems

The broad criteria previously described may be applied considering, on one hand GISs and, on the other hand SMSs. Figure 2 thus results as an adaptation of Figure 1.

6.   TAXONOMY OF STRATEGIES

Among several possible relationships among categories and subcategories illustrated in Figure 1, this study identifies four cases that

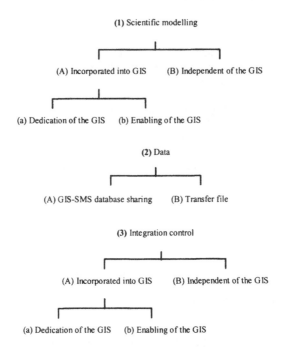

Figure 2. Taxonomy of criteria applied to GIS and SMS for developing SDSS.

Table 1. Taxonomic classes of strategies for developing SDSS.

Coupling class	Description	Label
loose	independent scientific modelling	1B 2B 3B
	data transferring	
	independent integration control	
close	independent scientific modelling	1B 2B 3A
	data transferring	
	incorporated integration control	
tight	independent scientific modelling,	1B 2A 3A
	data shared	
	incorporated integration control	
full integration	incorporated scientific modelling	1A 2A 3A
	data shared	
	incorporated integration control	
mixed	undefined	-

146

deserve attention (refers to Table 1 and Figure 3), these are described bellow.

The integration, or coupling, is loose when subsystems are independent entities which do not share but transfer data between themselves in the integration process and whose control is disjoint of any subsystem (Fig 3). The logical and physical proximity are minimum or null. According to Figure 2 the label used to express this level of integration is 1B 2B 3B.

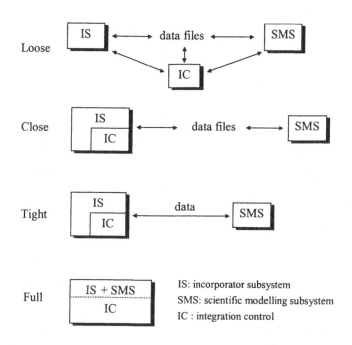

Figure 3. Classes of strategies for developing SDSS.

Loose coupling strategy allows integral reuse of existing subsystems, without need to develop a specific code to run integrated subsystems' tasks. Programming effort concentrates, basically, on mechanisms that implement the integration control, allowing less cost for developing. The main inefficiencies of such an approach refers to data storing and system performance levels. Transfer files set devices that duplicate data stored to be used by subsystems. The construction of these files requires translation operators besides routine operations of input/output. The execution of repeated sequences of data transfer through files may be slow and may need many operations to purge transferred data (Watkins *et al* 1996).

6.1.1   An instance of loose coupling applied to GIS-SMS subsystems

In the so called loose coupling ["loose" according to Goodchild *et al* (1992), Handcock (1995), Matson *et al* (1995), Turban (1995), Karimi (1997); "link" according to Fedra (1993) and Watkins *et al* (1996)] GIS and SMS are autonomous but integrated through files. The program, or set of interface programs, converts data among GIS and SMS file formats (Watkins *et al* 1996). GIS is used to geographic analysis, generate input files to SMS, read, analysis and display output files from SMS. Transfer files may be in ASCII or binary formats.

At link level, subsystems interface accesses GIS database, prepares files to be read by SMS, activates SMS, activates a program to format output file from SMS to be read for GIS and may call other programs to prepare GIS data structure to receive the results (Watkins *et al* 1996).

The WATERSHEDSS (Water, Soil, and Hydro-Environmental Decision Support System) is an example. The system aims evaluating impacts of individual nonpoint source pollution through information analysis of water quality and soil chemistry. Basically WATERSHEDSS integrate a GIS (GRASS - *Geographic Resource Analysis Support System*) with a SMS that implements a nonpoint source pollution model (AGNPS - *Agricultural Non-Point Source Pollution*). The integration is made through transfer file generator. The GIS is used to generate 22 input parameters required for AGNPS. The GIS and AGNPS spatial data model is raster (USDA 1997).

6.2   Close Coupling

Close coupling strategy differs from the loose coupling for it incorporates integration control to one of the subsystems (Fig 3). Despite of an incipient logical and physical proximity, close coupling is a move toward closeness. The label to express this level of integration is 1B 2B 3A.

Close coupling has the positive aspects of loose coupling with additional advantages is achieved by programming the mechanisms of integration control with high level languages, such as macros, owned of incorporator subsystem, under a user-friendly environment. Usually programming environment is such as to assist user for application development, including tools for visual interfaces. This means less programming effort as well as time for developing SDSS. The inefficiencies of loose coupling though remain with the additional disadvantage, at system performance interpreted languages are used (Dragosits *et al* 1995, Watkins *et al* 1996). At dedication level (low level languages), integration control codes are compiled to run-time language, and a decrease on system performance is unlikely. Dedication level may require a greater programming than enabling level due to the lack of a user-friendly environment.

6.2.1   An instance of close coupling applied to GIS-SMS subsystems

The functional and operational features in close coupling strategy ["close" according to Goodchild *et al* (1992) and Handcock (1995), "integrate" according to Watkins *et al* (1996)] of GIS-SMS are similar to loose coupling.

The MGE-MODFLOW is an instance of close coupling strategy. The integration results of close coupling at dedication level among a GIS MGE (*Modular GIS Environment*) with a SMS MODFLOW (groundwater flow simulation software). Integration control was developed through C language and its graphical user interface has dialog boxes, graphics commands and database query resources. The control allows input data from several fonts and translates to MODFLOW standard file. The system provides display and analysis tools of MODFLOW outputs. (Watkins *et al* 1996)

6.3   Tight Coupling

Tight coupling strategy takes advantages of positive aspects of close coupling and implements a closer logical proximity between subsystems. Logical proximity is not complete as well as physical proximity. At such an approach the subsystems are still independent, integration is controlled by incorporated mechanisms, but there is no transfer files (Fig 3). Integrated subsystems at tight coupling strategy share data through direct access into their data structure or memory. This operation may be executed by integration control with aid of database manager. Due to data sharing one would improvement on system performance and decrease of storing space. Tight coupling allows a better run-time for data formatting and re-formatting operations are significantly reduced (Watkins *et al* 1996). This is relevant when there is a massive quantity of data input/output.

Although it presupposes data sharing among integrated subsystems, the mechanisms of integration control represent artificial devices developed to allow interoperability at data flow level. Subsystems remain independent, developed at distinct time to meet requirements on different application domains and, thus, are suited for a set of distinct specifications. Each subsystem has a specific data structure oriented to its own model. SDSS developed under this strategy tends to present a polycotomic vision of an application supradomain, according to each subsystem vision on its own application domain. A problem of such an approach is the lack of flexibility to introduce the modelbase and the implementation of its associated concepts. This is illustrated by a number of SDSS already developed in which a timid approach of modelbase concepts may be observed. At such systems one may find the scientific modelling segment incorporated in a rigid way to support development and experimenting of new models that use already operational components.

According to Bennett (1997), SDSS should support the development and modification of scientific models and facilitate user interaction during a hydrological simulation event. User may wish to stop processing, evaluate results of partial analysis, check parameter values or visualise spatially processes. For they are independent, SMS usually run a determined sequence of procedures and then store the outputs. This sequential procedure obstructs the implementation of inference.

Time and spatial scale have been widely debated within the scientific community, both in relation to environmental modelling and spatial database. These are an ongoing debate that still deserves attention from researchers.

6.3.1 An instance of tight coupling applied to GIS-SMS subsystems

At tight coupling ["tight" according to Fedra (1993), Voris *et al* (1993), Matson *et al* (1995), Turban (1995) and Karimi (1997)] GIS and SMS are independent subsystems, integrated by control mechanisms incorporated into GIS at enabling level or dedication level. The input/output operations are always carried out directly on files of GIS database. It's mean that GIS data structure is shared (Handcok 1995).

An instance of this strategy is the MODFLOWARC system (Watkins *et al* 1996). MODFLOWARC system has the mechanisms of integration control incorporated into the GIS (Arc/Info) through enabling. The control activates program modules to read/write directly into GIS database. While spatial data matrix are stored into GIS database some control informations, such as time step and iteration parameters, are stored into ASCII files.

6.4 Full Integration

An important difference between full integration and tight integration strategies is the incorporation of the functional tasks, from one subsystem to the other subsystem. Physical and logical proximity are maximum because the incorporation is made by code sharing and data sharing, respectively, into only one of the subsystem. Full integration embrace architecture integration and semantic integration (Sage 1991). With respect to enabling and dedication levels the same observations made in description of scientific modelling criterion, are valid here.

As macros languages are designed to support .user with little programming experience, enabling allows high level of abstraction; that is, the user does not need to worry with details of internal storing. Implementation details are hidden from those developing an application, allowing more dedication to problem domain. For these reasons, the enabling level leads to full integration faster than the dedication level. For been interpreted macros may be problematic in complex applications due to the volume of procedures and processing time. Complex applications such as environmental modelling tend to require procedures to be grouped into structured modules. Excessive number of modules may impair comprehension and reusing.

As seen on the scientific modelling description the dedication level is the more demanding in software development and there is much involvement among the application and incorporator subsystem. When integration involve GIS and SMS, Handcock (1995) considers GIS as the main device of integration.

Although there are no experimental evidences on the scientific literature, the main positive aspects of such an approach would be:

i.   sharing data models: application data model is integrated naturally into GIS data model;
ii. sharing database: input/output data of the scientific models are stored into GIS data structure;
iii. efficiency and efficacy in data access: expected due to absence of transfer files and programs for data formatting and transferring;
iv. flexibility to develop complex applications;
v.   shared code allows:
   a. flexibility to implement modelbase concepts including management, experimentation and development;

b. flexibility to implement interruption devices on the scientific modelling process, to evaluate partial outputs and parameter values;

c. improvement of software performance associated to interactions among call and data-model management functions;

d. better user-computer interaction without intermediary languages interpreters;

e. reuse of code, according to software modelling;

f. complete access to GIS functionality;

vi. sharing same visual interface subsystem: user may be supported by GIS visual interface;

vii. scientific modelling can be seen as part of GIS tools;

viii. flexibility to incorporate new processes devices such as multicriteria decision making, knowledge base, intelligent agent, adaptive interface, neural network and virtual reality;

ix. support and documentation from the vendors (Goodchild *et al 1992)*;

x. access to scientific models is granted to all GIS users (Goodchild *et al 1992)*.

The negative aspects of full integration strategy at dedication level would comprehend:

i. more demanding on programming;

ii. costlier;

iii. scientific models might become ossified within GIS and changes that might be required by some users might not be needed by other users (Goodchild *et al 1992)*;

iv. difficult to persuade GIS developers to adopt the approach without pressure from the market (Goodchild *et al 1992)*.

Items *i* and *ii* are naturally expected in complex systems but they may not be that relevant if one is to develop an effective SDSS. As for the ossifying risk the system would need a flexible solution to support model development by different groups of users.

6.4.1 An instance of full integration applied to GIS-SMS subsystems

At full integration ["full integration" according to Goodchild *et al* (1992), "embed" according to Turban (1995), Watkins *et al* (1996)] SMS methods are completely inserted into GIS software. The most common approach in the full integration between GIS and SMS is enabling. Macros are developed within GIS environment and perform sequences of procedures such as database handling, objects selection, assigning values to objects, calling external programs, mapping algebra, and so on. Arc/Info is commonly used to implement enabling, for it has its own data handling language, named AML (Maguire, 1995). At the dedication level functions, routines, and algorithms are developed directly through GIS code. Less frequently this approach has been implemented open architecture. GIS developers have shown a trend to simplify their systems and modularity, abandoning large, complex and generalised systems. Nowadays various subsystems each of which simpler, more portable, more integrated and in a flexible architecture that allows full integration. The object-oriented paradigm has been used as a useful methodology to construct these subsystems.

It is important to consider Dragosits *et al* (1995) and Watkins *et al* (1996) on computer performance for SDSS under full integration. Two kind of systems were compared: GIS only (enabling level) and GIS + FORTRAN (loose coupling). The authors concluded that the system developed at enabling level has worse performance than loose coupling system because:

    i. interpreted languages as AML have worse performance than compiled languages, specially when complex algorithms are used;

    ii. matrix processing on GIS (Arc/Info) generates attribute tables and statistics;

    iii. iterative calculus required grid reading/writing;

    iv. some operator, map algebra functions, loop and matrix indexing on AML are less powerful than FORTRAN.

It may be added that in FORTRAN the iterations on matrix model were made widely at memory level without reading/writing operations of intermediary results.

These papers showed the full integration at enabling level might seriously impair system efficiency. More important are the resources present into incorporator subsystem to enable functions of the incorporated subsystem. These resources are limited due to the language used. Height level languages benefit inexperienced users but are limited to more simple applications. Low level languages require experienced programmers but allow complex applications development.

6.4.2 Mixed integration

Mixed integration does not represent a strategy as such, but is a term that expresses the methodology, which congregate diverse approaches, such as loose coupling, tight coupling and the use of uncoupled software. Matson *et al* (1995) presents a mixed integration methodology applied to the study of the interaction among land use management, hydrologic conditions, and pollutant fate and transport, so as to forecast their effects on water quality. The scheme proposed by this project joins various application softwares: GIS, data management, water quality models and statistic packages (Fig 4).

Aspects discussed on loose and tight coupling and mixed integration show a variety of ways to connect software applications. User interface has a fundamental role in isolating user from inumerous transfer operations and controls. It seems that the best approach to integrate software subsystems may be through intelligent interfaces (Rodrigues & Raper 1997).

7. CONCLUSIONS

Watershed may be seen as spatial units where spatially distributed phenomena occur as related to the cycle of water (hydrologic cycle). Such phenomena has been studied and large number of scientific models have been developed so as to simulate natural physical processes and supporting scientists and decision makers. These are the scientific modelling subsystems (SMS).

Spatial Decision Support Systems (SDSS) are computer-based systems functionally composed of data and scientific models coupled with views to supporting decision making processes. The appropriate handling of the spatial dimension is one of the main difficulties to make SDSS effective. The search for solutions has a number of systems in which two subsystems are detached: GIS and SMS. Presence of GIS is justified, essentially, for its

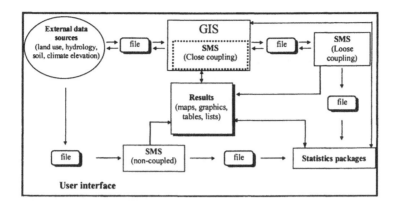

Figure 4. Integration of application software through mixed integration (adapted from Matson et al (1995).

capacity to collect, store and handle spatial data, while SMS for its formalism guided to an application domain. Integration GIS-SMS has been achieved by mean of distinct methodological approaches with their advantage and limitations.

The main factors that determine SDSS development approach were grouped into taxonomy of criteria whose roots are: scientific modelling, data and integration control. Such criteria were applied to analyse current methodologies for developing SDSS, thus resulting in a proposal taxonomy of strategies for developing SDSS. Five classes were identified and labelled: loose coupling, close coupling, tight coupling, full integration, mixed integration. These classes comprehend the majority of current methodologies and anticipate new technologies. These classes were discussed and it was suggested that full integration strategy at dedication level is the one most likely to show the greatest technical flexibility to provide effective SDSS.

REFERENCES

Anderson, A. H., 1998, The development of a 4-dimensional gis/cadd-based decision support system for managing environmental remediation projects. Watermodelling Resources. *[http://www.watermodelling.org/html/4d_gis_cadd.html] [Accessed in 03/17/1998]*

Barber, J. L., Lage, K. L., Carolan, P. T., 1994, Stormwater management and modelling integrating swmm and gis. URISA (1994). p.310-314. *[http://www.odyssey.maine.edu/gisweb/spatdb/urisa/ur94027.html] [Accessed in 10/29/1998]*

Bayas, M. R. C., 1995, *Caracterização da atividade de análise no desenvolvimento de software orientado a objetos.* Dissertação de Mestrado (Mestre em Engenharia). Escola Politécnica da Universidade de São Paulo.

Bennett, D. A., 1997. A framework for the integration of geographical information systems and modelbase management. *International Journal of Geographical Information Sciences* (11)4: 337-357.

Burrough, P. A., 1986, *Principles of geographical information systems for land resources assessment*. Oxford, New York.

Densham, P. J., 1991, Spatial Decision Support Systems. In *Geographical Information Systems: principles and applications* (D. J. Maguire, M. F. Goodchild and D. W. Rhind, eds.), Longman, New York, v 1. pp.403-412.

DeVantier, B. A., Feldman, A. D., 1993, Review of gis applications in hydrologic modelling. *J. Water Resources Planning and Management* (119)2: 246-260.

Dragosits, U., Place, C. J., Smith R. I., 1996, The potential of gis and coupled gis/conventional systems to model acid deposition of sulphur dioxide. Third International Conference/Workshop on Integrating GIS and Environmental Modelling. CD-ROM. January, 21-25, Santa Fe, New Mexico *[http://www.sbg.ac.at/geo/idrisi/GIS_Environmental_Modelling/sf_papers/dragosits_ulli/my_paper.html] [Accessed in 11/03/1998]*

Fedra, K., 1993, Gis and Environmental Modelling. In *Environmental Modelling with GIS* (M. F. Goodchild, B. O. Parks and L. T. Steyaert, eds.), Oxford, New York, pp.35-50.

Goodchild, M., Haining, R., Wise, S., 1992, Integrating gis and spatial data analysis: problems and possibilities. *International Journal of Geographical Information Systems* (6)5: 407-423.

Handcock, R. N., 1995, Model objects - managing an environmental model from within the gis. Part of M. Sc. at University of Toronto (1995). *[http://eratos.erin.utoronto.ca/grad/handcock/mosaic_files/esri/P366.HTM] [Accessed in 10/20/1998]*

Karimi, H. A., 1997, Interoperable gis applications: tightly coupling environmental models with GISs. International Conference and Workshop on Interoperating Geographic Information Systems, 3-4, December, Santa Barbara, CA. *[http://bbq.ncgia.ucsb.edu/conf/interop97/program/papers/karimi.html] [Accessed in 11/05/1998]*

León, L. F., Lam, D. C., Swayne, D., Farquhar, G., Soulis, R., 1998, Integration of a nonpoint source pollution model with a decision support system. *[httP://sunburn.uwaterloo.ca/~lfleonvi/artics/art25.html] [Accessed in 10/29/1998]*

Maidment, D., 1993, Gis and hydrologic modelling. In *Environmental Modelling with GIS* (M. F. Goodchild, B. O. Parks and L. T. Steyaert, eds.), Oxford, New York, pp.146-167.

Maguire, D. J., 1995, Gis application development. *[http://globe.geo.u-szeged.hu/arcinfo.htmls/ovmsproc95/p243.html] [Accessed in 09/14/1998]*

Mamillapalli, S., Srinivasan, R., Arnold, J. G., Engel, B. A., 1996, Effect of spatial variability on basin scale modelling. Third International Conference/Workshop on Integrating GIS and Environmental Modelling. CD_ROM. January, 21-25, Santa Fe, New Mexico. *[http://ncgia.ucsb.edu/conf/SANTA_FE_CD-ROM/sf_papers/mamillapalli_sudhakar/my_paper.html][Accessed in 06/13/1997]*

Matson, K. C., Stallings, C., Jennings, G. D., McLaughlin, R. A., Coffey, S. W., 1995, Watershed scale hydrologic modelling in a gis environment. GIS/LIS'95. *[http://www2.ncsu.edu/bae/people/faculty/matson/gislis95/w-shed_mod.html][Accessed in 10/20/1998]*

Pearson, J. M., Shim, J. P., 1994, An empirical investigation into decision support systems capabilities: a proposed taxonomy. *Information and Management* 27(1): 45-57.

Porto, R. L. L., Azevedo, L. G. T., 1997, Sistemas de suporte a decisões aplicados a problemas de recursos hídricos. In *Técnicas Quantitativas para o Gerenciamento de Recursos Hídricos* (R. L. L. Porto, ed.), UFRGS-ABRH, Porto Alegre (Brasil), pp.43-95.

Pressman, R. S., 1995, *Engenharia de Software*. Makron Books: São Paulo.

Rodrigues, A., Raper J., 1997, Defining spatial agents. To appear in Research Monograph to be published by Taylor and Francis, 1997. *[http://helios.cnig.pt/~armanda/html/Livro.html] [Accessed in 10/14/1997]*

Sage, A., 1991, *Decision Support Systems Engineering*. Wiley & Sons, New York.

Sprague Jr., R. H., 1991, Estrutura para o desenvolvimento de sistemas de apoio à decisão. In *Sistema de Apoio à Decisão. Colocando a Teoria em Prática* (R. H. Sprague Jr and H. J. Watson, eds.) Campus, Rio de Janeiro, pp.9-42.

Thomas, G., 1995, A complete gis-based stormwater modelling solution. *[http://spheroid.otago.ac.nz:808/esriproc/proc95/to050/p038.html] [Accessed in 06/18/1997]*

Turban, E., 1995, *Decision support systems and expert systems*. 4 ed. Prentice-Hall, Englewood Cliffs, NJ.

154

USDA Global Environmental Change Data Assessment and Integration Project, 1997, CIESIN, Research report on advances in spatial decision support system technology and application. *[http://www.ciesin.colostate.edu/USDA/Task%203%20Web/97T31.html] [Accessed in 06/25/1998]*

Voris, P. V., Millard, W. D., Thomas, J., 1993, Terra-vision – the integration of scientific analusis into the decision-making processes, *International Journal of Geographical Information Systems* (7)2: 143-164.

Watkins, D. W., McKinney, D. C., 1995, Recent developments associated with decision support systems in water resources. U. S. National Contributions in Hydrology 1991-1994, Reviews of Geophysics, Supplement, 941-948, July 1995. *[http://earth.agu.org./revgeophys/watkin00/watkinoo.html] [Accessed in 04/06/1997]*

Watkins, D. W., McKinney, D. C., Maidment, D. R., Lin M., 1996, Use of geographic information systems in ground-water flow modelling. *Water Resources Planning and Management* (122)2: 88-96.

UNDERSTANDING IS STRATEGIES IMPLEMENTATION:
A PROCESS THEORY BASED FRAMEWORK

Carl Erik Moe, Hallgeir Nilsen, and Tore U. Ørvik
Agder University College, Dept. of Information Systems
N-4604 Kristiansand, Norway

Key words: IS strategy, strategic planning, implementation, evaluation, process theory

Abstract: This paper addresses the issue of information systems (IS) strategy implementation. Previous work in this area has addressed the issue from a factor analysis perspective, identifying factors that are positively or negatively related to implementation.

Based on a discussion of research literature we argue that this concept cannot effectively be understood in a binary sense, i.e. implemented or not implemented, it should also be studied as a long term ongoing process.
The paper relates IS strategy implementation to previous work on how IT creates business value and suggests a process theory based framework for understanding and researching the extent and effectiveness of the IS strategy implementation process.

Introduction

In this paper, we examine the process of information systems (IS) strategy implementation. Development of IS strategies has been in focus among both academics (e.g. Boynton and Zmud 1987, Premkumar and King 1994a) and practitioners (e.g. Niederman et al. 1991) for years. The issue of IS strategy has ranked high in surveys of key management issues in IS (Brancheau and Wetherbe 1987; Watson et al. 1997; Morgado, Reinhard and Watson 1995) and it is a common belief that the businesses that succeed with IS strategy development may conceive substantial competitive advantages.

Most of the academic work have been directed at exploring ways to undertake the process of IS strategy planning, resulting in numerous planning methods, models and approaches, and related desirable process and content characteristics. According to Gottschalk (1998) relatively little work has however been done to explore the implementation of information systems strategies. The answers to such fundamental questions as: to what extent are IS strategies implemented, if not implemented, what reasons are evident; and if implemented, are they leading to the outcomes stated in the plan, are largely unknown.

Systems Development Methods for Databases, Enterprise Modeling, and Workflow Management,
Edited by W. Wojtkowski, *et al.* Kluwer Academic/Plenum Publishing, New York, 1999.

157

Based on previous work related to implementation issues and to evaluation of the business outcomes of IT, we will propose a process theory based framework that will be used in a forthcoming case study of the extent and effectiveness of IS strategy implementation.

We will first discuss the strategy concept and the strategy formation process. Then we will show how strategy implementation can be understood not just as a single action, but as a series of interrelated, time-consuming processes. Next we propose a framework for understanding the IS strategy implementation process. The final part of the paper proposes how this framework can be further developed and validated.

IS strategy: content and development

Historically strategy development has been understood as a deliberate and planned process (e.g. McLean and Soden 1977, King 1978), and IT consulting companies have been spending large resources on formation of structured methods for IS strategy development (e.g. Andersen Consulting's Method/1 and IBM's BSP). The focus has mainly been on a rational planning approach. Earl (1989) defines IS strategy as the long-term directional plan for how to use IT to meet business goals, by giving guidelines for how to exploit IT either to support business strategies or create new strategic options.

Other researchers have pointed to the need for understanding strategy development more as a mix between deliberate and emergent strategies (Mintzberg and Waters 1985) and question the value of formal planning (Mintzberg 1994). Several authors also describe strategy development as a learning process, a process of designing and deliberately adjusting or modifying a formulated strategy as reactions to the environment (e.g. de Geus 1988, Kenyon and Mathur 1993). Ciborra (1991) even points out that many of the classical examples of strategic information systems have been developed not as a result of a planned information systems strategy, but more in spite of such a strategy.

Some researchers claim that the most important part of the IS strategy is the project portfolio (Salmela et.al. 1996). Others state that IS strategies should be strategic, focus on outcomes, and not include projects (e.g. Earl 1989). Projects will however often be a part of the IS strategy. While some projects will be stated in the strategy, others will be established later to meet goals given in the strategy. The conversion from goals to projects is thus sometimes a part of the strategy development process and sometimes a part of the implementation.

Implementation of information systems strategies is found to be problematic in many cases (e.g. Earl 1993, Galliers 1994), in spite of large resources spent on the planning process. Nevertheless, IT investments continue to constitute an increasing rate of total investments and shifts in technologies highlight the need for making the right choices. Large corporations continue to spend a lot of resources on strategy development and it is important for research and practice to understand questions regarding implementation of these strategies. Our interest lies therefore more in the process of implementation of these strategies than in their development. In our work we propose a framework where IS strategy is treated as a dynamic concept including both the deliberate and emergent perspective.

158

Implementation

Although a lot of work has dealt with desirable features of the IS planning process and content characteristics of the plan, (e. g. Lederer and Sethi 1996, Earl 1989), according to Gottschalk (1998) relatively little is done that addresses the implementation issue directly.

A few have tried empirically to identify dimensions in the planning or strategy development process that are critical to implementation (e.g. Lederer and Sethi 1992, Earl 1993, Gottschalk 1998). Some present barriers to, or problems with IS strategy implementation and indicate that plans are often not implemented (Earl 1993, Galliers 1994). Others present factors that positively influence implementation (Bryson and Bromily 1993, Premkumar and King 1994b, Lederer and Salmela 1996, Gottschalk 1998).

Although directly related to implementation issues, this research still reflects a lack of consensus as to what constitutes implementation, especially when the term needs to be instrumentalised in some kind of factor analysis study.

The apparent ambiguity might be exacerbated by a conceptual difference in understanding the term as it relates to both implementing something relatively abstract like a plan, and some concrete (IT based) artifact. The studies on IS strategy implementation all instrumentralise the concept by studying some measure of completion of individual IT development projects identified in the plan and assessing the proportion of these that have been implemented. There is however also the issue of the whole being more than the sum of it's parts. Implementation of a (whole) plan might well be more than implementation of individual projects. A reductionistic approach allow for more accepted instrumentalisation, but it might well ignore the holistic underpinnings of the plan.

On the other hand information systems strategies can be viewed as consisting of different parts, with different levels of detail, different planning horizons, and different focuses. Earl (1989) treats IS-strategy as a collection of three different strategies or approaches, an Information Technology (IT) strategy, an Information Systems (IS) strategy and an Information Management (IM) strategy and also studies implementation of IT-strategies based on different approaches (1993). Other ways of classifying individual parts of the IS-plan are also conceivable. Such parts might well have very different implementation characteristics, and we believe identifying and studying different generic parts of the strategy might be of importance in understanding problems in implementation.

Information technology is only part of the IS strategy. Although the strategy usually defines new information technology that the organization should put to use, it also concerns the diffusion of that technology into the organization. Leonard-Barton (1988) addresses this aspect in her analysis of implementation characteristics by distinguishing between the adoption decision (related to IT) and the innovation response (by the organization and it's members). The response is manifesting itself over time, partially shaped by the technology and partially reshaping the technology itself. The implementation of the IS strategy is thus more than putting the technology to use. It can be understood as encompassing the whole diffusion process.

A dictionary definition of the verb implement (Crowter 1995) states that to implement is to "put into effect" or "to carry something out". These are activities that can hardly happen instantaneously. Implementation can thus be seen to have different aspects or dimensions, we suggest three such dimensions; time, quantity and quality.

Dimensions of implementation: time, quantity and quality
The time dimension is perhaps the most crucial of the three dimensions. The verb implement implies a process that will occur over time. Implementation refers to this process. Such a process view rises the question about when the process has been completed and implementation has occurred.

The concept of implementation has been discussed in the IS literature in a number of ways. Much has been related to the acceptance of new information technology. The work related to the technology acceptance model (TAM) done by Davis (1989) is a prominent example. The diffusion of innovation literature, initiated by the work of Rogers (1983), and followed up by others (e.g. Cooper and Zmud 1990, Lai and Mahapatra 1997), has also addressed implementation.

Rogers regards implementation as one step in the diffusion process: "Implementation occurs when an individual (or decision making unit) puts an innovation into use". The "use" concept however cannot be understood in a binary sense, there will certainly be degrees of use, especially when use is voluntarily. At what stage then do we have sufficient use for implementation to have happened? There might be an argument instead that implementation has occurred when the artifact is available for use, and that actual use is not an issue in this context. This is often the meaning that is implicit in much of the system development literature.

Rogers' definition is mainly related to technological artifacts, and the "use" issue is fairly clear when related to concrete objects. The issue becomes more complicated when we are addressing the implementation of the more abstract concept of a plan, namely the IS strategy plan which deals with the diffusion of IT in the organization. This means that the diffusion aspect must be a part of our understanding of implementation in this context.

Rogers recognizes that diffusion is strongly related to time. This is developed further by Kwon and Zmud (1987) into six stages: initiation, decision, adaptation, acceptance, routinisation and infusion, thus recognizing that diffusion is a process that will evolve and materialize gradually. The problematic nature of the time issue is accentuated when one tries to instrumentalise the implementation concept. At what stage is the plan "put to use", or following the dictionary: "put into effect" or "carried out". Conceivably this could be when the actions identified in the plan have been funded and initiated, when the identified outcomes have been realized, or at any intermediate stage.

From the above discussion follows that the implementation issue cannot be understood from a point in time perspective and that we need to understand implementation related phenomena as they unfold over time.

The quantity dimension is closely related to measuring problems when we are assessing the implementation process. Looking at different indicators of implementation in progress, there will usually be matters of degree. Are projects fully funded, or just partially? Is all the planned functionality available for use, or just some of it? And if technology is put to use - how much use is there?

The IS strategy plan will usually be divided into different activities, often organized as projects. Implementation of these might start at different times. This raises the question of how much of the plan is implemented or under implementation.

The quality dimension is also important when assessing different implementation indicators. If technology is available for use, how good is that technology and the information it provides? If use of the technology is evident, how effective is that use and finally, when assessing outcomes, what is the extent of quality related to those outcomes and how well do they conform to the outcomes envisioned in the plan?

Implementation outcomes

From the discussion above follows that implementation is not a simple construct. We have identified time and matters of degree, related to quantity and quality, as main issues in the implementation assessment process.

Although it is conceptually difficult to determine if implementation has (fully) occurred, it should however usually be possible to decide if implementation activities have stopped altogether or are still going on. When implementation is stopped short of being completed this could reflect either a lack of ability to carry through the intended work, or a realization that change in circumstances mandates changes in the planned activities.

Stopped or still in progress, in either case there is a continuum of possible outcome states. The main outcome states of the process can be portrayed as follows (figure 1):

Nothing <------------No implementation activity ------------->**Fully implemented**

    Almost nothing <----Implementation in progress ---------------> **Almost fully**

Figure 1: Possible implementation outcome states.

If implementation is still in progress, we will be somewhere along the continuum from almost nothing implemented to almost fully implemented. If there is no implementation activity we may have nothing implemented, we may be somewhere along the same continuum, or we may have full implementation. This outcome model should apply to both implementation efforts related to the strategy as a whole and to individual projects.

Based on the discussion of implementation issues above, we will argue for choice of theory and suggest a framework for implementation evaluation.

Research model

In the presentation of our research model, we briefly argue for the choice of process theory as a basis, and present the IT business value creation process model developed by Soh and Markus (1995). Our model is an extension of theirs, including separate stages for IT strategy planning and project funding.

Process theory

Variance theories are used to explain variations in the magnitude of a certain outcome. But these theories are not well suited when the outcome may or may not occur. In cases of outcome uncertainty, process theories have been shown to have distinct advantages over variance theories. In studies of IT and business value, it has been demonstrated that process theories have advantages when explaining what is happening, or not happening (Markus and Robey 1988).

Information systems strategy implementation is, as discussed in the implementation section, a time related process with uncertain outcome characteristics. Process theories should thus prove to be advantageous in the study of implementation. Strategy implementation is also clearly linked to the creation of business value, since this is the main rationale behind the strategic planning process. One process model that addresses the creation of business value is "The IT business value creation process model", presented by Soh and Markus (1995). We will use this model as our base model, but will extend it to include development and operationalisation of IS strategies and funding of strategy related activities.

The IT business value creation process model

The model of Soh and Markus (1995) depicts IT based value creation as consisting of three main processes, each converting some input to an output, where the output of one process is a necessary but not sufficient input to the next.

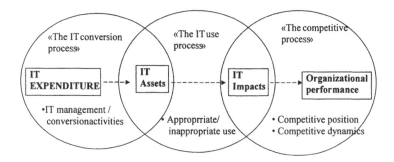

Figure 2: Soh & Markus

The implications of the model is best understood if it is read in a "reverse order", from the right to the left. For IT based organizational performance to materialize, there must be IT impacts evident, (created through appropriate IT use), and for IT impacts to be evident, there must be IT assets available. And for IT assets to be available, money must have been invested.

In the IT conversion process, IT expenditures are converted into IT assets. IT expenditures will however not always lead to IT assets, there are numerous examples of system development and technology acquisition projects gone wrong. This implies that IT expenditure is necessary, but not always sufficient for acquiring IT assets.

In the IT use process, available IT assets are converted into IT impacts through appropriate use. In the same way, availability of IT assets will not automatically lead to IT impacts. Appropriate use is necessary to generate IT impacts. What constitutes appropriate and inappropriate use, will depend on situation and context. According to Soh and Markus (op. cit.), many previous studies have assumed variance theory formulations on the form "greater IT use, leads to greater IT impacts". The perspective in the model is to see IT use as a probabilistic process that affects whether and how IT assets become IT impacts, rather than as an input variable in a necessary and sufficient relationship.

In the competitive process, IT impacts are hopefully converted to organizational performance. The IT impacts, or changes in the organization, may lead to improved performance. But, again, this is only a necessary, but not sufficient condition. External factors, mainly actions from competitors, will influence if certain impacts generate improved organizational performance. If the IT impacts generate customer value, but the same type of customer value is developed simultaneously by the competitors, changes in sales might not materialize, and the bottom line effect of increased organizational performance is competed away.

The IS strategy implementation model

To study the IS strategy implementation process we propose an extension to this process model of how IT creates business value. Soh and Markus (op. cit.) view information systems strategy development as part of the "IT conversion process", the first process in their model. For our purpose, assessing the extent and effectiveness of strategy implementation, we find that a clearer distinction between "IT conversion" activities and planning activities is necessary. Specifically we want to separate out activities related to IS strategy development from the "IT conversion process". We thus propose to extend the model as shown in fig. 3, below:

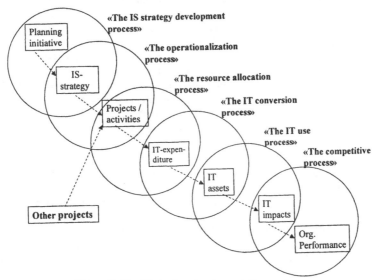

Figure 3: The IS strategy implementation model

This extension involves three new processes. The first is the IS strategy development process itself. The second process is the process of operationalising the outcome statements and general activities identified in the planning process into proposals for concrete projects and activities that can be considered for funding. The third additional process is called the funding process, in which funds are allocated and committed to projects and activities.

In keeping with the conversion view of the original model, the first of these processes converts a planning initiative into a formally documented and approved IS strategy.
The second process converts the strategy or elements of it into concrete project and task proposals. The third process converts proposals into IT expenditure which is the first element in the original model and the starting point for the IT conversion process. The model thus traces how the planning initiative is ultimately converted into organizational performance and links the information systems strategy to the creation of business value.

The three processes from the original model have been extensively researched and are fairly well understood, despite numerous problems in carrying these processes out in practice. The same can be said for the first two processes in our extended model. The general interaction, (in the form of implementation), between the IS strategy constructs and the IT value creation processes are however much less researched, as discussed in the implementation section.

The extended model should be read in the spirit of the original. It does not portray a process where one outcome necessarily leads to the next. At every stage it highlights elements that are necessary, but not necessarily sufficient for further conversion. Although the model is portrayed as basically a sequence of processes, a strict sequentiallity is not implied. That nonsequentiallity is often the case is quite evident in some of the research using structuration theory to explain the "IT use process" (DeSanctis and Poole 1994, Robey 1995). The main thrust of this work is that there is a mutual relationship between the IT assets and IT impacts, each influencing and shaping the other.

Using the same type of structuration thinking there may well be other intra and inter process relationships in the framework. As outcomes from processes becomes evident, previous processes might carry on and be reshaped by the emerging results. This underscores the possibility of intermediate implementation process outcomes influencing previous processes, including the IS strategy development process itself, and is consistent with the organizational learning perspective found in some of the IS strategy literature (Mintzberg and Waters 1985, de Geus 1988, Kenyon and Mathur 1993). Ideally there should be feedback links in the model from every process to the planning process, to inform this process about changes and results as the implementation process is unfolding, and in this way create a dynamic planning environment.

Applications of the model
A technological determinism perspective implies that a certain technology leads to certain outcomes. This deterministic view is often implicitly underpinning much of the IS strategy planning literature which is concerned with prescriptive methods and frameworks aimed at assisting management in formulating and implementing strategy (Earl 1989, Lederer and Sethi 1996). The present framework should provide guidance in assessing the implementation process of IS strategy based on such deterministic thinking. It should however also provide guidance in interpreting the ways in which

strategy forms and unfolds in practice, which we believe to be of a non-deterministic nature.

Some of the research mentioned earlier (Lederer and Sethi 1992, Earl 1993, Galliers 1994) states that some IS strategies or parts of IS strategies are not implemented or acted upon at all. We believe that not having funds allocated is often a crucial cutoff point for implementation or a main indicator of (temporarily) halts in the implementation process. By including the fund allocation process in the model we focus attention on this important activity. The model should inspire a study of the reasons behind "no funding" decisions, when such decisions are clearly stopping the implementation process.

As discussed earlier, an operationalisation of the strategic goals is often an integral part of the information systems strategy. Since that is not always the case however, we find it more useful to include this activity as a distinct process. It is well known that organizations start IT related initiatives and projects that are not part of a formal IS strategy even when such plans exist (Ciborra 1991). Such project initiatives also need to be operationalised in some kind of way. By separating out operationalisation as a distinct process and introducing the idea of projects and tasks entering the IS strategy implementation model from the "outside" we also facilitate a study of the projects and activities that are operationalised and funded without being a part of the formal strategy and possibly not even in line with the strategy.

When studying the implementation process, the framework should help us trace organizational outcomes or intermediate outcomes back to the plan, thus increasing our understanding of the IS strategy planning process and the strategy implementation. To facilitate this, there should be evaluation activities at every stage of the model. The model itself should also be helpful in assessing to what extent these evaluation activities are done and help us understand and organize the possible and desirable evaluation approaches.

Contemporary evaluation literature informs us about ways and means to evaluate the outcome from the various processes, and provides a number of measuring instruments that can be used at various stages in the implementation process (Delone and McLean 1992, Van Der Zee 1997, Garrity and Sanders 1998). Generally speaking, the further to the right in the model one moves the harder it is to instrumentalise measuring constructs and the harder it is to link the outcomes to the IS strategy. The field of information systems evaluation is still evolving and tries to address the problems related to evaluation of IS initiatives in many different ways. As new results become available the present framework should prove useful in organizing and applying emerging approaches.

The model highlights elements that are necessary, but not necessarily sufficient for further conversion. It should thus be useful in studying situations where the implementation process has stopped. Others have studied reasons for IS failure (Lyytinen and Hirschheim 1987, Sauer 1993), but little of this have been linked to the information systems strategy. By using the framework it should be possible to identify where in the process implementation abortions have occurred and then search for appropriate explanations.

Implications for research and practice

For research purposes the model must be validated. Being a process model this is best done through case studies. The model was initially developed to guide a forthcoming study of implementation of IS strategies in several Norwegian municipalities. This will be carried out as the first step in the validation effort.

Based on the model and through case studies we might also be able to develop variance theory type hypothesis which may be tested in a traditional empirical way.

The model itself may also be further developed. Coupling it with theory from the organizational learning field might show us how we can include learning through feedback and adjusting. In addition we believe that treating IS strategy as consisting of different generic parts or substrategies, may be important. Evaluating implementation of the different parts of information systems strategies may reveal even further insight into the implementation process.

The model can be used as a tool by practitioners to guide their evaluation of IS strategy implementation. Since we do not view implementation as a binary construct, evaluation should be carried out at every stage of the implementation process. Such studies can generate insight and learning which can facilitate more successful planning and implementation in later projects. In some cases such studies can also facilitate actions to revitalize stalled implementations or redirect the process.

References

Boynton, A.C. and Zmud, R.W. (1987) "Information Technology Planning in the 1990's: Directions for Practice and Research," *MIS Quarterly*, vol. 11, no. 1, pp. 59-71.

Brancheau, J. C. and Wetherbe, J. C. (1987) "Key Issues in Information Systems Management", *MIS Quarterly*, vol. 11, no. 1, pp. 23-45.

Bryson, J. M. and Bromiley, P. (1993) "Critical Factors Affecting the Planning and Implementation of Major Projects," *Strategic Management Journal*, vol. 14, no. 5, pp. 319-337.

Ciborra, C. (1991) "From thinking to tinkering: The grassroots of strategic information systems", *Proceedings of the eleventh International Conference on Information Systems*, 16-18 December, USA: New York, pp. 283-291.

Cooper, R. B. and Zmud, R. W. (1990) "Information technology implementation research: A technological diffusion approach", *Management Science*, vol. 36, no. 2, pp. 123-139.

Crowter, J. (ed) (1995) *Oxford advanced learners dictionary*, fifth edition, UK: Oxford, Oxford university press.

Davis, F. D. (1989) "Perceived Usefulness, Perceived Ease of Use, and User Acceptance of Information Technology," *MIS Quarterly*, vol. 13, no. 3, pp. 391-339.

Delone, E. H. and McLean, E. R. (1992) "Information System Success: The Quest for the Dependent Variable" *Information Systems Research*, vol. 3, no. 1, pp. 60-95.

DeSanctis, G. and Poole, M. S. (1994) "Capturing the Complexity in Advanced Technology Use: Adaptive Structuration Theory," *Organization Science*, vol. 5, no. 2, pp. 121-147.

de Geus, A.P. (1988) "Planning as Learning", *Harvard Business Review*, vol. 66, no. 2, pp. 70-74.

Earl, M. J. (1989) *Management strategies for information technology*, UK: Hertfordshire, Prentice Hall.

Earl, M. J. (1993) "Experiences in Strategic Information Planning", *MIS Quarterly*, vol. 17, no. 1, pp. 1-24.

Galliers, R. D. (1994) "Strategic information systems planning: myths, reality and guidelines for successful implementation", in: Galliers, R. D. and Baker, B. S. H. (eds), *Strategic Information Management*, UK: Oxford, Butterworth-Heinemann, pp. 129-147.

Garrity, E. J. and Sanders, G. L. (1998) *Information systems success measurement*, USA: Hersey, Idea Group Publishing.

Gottschalk, P. (1998) "Content characteristics of formal information technology strategy as implementation predictors", Ph.D. thesis, Henley Management College, Brunel University.

Kenyon, A. and Mathur, S.S. (1993) "The Designed versus Emergent Dispute", *European Management Journal*, vol. 11, no. 3, pp. 357-360.

King, W.R. (1978) "Strategic Planning for Management Information Systems", *MIS Quarterly*, vol. 2, no. 1, pp. 26-37.

Kwon, T. H. and Zmud, R. W. (1987) "Unifying the Fragmented Models of Information Systems Implementation", in: Boland, R. J., and Hirschheim, R. A., *Critical Issues in Information Systems Research*, UK, John Wiley and Sons, pp. 227-248.

Lai, V. S. and Mahapatra, R. K. (1997) "Exploring the research in information technology implementation", *Information and Management*, vol. 32, April, pp. 187-201.

Lederer, A. L. and Salmela, H. (1996) "Toward a Theory of Strategic Information Systems Planning", *Journal of Strategic Information Systems*, vol. 5, no. 3, pp. 237-253.

Lederer, A. L. and Sethi, V. (1992) "Root Causes of Strategic Information Systems Planning Implementation Problems", *Journal of Management Information Systems*, vol. 9, no. 1, pp. 25-45.

Lederer, A. L. and Sethi, V. (1996) "Key Prescriptions for Strategic Information Systems Planning", *Journal of MIS*, vol. 13, no. 1, pp. 35-62.

Leonard-Barton, D. (1988) "Implementation as mutual adaptation of technology and organization", *Research policy*, vol. 17, pp. 251-267

Lyytinen, K. and Hirschheim, R. (1987) "Information system failures: a survey and classification of the empirical literature.", *Oxford surveys in information technology*, 4, pp. 257-309.

Markus, M. L. and Robey, D. (1988) "Information Technology and Organizational Change: Causal Structure in Theory and Research.", *Management Science*, vol. 34, no. 5, pp. 583 - 598.

McLean, E.R. and Soden, J.V. (1977) "Strategic Planning for MIS", in: *Proceedings from the National Computer Conference 1976*, USA: New York, John Wiley, pp. 425-433.

Mintzberg, H. and Walters, J.A. (1985) "Of Strategies, Deliberate and Emergent", *Strategic Management Journal*, vol. 6, pp. 257-72.

Mintzberg, H. (1994) "The Fall and Rise of Strategic Planning", *Harvard Business Review*, vol. 72, no. 1, pp. 107-114.

Morgado, E. M; Reinhard, N. and Watson, R. T. (1995) "Extending the analysis of key issues in information technology management" In: *Proceedings of the sixteenth International Conference on Information Systems*, 10-13 December, Netherlands: Amsterdam, pp. 13-16.

Niederman, F.; Brancheau, J.C. and Wetherbe, J.C. (1991) "Information Systems Issues for the 1990's", *MIS Quarterly*, vol. 15, no. 4, pp. 475-500.

Premkumar, G. and King, W.R. (1994a) "Organizational Characteristics and Information Systems Planning: An Empirical Study", *Information Systems Research*, vol. 5, no. 2, pp. 75-109.

Premkumar, G. and King, W. R. (1994b) "The evaluation of strategic information system planning", *Information & Management*, vol. 26, no. 6, pp. 327-340.

Robey, D. (1995) "Theories that Explain Contradiction: Accounting for Contradictory Organizational Consequences of Information Technology", *Proceedings of the sixteenth International Conference on Information Systems*, 10-13 December, Netherlands: Amsterdam, pp. 55-64.

Rogers, E. M. (1983) *Diffusion of innovations*, USA: New York, Free Press.

Salmela, H; Lederer, A. L. and Reponen, T. (1996) "Prescriptions for Information Systems Planning in a Turbulent Environment", *Proceedings of the Seventeenth International Conference on Information Systems*, 16-18 December, USA: Cleveland, Ohio, pp. 356-368.

Sauer, C. (1993) *Why Information Systems Fail: A Case Study Approach*, UK: London, Waller.

Soh, C. and Markus, M. L. (1995) "How IT creates business value: a process theory synthesis" *Proceedings of the sixteenth International Conference on Information Systems*, 10-13 December, Netherlands: Amsterdam, pp. 29-41.

Van Der Zee, H. T. M. (1997) *In search of the value of information technology*, Netherlands: Tilburg, Tilburg University Press.

Watson, R. T; Kelly, G. G; Galliers, R. D and Brancheau, J. C. (1997) "Key Issues in Information Systems Management: An International Perspective", *Journal of Management Information Systems*, vol. 13, no. 4, pp. 91-115.

ON METHOD SUPPORT FOR DEVELOPING PRE-USAGE EVALUATION FRAMEWORKS FOR CASE-TOOLS

Björn Lundell[1] and Brian Lings[2]
[1]*Dept. of Computer Science, University of Skövde*
P. O. Box 408 S-54128 Skövde SWEDEN
E-mail: bjorn@ida.his.se

[2]*Dept. of Computer Science, University of Exeter, U. K.*
E-mail: brian@dcs.exeter.ac.uk

Key words: Evaluation framework, Pre-usage evaluation, Qualitative method, Method support, Organisational setting, CASE-tool

Abstract: In this paper we consider the issue of providing method support for evaluation of CASE tools. The importance of this issue is indicated by the high percentage of CASE tools which are purchased but rarely or never used. We consider the issue of CASE evaluation from the point of view of pre-usage evaluation, and report on an extensive study of related literature, the goal of which was to identify key issues in developing reliable frameworks for evaluation. A number of weaknesses were found in available method support for evaluation where the goal was successful adoption. These weaknesses are highlighted in the paper. We analyse our own method proposal for impact in each of the highlighted areas. It is found that the qualitative approach used in the method, and its emphasis on iterative, holistic refinement, offers significant methodological advantage over other extant approaches.

1. INTRODUCTION

It is widely recognised that evaluation is difficult, and at the same time is of fundamental importance in the deployment of information systems (IS) in organisations (Iivari, 1996; Nelson and Rottman, 1996; Symons, 1991; Walsham, 1993). An evaluation can be undertaken for a variety of reasons and many issues affect the degree of success of such an effort. One major factor concerns evaluation methods themselves. Symons (1991) claims that:

"there is no commonly accepted framework or methodology for information system (IS) evaluation" and that "developments in evaluation methods have not kept pace with the shifts in use of information systems which over time have widened the scope of their effects." (p. 205)

As pointed out by Symons, there are many stages in IS adoption at which evaluation plays a crucial role but in this paper it will be the stage 'before introduction', which we will characterise as *pre*-usage, on which we concentrate. "Before introduction, the ability satisfactorily to evaluate

technological innovations is an important factor in the decision to purchase."
(*ibid.*, p. 206)

One important example of 'technological innovation' relates to the adoption process for CASE-technology. The problem statement for one of the Software Engineering Institute's projects concerning CASE (SEI, 1995) contains the claims that the "introduction and use of integrated CASE technology is chaotic" (p. 1) and "evaluation techniques are unavailable" (p. 1).

The implications of this can be significant. Hardy *et al.* (1995) reports on mixed experiences from CASE tool users in a survey from the UK, noting that mistakes are still being made in the choice of tools, and Kemerer (1992) cites studies in which over 70 percent of purchased CASE tools are never used and one study suggesting as few as 5 percent are widely used.

1.1   Evaluation

According to Rossi and Freeman (1993), "[e]valuations are undertaken to influence the actions and activities of individuals and groups who have, or are presumed to have, an opportunity to tailor their actions on the basis of the results of the evaluation effort." (p. 46). With respect to IS evaluation, Hirschheim and Smithson (1988) argue that there has been a tendency by evaluators to adopt a 'technical' underlying perspective, thereby de-emphasising the important social aspects:

"Most IS evaluation has concentrated on the technical rather than the human or social aspects of the system. Part of the reason for this lies in the ontological beliefs of the evaluators -- that information systems are fundamentally technical systems (although they may have behavioural consequences). With such a belief, it is not surprising that only the technical aspects of IS are evaluated, especially considering the difficulty of evaluating social aspects." (p. 31)

More recently, when the same authors re-examined the same problem ten years later they concluded that "IS evaluation clearly remains a thorny problem" (Smithson and Hirschheim, 1998, p. 171)

Rossi and Freeman (1993) discusses the circumstances for evaluators who work in a social context, and notes that:

"Although some theorists in the field emphasize the scientific aspects of evaluation and others its ad hoc qualities, a pragmatic view sees evaluation as necessarily rooted in scientific methodology but responsive to resource constraint, to the needs and purposes of stakeholders, and to the nature of the evaluation setting." (p. 55)

Mathiassen and Sörensen (1996), in observing that "CASE tools are never introduced into an organizational vacuum", reinforce this view in reviewing the capability maturity model's "support to the management of CASE introduction", concluding that it is therefore "interesting and relevant, but limited in perspective" (p. 205).

Ramage (1997) discusses evaluation with respect to co-operative systems, i.e. systems which "are collections of people and organisations in cooperation facilitated by technology" (p. 769), and focuses on evaluation "of such systems in use" (p. 769). He suggests that any evaluation of such systems should "be designed with the multiple, possibly contradictory, interests of different stakeholders in mind" (p. 769), and be conducted as "an ongoing process" (p. 769).

Closely allied to the idea of evaluation is the notion of quality. Reeves and Bednar (1994) claims that quality, when seen as conformance to specification, can be counterproductive when human aspects are involved. It is therefore an alternative notion of quality, seen by Reeves and Bednar as the most complex definition of quality, which sets the ultimate goal for an evaluation. In the words of Swanson (1997):

> "Seeking to meet and/or exceed customer expectations in maintenance reflects a very different view of IS quality. Here the IS is not treated primarily as a software product ..." (p. 849)

This definition of quality is, according to Reeves and Bednar (1994), "an externally focused definition of quality" (p. 433) which is "responsive to market changes" (p. 435) and thereby useful in addressing what is offered in a fluid market such as, for example, the CASE-technology market.

1.2   Evaluation Frameworks

Symons (1991) emphasises the importance of what to consider in an evaluation (something referred to in the ISO standard (ISO, 1995) as an 'evaluation framework'):

> "The choice of criteria determines the content of evaluation, as much by what it excludes from consideration as by what is included." (Symons, 1991, p. 207)

Hirschheim and Smithson (1988) argue that the important phase of emphasising *what* to measure is often neglected in favour of a focus on *how* to measure:

> "it must first be decided exactly what to measure and why; unfortunately, parts of this simple relationship appear to have been forgotten in the desire to create ever more powerful techniques for evaluation." (p. 18)

Ramage (1997) acknowledges the difficulty of identifying *the* set of criteria to be used in an evaluation: "If we are evaluating according to some pre-defined criteria or objectives, then whose criteria?" (p. 777)

Churchman (1959) addresses, with respect to social sciences, the complexity involved in measurement. In particular he raises the issue of what he refers to as *specification*: the importance of knowing, and sharing a common understanding among stakeholders, of what is being addressed.

It is this aspect of sensitivity to the context of the particular organisational setting in which, and for which, an evaluation is to be undertaken, which makes the development of evaluation frameworks for quality decision making so difficult, and method support so critical.

1.3   The problem

Based on our conviction that development of an evaluation framework is of fundamental importance and a necessary pre-requisite for the success of any evaluation effort, we analyse method support for this task with respect to CASE-tool evaluation. To this end, we undertake a critical review of existing methods for CASE-tool evaluation with a specific emphasis on support for the task of establishing an evaluation framework which is to be used in a *pre*-usage CASE-tool evaluation. From the many reports of dissatisfaction when adopting CASE-technology in organisations, we consider this task to be critical.

The difficulty of integrating the 'softer' social and organisational requirements with the more technical demands of the technology, and what can realistically be expected from it with respect to current 'state-of-the-art' CASE-technology, into an evaluation framework that is considered relevant for a particular IS development organisation is therefore an important issue for the research community.

1.4   Overview of the paper

The rest of this paper is organised as follows. First, we report on a study of IS evaluation methods which we have conducted, leading to a categorisation which allows us to identify important issues with respect to current practice in CASE-tool evaluation (2). We then use these findings to analyse our method for establishing evaluation frameworks for CASE-tool evaluation (Lundell *et al.*, 1999) (3). Finally, we summarise our findings (4).

2.   CATEGORISATION OF CASE EVALUATION METHODS

Before analysing the literature on CASE evaluation, it is illuminating to consider what is understood by the term CASE itself. Dixon (1992) notes that definitions of CASE are "too broad to be useful". In our own study we have found significant variation in interpretation. Although a majority of authors refer to CASE as either 'Computer Aided Software Engineering' or 'Computer Aided Systems Engineering', some do not define what they understand by the acronym and others use one (or more) of the many combinations available in Figure 1.

Computer	Aided Automated Assisted	Software Systems [Information] Systems	Engineering Emergence

Figure 1. The acronym CASE

It is also useful to distinguish between *pre*-usage and *post*-usage evaluation. *Pre*-usage refers to activities that take place before a tool being evaluated is in real use in a specific organisational setting. Such activities include any organisational effort with respect to initial selection of a CASE tool, and is the main focus of this paper. *Post*-usage refers to reports of actual experience and use of CASE technology in an organisational setting. Such evaluation is therefore reflective (e.g. Orlikowski, 1993).

In fact, as noted by Le Blanc and Korn (1994), an author's underlying perspective with respect to CASE is often indicated by the adopted interpretation of the acronym. For example, it is apparent that the broader scope implied by the choice of the term 'Systems' is often preferred for studies in an organisational setting. For *post*-usage evaluations this usually implies that the perspective being taken is that of 'Information Systems' in the sense of the FRISCO definition (FRISCO, 1998, p. 73), incorporating human actors and the organisation by which the formal software system has been adopted. It is therefore important to discriminate between these situations in any study of evaluation methods.

2.1 Evaluation and selection

The reference framework we have adopted for categorising the various approaches to CASE evaluation is based on that used in the ISO standard (ISO, 1995, p. 6). It has two aspects, as represented in Figure 2, these concerning process and product.

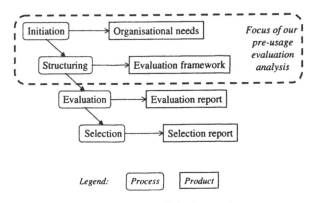

Figure 2. Evaluation and selection overview

With the distinction between *pre*-usage and *post*-usage evaluation in mind, a literature analysis was initiated using a qualitative approach. In the initial analysis, in which sources were considered rather holistically, it soon became clear that this binary distinction was not discriminating enough, and therefore additional categories were introduced in the coding scheme. The continuation of our analysis allowed us to identify which of the defined processes and products were being targeted by each report. In practice it was found more illuminating to also identify the context in which a proposed method for a process was to be applied (for example, an evaluation method may be designed for *post*-usage and for application in a specific adoption environment). Similarly, for product it was necessary to discriminate in terms of intended use of product (for example, an evaluation framework to be promoted for general use). If a source has a product as output, then the process used to generate that product was identified; if a source has a method for a process as output, then the product produced by the method was classified as above. Section 2.2 gives more detail of the refinements actually used, and section 2.3 presents some results of the study.

2.2   Study of CASE Evaluation Methods

In all, 39 academic publications (including conference and journal papers, books and technical reports) were consulted in the study. Its primary aim was to investigate method support offered for CASE tool evaluation, and particularly processes and products of the first two phases of Figure 2. To this end we initially classified the approach used in, or suggested by, each paper according to the figure supplemented by the categories *pre-* and *post-*usage. Our goal was to consider current practice in the light of the issues raised in section 1, and thereby to identify where that practice might be lacking.

Among the sources which we have classified as being concerned with *pre-*usage, it is possible to observe four primary concerns:
– Systematic methods
– Checklists
– Product reviews
– Experience reports

A *systematic method* embodies a systematic 'way of working' during an evaluation and typically also presents, as part of the method, a *checklist*. Examples include du Plessis (1993), ISO (1995), Le Blanc and Korn (1994), Zucconi (1989). These, together with their associated checklists, are of fundamental importance to the study.

A *checklist* contains a set of factors (or criteria) to be used in an evaluation. A source primarily concerned with checklists in general does not include significant advice on how to perform an actual evaluation but is intended to be generally applicable in evaluation efforts. Some papers describe factors at a rather high level of abstraction (Miller's (1956) "7 +/- 2 rule" seems to be an inspiration), while others include more detailed criteria. Some present rather extensive checklists which are hierarchically organised, perhaps into several levels of abstraction (e.g. Dixon, 1992). We notice that most of these sources emphasise that the contents of these lists are to be taken as *potentially relevant* aspects to be considered in the evaluation *if found appropriate* for the evaluation task at hand. The method by which aspects are to be classified as appropriate is usually not elaborated. The checklist approach is exemplified by Reiner (1992).

A *product review* is based on a systematic analysis of a CASE-tool, independently of any specific or general organisational requirements. It will implicitly address the question: 'how good is this CASE-tool?' in an 'objective' manner. As this is done independently of any organisational setting, the basis for these analyses is often (implicitly) extractable into a general checklist, although a detailed product analysis is the primary goal. An example is Schumacher (1997).

An *experience report* in this context refers to experience of a tool, not of its use in an organisational setting (since such a situation would then have been classified as *post-*usage). An evaluation effort may be based either on 'real' (e.g. Kudrass *et al.*, 1996) or 'hypothetical' (e.g. Strange, 1990) requirements from an organisation, and is intended to produce a general report on a tool. For clarification of the difference between this emphasis and the rest of the framework, we note that it is the lack of organisational *experience* which distinguishes it from the category *post*-usage and it is the existence of organisational input which separates it from a *product review*.

Among the many *post-usage* evaluations, we observe an interesting aspect with respect to emphasis. In the majority of sources (particularly recent sources) from this category we observe the broader goal of providing support for the *system* to be designed, not the *software* as a "separate entity". For example, King (1996) emphasises that "CASE *research* is of limited value if context is ignored" (p. 174). It is not therefore surprising that the majority of *post-usage evaluations* use a qualitative approach (Cronholm, 1995; King, 1995, 1996; Orlikowski, 1993).

2.3   Study Analysis

With the above issues in mind, papers were analysed in terms of process and product. For a process, we were interested in the method either used or suggested for that process, and particularly the context in which the method was (or is to be) applied. We classified context as one of: current CASE technology; general organisational requirements; specific adoption environment. For a product we were interested in the goal for future use of that product. We classified each goal as either general usage or usage specific to one organisational adoption environment. Overall, most *pre-usage* sources were found to offer little in terms of method support, but instead concentrated on products, primarily evaluation frameworks (checklists) and evaluation reports (product reviews and experience reports).

For example, Le Blanc and Korn (1994) can be classified as a *pre-usage* study. It reports on using a structuring method based in current CASE technology with a product (a checklist) intended to be general, for use in any organisation. In fact the overall evaluation process envisaged in this source is explicit, and close to that of Figure 2. In common with many other such sources, there is an assumption of organisational needs feeding in to the initiation and structuring processes, but detailed method support – particularly for this complex task – is at best very high level. By contrast, Cronholm (1995) can be classified as a *post*-usage study using a qualitative method in the context of several organisations. Its goal was to produce a general evaluation framework grounded in the needs of organisations. This source exemplifies *post*-usage approaches in that emphasis is placed on organisational need rather than current technology. How current technology is brought into the equation is usually not elaborated.

A number of issues were raised by the study which together, we believe, help to characterise the nature of the current problem with CASE-tool evaluation. We discuss each briefly below.

Firstly, there is the tension identified above between grounding any study in current CASE technology and grounding it in organisational need. There is strong evidence from the study that *pre*-usage studies are based on CASE technology and at best acknowledge the need for organisational input (e.g. Le Blanc and Korn, 1994). Where it is acknowledged, there is an underlying assumption of two frameworks which can somehow be merged. On the other hand, *post*-usage studies are firmly based in organisations, some in a single organisation and concerned with evaluating the level of success of an adoption process. The importance of the state of the art in CASE technology may be acknowledged if the goal is to produce a generally usable output, but many studies are for the purpose of reporting and so are entirely reflective.

Secondly, and particularly for *pre*-usage evaluation, there is a general lack of detailed proposals for method support for the various processes. For example, not only is there an assumption of merging as above, but there is no suggestion as to how such merging should be undertaken. Just as serious for current practice is a lack of guidance for future use of products, given the acknowledgement that future use implies refinement. Just how such refinement should proceed is left unremarked (see for example ISO (1995, p. 25)).

Thirdly, a lack of method support is evident at a meta-level also, in the sense that many sources assume a sequential progression through an evaluation with outputs which are evidently not mutually independent (for example, du Plessis (1993)). This partly relates to the underlying assumptions on merging, but is broader – perhaps more closely related to the waterfall models of development. Particularly when organisational requirements are given prominence, explicit cycles of refinement can be anticipated but are not evident in the sources studied.

Finally, the organisationally grounded studies identify another key issue: that of the "culture gap". King (1997) analyses "culture gaps in a CASE-based IS development situation" (p. 325) and, based on the work of Orlikowski (1993), claims that:

> "culture gaps can occur *between* IT professionals, where CASE tool suppliers fail to understand the work, motivations and expectations of their customers, the system developers." (King, 1997, p. 325)

Such gaps are also evident in the interpretation of general frameworks – for example checklists – for evaluation. Before a checklist can be used (or refined) there must be a common understanding amongst participants of the underlying assumptions of the concepts used. In fact, transfer of results is generally considered a critical facet of any qualitative study for this reason.

In section 3 we review our suggested method (Lundell *et al.*, 1999) for the Initiation and Structuring processes of evaluation in the light of these identified problems. It is unusual[1] in that it targets *pre*-usage in a real organisational setting.

3.   PRE-USAGE EVALUATION IN AN ORGANISATIONAL SETTING

In this section we briefly analyse a method proposed by Lundell *et al.* (1999) in the light of the above study, hereafter referred to as the method. The work in question aims to offer method support for the selection process (and aspects of the initiation process) for evaluation in a *pre*-usage situation. The method is to be applied in a specific adoption environment, with the

---

[1]   In our literature analysis we found no other method with an emphasis on integrating organisational and technical considerations for development of an evaluation framework.

goal of establishing a CASE-tool evaluation framework for that adoption environment. It is based on a qualitative research approach informed by Grounded Theory (Glaser and Strauss, 1967; Glaser, 1978, 1992; Starrin *et al.*, 1997). Figure 3 presents an overview of the method.

An inherent characteristic of the method is that it does not assume an evaluation framework which has been developed *a-priori*, but rather grounds it in the organisational setting. Previously developed frameworks can be used as (external) input data for the method, but no special status is afforded them. In particular, the approach stresses the importance of a shared understanding among involved stakeholders for the emerging concepts in the framework and their interrelationships (Lundell and Lings, 1997, 1998).

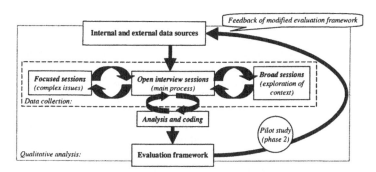

Figure 3. An overview of the method

We proceed by considering the method in the light of each of the issues raised by the study reported in section 2.

There is an assumption that evaluation frameworks encompass aspects both of organisational need and current technology; the first ensures relevance and the second realism. It is implicit in most of the work considered that these aspects can be dealt with separately, and an evaluation framework developed which incorporates both. However, concepts resulting from analysing current technology cannot simply be incorporated with concepts based on organisational need into a single framework. The method addresses this problem essentially by avoiding it, concentrating instead on making the method holistic. In other words, the first identified issue is addressed in the approach to the second. This is done in the following way.

An evaluation framework is developed in the method by an iterative refinement process which receives organisational input during broad, organisationally grounded sessions, and technology input during certain focused sessions (these taking the form of pilot CASE-tool evaluations).

This makes the overall process holistic, whilst keeping the goal at any point in the process simple. The evaluation framework thereby emerges from the alternating consideration of organisation and current technology. Broad organisational needs can be refined into realistic requirements of CASE tools, and possibilities offered by case tools can help to shape an organisational understanding of possibilities and as yet unexpressed needs. This does address the issue of outputs not being mutually independent, as so far this implies only one output (an evaluation framework encompassing organisational need but relevant to an evaluation of current technology). However, the issue of future use is not implicitly addressed.

In fact, the place of the technology-based sessions within the method is critical. Because of the holistic nature of the process, initiation and structuring are given a very particular interpretations and the outputs are distinguished in the method. Firstly, the nature of the iteration proposed is alternation of the first two phases identified in Figure 2. Secondly, the output of phase 1 (initiation) is maintained purely as an expression of organisational need. This need may, through successive cycles, be refined: both from interview sessions within the organisation and from new perspectives brought from the study of CASE technology. Technical input will, however, only be used to broaden debate not to enforce a degree of realism on current expectations. The resulting expression of *organisational need* will therefore be technology independent and in this sense reusable; if new advances are made in the technology, then this will be a good basis for refinement into an evaluation framework based on the new technology. It is the *evaluation framework* which represents the output of a holistic process. It is successively refined also, but is grounded both in organisational need and current technology (through pilot evaluations).

In these ways it can be asserted that the method does indeed address the open issues identified in the study of the CASE evaluation literature and so may be seen to have the potential to improve the quality of evaluation practice.

In Lundell *et al.* (1999) we report on experiences from a field-study using this method in a complex application domain. In that study, the advantage of developing an evaluation framework in this way was found to be very beneficial by the company concerned. The study also noted, however, that a belief in the method used might be seen as a pre-requisite for success. Further field studies are needed before stronger claims for the method can be made, something which is currently being undertaken.

4.   CONCLUSIONS

In this paper, we have categorised approaches to the evaluation of CASE-technology, and specifically focused our analysis on *pre*-usage evaluations designed to assist in quality decision making with respect to tool adoption. The distinct separation of approaches to *pre*- from *post*- usage which we found in the study is in line with the findings of Jarzabek and Huang (1998, p. 94):

"When we initially evaluate a CASE-tool for possible adoption in a company, we tend to examine how well a CASE tool supports technical details of a method we use. However, when we put a CASE tool into use, the evaluation perspective changes, becoming holistic."

The analysis of documented approaches to CASE-tool evaluation identifies a lack of method support for the important task of establishing evaluation frameworks within this process. Several specific weaknesses in current approaches are described. In brief, these relate to:
- (implicit) assumptions about the ability to separate consideration of organisational needs from those of current technology when composing an evaluation framework
- lack of method support for considering both aspects within an evaluation, and for adapting pre-existing frameworks
- the sequential nature of the assumed overall process, with the implicit mutual independence of the products of each component process
- the lack of acknowledgement of the importance of 'culture gap' in establishing frameworks and reusing published frameworks

We have then investigated the recent proposal for method support in this area by Lundell *et al.* (1999) which, unusually, takes a qualitative approach and explicitly introduces iterative refinement into the process of evaluation framework development. The approach is shown to have significant benefits with respect to each of the above.

ACKNOWLEDGMENTS

The authors are indebted to Per-Ola Gustafsson, Skövde Systemutveckling AB for his contribution in a previously undertaken field-study using the proposed method.

REFERENCES

Churchman, C.W., 1959, Why Measure?. In *Measurement: Definitions and theories* (Churchman, C.W., and Ratoosh, P., eds), John Wiley & Sons, New York.

Cronholm, S., 1995, Why CASE Tools in Information Systems Development? – an empirical study concerning motives for investing in CASE tools, In *Proceedings of the 18th Information Systems Research Seminar in Scandinavia: IRIS 18 – "Design in Context"*, (Dahlbom, B., Kämmerer, F., Ljungberg, F., Stage, J., and Sörensen, C., eds.), Gothenburg Studies in Informatics, Report 7, pp. 131-144.

du Plessis, A.L., 1993, A method for CASE tool evaluation. *Information & Management*, 25(2): 93-102.

Dixon, R.L., 1992, *Winning with CASE: Managing Modern Software Development*. McGraw-Hill, New York.

FRISCO, 1998, A FRAMEWORK OF INFORMATION SYSTEM CONCEPTS: The FRISCO Report (Web edition) (Falkenberg, E.D., Hesse, W., Lindgreen, P., Nilsson, B.E., Oei, J.L.H., Rolland, C., Stamper, R.K., van Assche, F.J.M., Verrijn-Stuart, A.A., and Voss, K., eds.), IFIP December 1996, ftp://ftp.leidenuniv.nl/pub/rul/fri-full.zip *(as is: 1st Nov 1998)*.

Glaser, B.G., and Strauss, A.L., 1967, *The Discovery of Grounded Theory: Strategies for Qualitative Research*, Weidenfeld and Nicolson, London.

Glaser, B.G., 1978, *Advances in the Methodology of Grounded Theory: Theoretical Sensitivity*, The Sociology Press, Mill Valley, California.

Glaser, B.G., 1992, *Basics of Grounded Theory Analysis*, Sociology Press, Mill Valley, California.

Hardy, C., Stobart, S., Thompson, B. and Edwards, H., 1995, A Comparison of the Results of Two Surveys on Software Development and the role of CASE in the UK, In *Proceedings Seventh International Workshop on Computer-Aided Software Engineering* (Muller, H.A., and Norman, R.J. eds.), IEEE Computer Society Press, pp. 234-238.

Hirschheim, R. and Smithson, S., 1988, A critical Analysis of Information Systems, In *INFORMATION SYSTEMS ASSESSMENT: Issues and Challenges – Proceedings of the IFIP WG 8.2 Working Conference on Information Systems Assessment* (Björn-Andersen, N., and Davis, G.B. eds.), North-Holland, pp. 17-37.

Iivari, J., 1996, Why Are CASE Tools Not Used?. *Communications of the ACM*, **39(10)**: 94-103.

ISO, 1995, *Information Technology – Guidelines for the evaluation and selection of CASE tools*, ISO/IEC JTC1/SC7/WG4, ISO/IEC 14102:1995(E).

Jarzabeck, S., and Huang, R., 1998, The Case for User-Centered CASE Tools, *Communications of the ACM*, **41(8)**: 93-99.

Kemerer, C., 1992, How the Learning-Curve Affects CASE tool Adoption, *IEEE Software*, **9(3)**: 23-28.

King, S.F., 1995, Using and Evaluating CASE Tools: from Software Engineering to Phenomenology, Ph.D. thesis, Warwick Business School, University of Warwick, Coventry.

King, S., 1996, CASE tools and organizational action, *Information Systems Journal*, **6(3)**: 173-194.

King, S., 1997, Tool support for systems emergence: A multimedia CASE tool, *Information and Software Technology*, **39(5)**: 323-330.

Kudrass, T., Lehmback, M., and Buchmann, A., 1996, Tool-Based Re-Engineering of a Legacy MIS: An Experience Report, In *Advances Information System Engineering, 8th International Conference, CAiSE'96* (Constantopoulos, P., Mylopoulos, J., and Vassiliou, Y. eds.), Springer, pp. 116-135.

Le Blanc, L.A., and Korn, W.M., 1994, A phased appraoch to the evaluation and selection of CASE tools, *Information and Software Technology*, **36(5)**: 267-273.

Lundell, B., and Lings, B., 1997, Evaluation and selection of CASE-tools within the ISO framework: qualitative issues, In *Proceedings of the 8th Australasian Conference on Information Systems* (Sutton, D.J. ed.), School of Information Systems, University of South Australia, Adelaide, pp. 243-254.

Lundell, B., and Lings, B., 1998, An Empirical Approach to the Evaluation of CASE-tools: Method Experiences and Reflection, In *Third CAiSE/IFIP8.1 International Workshop on Evaluation of Modeling Methods in System Analysis and Design: EMMSAD'98* (Siau, K. ed.), University of Nebrasca-Lincoln, pp. N:1-12.

Lundell, B., Lings, B., and Gustafsson, P.-O., 1999, Method support for developing evaluation frameworks for CASE tool evaluation, In *1999 Information Resources Management Association International Conference – Track: Computer Aided Software Engineering Tools*, Hershey, Pennsylvania, 16th-19th May 1999 *(accepted for publication)*.

Mathiassen, L., and Sörensen, C., 1996, The capability maturity model and CASE, *Information Systems Journal*, **6(3)**: 195-208.

Miller, G.A., 1956, The magic number seven, plus or minus two: Some limits on our capacity for processing information, *Psychological Review*, **63**: 81-93.

Nelson, A.C., and Rottman, J., 1996, Before and after CASE adoption, *Information & Management*, **31(4)**: 193-202.

Orlikowski, W.J., 1993, CASE Tools as Organizational Change: Investigation Incremental and radical Changes in Systems Development, *Management Information Systems Quarterly*, **17(3)**: 309-340.

Ramage, M., 1997, Developing a methodology for the evaluation of cooperative system, In *Proceedings of the 20th Information Systems Research In Scandinavia (IRIS 20)* (Braa, K., and Monteiro, E. eds.), Conference Proceeding Nr. 1, Department of Informatics, University of Oslo, Norway, pp. 769-789.

Reeves, C.A. and Bednar, D.A., 1994, Defining quality: alternatives and implications, *Academy of Management Review*, **19(3)**: 419-445.

Reiner, D., 1992, Database Design Tools, In *Conceptual Database Design: An Entity-Relationship Approach* (Batini, C., Ceri, S. and Navathe, S.B. eds.), The Benjamin/Cummings Publishing Company Inc., Redwood City, California, pp. 411-454.

Rossi, P.H., and Freeman, H.E., 1993, *Evaluation: A Systematic Approach*, SAGE Publications, Newbury Park, London.

Schumacher, R., 1997, Erwin/ERX 3.0, *DBMS*, **10(11)**: 31-32.

SEI, 1995, CASE Environment Project Description. Slide presentation January 1995, Software Engineering Institute, Carnegie Mellon University, Pittsburgh, http://www.sei.cmu.edu/legacy/case/case_desc.ps.Z *(as is: 26th October 1998)*.

Smithson, S., and Hirschheim, R., 1998, Analysing information systems evaluation: another look at an old problem, *European Journal of Information Systems*, **7(3)**: 158-174.

Starrin, B., Dahlgren, L., Larsson, G., and Styrborn, S., 1997, *Along the Path of Discovery: Qualitative Methods and Grounded Theory*, Studentlitteratur, Lund.

Strange, G., 1990, International CASE Tool Selection and Evaluation, In *CASE on Trial* Spurr, K., and Layzell, P. eds.), Wiley, Chichester, pp. 81-93.

Swanson, E.B., 1997, Maintaining IS quality, *Information and Software Technology*, **39(12)**: 845-850.

Symons, V.J., 1991, A review of information systems evaluation: content, context and process, *European Journal of Information Systems*, **1(3)**: 205-212.

Walsham, G., 1993, EVALUATION, In *INTERPRETING INFORMATION SYSTEMS IN ORGANIZATIONS* (Walsham, G.), John Wiley & Sons, Chichester, pp. 165-185.

Zucconi, L., 1989, Selecting a CASE Tool, *ACM SIGSOFT: Software Engineering Notes*, **14(2)**: 42-44.

ASSISTING ASYNCHRONOUS SOFTWARE INSPECTION BY AUTO-COLLATION OF DEFECT LISTS

Fraser Macdonald, James Miller, and John D. Ferguson
Dept. of Computer Science, University of Strathclyde
Glasgow G1 1XH, U. K.

Key words: ASSIST, software inspection, auto-collation, computer support

Abstract:

Software inspection is well known as an effective means of defect detection. Recent research in inspection has considered the application of tool support, with the intention of further improving inspection efficiency. An expensive component of the inspection process is the group meeting, where individual defect lists are collated into a single master list. Performing this collation automatically could reduce the length of the group meeting, or contribute towards its replacement by an asynchronous or meetingless inspection. This paper describes a means of automatically collating multiple defect lists into a single list, while removing duplicates. An experiment investigating a simple technique is also described, making use of defects lists generated by subjects in previous controlled experiments. This experiment demonstrates that an automatic collation mechanism is feasible and provides a sound basis for further work in this area.

1. INTRODUCTION

Software inspection is a method for statically verifying documents. It was first described by Michael Fagan (1976). Since then there have been many variations and experiences described, but a typical inspection involves a team of three to five people reviewing and understanding a document to find defects. The benefits of inspection are generally accepted, with success stories regularly published. In addition to Fagan's papers describing his experiences (Fagan 1976, 1986), there are many other favourable reports. For example, Doolan (1992) reports industrial experience indicating a 30 times return on investment for every hour devoted to inspection of software requirement specifications. Russell (1991) reports a similar return of 33 hours of maintenance saved for every hour of inspection invested. This benefit is derived from applying inspection early in the lifecycle. By inspecting products as early as possible, major defects are

Systems Development Methods for Databases, Enterprise Modeling, and Workflow Management,
Edited by W. Wojtkowski, *et al.* Kluwer Academic/Plenum Publishing, New York, 1999.

caught sooner and are not propagated through to the final product, where the cost of removal is far greater.

Unfortunately many organisations are failing to reap the significant benefits, due to their failure to implement a full and rigorous inspection process. One solution to the problem of rigour lies in providing computer support for inspection, and to this end a number of prototype tools have been developed. A comprehensive review of these can be found in Macdonald (1995) and (1996). These tools allow the inspection team to browse and annotate the product (the document under inspection) on-line, and may support discussion of defects during team meetings. Apart from this, the features provided by these systems vary. For example, ICICLE (Shaw 1971, Brothers 1990) makes use of the UNIX tool lint and ICICLE's own rule-based static debugging system to detect certain types of defects. These defects can then be analysed for correctness during the inspection. Scrutiny (Bull 1994, Gintell 1993) provides support for both individual defect detection and group discussion of defect lists. It also provides limited support for holding a distributed meeting. CSRS (Johnson 1994), on the other hand, was designed with the intention of minimising the synchronous meeting element of the inspection, and therefore implements an almost completely asynchronous inspection process.

Although existing systems present innovative approaches to supporting software inspection, in general they suffer from a number of shortcomings. Primarily, they support only a single, usually proprietary, inspection process. For example, Scrutiny implements the inspection process used by Bull, since user acceptance of the tool depended on it performing inspections that counted as part of the development process. These tools also only support inspection of plain text documents, while today's software development environments produce a number of different document types, from plain text to postscript and other graphical formats. Support for inspection of all of these formats is desirable. There is also limited support for defect finding aids such as checklists, while new meeting technologies such as videoconferencing and electronic whiteboards, along with asynchronous meeting arrangements, are still at an early stage of development. Finally, although collection and analysis of inspection data is deemed to be desirable for process improvement, existing tools perform little of such analysis.

Given the limitations of existing tool support, the authors have been working to design and implement a second generation inspection support tool which embodies many important lessons learned from first generation tools, as well as tackling perceived weaknesses. This system is known as ASSIST (Asynchronous/Synchronous Software Inspection Support Tool). This paper provides an overview of the current system to support the inspection process and proceeds to describe the automatic collation of defect lists. It concludes by discussing an approach to estimating the effectiveness (in terms of defect removal) of the current inspection process.

2. THE ASSIST SYSTEM - OVERVIEW

The ASSIST system shares many features found in other systems providing support for software inspection, along with a number of unique features:

- Support for ALL types of inspection processes - ASSIST incorporates a specialised inspection process description language (IPDL), which can be used to model all current types and modes of inspection, and many new forms of inspection processes. The system uses these process descriptions to enforce the rules of the inspection. The modelling language implementation is highly optimised to provide a simple and easy to use interface, e.g. most current inspection processes can be described in around one hundred lines of Structured English.

- Support for different types of documents - ASSIST incorporates two different components to achieve this objective. Firstly, ASSIST has an open-architecture, with regard to document type, allowing any document type-specific browser to be introduced into the system. Currently ASSIST has a default text browser, a generic code browser, a C++ browser and several others. It can also make use of Netscape to display HTML and Ghostview to display Postscript files. Secondly, ASSIST has a flexible mechanism for defect annotations, allowing annotations at virtually any scale, from individual letters or words up to paragraphs, sections and whole documents.

- Active Checklists Facilities - several leading researchers (Miller 1998, Parnas 1985, Porter 1995) have suggested that making the inspector actively interact with the documents, rather than simply passively reading the documents, produces improved detection rates. ASSIST provides mechanisms where the inspector can interact with the checklist to provide an initial starting point in providing automated support for this idea.

- Automatic collation of individual inspectors lists - again several leading researchers (Votta 1993, Porter 1995 and 1997, Miller 1998) have suggested that inspection models without meeting components, may offer advantages in many situations. ASSIST provides support for this idea by providing a tailorable mechanism that enables the moderator to automatically collate each inspectors defect list into a single master defect list.

- Indexing - ASSIST fully supports the automatic indexing of any term in an inspection document, including document-type specific stop-lists and stemming procedures. The built-in cross-referencing facilities then allow the user to easily switch between related areas both within and across documents.

3. ASYNCHRONOUS AND SYNCHRONOUS SOFTWARE INSPECTION

The traditional, synchronous, inspection process has a period of individual defect finding followed by a group meeting where individual defect lists are merged into a single master list. This meeting is expensive to set up and run, requiring the simultaneous participation of three or more people. Some would argue that their cost is unjustified and they should be replaced altogether. For example, Votta (1993) presents evidence that meeting costs outweigh their benefits and suggests their replacement by depositions. Proponents of group meetings, on the other hand, contend that the benefits of group meetings are not easily quantified. Education of new inspectors is one quoted benefit, while synergy, found in many small group situations (Shaw 1971), is another.

With the asynchronous or meetingless approach, (Mashayekhi 1995) inspectors individually go through the document looking for defects. Following this, each inspector circulates a copy of their own defect list to the other inspectors and the leader for review. The review allows participants to pick up on any defects they failed to identify and can be carried out using any communication mechanism e.g. email, bulletin board, or any other medium. The asynchronous discussion associated with the review period mimics a face-to-face meeting, and is thought by many to offer improvements in a number of ways:

- The inspectors can explain themselves in whatever detail they think is required to ensure clarity, including references to the original document.

- We no longer have the situation suggested by Votta, where members of the review team are simply waiting for the meeting to finish; contributions can be made at any convenient time.

- An inspector does not have to wait for someone to stop talking and risk losing their train of thought.

- Conversations can now proceed concurrently, allowing separate threads of discussion to proceed simultaneously.

4. AUTOMATIC DEFECT LIST COLLATION

To reduce the effort required at formal inspection meetings, Humphrey (1989) recommends that the moderator should perform the collation before the meeting. The collated defect list then becomes the agenda for the meeting. While not particularly demanding, collating defect lists can be time consuming. Hence, one means of increasing the efficiency of a traditional inspection is to provide support for this collation. With the asynchronous meetingless approach to inspection, some mechanism for collating defects forms an essential ingredient. This paper describes a simple system to explore the feasibility of the automatic defect list collation (auto-collation), implemented as part of ASSIST. The system is designed to allow multiple lists of issues or defects to be combined into a single list with duplicate entries automatically removed.

It is highly unlikely that duplicate defects from different inspectors will be identical, hence some form of approximate matching must be used. A defect or issue in ASSIST has several components:

- The title of the item.

- The document in which this item occurs.

- The position within the document where the item occurs.

- A free-form text description of the item.

- Up to three levels of classification.

When comparing items for duplicates, document name and position are considered together as the position, the title and text description are considered together as the content of the issue, and the classification is considered on its own. The mechanism used is to score items on their similarity in each of these facets. If two items match with a score above a threshold, one of the items is discarded.

In terms of position, the closer the physical locations of the two items, the higher the score is given, with 0 representing no match and 1 representing identical positions. Items occurring in different documents are given a score of 0. Positions in ASSIST are given as one or more integers, separated by points if necessary. For example, a simple line number is given as 23. A specific character on that line may be 23.12. The scheme can be extended to as many levels as required, and its hierarchical nature allows it to be used for many document types. For example, chapter and section numbering in English documents follows this scheme. When performing a comparison, each component of the position is weighted according to its importance, with leading numbers being more significant.

When comparing the contents of two items, the first step is to generate a list of words occurring in each item. Each word is then checked against a stop list of information-free words (such as "a" and "the"), and discounted if it appears, since they contribute little to the meaning of the contents. If it does not appear in the stop list it is stemmed to find the root. Stemming allows words that are related but not identical to be linked (Harman 1994). For example, *calculation* would also reference *calculates* and *calculated*. The stemming algorithm used is derived from Porter (1997). The two word lists are compared and the number of common words found, expressed as a fraction of the total number of words in the smaller list, gives a score of between 0 and 1. The more words that are common to both items, the higher the score.

The three classification levels also contribute to the similarity score, with the levels scaled in the ratio 2:1:1. This allows one classification factor to be given importance over the other two. Each classification level in one item is checked with the corresponding level in the other, and a simple binary decision made whether they match or not. Checking all three levels gives a score between 0 and 1.

To allow more flexibility, the implementation of auto-collation in ASSIST allows the setting of the Contents, Classification and Position values to indicate the relative importance of each facet when calculating the similarity between two items. The total of all these factors must sum to 1, hence increasing (decreasing) one factor decreases (increases) the others. The similarity value is calculated by multiplying the individual values by the appropriate factor, adding them together, then scaling to give a value between 0 and 1. Finally, the similarity value for two items must be compared with a value used to determine whether the items are sufficiently similar to be declared duplicates. This value is the Acceptance Threshold, and it can be defined by the user. The higher the threshold value, the more similar two items must be to be declared duplicates. Too high a threshold, however, will result in no matches being made.

An initial investigation into the use of auto-collation proved encouraging, and it appears to be an efficient way of removing duplicate items. As can be imagined, however, the set of values used as factors has a huge effect on the result of the auto- collation, with the outcome ranging from discarding virtually all items to performing no removals. Hence, a more rigorous experiment was performed to determine the ranges within which these factors are most usefully set. The next section describes the method used to perform the experiment.

5. EXPERIMENTAL METHOD

Two controlled experiments comparing the effectiveness of paper-based inspection with that of tool-based inspection had already been carried out

(Macdonald 1998). During these experiments, subjects made use of ASSIST to inspect two C++ programs, storing their defect lists on-line. It was decided to use these lists as material for the auto-collation experiment as they reflect real usage. 43 lists were available from the first experiment, while 48 lists were available from the second experiment. The items in each individual list were tagged with a number indicating the identity of defect or 0 if the defect was a false positive. Auto-collation was then applied to groups of three or four lists, as determined by the group allocation from the appropriate experiment, using a number of factor settings. The factor settings used were: -

- Content: 0.05 - 0.95, in steps of 0.05.

- Position: 1 minus the current Content setting.

- Classification: always set to 0, since subjects were not asked to classify their defects.

- Threshold: 0.05 - 0.95, in steps of 0.05.

For each setting of the content factor, the position factor was set appropriately and auto-collation applied for each threshold value. For each group of lists the optimal defect list which could be produced was calculated, consisting of the set of all defects found in all lists minus duplicates and false positives. This allows the percentage of correct defects in the list produced by auto-collation to be measured. The worst case list for each group was calculated as the concatenation of all lists minus false positives. Performance of duplicate removal can then be measured as the number of duplicates in each auto-collated list, expressed as a percentage of the total duplicates in the list. The data for both programs was grouped together, however the data for both experiments was treated separately. This is because the defect positions for the first experiment consist only of line numbers, while those in the second experiment consist of line numbers and character positions, due to the different browsers used to display the code.

6. RESULTS

Figure 1 shows the performance of auto-collation in terms of the percentage of correct defects in the generated list, applied to defect lists from Experiment 1. Overall, the performance is very stable, with the graph having three distinctive areas. With low thresholds ($<=0.25$) most items are discarded, as expected, since items easily pass the similarity test. With high threshold values ($>=0.6$), most items are kept, since the similarity test is harder to pass. In between these threshold values, there is steady increase in the percentage of correct defects, with the rate of this increase rising with larger content factors. Note that the lowest value for each setting of the content factor is always just above zero, since one defect must be put in each list to begin the auto-collation.

The average percentage of duplicates left in each list for various values of content and threshold is shown Figure 2. This graph shows a similar trend to Figure 1, although the threshold values corresponding to the three distinctive areas are higher than their counterparts, at around 0.45 and 0.65. The content factor setting has an effect here similar to that in Figure 1 : as it increases so too does the rate of increase in the percentage of duplicates remaining.

To find the optimal settings for the content and threshold factors, the percentage of correct defects in the list must be compared with the percentage of duplicates remaining. The optimal settings occur when all the correct defects are present, while as many duplicates as possible are removed. Figures 3, 4 and 5 show the performance in terms of correct defects and duplicates for varying threshold factors for content factors of 0.05, 0.5 and 0.95, respectively. For a content factor of 0.05 (Figure 3), 100% of defects occur at a threshold of 0.7, but so also do 100% of duplicates. The threshold value of 0.65 is interesting, since less than 10% of duplicates occur, but only around 95% of correct defects occur. This presents an interesting dilemma: is it more efficient to remove 90% of duplicates, even if it means losing one or two real defects? When the content factor is set to 0.5 (Figure 4), the 100% region for defects occurs at a threshold of 0.6. At this threshold around 65% of duplicates occur. At a content factor of 0.95 (Figure 5), 100% of defects occur at a threshold of 0.55. At the same threshold only 65% of duplicates remain. Overall, setting the threshold to around 0.6 along with a contents factor of at least 0.5 appear to give the best results, removing about a third of duplicates, with no loss of defects. If the loss of one or two defects is acceptable then the contents factor can be reduced and the number of duplicates removed increased significantly.

The graphs of performance for the data from Experiment 2 are very similar to their Experiment 1 counterparts. Figure 6 shows the average percentage of correct defects in each list versus the content and threshold settings. Again, three distinct regions appear in the graph. The lower region, with threshold being less than 0.2, is where virtually all defects are discarded. The upper region, above a threshold of 0.6, is where all defects are kept. The middle region provides a steady increase in the number of defects retained, with the rate of increase rising as the value of the content factor rises. Figure 7 shows the average percentage of duplicates for the data from Experiment 2. Once again, the graph has a similar form, with the boundaries occurring at threshold values of 0.35 and 0.6.

Figures 8, 9 and 10 show the average percentage of defects remaining and the average percentage of duplicates for contents factors of 0.05, 0.5 and 0.95 respectively. These graphs follow their counterparts for the Experiment 1 data very closely. Considering Figure 8, if the criteria applied is that 100% of correct defects must occur then these settings are obviously not useful, since 100% of duplicates occur for any threshold setting where 100% of defects occur. If, however, the condition is relaxed slightly, with around 99% correct being acceptable, then the threshold value of 0.65 allows rejection of almost 60% of duplicates. Since the 99% value is an average, in most cases 100% of all defects are being found with just a single instance of perhaps one defect being lost. In comparison with the same values for the Experiment 1 data, slightly fewer defects are lost, although many more duplicates remain. The difference is almost certainly due to the more accurate defect positioning allowed in the second version of ASSIST, where differences in character position, not just line number, have an effect. In Figure 9, when the 100% criterion for correct defects is fulfilled, less than 5% of duplicates have been removed. Relaxing the criterion slightly gives just over 20% of duplicates removed at a threshold of 0.6. This is a slightly worse result than for the Experiment 1 data, both for correct defects and duplicates remaining. Figure 10 follows the pattern observed so far. At a threshold value of 0.55, it provides slightly worse results than the

Experiment 1 data in terms of correct defects and duplicates remaining. Again, for acceptable duplicate removal performance, a small loss of real defects has to occur.

One reason why the data for Experiment 2 provides slightly worse results than that from Experiment 1 is the difference in the way in which defects positions are represented. In Experiment 1 only line numbers are used, while both line number and character positions are used in Experiment 2. This more accurate positioning system exaggerates small differences in defect positions. For example, when using line numbers only, two defects on the same line have identical positions. When using both line number and character position, two defects on the same line may have positions such as 32.0 and 32.4. Some subjects may consider the defect to occur at the start of the line (even if there is blank space at the start of the line), while others may mark its position exactly. Instructing subjects on a uniform method of deciding defect positions would help reduce this variability. The effect of the positioning strategy should be reduced as the contents factor is increased, since the value of the position factors decreases at the same time.

One factor which may generally have had an adverse effect on the results of the auto-collation experiments was the variability in spelling among subjects. Misspelling of long words such as 'initialisation' was commonplace. Such misspellings reduce the effectiveness of word matching between items, stop list matching and stemming. The result is a reduction in the probability of two items being declared duplicates, thereby reducing the effectiveness of auto-collation. Although misspellings could have been fixed before the experiments were carried out, this would not have reflected real usage. The obvious solution to this problem is provision of a spelling checker within the tool. Subjects also tended to use varying terminology. In an industrial setting terminology would probably be more consistent, which would also improve performance.

7. SUMMARY AND CONCLUSIONS

An essential component of any software inspection process is the collation of defects into a single master list. In the synchronous or meeting orientated approach, this process can consume considerable effort at the meeting with an associated cost. In the asynchronous or meetingless approach, the collation of defects again has a significant cost. In an attempt to reduce this cost, a system for automatically performing this collation has been developed as part of a prototype software inspection support tool. The system was evaluated by performing auto-collation on real defect lists obtained from previous experiments comparing tool-based and paper-based inspection.

With the strict criterion of absolutely no loss of defects, auto-collation can reliably remove 10%-35% of duplicates. If the criterion is relaxed slightly, with one or two losses being acceptable, the rate of removal can be as high as 60%-90%. The question of whether the defect loss is acceptable to reach such removal rates is a difficult one to answer, and depends on the context in which the system is being used and the relative costs and benefits.

Finally, it is believed that this initial feasibility study, despite the early stage of development, demonstrates that the auto-collation of defects lists is a viable process. Further, additional performance can be obtained by:

- linking the auto-collation process with spelling, thesaurus and grammatical aids, which are currently found in current text processors.
- constraining the language used by inspectors; for example, standard definitions of defects.
- more advanced approximate matching techniques.

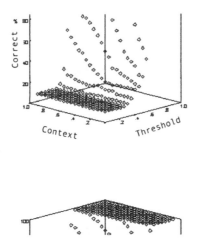

Figure 1. Average percentage of correct defects in collated lists for each value of content and threshold (Experiment 1 data).

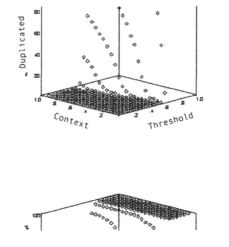

Figure 2. Average percentage of duplicates in collated lists for each value of content and threshold (Experiment 1 data).

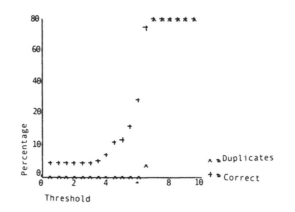

Figure 3. Average percentage of defects remaining and average percentage of duplicates in collated defects lists for contents factor of 0.05 (Experiment 1 data)

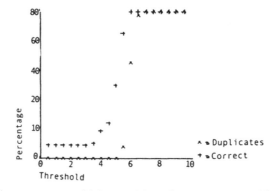

Figure 4. Average percentage of defects remaining and average percentage of duplicates in collated defects lists for contents factor of 0.5 (Experiment 1 data)

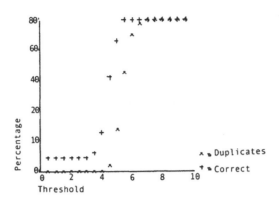

Figure 5. Average percentage of defects remaining and average percentage of duplicates in collated defects lists for contents factor of 0.95 (Experiment 1 data)

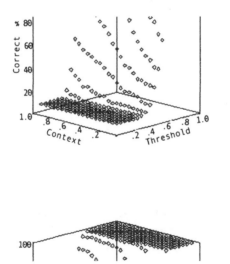

Figure 6. Average percentage of correct defects in collated lists for each value of content and threshold (Experiment 2 data)

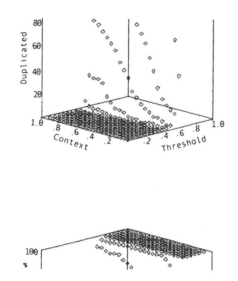

Figure 7. Average percentage of duplicates in collated lists for each value of content and threshold

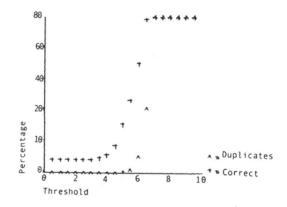

Figure 8. Average percentage of defects remaining and average percentage of duplicates in collated defects lists for contents factor of 0.05 (Experiment 2 data)

194

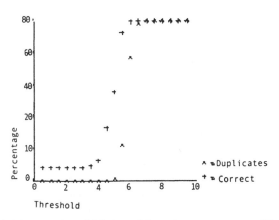

Figure 9. Average percentage of defects remaining and average percentage of duplicates in collated defects lists for contents factor of 0.5 (Experiment 2 data)

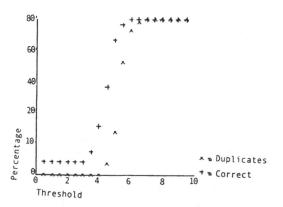

Figure 10. Average percentage of defects remaining and average percentage of duplicates in collated defects lists for contents factor of 0.95 (Experiment 2 data)

REFERENCES

Brothers, L. R. Sembugamoorthy, V., and Muller, M., 1990, ICICLE: Groupware for code inspections. In *Proceedings of the 1990 ACM Conference on Computer Supported Cooperative Work*, pages 169-181

Bull, HN, 1994, Information Systems, Inc., U.S. Applied Research Laboratory. *Scrutiny User's Guide*

Doolan. E.P., 1992 Experience with Fagan's inspection method. *Software-Practice and Experience*, 22(2):173-182

Fagan. M.E., 1976 Design and code inspections to reduce errors in program development. *IBM Systems Journal*, 15(3):182-211

Fagan. M.E., 1986, Advances in software inspection. *IEEE Transactions on Software Engineering*, 12(7):744-751

Gintell, J.W., Arnold, J., Houde, M., Kruszelnicki, J., McKenney, R. and Memmi. G.,1993, Scrutiny: A collaborative inspection and review system. In *Proceedings of the Fourth European Software Engineering Conference*, September

D. Harman. Automatic indexing. In R. Fidel, T. B. Hahn, E. M. Rasmussen, and P. J. Smith, editors, *Challenges in Indexing Electronic Text and Images*, chapter 13, pages 247-264. Learned Information, Inc., 1994.

Humphrey. W. S., 1989, *Managing the Software Process*, chapter 10, pages 171-190. Addison-Wesley

Johnson. P.M., 1994 An instrumented approach to improving software quality through formal technical review. In *Proceedings of the 16th International Conference on Software Engineering*, May

Macdonald, F. and Miller, J., 1997, A software inspection process definition language and prototype support tool. *Software Testing, Verification and Reliability*, 7(2):99-128, June

Macdonald, F. and Miller, J., 1998, A comparison of tool-based and paper-based software inspection. *Empirical Software Engineering: An International Journal*, 3(3), Autumn

Macdonald, F., Miller, J., Brooks, A., Roper, M., and Wood, M.,1995,. A review of tool support for software inspection. In *Proceedings of the Seventh International Workshop on Computer Aided Software Engineering*, pages 340-349, July

Macdonald, F., Miller, J., Brooks, A., Roper, M., and Wood, M.,1996,. Automating the software inspection process. *Automated Software Engineering: An International Journal*, 3(3/4):193-218, August

Mashayekhi, V., 1995 *Distribution and Asynchrony in Software Engineering*. PhD thesis, University of Minnesota, March

Miller, J., Roper, M., and Wood, M., 1998, Further experiences with scenarios and checklists. *Empirical Software Engineering.*, 3(1): 37-66

Parnas, D.L.,and Weiss, D.M., 1985, Active design reviews: Principles and practices. In *Proceedings of the Eighth International Conference on Software Engineering*, pages 132-136, August

Porter, A.A., Votta, L.G., and Basili, V.R., 1995, Comparing detection methods for software requirements inspections: A replicated experiment. *IEEE Transactions on Software Engineering*, 21(6):563-575, June 1995.

Porter, A., Johnson, P., 1997, Assessing software review meetings: Results of a comparative analysis of two experimental studies. *IEEE Transactions on Software Engineering*, 23(3):129-145

Porter, M.F., 1980, An algorithm for suffix stripping. *Program*, 14(3):130-137

Russell, G. W., 1991 Experience with inspection in ultralarge-scale developments. *IEEE Software*, 8(1):25-31, January

Shaw, M. E., 1971, *Group Dynamics: The Psychology of Small Group Behaviour*. McGraw-Hill

Votta, L. G., 1993, Does every inspection need a meeting? In *Proceedings of the First ACM SIGSOFT Symposium on the Foundations of Software Engineering*, pages 107-114, December

META-MODELING:
THEORY AND PRACTICAL IMPLICATIONS

Milan Drbohlav
Department of Information Technologies
Prague University of Economics, W. Churchill Sq. 4,
130 00 Praha 3, CZECH REPUBLIC

Key words: model, meta-model, meta-meta-model, level of inquiry,
modeling dimensions, concepts, CAME, meta-database

Abstract: There is an obvious need to integration (data integration as well
as functional integration) of current methods in different
engineering disciplines. Meta-modeling constitutes foundation,
which can satisfy these requirements. It is called a new approach
to adapting methods into various environments (for example
CASE). Without precise and accurate specification of a method
using explicit and clear meta-model, this adaptation is not
possible.
Meta-modeling is the process of specifying the requirements to
be met by the modeling process or establishing the
specifications which the modeling process must fulfil. It also
uses its own methods and tools which, in turn, can be described
on one level higher, on meta-meta-modeling level. This is the
highest relevant abstract level (thinking about modeling), which
describes only structural aspects of meta-modeling and for
which is typical its auto-reference property.
The meta-modeling theory has its implication in the
development of tools or toolkits without defined meta-model.
Users can configure the meta-model of the tool.

1. INTRODUCTION

Many human activities have been transformed from an art to an
engineering discipline in the last ten years (for example analysis, design and
implementation of information systems or management of processes in the
domain of informatics). The elementary condition for such transformation is
improvement of our understanding and formalization of the knowledge that
is used, captured, transferred, and transformed while performing mentioned
activities (see Brinkkemper 1990). This can be achieved by developing
notations, approaches and theories that are concerned with these activities
and also by developing structure and formal presentation of their outcomes.
This growing and intensively researched area is called meta-modeling and

the principles and systematic of conducting meta-modeling method engineering.

While talking about meta-modeling, we have to mention at least three sets of issues:

1. An ontology of meta-modeling: what does it mean, what kinds of meta-modeling "dimensions" are there,
2. Meta-modeling process: how to perform meta-modeling, how to recognize and/or build meta-models based on application/business requirements, etc., and
3. Meta-models implementation: how to represent meta-models, what additional notational constructs are needed, what kind of support does modeling tools should provide, etc.

All three issues are pretty wide research areas and there is no space to discuss them in this paper. The paper addresses the following:

- what meta-modeling really is and what are the differences between modeling, meta-modeling and meta-meta-modeling
- why does meta-modeling recently get so much attention
- what are the main uses of meta-modeling
- what are the key components of a meta toolkit as an implication of meta-modeling theory

2.   LEVELS OF INQUIRY

Many people talk about meta-modeling in the context of information system development methodologies, especially in the context of object oriented methodologies and object oriented programming languages (Artsy 1995). The reason for this is that the most tangible meta-modeling achievements were historically made by work on reflection in programming languages, also referred to as computational reflection. These efforts have identified two kinds of reflection[1]:

- Structural Reflection: can be defined as the "ability of a language to provide a complete reification of both the program currently executed as well as a complete reification of its abstract data types" (DMRG 1997).
- Behavioral Reflection: can be defined as "the ability of the language to provide a complete reification of its own semantics (processor) as well as a complete reification of the data used to execute the current program" (DMRG 1997).

However, meta-modeling is obviously an approach useful also in other engineering disciplines. In the simplest form we can say that meta-modeling is an activity that produces meta-models (Rumbaugh 1995), an activity that is closely related to modeling. One possibility off explaining differences between modeling and meta-modeling is considering levels of inquiry or of expertise that can be identified when dealing with a problem. These levels influence the definition of the problem and the types of suggested solutions. We can distinguish four levels of inquiry (as shown in Table 1 – source Drbohlav 1998).

The first level is the level of implementation, where people (employees, citizens) participate in activities involving real world problems. In management terminology this represents the operational level of the traditional organization's hierarchy. This level always implements methods and procedures, which originate at the higher level of inquiry. In software

---

[1] this concept corresponds to the concept of two meta-models types mentioned later

Table 1. Levels of inquiry - summary

Level of inquiry	Principle	Corresponding management level	Position of the level in software development process
level of implementation	utilization of methods and procedures	operational	performing particular phases
modeling	formulation of methods and procedures	tactical	designing a methodology
meta-modeling	design principles of methods, generalization	strategic	specification of types of models, specification of entity types, of relationships types, their assignment to specific models
meta-meta-modeling	description of meta-modeling	generalized practices, experiences, knowledge	specification of general modeling principles

development process this level corresponds to a running developed (analyzed, designed and implemented) system, which supports some processes.

At the second level of inquiry, experts deal with problems of finding solutions to the problems encountered at the implementation level. People at this level are occupied with the identification of the problems discovered at the first level, the failure diagnostic, and the formulation of methods and procedures by which the various activities at the implementation level can be carried out. Usually, understanding and solving the problem at this level requires the formulation of a model. That is why the level is called the modeling level of inquiry. Modeling is thus the process of converting our perceived view of reality into a formal representation. Traditionally, this level is called the tactical level of the enterprise. Analyses and design phases of a software development process are typical examples of this level of abstraction. Analysts and designers try to "formalize" the reality using appropriate set of methods and procedures (recommended by chosen development methodology). They develop "methodology" for using such formalization in form of user guides.

At the third level, we find other specialists whom work on the generic problems of methodologies, regardless of their origin – in other words they try to generalize. At this level, people deal with the design of the methods and approaches used at all lower levels of inquiry. Because of the degree of abstraction required solving the type of problem offered at this level, it is called the meta-modeling[2] level of inquiry. In the traditional management hierarchy, this is called the strategic level. This level of inquire does not have direct equivalent in software development process. However, it should be used for example while developing concrete methodology. In this case it is necessary to describe:
– supported types of models

---

[2] The prefix "meta" originates from Greek and means "about" – in our case "meta-modeling" means "about modeling.

- types of entities and allowed relationships between them, graphical symbols and semantic of the model have to be described for every type of supported model

The fourth level of inquiry is the most abstract level. Meta-modeling also uses its own methods and tools which, in turn, can be described on one level higher, on meta-meta-modeling level. Theoretically this cycle can run ad infinitum. Whether there is a meta-meta-meta-modeling level or not, it is an academic question. In fact some authors (van Gigch 1991) do not speak about fourth level of inquiry at all. From the philosophical point of view, the meta-meta-modeling level does not make sense and the meta-modeling level is the highest level of inquiry.

But there is a good argument for distinguishing the meta-meta-modeling level. When we talk about architecture of a meta tool, i.e. a tool, which can describe concrete method or which, can generate a tool that supports concrete set o methods, than we are talking about formulation of a meta-meta-model. Meta-meta-modeling level captures the structural information of meta-models, not their behavior – though, of course, the meta-models can capture the behavior of user models. The most interesting property of the meta-meta-modeling level is auto-reference. This feature means, that the meta-meta-model (output of meta-meta-modeling process) can be described using the same concepts (constructs, notations) as the meta-model (output of meta-modeling process).

I present Figure 1 for better understanding of my view of particular levels of inquiry:

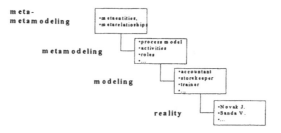

Figure 1. Example of different levels of inquiry (see vanGigch 1991, Drbohlav 1998)

3. META-MODELING

Meta-modeling is the process of specifying the requirements to be met by the modeling process or establishing the specifications which the modeling process must fulfill. So meta-modeling is two abstraction levels "removed" from the reality – it takes place one level of abstraction and logic higher than the standard modeling process. Clearly, no modeling is possible without some sort of explicit or implicit meta-modeling.

The output of meta-modeling process – called meta-model – captures information about the concepts, representation forms and use of a method. In other words, a meta-model is a conceptual model of a method, which describes the concepts in the method and their relationship to each other.

For example, a meta-model of process model defines, what are the key concepts of the model (activities, roles, states, and documents), the way of representation of each concept (graphical representation) and the stages of creating the model.

Smolander distinguishes three dimensions in modeling methods (Smolander 1990). The first dimension deals with the two abstraction levels needed to build up a model hierarchy: the meta-model and its model instantiations. This we call a type-instance dimension in meta-modeling. Types included in a meta-model determine what one can describe or observe on the instance level while using the method. The second dimension makes a difference between the conceptual structure of a method and its representational form - so called conceptual-representational dimension[3]. This helps to achieve "representation independence", i.e. that the method knowledge can be conveyed in varying representations covering graphical, matrix or textual form. The third dimension recognizes time as an important feature of meta-modeling and thereby introduces dynamics into the modeling domain. This is called the static-dynamic dimension of meta-modeling and it permits us to distinguish between meta-datamodels and meta-activity models, where only the latter involve time. Meta-datamodels describe the static aspects of specific method like its concepts and terms, notation or representation of the method. These models declare <u>what</u> a method allows one to perceive. Meta-activity models capture the process (dynamic) aspects of specific method like its stages, tasks or steps. They specify <u>how</u> a method should be followed, i.e. how and in what order of tasks (in time) a method produces its products (representations).

Presented three dimensions allow us to discern three key concerns in method engineering:

1. what knowledge is conveyed, stored and manipulated while using the method
2. how the knowledge is represented for different users of models
3. how the method knowledge is produced and consumed in practice; moreover, the dynamics allows us to analyze the dynamics of the method evolution and its modifications in conceptual/representational dimensions

Meta-modeling has been around for at least 10 years, but it has gained so much attention just recently. Commonly mentioned reasons are (see Tolvanen):

– its time has come; data and functional integration is something that is seriously getting attacked now and meta-models are the foundation for integration (even if they are not always called meta-models)
– the advent of meta-CASE tools; earlier-generation CASE tools would usually support only one methodology, later more than one. Meta-CASE tools are tools that do not have a fixed meta-model, instead, the user to support any meta-model can configure them. Thus meta-model can be used to directly drive tool support.
– the increasing availability of meta-model-driven technologies and standards; some of these include repositories, data exchange between modeling tools, and standards for it

---

[3] has been recognized since the ANSI/SPARC proposal for a three-level data base architecture

I am pretty sure, that the most important reason is the first one – the integration requirements. And these requirements are perceptible not only in the area of development tools (the goal is to interchange data among different models being developed), but also in all engineering areas - for example in the area of development of application packages or systems integration as well. Too many people interfere with problems, the source of which can be called lack of meta-model knowledge about the used method. Without a precise and accurate specification of a method, as would be expressed in an explicit and clear meta-model, it is difficult to understand, accept, and adhere to the method, inter-operate between tools implementing it, let alone inter-operate with tools implementing other methods.

Very few tools implementing a method do have an explicit meta-model, and even fewer publish it. Without such a model, the tool's user cannot know precisely how accurately the tool views or implements certain concepts. Furthermore, even when the tool has an explicit meta-model, but the tool is not model-driven, it is inflexible to change whenever the specific method evolves.

Even when there are meta-models for specific method and tools, they usually are expressed in very particular notations, which differ not only graphically (boxes vs. clouds, for example), but also semantically. Further more, they would use similar names to mean different things, or different names to mean similar concepts. These certainly are barriers for interoperability: how could you pass design information among tools if they cannot understand each other's concepts?

Similar to meta-modeling a system and an application, there is a problem of how to represent meta-models for methods and tools. Which meta-meta-modeling concepts can express such diverse notations and different meanings?

The need for meta-modeling demonstrates the following example concerning the role of Systems Integrator: a systems integrator might have to integrate multiple software and hardware components in a way that they more or less appear like one integrated system. In order to do this, the systems integrator needs to pay careful attention which component supports which concepts, and how a particular concept of a particular component relates to which other concepts of another component. This analysis provides a list of concepts and their relationships, i.e. a meta-model. This meta-model can then be used to decide which data is kept where, and how components are supposed to talk to each other.

Overall, the main uses of meta-models today are:

- as conceptual schemas for repositories that hold software engineering and related data
- as conceptual schemas for modeling tools such as CASE tools
- to define modeling languages, such as for object analysis and design
- as part of technology (together with a transfer mechanism such as a file format[4]) that allows interoperability of modeling tools
- as a tool to help understand the relationships among concepts in different modeling languages

---

[4] for example, the CDIF standard committee of EIA has been developing standards for an integrated meta-model supporting information interchange between development tools. The committee has published meta-meta-modeling framework for meta-modeling and extensibility, a transfer format with encoding/decoding rules, and a few subject areas defining the semantics of the interchanged models.

4.   COMPONENTS OF A META TOOLSET

In the last part of the paper I like to outline key components of a meta toolkit as an implication of the described basics. The architecture of this meta toolkit is being developed at Prague University of Economics – Department of Information Technologies – under code name MaTeS (Vorisek & Stanovska 1994, Drbohlav 1999).

Generally we can say MaTeS is a CAME (Computer Aided Method Engineering) toolkit which should dispose of following properties:

1. Acceptance of all three described modeling levels: modeling, meta-modeling and meta-meta-modeling

    We suggested a meta-meta-model, which covers basic modeling principle. These principles must be fulfilled at lower abstraction and logic levels. We formulated a meta-meta-datamodel (expressed using an ERD diagram) at meta-meta-modeling level and methodology, how to recognize and build meta-models.

    One of the main goals is to develop an architecture, which does not have built-in meta-model and which allows formulating of different meta-models; a toolkit with such architecture can be used to model different methodologies.

    Different types of users (see Figure 2) can use particular tools. We distinguish tree types of toolkit end users:

    -   metadesigners: designers who work on meta-modeling level; they are responsible for designing all concepts of described methodology, that is they specify:
        -   entities (entity types), their attributes and relationships among them
        -   particular types of models, which can be developed at the lower (modeling) level
        -   activities, which can be run over a specific model, over all (or selected) instances of all (or selected) entities or over all (or selected) relationships; one group of these activities constitute checking of consistency rules

            All these concepts (the meta-model) are stored persistently in a meta-database.

            In practice, metadesigners should be high skilled experts, who are able, based on experiences, practices and knowledge, to formulate a meta-model, which is most suitable for a specific case.
    -   modelers – implementants: the challenge for modelers is to model the focused system, that is to fill the designed meta-database with particular instances of entities and their relationships; this role can be fulfilled by skilled consultants and by end users as well; the output of their activities is filled and consistent meta-database (different models)
    -   end users: people who use developed models to support decision making, development of an information system, etc.

2. Graphical representation; today this is the basic modeling requirement, which is useful not only for end users, but for experts as well; as explain earlier, meta-modeling level represents a higher level of abstraction, where textual description has not enough explanatory power

3. Ease of use and easy of gathering information; while developing the
    MaTeS´s architecture we think about end users, who are not skilled in
    modeling technique at all; so we try to hide the complexity and different
    possible uses of particular tools and thus increase likelihood of ease use
    by not skilled users
4. Ability to inter-operate with other tools and software products via defined
    interfaces

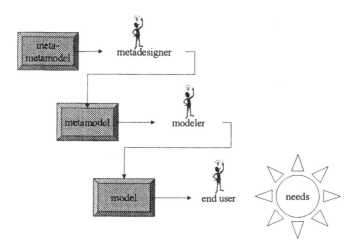

Figure 2. Types of MaTeS toolkit end users (Drbohlav 1999)

The MaTeS toolkit consists of following tools:
1. meta-database designer
    based on meta-meta-model this tool enables the development of all meta-
    models concepts, that is entities, attributes, relationships among entities,
    different types of models and activities; the main output of this model is
    the defined structure of a meta-database; reengineering of existing meta-
    models (i.e. meta-databases) is possible; users of this tool are
    metadesigners
2. meta-database editor
    the challenge of this tool is to fill the meta-database structure with
    instances of entities and their relationships; users of this tool are
    modelers and they, based on defined meta-model, develop different types
    of models; the process of model development can be described as
    development of multiple level net of instances of particular entities and
    their relationships;
    this tool is used to specify analytical appliances over models; user
    defined analytical appliances are supplement to predefined analytical
    appliances like what-if-analysis, analysis of hierarchical causality, etc.
3. meta-database viewer
    meta-database viewer is a typical tool for end user; it allows to navigate

across different models and entity instances; it is designed for running all the defined analyses as well

4. meta-database administrator
   this tool should offer administration function for all other tools; first of all in two areas:
   administration of users and access rights
   replication of meta-database for purpose of off-line activities

5.   CONCLUSIONS

Let me conclude with the following statements:

1. A common approach for describing modeling structures employs physically and conceptually different modeling levels.
2. The meta-modeling process is useful in wide area of engineering discipline, not only in software development process.
3. The meta-modeling level, which is the object of methodology engineering discipline, can be described at the meta-meta-modeling level. I believe, this is the highest abstract level, which makes sense and which disposes of auto-reference property.
4. The meta-meta-modeling level is important while constructing a meta tool, i.e. a tool without pre-defined meta-model.
5. Architecture of a meta toolkit was developed as implication of meta-modeling theory. The toolkit consists of following types of tools:
   - definition tools (to define meta-model and different models)
   - evidence tools (to fill meta-database with instances of entities and their relationships)
   - query and presentation tools (to perform analysis over defined models and present their results)
   - administration tools

6.   REFERENCES

ANSI: The ANSI/X3/SPARC DBMS Framework report of the Study Group on Database Management Systems, Information Systems 3, Pergamon Press, pg. 173-191

Artsy, Yeshayahu Shaike, 1995, Meta-modeling the OO Methods, Tools, and Interoperability Facilities. OOPSLA '95 Workshop on Meta-modeling

Brinkkemper, S., 1990, Formalisation of information systems modelling. PhD-Thesis, Katholieke Universiteit Nijmegen, The Netherlands

Chen, P., 1976: The entity-relationship model – toward a unify view of data. ACM Transactions on Database Systems, 1 , 9-36

DMRG group, 1997, Method Engineering Encyclopaedia. Technische Universiteit Twente, The Netherlands, http://wwwis.cs.utwente.nl:8080/dmrg/MEE/index.html

Drbohlav, Milan, 1998, Implementation of a Metainformation system. Research paper, Prague University of Economics, Prague

Drbohlav, Milan, 1999, MaTeS – a meta-CASE tool. Research paper, Prague University of Economics, Prague

van Gigch, J.P., 1991, System design modeling and meta-modeling. In Plenum Press, New York

Mohammed, Michael, 1995, Meta-modelling Proposal. OOPSLA '95 Workshop on Meta-modeling

Rumbaugh, J., 1995, What is a Method?. Rational Software Whitepaper, http://www.rational.com/uml/resources/whitepapers/index.jtmpl

Smolander, K., Marttiin, P., Lyytinen, K., Tahvanainen, V.-P., 1990, MetaEdit – a flexible graphical environment for methodology modeling. The next Generation of CASE-tools, Nordwijkerhout, Netherlands

Tolvanen, J.-P., Lyytinen, K., Flexible method adaptation in CASE, The Meta-modeling Approach. University of Jyväsjykä, Finland, http://iris.informatik.gu.se/sjis/vol5/tolvanen.shtm

Vorisek J., Stanovska I., 1994: Metainformation System – Tool of Systems Integration, Systems Integration '94, conference proceedings, Prague

Wakeland, Wayne, 1990, Meta-modeling Aspects of Model Conceptualization. ISSS 1990, conference proceedings

CRITICAL SYSTEMS THINKING AND THE DEVELOPMENT OF INFORMATION TECHNOLOGY IN ORGANISATIONS: ANOTHER PERSPECTIVE

Diego Ricardo Torres Martinez
Department of Systems Engineering
Pontificia Universidad Javeriana, Carrera 7 Numero 40-62,
Santa Fe de Bogota, COLOMBIA

1.   INTRODUCTION

Most of the literature[1] and the practice of the development of IT in organisations seem to be focused upon the technical part [Davies *et al*, 1989: 61]. It focuses attention on the technological devices in isolation; where computer science plays an important role. It is concerned with the technical hardware development and software engineering, rather than with the relationships between IT and for example, individuals and the organisation itself. Also it is clear that an historical overview of the evolution of IT in organisations allows us to have a better understanding of the present role of IT in organisations. The significance that the computer itself is a radical invention[2] makes clear the social construction of IT. Furthermore, it is clear that IT is a strategic resource and there exists an important concern about how IT can fit into and support organisations. The history of IT displayed the relationship between the enabling effects

---

[1] Davies *et al* (1989), argue that the literature of IS development is dominated by a rationality which is rooted in logical positivism and functionalism, which suggest that data flow design is more important than the nature of information (pg. 61) as for example. In addition the predominant view of IS is equate of computing, focusing in technical hardware development and software engineering, which concern of computing science. Winograd and Flores (1986), show how in the literature on computers and decision making, a wide range of human activities and concerns are subjected to an rationalistic analysis given more importance to the technical part of IT development (pg. 20)

[2] Hughes (1987) shows how the history of evolution of a large technological system, for instance IT, can be presented in the phases in which the activity named predominates: invention, development, innovation, transfer and growth, competition and consolidation (pg. 56). Development is the phase in which the social construction of technology becomes clear. Innovation clearly reveals technologically complex systems, as they combine the invented and developed physical components into a complex system consisting of manufacturing, sales and services facilities (pg. 64). The transfer of technology, though, which can occur at any time during the history of the technological system [pg. 66], is clearer after the innovation phase. Technological systems, even after growth, competition and consolidation, do not become autonomous; they acquire momentum (pg. 76).

Systems Development Methods for Databases, Enterprise Modeling, and Workflow Management,
Edited by W. Wojtkowski, *et al.* Kluwer Academic/Plenum Publishing, New York, 1999.

of IT and organisational change[3]. On the other hand, it is admitted that information technology is developing very rapidly. An important characteristic concerned with the relationship between the very rapid development of technology and the practice of this technology in an organisation, is that continuous advances in hardware and software permit users to discover new practices in the way of developing their tasks. There are many existing factors associated with its use that have affected both the development and the completion of tasks in organisations. Checkland (1998) noted how the introduction of IT can have very profound effects on both the individual and organisations (pg. 6). It seems to be that the evolution of IT has presented a diffusion of an organisation's many features, which deserve an understanding of IT not only from a merely technological point of view. For example, IT has affected the processes of communication within organisations and between them and processes of decision making are currently supported by IT tools. The consequences of its use have implications beyond the merely technical and its understanding should involve its interrelationship with human activity and the social and political aspects of organisations.

In summary, IT contains messy, complex, problem-solving components. In IT, complexity is present in various shapes and forms and this poses a variety of difficulties that need to be addressed. These issues are key features in removing common assumptions made by most of the persons involved in the development of IT in organisations; that the use of IT is only a technical matter. Repeatedly in organisations, managers can be motivated to employ information technology, but they do not know how it can help the organisational business. Furthermore, here it is important to stress that managers may consider those themes as merely technical, leaving it to technical areas, which them may become, one of the principal problems. Due to the possible solutions which can yield only under technical issues rather than social and political issues [Bloomfield and Danieli: 29] (which will be dealt with later), and have a great influence over the limited vision of the development IT in an organisation.

This article explores the development of IT in organisations from hard, soft and critical systems thinking traditions respectively. In addition, the role of IT in organisations and the social and technical interaction are subjects under analysis. This article primarily attempts to elaborate an insight of the traditional computer systems analysis from a hard system approach, as it has been widely applied in order to enhance development of IT in organisations. This approach is founded in a positivistic and reductionistic tradition. Then, the article will discuss the contribution of the soft tradition, as it presents a fundamentally different view of human action. Finally, the article will introduce the new tradition: Critical Systems Thinking (CST), and discuss how this approach can enrich the development of IT in organisations. This section will discuss the benefits to the development of IT in organisations of this approach, through a reflection of its three commitments: Critical awareness, Emancipation and Methodological Complementarism, and the theory and practices of the development of IT in organisations.

---

[3] Tapscott and Caston (1993), present the paradigm shift in IT and how it relates to other changes in the world. For instance, the technology paradigm parallels the other shifts. Like the new enterprise it is open and networked. It is modular and dynamic-based on interchangeable parts. It technologically empowers, distributing intelligence and decision makers to users (xiii).

2. COMPUTER SYSTEMS ANALYSIS (CSA) AND ITS CRITIQUE

It is probably correct to say that the history of IT implementation in organisations has not provided the quick success, or gains efficiency that was anticipated. Actually, organisations find a competitive environment in which IT is considered very helpful and managers keep the hope of being more competitive by using it. But, there is not a causal relationship between the development of formal IT in the organisation and the improvement of organisational performance [Lyytinen, 1987: 4]. Some of the possible causes are argued to be the approaches undertaken, which did not take into consideration the real informational needs of the organisation and/or the methodologies used by engineering systems which are not powerful enough [Bloomfield & Best, 1992: 534].

The reason for all these problems surrounding the development of IT in organisations could be focused on a limited spectrum of development issues. Technology and engineering are interested in action directed to a defined end, and whether it is successful or not [Checkland, 1978: 100]. This fact has been depicted clearly by the picture of systems engineering, which is more focused on achieving a good design rather than analysing the way they went about achieving it [ibid. 101]. The picture of systems engineering is drawn up by conceiving, designing, evaluating and implementing a system to meet certain defined needs [ibid. 102]. Thus, the core of CSA is the single idea that a real-world problem can be formulated in the way of defining both the actual and desired situation and then proposing and selecting the best way or alternative solution. The way of getting from the current situation to the desired situation, in order to meet the objective desired [ibid. 108]. Checkland argued that the belief that real-world problems can be formulated in this way is the distinguishing characteristic of all hard systems thinking.

According to all explained above, it seems to be that the perception by analysts of the system (problem situation) in study, can be expressed by a positivist point of view. The knowledge about the system can be an explicit statement of laws and facts that are positively corroborated by measurement [Lyytinen, 1987: 9]. So, it is possible to suggest that there are two main weaknesses of the CSA methodology. They are as follows:

Firstly, CSA and the different methods and techniques associated with it are concentrated more on the design of both the applications [Lyytinen, 1987: 4] and the implementation of IT in an organisation, which merely focus on technical issues, ignoring some factors that influence its use. For instance, social and political matter of the organisation, changing in the organisational environment, etc. Davies and Wood-Harper (1989) argue that most of the current methodologies are concerned with computing science, where the 'framework is essentially technical in nature and is concerned with the technical aspects of work or data flow' (pg. 63). The principal view related in information/data as an entity and the world of computer programming. That is, when information is treated as an entity which operates in logical form within the technological necessity of fulfilling the organisational goals. In relation to the systems life cycle, a computational view focuses the problem in structuring the concept of the problem to be capable of representation and manipulation of the information through a computer-based technology (pg. 62). We can appreciate that these views offer little consideration to the political and social elements of the development of IT in an organisation.

Secondly, the understanding of both the problem situation and its possible solution is made from both a positivist and a reductionist point of view. It is an inadequate conceptual base to deal with the complexity surrounding the development of IT in organisations. Reductionism (both in science and in design), 'is not able to deal with the holistic transcendental character of phenomena' [Fuenmayor and Lopez, 1991: 446]. Structured methods address technical complexity, using tools of reductionism, rather than recognising sociotechnical complexity [Vidgen, 1997: 21]. This position neglects the need for a widely systemic approach to deal with the problem situation. Winograd and Flores (1986), claim that a rationalistic tradition led to many of the problems created by the use of computers, as it presents a blindness to the nature of human thought, leading to a broad misunderstanding of the role that will be played by computers (pg. 8).

Some of the consequences of these issues, as expounded earlier, are reflected in the failure to meet the needs of management or organisational requirements; designs made just to replace current processes or tasks, rather than reach a possible improvement of the task or process [Avison, 1992: 21-26]. The solution may focus on just one functional area or subsystem with no interrelationship with the rest of the organisation or without providing the possibility of future interrelationships, given inflexibility in the IT systems developed.

Summarising, the limitation of structured methodologies to deal with IT complexity in organisations has caused most of the failures in its implementation. The principal characteristics of these failures lie in the application of the scientific paradigm, as a ground of the IS development, which neglects dealing with human situations in an organisation. On the other hand, an organisation is better conceptualised via human activity systems, acknowledging shared human characteristics such as beliefs, values, feelings, motives, aspirations, relationships, cultural norms, etc. Therefore, it is important to explore the relationship between IT and human activity in the development of IT in an organisation. The next section will present an insight of the development of IT in organisations from a 'soft' tradition point of view, as it presents a fundamentally different view of human action.

3.   SOFT SYSTEMS APPROACH AND THE DEVELOPMENT OF INFORMATION TECHNOLOGY IN ORGANISATIONS.

The Soft systems tradition has a provenance in Churchman's work, an enquiring system (1971). Significant characteristics of Churchman's work lie in the emphasis of participation ('sweeping in'), and a recognition that any world view is restricted. He argues that no optimal design of a part of a system is possible without prior knowledge of the 'whole' system [Churchman, 1971: 42]. For him, systems are not real entities existing 'out there' waiting to be identified [Flood & Jackson, 1991b: 126]. Rather, systems are whole systems judgements. This suggests a process of enquiry, which focuses on taking into account different people's perceptions of the problem situation or systems. This fact allows a process of contrasting ideas and debate in order to understand the situation problem. This general framework led Checkland to conclude that there are two paradigms in systems thinking, the hard and soft paradigm. As it was explained earlier, in the hard paradigm, the real word is assumed to be systemic and the methodologies that are used to investigate that reality are systematic [Flood & Jackson, 1991a: 170]. Soft paradigm sees the world as problematical but the process of enquiry into it may be systemic [ibid. 170]. Checkland and Scholes (1981) define systemic as the properties, which refer to the whole and are meaningless in terms of the parts that make up the whole (emergent properties). Fundamentally the soft tradition portrays a different view of human action. It rejects the goal-seeking model of human behaviour. Therefore, the core of the 'soft' tradition concerns a debate about possible courses, which might be followed and the relationships they will affect [Checkland, 1998: 47].

3.1 Some Consideration of the Use of SSM in the Development of IT in Organisations

Information Systems should be concerned with a broader perspective [Checkland and Scholes, 1981: 53-58], rather than being concerned with just developing the specification of a perceived problem. IS should involve human activity, with attention to purposeful action which the IS serve [ibid. 307]. They argued that in order to create an IS, it is necessary first of all to understand how the people in the situation conceptualise their world. Secondly, to find out the meaning they attribute to their perception of the world and hence understand which action in the world they regard as sensible purposeful action, and why. These latter points are incomes to build the human activity systems. The boundary of an IS must be based upon human activity and should have a designer guarantee, with a proactive user's participation, that users can interpret that manipulated data into an attribute organisational meaning.

Now, one can appreciate that dealing with Soft Systems Methodology - SSM[4] requires a different understanding of the notions of systems by the practitioners and the stakeholders of the systems. It could be said that analysts require experience; they must be aware of this fact, addressing activities that permit both the analyst and persons involved in the project to internalise these concepts. On the other hand, methodologies of expression, such as the rich picture, provide a useful starting point for the identification of stakeholders and for gaining a pre-understanding of the problem situation [Vidgen, 1997: 45]. The construction of many root definitions may address the problem that clients and owners do not necessarily have a clear understanding of the problem situation [Jayaratna, 1991: 66]. SSM provide all concerned, including the analysts, opportunities to understand and to deal with the problem situation. The analysts are perceived as being involved in the problem situation [Avison, 1992: 249]. Furthermore, as the core of the 'soft' tradition concerns debates about possible courses of insight into the problem situation, given recommendations and taking action, SSM users need to have a high degree of interpersonal skills [Jayaratna, 1991: 67].

Vidgen (1997) argues according to his experience using SSM in action research, that SSM has been used to conceptualise the problem situation on the grounds that it provides an approach for addressing pluralism and complexity. However, SSM does not provide adequate guidance for addressing the role of technology and the specification of requirements that would inform the design and implementation of a computer-based

---

[4] Checkland & Scholes (1981) define Soft Systems Methodology (SSM) as a methodology that aims to bring about improvement in areas of social concern by activating a learning cycle for the people involved in the situation, which is ideally never-ending. The learning takes place through the interactive process of using systems concepts to reflect upon and debate perceptions about the real world. Systems ideas are used to organise thought about a problem situation [Vidgen, 1997: 23], which permits the analyst to focus on problem formulation and what needs to be done rather than problem solving.

tool. Vidgen's appreciation can be supported by Avison's (1992) ideas, who argues that 'SSM provides tools for using in particular situations at particular times in the development of IT in organisation' (pg. 249). One possible way that SSM can be fitted into IT development in an organisation is by using it as a 'front end', before proceeding to the 'hard' aspects of systems development (pg. 250). Avison & Wood-Harper (1990) draw upon SSM in the early part of the system definition process and combines the rest of the project with structured methodologies. They argue that SSM concerns analysis, whereas the hard methodologies tend to emphasise design, development and implementation.

In summary, SSM, which is based in an organised set of principles, which guide action in trying to manage real-world problem situations, is in my understanding a potential alternative to addressing the development of information technology in an organisation. The fact that SSM is systems-thinking based and is applicable to taking purposeful action to change real situations, its use permits to practitioners to have a better understanding of the problem situation. SSM deploys first of all a holistic view of the problem situation, and secondly it needs to generate a deeper process of analyses. An intellectual reasoning is required in order to build the relevant systems of the problem situation. But applying SSM to the development of IT technology does not present a clear solution to the gap existing between the conception of the problem situation and the construction of an application, which at the end of the day, in the eyes of most of managers and analysts is the palpable solution. However, it is true that implementation is to a lesser extent not as technical as design and construction, as it also concerns social matter; and that the specification may be supported by formal methods in order to maximise computer based-technology resources. It seems that the complexity of the development of IT in organisations is regarded in the complexity between the understanding of the problem situation and the design of the IT system, i.e. bridging from social analysis to IT design. The Multiview methodology (Avison & Wood-Harper, 1990) differentiates IS issues between what is organisational in nature and what demands technical or computing approaches are needed.

4. CRITICAL SYSTEMS THINKING AND THE DEVELOPMENT OF INFORMATION TECHNOLOGY IN ORGANISATIONS.

Although a positivistic and reductionistic tradition has been the basis for a great deal of technological progress, the limitations of structured methodologies to deal with IT complexity in organisations have caused most of the failures in IT implementation. The scientific paradigm as the framework for IS development neglects the human aspect of an organisation. Organisations are better conceptualised as systems. The complex relationship between an individual's behaviour and their role within a system defines the energy and dynamism of the organisation. The hard tradition sees the organisation as a goal-seeking entity. Thus, the role of IT (focused merely on information) is to aid decision-making. This conceptual model sees human and organisational behaviour as only decision making and problem solving in pursuit of goals. This is limiting in its definition of behaviour. Therefore, a different method of dealing with these kinds of problem situations is necessary. An insight to the contribution of 'soft' tradition methodologies to the development of IT in organisations was explored, as it presents a fundamentally different view of human action. Soft systems methodology (SSM) presents a broader perspective concerning information systems, which goes beyond merely developing the specification of a perceived problem. SSM involves human activity, with attention to purposeful action, which the information systems serve. SSM provides all concerned, including the analysts, opportunities to understand and to deal with the problem situation, as the core of the 'soft' tradition concerns the debate about possible sources of insight into the problem situation. The fact that SSM is systems-thinking based and is concerned with taking purposeful action to change real situations permits practitioners to have a better understanding of the problem situation. Thus, SSM is a potential alternative to addressing the development of information technology in an organisation. However, it appears that the complexity of the development of IT in organisations is regarded in the complexity between the understanding of the problem situation and the design of the IT system, i.e. the bridge from social analysis to IT design. Thus, a new perspective of the development of IT in organisations seems to be necessary in order to deal with these kind of projects. In summary, in the development of IT in organisations, the positivist and reductionist paradigm that underpins the hard tradition limits the domain of it's applicability. And the soft tradition neglects dealing with the technical context of the IT development in organisations. The complexity surrounding the development of IT in organisations presents a variety of contexts within problem situations, which first of all, they are present in an inextricably interlinked way and secondly, they belong to different domains (hard and soft). It seems to be then, that the domain of the development of IT in organisations is neither from a hard systems tradition nor from a soft systems tradition.

The principal purpose of this section is to introduce the new tradition: *Critical Systems Thinking (CST)*, and see how this approach can enrich the development of IT in organisations. This section will discuss the benefits to the development of IT in organisations of this approach, through a reflection of its three commitments: *Critical awareness, Emancipation* and *Methodological Complementarism*, and the theory and practices of the development of IT in organisations.

The emergence of Critical Systems Thinking (CST) has brought a new tradition of thinking to management and systems sciences. Flood and Jackson (1991b) claim that CST shares the soft tradition critique of hard approaches, but generates the possibility of good use of such approaches through a reflection of the context in which they would be employed (pg. 1). CST possesses three fundamental commitments[5]: *Critical awareness, Emancipation* and *Methodological Complementarism*. These commitments are summarised by Flood and Jackson (1991b: 2) and Flood and Romm (1996: 81), as follows:

Critical awareness: To critique or reflect on the relationship between different organisational and societal interests and the dominance of different theories and methodologies

Emancipation: To develop system thinking and practice beyond its present conservative limitations and, in particular, to formulate new methodologies where the operation of power prevents proper use of systems approaches.

Methodological Complementarism: To reveal and critique the theoretical (ontological and epistemological) and methodological bases of systems approaches, and to reflect on the problem situations in which approaches can be properly employed and to critique their actual use.

---

[5] Five commitments or interrelated intentions were identified by Jackson (1991): critical awareness, emancipation, Complementarism at the methodological level and at the theoretical level and social awareness. As methodologies embody theoretical assumptions, the two forms of Complementarism was reduced to methodological pluralism. Social awareness becomes an implicit part of the commitment to emancipation.[Midgley, 1996: 11].

Midgley (1996), who realised that there were different definitions of CST, argues that 'CST can be seen as an evolving debate around a set of themes that are considered important by some systems practitioners' [Midgley, 1996: 12]. Based on this statement, the following sections will present the principal features outlined by some interpretations of CST in order to present a reflection of the implications of the three commitments for the development of IT in organisations.

4.1.1   Critical Awareness Commitment

Midgley (1996) shows critical awareness to be in support of the commitment to methodological pluralism and of the commitment of emancipation. It is primarily based upon an understanding of the strengths and weaknesses and the theoretical underpinning of available methods. Secondly, critical awareness aids in our attempts to understand both the context of application and the possible consequences of using various methodologies. Finally, it provides a close examination of the assumptions and values entering into the design of the solution. According to Midgley the support of the commitment to methodological pluralism relates two critical issues: critical thinking about methodology - development of effective metatheories, and critical use of the methodology or methodologies selected, where the focus of critique is the context of application. The support to the commitment of emancipation is, in fact, a statement that power relations can be understood and improvement defined within an ethical framework. The following sections will utilise critical awareness to present both methodological complementarism and emancipation. These will be examined to help reflect the applicability of CST in the development of IT in organisations. To do this, the principal features of both commitments will be outlined and the relationship between each commitment and the development of IT in organisations will be discussed.

4.1.2   Methodological Complementarism Commitment.

Methodological Complementarism suggests reflection upon the variety of different contexts present in a problem situation and its relationship with the range of approaches that are available to undertake it. That is, questioning the methods, practices and theories [Schecter, 1991]. In this sense, there is not one single approach which can deal with all the problem situations and all systems approaches are able to be used, through reflecting upon the strengths and weaknesses of hard, soft and emancipatory approaches in order to address a problem situation.

Jackson and Keys (1984), through a systematic analysis of problem contexts and the identification of the methodologies most suitable for the different contexts, provide a 'system of systems methodologies'. This is based on the idea that methodologies from different paradigms make particular assumptions about the contexts within which they will be used [Mingers, 1997: 492], which support the direction of intervention in the real world. This suggests a complementary manner of addressing the complexity of the problem situation [Midgley, 1996: 13]. In other words, it aligns 'problem situation with problem-solving' [Gregory, 1996: 39]. The problem contexts are constituted by both the level of complexity involved in understanding the system and the situational context within which the decision maker finds himself. The decision maker's role revolve around the maintenance and efficiency of a system. The system and the decision maker are affected by changes in the nature of that context. The management and improvement of the system relates to two distinct, but interconnected processes: planning (development), and control (evaluation). Based on this framework, Jackson and Keys expose a classification of problem contexts in connection to solving methodologies, arguing that the variety of methodological assumptions allow for the decision maker to address the variety of problem contexts and use the most applicable methodology for that context [Midgley: 1996: 13]. Flood and Jackson (1991b: 326), design a metatheory of systems methodologies for use as a tool provided by total systems intervention[6] (TSI) to help with the choice of an appropriate systems based intervention methodology to fit specific organisational contexts. This metatheory claims that different whole methodologies may be used within the same intervention to deal with different issues or to provide different points of view [Mingers and Brocklesby, 1997: 492]. According to Midgley, the focus of the two latter works is on how methodologies can most appropriately be used in practice [Midgley, 1996: 13]. Mingers[7] (1997) presents a third combination. This is when methodologies are split or partitioned into components and they are combined together to construct an ad-hoc multimethodology suitable for a particular problem situation.

Flood and Romm (1996) express Methodological Complementarism as an attempt to preserve diversity in theory and methodology. This became a discussion about the possibility of the joining of methodologies (commensurability) at both theoretical and practical levels [Flood and Romm, 1996: 83]. Commensurability is defined in terms of being measurable by a common standard. On the contrary, incommensurability suggests that there is not a common standard, so there is no possible way of comparing theories. This has been criticised of isolationism - the belief that only their version is the truth, i.e. the use of a single methodology in all problem-solving circumstances. Flood and Romm argue the possibility of choice because there is a diversity of management, i.e. management take into account wider matters of contemporary concern. This is an ongoing theoretical debate. As a consequence of this, choice-making is totally desirable, providing it is dynamic and widely informed in order to support the choice. The next section develops a reflection between methodological complementarism and the theory and practices of the development of IT in organisations.

4.1.2.1   Methodological Complementarism Commitment and the Development of IT in Organisations

Throughout the history of the development of IT in organisations, the system life-cycle has been a stable feature, which presents a logical structure of the phases which the designer or developer of the system has to plan, control and finally carry through. This methodology can be summarised in a basic form, as a consequence of three main consecutive phases: analysis, construction (design and program codification) and implementation. On the side of designers, by completing the practice in these stages, they are dealing with both the richness of the classical problems of the organisational design and management and the problematic surrounding the construction of the system itself.

The development of IT in organisations is involved in many of the classical problems of an organisation, where relations of power and political aspects of organisations and their environment are present. Regarding Churchman's dialectical inquiry (1971) one can appreciate that the development of IT in organisations is teleological. It has a set of goals, which are related with the IT implementation in its

---

[6] Flood and Jackson (1991b: 321) defined TSI as an approach to planning, designing 'problem-solving' and evaluation.

[7] Mingers (1997) wonders the nature of the relationship between multimethodology and CST. He argues that multimethodology has much in common with others CST such as TSI, which delivers its own set of commitments.

specific context. It seems that these sets of goals have a strong relationship with the areas of knowledge involved; that is, in the domains of IT and management science. In addition, although these kinds of projects tend to serve the interest of the persons who pay the designer; in the analysis phase, participation and interrelation between analysts and clients is necessary. Ideally, within a participatory approach, everyone should be involved in the decision making process. However, frequently in the practice the decision maker is defined by those who exercise power, who are ultimately important when dealing with the creation of participatory fora, as they are, at the end of the day, who the designer or developer wants to serve.

In the development of IT in organisations several components are present. They could be as follows: Culture, values and policies of consultant firm (if this is required). Culture, values and policies of the organisation as such; different knowledge involved (e.g. IT, organisational and management knowledge). IT itself, which is represented by

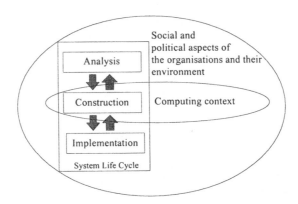

Figure 1. Relationship between the system life cycle and its contents

the suppliers and finally the needs and skills of the persons involved in the project and affected by the implementation. On the other hand sometimes the IT environment exercises a pressure upon technical decision making, e.g. acquiring a computer-based determined technology just for the sake of being the most sophisticated and modern. Relating to the construction of the IT systems, the analysts who design and construct the systems base themselves on formal methods. In the IT community and computing context, formal notation and measures oriented to technical matters are relevant in order to produce quality software.

According to the above, one can appreciate that a variety of contexts are present in the development of IT in organisations. And the principal activities of the three main phases of the system life cycle are related to a specific context. That is, the analysis phase deals with the richness of the real world depicted in an organisation; the view of the phase of construction is concerned more with a computational context where formality is relevant. And finally, in the implementation phase the implications of the decision-making done are appreciated. The following graph represents the interrelationship between the systems life cycles and their contexts.

This view allows us to appreciate the complexity surrounding the IT development in organisations. There exist a variety of complex contexts with different agents involved, associated with the main stages of the system life cycle. In order to deal with this complexity, Methodological Complementarism could be useful. The previous sections have been concerned with the problematic of the development of IT in an organisation from hard and soft systems traditions. The nature of the hard thinking approach both shapes and limits structured methodologies, as it neglects dealing with human situations in an organisation. A broader perspective, which involves human activity and its relationship with the purpose of the development of IT in organisations, was presented through an understanding of the soft tradition and its contribution to this theme. But this tradition does not present a clear solution to the gap existing between the conception of the problem situation and the construction of an application. The domain of the development of IT in organisations is best approached neither from a hard systems tradition nor from a soft systems tradition. Hard and soft traditions provide a means to address the development of IT in organisations from two different paradigms, which are based on empiricist and hermeneutic methodological foundations respectively. Methodological Complementarism, through a critique of the theoretical and methodological bases of systems approaches, supplements a reflection on the context situation in which the IT development is immersed. This proposal allows us to combine these paradigms to build a temporarily suitable approach for the particular problem situation. Methodological Complementarism, presents an interesting approach to deal with the complexity of these kind of projects, and allows application of an appropriate systems based intervention methodology to fit both the specific organisational context and the computing context. For instance, Avison (1992) shows a comparison based on the scope of current methodologies and the coverage of the systems life cycle, where one can appreciate the strengths and weaknesses of some methodologies dealing with the different stages of the systems life cycle.

4.1.3   Emancipation Commitment

The critical concern of communicative distortions, false consciousness and other ideological distortions, allows a recognition of coercive contexts [Schecter, 1991: 218], Therefore, new developments in methodologies focus on 'improvement'[8] [Midgley, 1996]. Improvement, which is defined dynamically and locally, takes issues of power into account. According to Jackson (1991) CST is dedicated to human emancipation and thus is concerned with improvement. For instance, Critical Systems Heuristics is classified as an explicit emancipatory method[9] for the critique of coercive situations,

---

[8] Emancipatory practice in CST has taken into account the Operational Research (OR) movement. Schecter (1991: 219) argues that emancipatory OR methods must: serve the oppressed, striving liberation and social justice, support all those who are involved in the problem situation and secure a process of transforming oppressive social systems.

[9] Ulrich (1996b) claims that critical heuristics does not understand itself as a self-contained 'method' of planning. Rather it seeks to complement and to changes others approaches in order to emancipate citizens from those who practise the approaches.

which aims at promoting a practice of systems thinking [Ulrich, 1996a: 177]. It is designed to help people think critically about the systems design, which affects them [Schecter, 1991: 218].

4.1.3.1   Critical Heuristics of Social Systems Design

Ulrich (1991) shows Critical Heuristics as an approach to both systems thinking and practical philosophy. This methodology provides a support to the problem of practical reason (pg. 104) - the claim to secure improvement and to be rationally justifiable. For Ulrich, the principal problem with applied science lies in the 'normative content'. This is value judgements (recommendations for actions, design, models, planning standards), and also the practical consequences and side effects of the scientific proposition in question for those who may be affected by their implementation. He critiques the 'ideal' models of practical discourses (e.g. Habermas' Ideal Speech situation), as it does not take into account that the reach of an argument is related to its premises and makes conclusions that it cannot question and justify any further. He called it the 'inevitability of argumentation break-offs' (pg. 104). Ulrich bases this concept on the concept of boundary judgement of systems science. Boundary judgements represent the

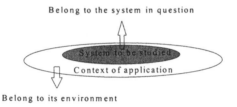

Figure 2. Justification break-offs as Boundary Judgments

assumptions that whenever we apply the systems concept to the real world, one must take into account what belongs to the systems and what belongs to its environment (pg. 106). The next graph attempts to show the argumentation break-off as boundary judgements. This graph is adapted from Midgley *et al* (1998). What belongs to the system in question relates to the whole systems judgement. The designer's boundaries concerning the system in study, how it can be improved and what falls outside the reach of this effort are helpful to define the problem situation. What belongs to the environment concerned within the context of application, justifies the normative implication of a design for those affected by its effects.

Ulrich argues that in order to deal with the problem of boundary judgement, a merely technical modelling point of view is inadequate. On the contrary, he argues that we cannot understand the meaning of the model if we do not understand the model environment, and that all the citizens, involved in the process or not, '*ought*' to be regarded as being part of the context of application. Critical Heuristics presents the measure of the rationality of a systems design by the degree to which it renders the justification break-offs enabling both those involved in and those affected by the design to reflect on the validity and legitimacy of those break-offs [Ulrich, 1991: 107]. Critical Heuristics presents twelve boundary questions split into four groups; each group contains three kinds of categories. The four groups ask for the value-basis, for the basis of power, know-how and legitimisation respectively. The three categories of each group relate social roles of the involved or the affected, the role-specific concerns and key problems or issues in determining the necessary boundary judgements relative to the two previous categories. Ulrich shows that the critical relevance of these categories and corresponding boundary questions may best be seen by contrasting each *'ought'* with the pertaining *'is'* judgement [Ulrich, 1991: 108-109]. This emancipatory methodology enables the designer to acquire a better understanding of the meaning of boundary judgement. Ulrich (1991) claims that it enables us to learn to see through the 'objective necessities' of the others (pg. 112). In this sense, the use of the Critical Heuristics approach concerns ethical issues. Analysts will be aware of the social and political implications of IT implementation for those who are involved in the problem situation and those who will be affected by it. The applicability of the boundary judgements argument in problem situations concerning the implementation of IT in organisations will now follow.

4.1.3.2   Improvement and the Development of IT in Organisations

Designers and developers of IT in organisations, for instance IT consultants, claim to have technical expertise but they are far from being mere technical experts, so they stress when speaking about organisations and IT. Bloomfield and Danieli (1995) draw some stratagems deployed by IT consultants as follows: first of all, making themselves indispensable to clients, creating a niche and persuading clients that they are within it, they seek to portray themselves as an obligatory passage point if IT implementation is desired. Secondly, the consultants' profile has to be maintained during the consultancy engagement. These stratagems are constitutive of the role of consultancy practice. IT consultants represent the category of change agents. They manage a special vocabulary that seems to be the same as the language of a manager, which managers believe adds value to the organisation. But rather, this organisational knowledge expressed by IT consultants shows that problems could be redefined in terms of existing IT solutions [Bloomfield and Best, 1992: 544]. The problems are formulated and the solution (pre-established) is proposed.

Such an assumption that IT consultants make interventions in the organisation with a pre-established solution which could create bias due to the fact that the problematic situation under study could be understood only from an IT perspective. That is, the intervention would focus on adapting problem issues to pre-established IT solutions, which could be the result of previous experiences in other companies. From improving the organisation as a system, *boundaries* of analysis take crucial importance [Midgley *et al*, 1998]. From the context shown above, improvement of the organisation would not clearly be seen, because the boundary of the system could be limited by prior assumption that solution is regarded only in the IT domain. In order to understand the meaning of the system to be studied by analysts, an insight of its contexts of application is needed. This means that an increase in the scope of the boundary is necessary to take into account all those who are involved and affected by the implementation of IT. Obtaining the objective necessities of these agents let to analysts to have a wider view of the problem situation but also come into view the implications of the decisions made in the context of application. The need for considering which aspects of the problem situation should be taken into account, and which should be left out, in order to approach the situation adequately is particularly difficult to work. In addition the meaning of the system can be affected by the positivist conception of the way of tackling these kinds of problem situations. When the principal concern is focused on the '*is*' question rather than the '*ought*' question, this is symptomatic of a limited boundary judgement, and limits the understanding of the problem situation. In this sense, the solution provided could be biased, as it doesn't include the real needs of the organisation and the implications of the design. This position disallows reflection by those who must be involved in or are affected by the system in study of the problem situation. Furthermore, in the practices of the development of IT in organisations a rational justification of the meaning of the system to be studied causes a limited participation of those who are involved in the problem situation, as the formulation of the '*is*' question may be addressed to particular agents. Those who can exercise power do so with the consent of the designer or developer.

Information Systems development is a multidisciplinary area [Gilligan, 1998]. Rather than solely concerning itself with the implementation of computer based information systems, disciplines as such as the humanities, management science, sociology and accountancy are needed. The evaluation of the achievements of these kind of projects in organisations has been addressed throughout the life cycle of the software and commonly includes factors for measuring financial and technical matters, rather than social and organisational aspects of the problem situation, even though they are very important.

4.2   A Reflection of the Development of IT in Organisations from a Critical Systems Thinking Perspective.

The practice of development of IT in organisations might deal with coercive contexts where communicative distortions, false consciousness and other ideological distortions are present. However, in the practice of analysts, emancipation may be a utopia - something theoretical and separate of the view that they have of the real world of the development of IT in organisations, which could be biased by their theoretical foundations of IS. The next part of this section concerns a reflection of the problematic surrounding IT development in organisations from a critical thinking perspective.

Although there exists a clear dependency of IS on the organisational and societal domain, it seems that IT tends to affect a complex of conditions which defines a sphere of life in modern society. For example, Klein and Hirschheim (1987:279-283), show a relationship between some kind of social attitudes and IS. Attitudes have changed regarding work and its relationship to the family, authority and power, the nature of organisation, planning, learning, knowledge, information, science, technology and perception of reality, and so on. These are some examples that depict the relationship between the use of IT and social attitudes. On the other side, the development of IT in organisations is composed of *means* (methods, methodologies, approaches utilised to achieving IS ends); e*nds* (that dominate the organisation); and *social relations*: either a focus to distribution of equalities/inequalities in terms of privileges and power [Kendall & Avison, 1993: 6], or a possible mechanism of domination. The two latter appreciation of the implication of the use of IT in social domain show us the presence of power relations in IT development. For instance, within the process of IT development in organisations among systems professionals and users, power is being exercised [Bloomfield and Best: 1992]. Such kinds of power can be manifested throughout the relationship between each party in the IS development process. This relationship is based on mutual negotiation, which in terms of Habermas' view, would be possible only in the context of an ideal speech situation which could be distorted by power or for example by self-interest of any party. Now both parties, systems professionals and IS users, have their own knowledge of the problem situation, which supports both their world picture of the problem situation and the way of tackling this problem situation. On the one side, is the professional knowledge and skills concerning IT. And on the other side, users that, although their knowledge of the IT domain may be limited, their knowledge of the organisation is better. Here is presented a narrow relationship between power and knowledge, which could be reflected at the moment of the intervention by the analysts.

Habermas' theory of Communicative Action has been applied in IS[10]. Habermas' Communicative Action Theory is based on an undistorted communication that he calls the ideal speech situation, which permits rational communication - All participants have the same chance to choose and to apply speech acts [Habermas, cited by Alvesson, 1996:144]. IS could play the role as a facilitator to provide opportunities for participation. IS has the potential to support communication among the organisation's members. But at the same time, it affects the ideal speech situation. Somehow, the Internet has facilitated individual communication across interest groups by way of Usenet. But in addition IS involves many of the classical problems of organisational design and management [Lyytinen, 1992: 163], drawing the structure, goals, norms and laws of the organisation. Moreover, IS could support processes of dialogue amongst different organisational agents (clients, stockholders, employees, etc.) through use of, e.g. a corporate Intranet. Now, from the point of view that an organisation can be understood as structures of distorted communication [Alvesson, 1996: 150], IS would be seen as an element that converges the process of communication towards a communication distorted by the possible relations of power that the Information Systems themselves embody.

Some of the sources of distortion are as follows: first of all, it is possible to identify someone who has better access to certain information based on the power of his/her position in the organisation. This inequality of access to information (through the IS) merely reflects the structure and division of labour of the organisation, so the definition of the boundary changes according to the user's position. Secondly, knowledge of the IT domain could create an inequality to those who have cognitive limits. These cognitive limits would be shifted with training, but in some communities their idiosyncrasies affect this communication [Forester, cited by Alvesson, 1996: 151]. For example, some farmers in Colombia reject the use of IS to support their production, due to the fact that they think this kind of technology does not concern the nature of his/her work. But they have to deal with, for example agricultural institutions, that use IS to control and manage their position in the global market. Finally, information needs reflected on the IS are given by just choice agents of the organisation, who draw a monopolistic creation of needs, maintaining the old organisational structure, which does not permit a real consensus in the origin of the system in order to accept social norms within the organisation. Thus, it is impossible to assume that the actions or practice implied by the speech act are consensually valid.

---

[10] See Kalle Lyytinen [Lyytinen: 1992], who present how primary action of Habermas' theory: instrumental, strategic, communicative and discursive, are prominent in IS, which has served as means of raising research's awareness of the ontological, epistemological and ethical dimension of IS. And, John Mingers [Mingers: 1997], who considers pragmatic aspects of Habermas' theory to draw the relationship between information and meaning.

Another perspective of power relations and IT could be based on Foucault's concept of power, which presents a different conception of power to Habermas' concept of power. But both seem to be prevalent in modern society and are frequently in discussion in contemporary academic studies. Foucault, with a conception more of all pervasive power, argues that the individual is relatively recent object of knowledge [Townley, 1994: 83]. To know individuals as a means of control and method of domination (Discipline). This identification of the individual can be achieved through *examination*, which provides the basis for judgement and measurement [ibid. 83]. Examination is a disciplinary process, which make the individual an object of both knowledge and power [ibid. 86]. It can be reached through looking closely or analytically, and judging according to a rule. Now, IS could support the link between power and the construction of knowledge. In addition, IS presents information processing capabilities - information processing tools, as one of its major roles in organisations, allowing features in monitoring and recording performance of employees and organisational units [Lyytinen, 1992: 161]. IS provides a means to direct attention to important aspects of organisational performance and reduce possibilities for covert action. IS providing mechanism to evaluate and measure the performance of the organisational agents. In this regard IS portray a mechanism of power. It is widely accepted that organisational standards, policies, systems of measure are embedded within IS, with which individuals started to be known in terms of these marks according to their performance achieved. Therefore, IS could be see as a mechanism through which individuals may be view, judged and compared. That in term of Foucault view, it constitutes the individual as effect and object of knowledge and power [Townley, 1994: 85]. Appreciating the meaning and effects of IS, as a technology of control may call for a consideration of a wider context in which they could be employed.

In these two appreciation of power relations and IS, it is very important to be aware that there are problems regarding who defines the systems [Robb, cited by Romm, 1997: 25], and the purpose for which it is created, as well as the aims of those who control them. IS are seen as a locus of information processing and as such could be understood that it would not have any power. However, Information Systems are dynamic, they express what the organisation wants to be, i.e. Information Systems are legitimating the organisation, reflecting its energy, problems, etc. In this sense, 'information moves from a structured data allowing information to be entified' [Boland, 1987: 373], to be treated as open to ongoing discursive repackaging, regarding information as meaningful rather than factual [Romm, 1997: 34]. Now, if IS validates the organisation, the organisational power-relations would be present and furthermore, IS could be an able to influence this power- relationship.

An emancipatory approach requires changes in the epistemological view of the development of IT in organisations by all those who are involved in these kind of projects. The prevalent role of the analysts must be changed to make rich their knowledge about the problematic surrounding IT. Insofar, it has appeared that analysts are able to both understand the epistemic principles of his/her role as such and the epistemic principles of the methodologies, the way of tackling the problem situations and being conscious of the implications of this. Analysts could be, in terms of critical theory, free of the self-delusions and self-imposed coercion of prevalent ideology that underpin the development of IT in organisations, and then be a facilitator to induce this process of emancipation in organisations. The positivist point of view may omit issues and values causing a loss of guidance to tackle the richness of the classical problems of the organisational design and management. Thus, human action, and the social and political aspects of the organisation would gain more importance. The predominant view of the development of IT in organisations focused in technical matters should be one of the first ideological constraints to be contended by analysts, but also by the people in charge of promoting IT in organisations. The manager's role should be more proactive.

It is commonly accepted that manager participation in the development of IT in organisations can be split in two big areas. Firstly, in both the conception and decision making about starting IT projects and at the end of the project in order to both approve and pay according to the results. Secondly, as a user of the system in which the manager is embedded into the methodology used as a client of the part of management systems information. [Mike Harry, 1998]. But the manager's role should be more active and not over-burdened with responsibility to ensure a good decision just on the knowledge of the analysts, what analysts say should be a starting point for interpretation and not the final word. Thus, the manager's role: act proactively in managing IT change within their own organisation [Kendall and Avison, 1993: 14]. Managers must act to recognise their weaknesses and ask for assistance to increase their cognitive limits in the IT domain. But rather they must ensure a process of knowledge and understanding of the problem situation, so that improvements become apparent to the people involved. A participatory democratic approach will be useful to reach the needs of the involved people. On the other side, IT is portrayed as indispensable to develop organisational activities, originating in a dependence of the organisation with IT. For example, some IS researchers [Dutta: 1997, Stein & Zwass: 1995] see the use of IS as knowledge management indispensable to guide current activities in organisations. Insofar as IT has been implemented in organisations, it has left IT as something completely necessary to support the daily operations of the company. In some cases, this dependency is reflected in the operation of a company and employee performance, which is drastically affected when its computer base technology and/or information systems infrastructure is down. But, IT is perceived as a merely technical concern, only involving analysts, leaving it to technical areas. This predominant ideology disallows a better understanding of the use of the IT in organisations. In this sense, an emancipatory approach provides the concerned people with a position to determine their true interest about IT and their relationship with the organisation. The importance of acquiring this broader knowledge is necessary to support the aims and ideals of emancipation.

The development of IT in organisations links issues which are presented in social and political aspects of the IT implementation, where power relations and ethical issues have to be taken into consideration. In addition, social and political aspects taken a great importance in order to have a better understanding of the phenomena.

5.   CONCLUSIONS

The purpose of this section is to bring together the principal subjects of the development of IT in organisations. The subject matter of the article concerns an exploration of the problematic surrounding the development of IT in organisation and consequently the understanding of both the role of IT in organisations and the social and technical interaction between the technology and the organisation. The framework for the analysis is based on the main issues of the systems thinking tradition and the development of IT in organisations.

In the development of IT in organisations complexity is present in various shapes and forms and this poses a variety of difficulties which need to be addressed. Hence, one of the major assumptions made by most of the persons involved in the development of IT in organisations - that the use of IT is only a technical matter -, must be removed. On the contrary, some of the different issues are concerned with the computer based-technology, both hardware and software, human and organisational behaviour, the social construction that IT itself presents and the different rational processes that are undertaken, and the tools and techniques that guide the solution of the problem situation. These issues, rather than being involved in the problem situation in an isolated way, are in fact strongly interlinked.

In this sense, the domain of the development of IT in organisations is best approached neither from a hard systems tradition nor from a soft systems tradition. Hard and soft traditions provide a means to address the development of IT in organisations from two different paradigms, which are based on empiricist and hermeneutic methodological foundations respectively. The argument here is that the understanding of the development of IT in organisations requires more than a technical approach.

Rather the use of alternative paradigms enriches its understanding. The value of social and political aspects takes relevance. The positivistic and reductionistic tradition has been the basis for a great deal of technological progress, but a positivistic analysis of problems surrounding the development of IT in organisations is focused on a limited spectrum of development issues. The limitation of structured methodologies to deal with IT complexity in organisations has caused most of the failures in its implementation. The principal characteristics of these failures lie in the application of the scientific paradigm, as a basis of IS development, which neglects dealing with human situations in an organisation. However, it does not mean that technical and formal methods lose significance in the development of IT in organisations. Technical qualification by analysts is important in the computing context in which the stage of construction of IS is immersed. But also an insight of the use of the IT and its implications are needed. Then, an understanding of the development of IT in organisations should be concerned with a broader perspective. Rather than being concerned with just technical matter it should involve social and political aspects. The process of inquiry of the soft tradition, where contrasting ideas and debate are relevant in order to understand the problem situation, portrays a different view of human action in the development of IT in organisations. Thus, applying SSM a different understanding of the notions of systems by practitioners and stakeholders is required. SSM has been used to conceptualise the problem situation on the grounds that it provides an approach for addressing pluralism and complexity, giving a better understanding of the problem situation. But applying SSM to the development of IT does not present a clear solution to the gap existing between the conception of the problem situation and the construction of an application.

In summary, in order to enrich the understanding of the development of IT in organisations, the predominant conceptualisation that the role of IT is based only on technical matters should be one of the first ideological constraints to be contended by analysts. Thus, human action, and the social and political aspects of the organisation need to gain more importance in order to understand the social and technical interaction. That is, break the dichotomy between what is science and what is politics in order to have a proper understanding of social and scientific change [Callon et al, 1986]. Furthermore, the manager's role in these kinds of processes should be more proactive. An emancipatory approach through a process of knowledge and understanding of both the role of IT and the problem situation is needed. So, improvement become apparent to the people involved. In this sense, an emancipatory approach provides the concerned people with a position to determine their true interest in IT and its relationship with the organisation.

Next table shows a synthesis of the subject matter of the article and the main issues of the systems thinking tradition and the development of IT in organisations.

Systems Thinking Tradition: Framework for Analysis	The development of IT in organisation	The understanding of the IT role in organisations	The social and technical interaction.
Hard	Its principal concern is	Fitting technology to	Seek a distinction

Systems Thinking Tradition: Framework for Analysis	The development of IT in organisation	The understanding of the IT role in organisations	The social and technical interaction.

	in design and its process of inquiry is based on systematic methodologies. Human and organisational behaviours are seen as decision making and problem solving in pursuit goals.	support decision making in pursuit of goals . IT is used to gain control of a socio-economic environment.	between science and politics.
Soft	Concerns analysis, addressing pluralism and complexity. Its process of inquiry is based in contrasting ideas and debate, in order to interpret the world.	IT is a part of interpreting the world, via creation and sharing of meaning to legitimate social action. Furthermore IT makes sense in the relationships that the IS will affect.	The understanding of the social and technical interaction leads to interpretation of IT in the world.
Critical	There is not single approach which can deal with all the problem situations. Requires normative consensus as a mechanism against distorted communication. A process of knowledge and understanding about IT and its organisational relationship is necessary, and a	IT serves technical, practical and emancipatory interests. The technical and practical role of IT are the same as that portrayed by the hard and soft tradition respectively. The emancipatory approach provides	IT and social relations are interrelated. The construction of IT embody social and political aspects. For instances power relations are presented.

Systems Thinking Tradition: Framework for Analysis	The development of IT in organisation	The understanding of the IT role in organisations	The social and technical interaction.
	reflection by the agents involved in the problem in order to improve, control, mutual understanding and emancipation.	the concerned people with a position to determine their true interest in IT and their relationship with the organisation. IS could converge the process of communication towards a communication distorted, distributing unequal normative power.	

6. BIBLIOGRAPHY

Alvesson, M., 1996, *Communication Power And Organisation*, Walter de Gruyter, New York.

Avison, D. E., and Fitzgerald, B., 1992. *Information Systems Development: methodologies, technologies, tools,* 2nd Edition, MacGraw-Hill, London.

Avison, D. E., and Horton, J., 1992. Evaluation of information systems, *AMS Discussion Paper in Accounting And Management Science* 31.

Avison, D. E., and Wood-Harper, A. T., 1990. *Multiview: An Exploration In Information Systems Development,* Blackwell Scientific Publications, Oxford.

Barry, H.,1996. *Discourses of power: from Hobbes to Foucault*, Blachwell, Oxford.

Bloomfield, B. P., and Best, A., 1992. Management consultants: systems development, power, and translation of problems, *Sociological Review* 40 533-560.

Bloomfield, B. P., and Danieli, A.,1995. The role of management consultant in the development of information technology: the indissoluble nature of socio-political technical skills, *Journal Of Management Studies* 32 23-46.

Boland, R. J.,1987. The in-formation of information systems, In Boland, R. J., and Hirschheim, R. (eds.), *Critical Issues in Information Systems Research,* John Wiley Information Systems Series, London.

Boland, R. J., and Hirschheim, R. A.,1987. *Critical Issues in Information Systems Research*, John Wiley Information Systems Series, London.

Callon, M., Law, J., and Rip, A.,1986. How to study the force of science. In Callon, M., Law, J., and Rip, A (eds.), *Mapping The Dynamics Of Science And Technology: Sociology Of Science In The Real World.* Macmilland, Basingstoke.

Checkland, P., 1978. The origins and nature of 'hard' systems thinking, *Journal of Applied Systems Analysis* 5 99-109.

Checkland, P., and Scholes, J., 1981. *Soft Systems Methodology In Action,* John Wiley & Sons, Chichester, New York, Brisbane, Toronto, Singapore.

Checkland, P., and Holwell, S., 1998. *Information, Systems And Information Systems –
Making Sense Of The Field,* John Wiley & Sons, London.

Churchman, C. W., 1971. *The Design Of Inquiring Systems: Basic Concepts Of Systems And
Organisations.* Basic Books, Inc. Publishers, New York.

Ciborra, C. U., 1987. Research agenda for a transaction costs approach to information
systems. In Boland, R. J., and Hirschheim, R. (eds.), *Critical Issues in Information
Systems Research,* John Wiley Information Systems Series, London.

Davies, L. J., and Wood-Harper, A. T., 1989. Information systems development: theoretical
framework, *Journal Of Applied Systems Analysis,* 16 61-73.

Dutta, S., 1997. Strategies for implementing knowledge-based systems, *IEEE Transactions on
Engineering Management* 44 79-90.

Flood, R., 1991. Redefining management and systems sciences. In Flood, R., and Jackson, M.
(eds.), *Critical Systems Thinking: Directed Readings,* John Wiley, Chichester.

Flood, R., and Jackson, M., 1991a. *Creative Problems Solving: Total Systems Intervention,*
John Wiley, Chichester.

Flood, R., and Jackson, M., 1991b. *Critical Systems Thinking: Directed Readings,* John
Wiley, Chichester.

Flood, R., and Romm, N., 1996. Diversity management, theory in action. In Flood, R., and
Romm, N. (eds.), *Critical Systems Thinking: Current Research And Practice,* Plenum Press,
New York and London.

Fuenmayor, R., and Lopez, G., 1991. The roots of reductionism: a counter-ontoepistemology
for a systems Approach, *Systems Practice* 4 419-448.

Geuss, R., 1981. *The Idea Of A Critical Theory. Habermas And The Frankfurt School,*
Cambridge University Press, London.

Gilligan, J., 1998. Back to the marketplace: mindfulness in information systems. *PARS II
Second Conference And Workshop On Philosophical Aspects Of Information Systems:
Methodology, Theory, Practice And Critique,* July.

Gregory, W., 1996. Dealing with diversity. In Flood, R., and Romm, N. (eds.), *Critical
Systems Thinking: Current Research And Practice,* Plenum Press, New York and London.

Hirschheim, R., 1985. *Office Automation: A Social And Organisational Perspective,* Wiley,
London.

Hirschheim, R., and Smithson, S., 1988. A critical analysis of information systems evaluation.
In Bjorn-Andersen, N., and Davis, G. B. (eds.), *Information Systems Assessment: Issues
and Challenges,* North-Holland, Amsterdam.

Honderich, T., 1995. *The Oxford Companion To Philosophy,* Oxford University Press,
London.

Huber, P. J., 1990. A theory of the effects of advance information technology on
organisational design, intelligence, and decision making. In Fulk, J. , and Steinfield, C. (eds.),
Organisations and Communication Technology, Sage publications, London.

Hughes, T. P., 1987. The evolution of large technological systems. In Bijker, W. E., Hughes,
T. P., and Pinch, T. F. (eds.), *The Social Construction. New Directions In The Sociology
And History Of Technology.* MIT press.

Jackson, M., 1991. The origins and nature of critical systems thinking, *Systems Practice* 4
131-149.

Jackson, M., and Keys, P., 1984. Towards a system of systems methodologies, *Journal Of
The Operational Research Society* 35 473-486.

Jayaratna, N., 1991. Systems analysis: the weak link in the systems development process?.
Journal of Applied Systems Analysis. 18 61-68.

Kelly, G., Watson, R., and Brancheau, J., 1994. Key issues in information systems
management: an international perspective, Warwick Business School Research Bureau
134.

Kendall J. E., and Avison D. E., 1993. Emancipatory research themes in information systems development: human, organisational and social aspects, *Discussion Paper in Accounting & Management Science,* University of Southampton. Discussion Paper 58, February.

Klein, H., and Hirschheim, R., 1987. Social change and the future of information systems development. In Boland, R. J., and Hirschheim, R. (eds.), *Critical Issues in Information Systems Research,* John Wiley Information Systems Series, London.

Kling, R., 1987. Defining the boundary of computing across complex organisations. In Boland, R. J., and Hirschheim, R. (eds.), *Critical Issues in Information Systems Research,* John Wiley Information Systems Series, London.

Lloyd, E., and Ralph, E., 1987. The concept of organisation mind, *Research In The Society Of Organisation,* 5 135-161.

Lyytinen, K., 1987. A taxonomic perspective of information systems development: theoretical constructs and recommendations. In Boland, R. J., and Hirschheim, R. (eds.), *Critical Issues in Information Systems Research,* John Wiley Information Systems Series, London.

Lyytinen, K., 1992. Information systems and critical theory. In Alvesson, M., and Willmott, H. (eds.), *Critical Management Studies*, Sage, London.

Midgley, G., 1996. What is this thing called CST?. In Flood, R., and Romm, N. (eds.), *Critical Systems Thinking: Current Research And Practice,* Plenum Press, New York and London.

Midgley, G., Munlo, I., and Brown, M., 1998. The theory and practice of boundary critique: developing housing services for older people, *Research Memorandum: The Centre For Systems Studies,* University of Hull 16.

Mike, H., 1998. Interview about role of the manager in process of implementation of information technology, February.

Miles, R. K., 1985. Computer systems analysis: the constraint of the 'hard' systems paradigm. *Journal of Applied Systems Analysis* 12 55-65.

Mingers, J., 1997. The nature of information and its relationship to meaning. In Winder, R. L., Probert, S. K., and Beeson, I. A. (eds.), *Philosophical Aspects of Information Systems.* Taylor & Francis, London.

Mingers, J., and Brocklesby, J., 1997. Multimethodology: towards a framework for mixing methodologies. *The International Journal Of Management Science.* 25, 489-509.

Morgan, G., 1986. *Images of Organisations,* Sage, Berveley Hills.

Oliga, J. G., 1991. Methodological foundations of systems methodologies. In Flood, R., and Jackson, M. (eds.), *Critical Systems Thinking: Directed Readings,* John Wiley, Chichester.

Robb, F., 1997. Some philosophical and logical aspects of information systems. In Winder, R. L., Probert, S. K., and Beeson, I. A. (eds.), *Philosophical Aspects of Information Systems.* Taylor & Francis, London.

Romm, N., 1997. Implications of regarding information as meaningful rather than factual. In Winder, R. L., Probert, S. K., and Beeson, I. A. (eds.), *Philosophical Aspects of Information Systems.* Taylor & Francis, London.

Schecter, D., 1991. Critical systems theory in the 1980s: a connective summary. In Flood, R., and Jackson, M. (eds.), *Critical Systems Thinking: Directed Readings,* John Wiley, Chichester.

Stafford Beer, 1988. *The Heart of Enterprise*, John Wiley & Sons.

Stein, E. and Zwass, V., 1995. Actualising organisational memory with information systems, *Information Systems Research* 6 85-117.

Symons, V., and Walsham, G., 1988. The evaluation of Information Systems: a critique, *Journal Of Applied Systems Analysis* 15 119-132.

Tapscott, D., and Caston, A., 1993. *Paradigm Shift, The New Promise of Information Technology,* McGraw-Hill, London.

Townley, B., 1994. *Refraining Human Resource Management - Power, Ethics And The Subject Of Work*. Sage, London.

Tudor, D. J., and Tudor, I. J., 1995. *Systems Analysis and Design: A Comparison Of Structure Methods*. NCC Blackwell, London.

Ulrich, W., 1991. Critical heuristics of social systems design. In Flood, R., and Jackson, M. (eds.), *Critical Systems Thinking: Directed Readings,* John Wiley, Chichester.

Ulrich, W., 1996a. Critical systems thinking for citizens. In Flood, R., and Romm, N. (eds.), *Critical Systems Thinking: Current Research And Practice,* Plenum Press, New York and London.

Ulrich, W., 1996b. A primer to critical systems heuristics for action research, *Research Memorandum: The Centre For Systems Studies,* University of Hull, March.

Vidgen, R., 1997. Stakeholders, soft systems and technology: separation and mediation in the analysis of information systems requirements, *Information Systems Journal.* 7 21-46.

Walsham, G., 1993. *Interpreting Information Systems In Organisations,* John Wiley & Sons, Chichester.

Winograd, T., and Flores, F., 1986. *Understanding Computers And Cognition,* Addison-Wesley. Ablex, Nowood, N.J.

INFORMATION SYSTEMS AND PROCESS ORIENTATION: EVALUATION AND CHANGE USING BUSINESS ACTIONS THEORY

Ulf Melin[1,2] and Goran Goldkuhl[1,2,3]
[1]Dept. of Computer and Information Science, Linköping University SE-581 83 Linköping, Sweden

[2]Centre for Studies of Humans, Technology, and Organization, Linköping University SE-581 83 Linköping, Sweden

[3]Jönköping International Business School, Sweden

Key words: Information Systems, evaluation, information systems development, organisational change, business processes, case studies, business action theory

Abstract: The first purpose of this paper is to outline suitable features of, and strategies for developing and purchasing Information systems (IS) in organisations striving for increased process orientation, based on the results from two in-depth case studies and related theory. The results are based on an evaluation of IS and business processes using Business Action Theory (BAT). The second purpose of the paper is to present experiences using BAT as an approach for evaluation of IS and business processes in an organisational context. In analysing data BAT helps to focus on the character and the relations between different critical business actions in a process dimension. The need for information on IS at different departments and a higher integration level was also highlighted when using BAT in studying critical business actions.

1. INTRODUCTION

In order to increase competitive advantage of organisations a common approach today is to become more process oriented. We can identify process orientation in change concepts such as Total Quality Management (TQM) and Business Process Reengineering or Redesign (BPR) (e.g. Imai, 1986; Davenport, 1993; Hammer, 1990; Hammer and Champy, 1993). Information technology or information systems (IS) is often proposed as a corner stone and enabler for organisational change in these concepts, particularly in the BPR approaches. The role of technology is to cut costs, support teamwork, shorten cycle-times, improve efficiency, support improvement, and/or innovation (Iden, 1994). BPR approaches also seem to imply a so-called clean slate approach to change, even concerning IS — however, the system legacy in organisations is an economic reality (due to large monetary and competence investments), that can not be ignored. IS developed for a more hierarchical, departmental focused, organisation can be a part of the system legacy, and may be an obstacle for increased process orientation. A diametrically op-posed view of IS (compared with the BPR concepts formal-rational perspective) is based on the web models developed by Kling and Scacchi (1982). Web models view IS as complex social objects con-strained by their context, infrastructure and history.

1.1 The Need for Evaluation of Information Systems and Business Processes

In order to change existing IS in the appropriate direction (to support and enable organisational change), or to replace them, there is a need for an evaluation of IS and business processes in organisations. There are many evaluation techniques or approaches, but no commonly accepted approach to perform evaluation. Symons and Walsham (1988) identify four groups of evaluation approaches with different content and focus: cost/benefit analysis, value analysis, decisions analysis, and management value added. These four approaches have been rejected since they are to narrow and insufficiently cope with social issues. The authors argues that there is a need for evaluation of both IS and organisation, and not independently (based on e.g. Kling (1987) and Hirschheim (1985)). They also argue for an interpretative approach to evaluation, based on four supplementary conceptual frameworks. These four are formal-rational, structural, interactionist, and political perspective. All these perspectives seem, however, to have an intra-organisational focus without an explicit view of the customer, and how to interact with and satisfy the customer.

We follow Symons and Walsham (ibid.) for an evaluation of both IS and organisation, together with an interpretative approach. However we reject that a conceptual framework only should have an intra-organisational focus. Instead such a framework should also include the business interaction with customers. For these reasons we choose to use Business Action Theory (BAT) (Goldkuhl, 1996; 1998) as a conceptual framework in evaluation of IS and business processes.

1.2 Research Questions and Purpose

The IS are almost taken for granted be a supportive force in several approaches to change — but does the IS have the features that are needed to meet the challenge of the "new", process oriented organisation? Do IS developed for a hierarchical organisation support a process oriented one? Is the IS managed in the appropriate way to support the expected organisational change? Is Business Action Theory a useful approach to perform evaluation and change of IS and business processes?

The first purpose of this paper is to outline suitable features of and strategies for developing and purchasing IS in organisations striving for increased process orientation, based on results from two in-depth case studies. The results are based on evaluation of IS and business processes using BAT.

The second purpose of the paper is to present experiences using BAT as an approach for evaluation of IS and business processes in an organisational context.

The next section of this paper describes BAT. Section three then contains research and evaluation strategy and the results from applying BAT in two case studies: the paper-mill and the manufacturing company. Conclusions are then presented, including important results from case studies and experiences from using BAT.

2.   BUSINESS ACTION THEORY

In many BPR-approaches there is an emphasis on viewing business processes as transformations of input to output. Hammer and Champy (1993, p. 35) define the notion of business process in the following way: "A collection of activities that takes one or more input and creates an output that is of value to the customer". Davenport (1993, p. 5) gives a partially similar, but more exhaustive definition: "In definitional terms, a process is simply a structured measured set of activities designed to produce a specified output for a particular customer or market." [...] "A process is thus a specific ordering of work activities across time and place, with a beginning and an end, and clearly identified inputs and outputs: a structure for action."

Reading these two definitions, one can see the basic idea that a process is conceived as a set of ordered activities that takes some input and transforms it to some output for a customer. We call this view a transformation perspective on business processes. This view has been challenged by Keen (1997). He argues for a co-ordination view to replace the transformation view. Co-ordination of people is considered as the essence of performing processes. We totally agree with Keen that co-ordination aspects are left out in many BPR approaches (cf. the above definitions) and that such aspects should be put into foreground when studying processes. However, we do not think that a transformation view should be totally abandoned: it should be kept, but put within a co-ordination view.

There are existing approaches to the design of business processes taking a co-ordination perspective. The most famous approach of this kind is probably Action Workflow (e.g. Denning and Medina-Mora, 1995). In Action Workflow there is a focus on the interaction between the two (generic) roles: performer and customer. Four different phases are described on how the two parties accomplish a "business transaction". After a preparatory phase the two parties come to an agreement on what is to be performed. After that the performance (by the performer) occur and the process is ended by an acceptance made by the customer.

In the spirit of viewing business processes as co-ordination and transformation BAT has been formulated (Goldkuhl, 1996; 1998). According to BAT a business process is divided into six generic phases:
1. Business prerequisites phase
2. Exposure and contact search phase
3. Contact establishment and negotiation phase
4. Contractual phase
5. Fulfilment phase
6. Completion phase

The Business Action Theory describes a business process in terms of the interaction between two generic business roles: A customer and a supplier. The different phases and the interaction between the two roles are described graphically in figure 1.

The first phase includes the establishment of prerequisites, on both sides, for business interaction. The supplier must have ability (a know-how and a capacity) to perform business. The customer has operations where there exist some lacks and needs. In the second phase there is a contact search of each side, which includes exposure of business interest. When making contact, the two parties can start negotiating (phase 3). This communication can be described as proposal stating. Bids and counter bids are made. The desire and demand of the customer are expressed.

SUPPLIER   CUSTOMER

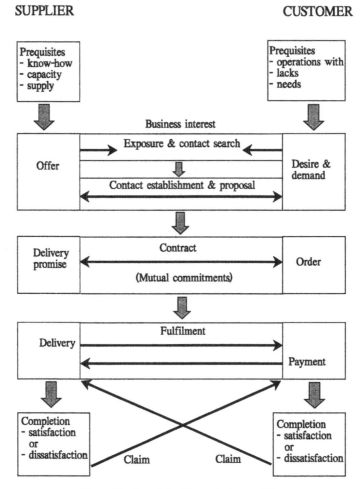

Figure 1 Business Action Theory: A phase model

236

The supplier can make offers (of standard or customised character). The negotiation can be terminated (which ends up the business processes) or transferred into a contractual phase. This is the fourth phase.

Customer and supplier come to an agreement concerning the business transaction. The contract involves mutual commitments. These different commitments must be fulfilled; otherwise the contract is broken. This fulfilment is made in the fifth phase: The supplier delivers and the customer pays. If not satisfied with the delivery, the customer can make a claim. The supplier is asked to make some modification in the delivery. Correspondingly, the supplier can make payment claims towards the customer if needed. This is the sixth, and last phase which involves assessments of the fulfilment leading to satisfaction or dissatisfaction.

BAT has thus a focus on generic business actions, both communicative and material actions. It involves the interactive co-ordination between customer and supplier. The fulfilment phase of the supplier usually consists of a transformation of input (raw material) to output (finished products). The BAT model is a symmetric model giving full account to both customer and supplier, which is seldom done in the marketing literature (confer critique in Axelsson and Easton, 1992; Goldkuhl, 1998). Many marketing theories seem to take the perspective of the supplier viewing at the customer (e.g. Kotler, 1994).

The theoretical sources behind BAT are of two kinds: Language action theories (e.g. Searle, 1969; Habermas, 1984) and business relationship theories (e.g. Axelsson and Easton, 1992; Gummesson, 1996). The language action background is used to identify and characterise different communicative actions. Business relationship theories are needed to give the proper business context, including interaction of the two roles.

The main purpose of BAT is to describe and explain business interaction. But this is not the only purpose. BAT can also be used as a theoretical lens in organisational change when developing business processes. The theory can be used as an interpretative framework when reconstructing, evaluating and redesigning different business processes. In such change situations it should be supplemented by suitable development methods (Goldkuhl, 1996; Lind and Goldkuhl, 1997). BAT should thus not be seen as a development method. The Action Workflow approach mentioned above is on the contrary an integration of theory and method into one unified whole. There are other important differences also, although BAT and Action Workflow partially share the same theoretical basis in language action theories. A comparative review of BAT and Action Workflow is made in Goldkuhl (1996) and Verharen (1997).

3. CASE STUDIES: EVALUATION OF INFORMATION SYSTEMS AND BUSINESS PROCESSES

This chapter contains the research and evaluation strategy of two case studies using BAT for understanding information systems and business processes. Information systems features and strategies for development and purchasing are also presented as a result from using BAT.

3.1   Research and Evaluation Strategy

In the two cases studies an overall qualitative research approach is applied (e.g. Yin, 1994). Two in-depth case studies has been performed with the use of interviews, studies of documentation and "face-to-face" studies of IS as methods for data collection (cf. Denzin and Lincoln (1994)). Interviews are conducted with people from different departments and from several hierarchical levels in the companies. The choice of informants has been made in order to get a "rich picture" of the IS and organisation and catch different perspectives of the phenomena (perceived quality of IS and business processes). IS/IT-managers, business development projects managers, system developers, and daily information system users in e.g. sales- and marketing, production planning, and production are represented in the group of informants. This approach is chosen in order to triangulate data (Denzin, 1978), to capture social meanings held by the actors within the organisation concerning IS and business processes, and also to discover new meanings.

The two anonymous cases, the paper-mill and the manufacturing company, are both striving for an increased process orientation in combination with changes/investments in IS. Some aims of the two Swedish companies are to increase quality, to become more customer oriented, to implement continuous improvements in the daily work, and make a more cost efficient use of IS. The two companies have, of course, both similarities (directions of organisational change, goals etc.) and differences (in size, production processes, products, markets and IS legacy).

The paper-mill is a rather small producer of paper and has a market strategy to be a flexible supplier of goods. Their competitive edge is to make fast adaptations of their production equipment in order to satisfy customer needs. The adaptations in the production equipment must, according to the company, be supported by a mental and competence capacity change in the staff to be successful. The IS in the paper-mill is often developed in-house, but some standard-packages for payment of salary exist as well.

The manufacturing company is, under Swedish circumstances, a large company. The company has, for many years, been an actor on a stabile market, but, for a couple of years, has been trying to expand its market. In its efforts the company meets competition, and wishes therefore to become more business oriented. The company was, when the case study was performed, implementing a large standard-package mainly for financial use. Other IS at the company are often developed in-house, during several years, and by support from several different subcontractors. The systems are often based on different platforms, disparate operative systems etc. that makes communication between them difficult.

The use of BAT in our research can be characterised as a guide for and "checklist" for both questioning and analysing empirical material. The approach has, however, not been excluding open questioning in the study. Such open questioning is necessary in order to generate not "pre-defined" concepts and categories, from the empirical field. The open questioning and "open use" is made in order to (1) have the possibility to discover phenomena that are not covered by the guide and checklist, and (2) in order to identify potential improvements for BAT.

This approach can be compared with the open and axial coding of the Grounded Theory approach (e.g. Strauss and Corbin, 1990). In the data collection phase we apply an open questioning. BAT is used as a checklist in this phase. The initial analysis of data is performed as an open coding. The next step in data analysis is to perform axial coding. The axial coding in this case is guided by BAT as an action oriented frame of reference for questioning and interpretation.

3.2 Understanding Information Systems and Business Processes Using BAT

The evaluation of IS in the two companies was made using BAT (figure 1). The different phases in BAT consist of several critical business actions. A selection of identified critical business actions performed by the suppliers in the case studies are presented below, and related to different phases of the BAT framework.

3.2.1 Exposure of Production Capacity

One critical business action that has been studied is the suppliers' possibility to identify and expose their production capacity. This critical business action is related to the contact and establishment, and negotiation phase in BAT. In the paper-mill a market assistant handles many questions from both established and potential customers concerning when the company has scheduled the production of a certain product, the time of delivery for a certain product etc. In this phase we have identified that the answers to the customers are sometimes unclear and/or delayed in time. The IS-support in performing these kind of critical business actions is weak. This has several negative effects, e.g. on the quality concerning the suppliers' external communication with the customer. The only information that is possible to access is based on historical production figures that of course are relevant, but not in this particular situation. The IS in the paper-mill's sales- and marketing unit is not integrated with the systems for production planning and production. Of course this information can be, and is, gathered by using the "manual" IS, but this is often time consuming and inefficient in this situation, "face-to-face" with the customer.

3.2.2 Presenting an Offer

When the manufacturing company puts together an offer to customers, there are several departments and actors involved, due to the high complexity in their product. The offer is a part of the exposure made by the supplier, and related to the contact and establishment and negotiation phase. The work is done by a so-called offer-team, with members from different departments. Putting together an offer is a time and resource consuming activity, even though the company has implemented a PC-network to support these tasks. The time from the customers tender for an offer to the company's answer to it can however be shortened. One problem with the PC-network is that the network is isolated from the rest of the organisation's IS-resources. One effect of isolation, that was highlighted when using BAT, is that information can not be efficiently moved from the team members' ordinary workstation to the offer-team network, in order to support a short cycle-time from question to answer.

3.2.3 Converting Order Commitments to Production and Product Specifications

In converting an order commitment (e.g. to produce a certain product to a certain customer) to a product specification as a basis for production planning and production, distortion can occur. The distortion is explained by the fact that the sales- and marketing department's information system and the information system for production planning is incompatible. The paper-mill has however tried to solve this problem with an in-house built "bridge" between the two systems. This conversion of order and product data is however not reliable, distortions occur that makes it hard for users to trust the information.

Sometimes when conversion failures are not identified, the original commitment to the customer (order) is changed and not identical with the production and product specification. A faulty production specification can lead to a low delivery quality. This critical business action is related to the link/interface between contractual phase and the fulfilment phase in BAT.

3.2.4 Producing Goods (Based on Order Commitments Out of Date)

Another problem related to the order information is based on the fact that the paper-mill has an unstable IS-structure, which makes data transmission from distributed IS located at different order- and sales offices and the central order database not transparent. When the situation occurs that a customer change the order specification, the-paper mill promises to change the order specification in the information sys-tem, and believes that they have made a new commitment, but the data transmission fails. The failure is then not automatically recognised by the system user, and the old commitment (not updated order information) could still be a base for production. A faulty production specification (based on the old commitment) can, again, lead to a low delivery quality. This critical business action is related to the link/interface between the contractual phase and the fulfilment phase in BAT.

3.2.5 Delivery Promises/commitments

When the paper-mill promises a specific customer to deliver a certain amount of goods, at a certain time, they use "week X" as the time of delivery. This is done in spite of the fact that the customers usually wish to have a particular "day Y" as time of delivery. In this case no particular IS is involved. This problem is related to the uncertainty of the paper-mill's future capacity to deliver goods at a certain time (a commitment to deliver at a particular day). This uncertainty is related to the fact that the paper-mill does not follow up their delivery quality (delivery of the agreed product, on agreed time). Delivery promises are given in the BAT contractual phase and should be followed-up in the completion phase.

3.2.6 Giving Information Concerning Ordered Products' Status

When a contract is established with a specific customer, the paper-mill's customers usually ask questions concerning where the ordered goods are in the production line, when it is going to be delivered, and so on. Again the market assistant in the paper-mill is assigned to answer this kind of questions. However it is not easy to get access to information of specific products and their status in the productions process. This information is stored in the production units database, and not possible to access through the IS in the sales- and marketing unit. The ways to get access to this kind of information is, also in this phase, time consuming and dependent on certain peoples (the production planners) presence. These kind of critical business actions are a part of the fulfilment phase in BAT.

3.2.7 Invoice Customers

Invoice customers are another important business action that are part of the fulfilment phase. The paper-mill's customers can at the same time have several different terms of payment due to different products. The information system and database for customer information, however, does not support this "one-to-many" relationship between customer and terms of payment. This restriction has the effects that (1) the system users has to construct fictitious customers, based on the "real" customers in order to allow one customer to have different terms of payment, and (2) the problem of possible information inconsistency when a customer changes addresses. Several updates of a customer's address have to be performed in order to avoid invoices (and even offers) being sent to wrong addresses.

3.2.8 To Improve Operations and Meet Customer's Changing Demands

The suppliers' knowledge of their customers' satisfaction or dissatisfaction is important in the completion phase in order to improve operations and meet customer's changing demands. The paper-mill and the manufacturing company do not, in a systematic way, follow-up or measure customer satisfaction or dissatisfaction. The only information concerning customers level of satisfaction is, ideally, "ad-hoc" collected by sales representatives. In order to make the best use of this kind of information it should be spread around the company, which is not done. However the paper-mill has a well-developed way of communicating with some of its customers when performing so called study tours. The objectives in performing these kind of activities is for the production workers to better understand why customers demand that certain quality measures are more important than others.

3.3 Information Systems Features and Strategy Related Results

The results presented below are based on case study research and can be viewed as a result from using BAT as a guide in analysing data together with an inductive research approach (the open questioning described in section 3.1).

3.3.1 Suitable Features of Information Systems in a Process Oriented Organisation

The results from the case studies show, among other things, that the IS should be able to support ex-change of information along processes, with other IS (e.g. IS traditionally related to other departments or functions) in order to support communication between actors in different departments in the organisation. The problems with IS that is not sufficiently integrated, are a well-known fact in the case studies, as well as in the literature (e.g. Keen, 1991, 1997; Tapscott and Caston, 1993). These aspects seem to be stressed when applying a process perspective on organisations including the focus on creating customer value, instead of focusing on single functions or departments activities only. BAT has been an instrument in our studies for identifying these process-related weaknesses with a point of departure from the performance of critical business actions.

We have also found that it is important that the IS has a potential to change (flexibility) based on what is technically possible and economically satisfactory. It is also important that the IS has an overall good correspondence with the business, irrespective of whether changes are initiated from technical progress or changes/improvements in business needs motivated by e.g. changes in work procedures or routines.

3.3.2 Suitable Strategies for Information Systems in a Process Oriented Organisation

In our study we also included questions concerning suitable IS/IT-development and purchasing strategies in a process oriented organisation. We have found that it is important to uncover different ideals (starting-points and objectives) before and during a parallel development of IS and organisation. In the manufacturing company this was clear when two different project groups worked with, on the one hand the organisational change (development of future process maps including work-flows etc), and on the other hand the development/ adaptation of an standard package. These two groups did not communicate with each other about existing and important decisions on designing future workflow and processes. Of course different perspectives and ideals can generate change, but a far too unclear apprehension of conflicts concerning different ideals appears to obstruct resource-efficient development. On the basis of the earlier discussion we claim that it is important to stimulate a change of perspectives in development of IS between technical possibilities of IS and businesses needs.

We also identified that it is important to be aware of the fact that planned strategies for development and purchase of IS are not necessarily identical with the realised ones. There are frequent differences between interpretations of planned strategies and often a set of developing strategies (according to Mintzbergs (1991) terminology) in the empirical material. The two organisations existing IS/IT-strategy (the realised strategy) is more of a consequence of different, so called "technical", decisions on the operative levels of the organisations. The sum of these often divergent decisions then constitutes the realised IS/IT-strategy. In order to reduce this, it is important to make decisions about IS/IT-development from a point of view that is beyond the domain of single departments or organisational functions (regarding the six generic phases of a business process according to BAT can help avoiding single functional view). By doing this the risk for decisions that generates isolated "islands of information systems" (see Melin and Ritschel, 1998; Tapscott and Caston, 1993) can be reduced.

4. CONCLUSIONS

In analysing data, BAT helps to focus on the character of, and relations/interfaces between different critical business actions in a horizontal (process) dimension, instead of a more traditional departmental or functional dimension. Using generic business actions and phases from BAT facilitates interpretation, comparison and evaluation of different cases.

The need for information from different departments' IS and a higher integration level was highlighted when using BAT in studying different critical business actions. The IS should be able to support actors critical business actions, e.g. to inform the customer about the suppliers manufacturing capacity, possible commitments, commitments made, time for fulfilment of an order, above. The IS support should contain information that is needed to perform these actions. No matter where the information is originally stored in the information systems of the organisations IS portfolio. The IS could also be designed to automatically perform some critical business actions. In the studied companies, the existing IS-support performing the critical business actions was weak in several situations. The IS in the paper-mill's sales- and marketing was not integrated with the systems for production planning and production.

A question that one could ask oneself is if it would have been possible to detect these features without using the BAT framework in evaluation? Using other evaluation approaches, e.g. the intra-organisational approaches mentioned in section one above could possibly result in knowledge of this kind. But it is accidental if this would be the case. The use of BAT provide direct attention towards features of IS in relation to business processes, business actions, and inter-organisational interaction.

BAT can be used, not only for evaluation, but also as a conceptual instrument in redesigning business processes and IS. BAT can also be used for redesign of specific critical business actions, and whole processes, consisting of several actions performed in concert. This is important in order to reach an organisation with more clear business actions and well worked-out business logic, both internally and in contact with suppliers and customers. In the case of the paper-mill the results from our BAT evaluation served as a basis for development of new IS.

In order for IS to support a process oriented organisation we have also identified (section 3.3) that it is important to give a broad-minded view of different types, and interpretation of IS-strategies and their status in the organisation. A change of perspectives in development of IS, between technical possibilities of IS and business needs is also important.

REFERENCES

Axelsson, B, and Easton, G. (eds., 1992). *Industrial networks. A new view of reality*, Routledge, London.

Davenport, T.H. (1993). *Process innovation. Reengineering work through information technology*, Harvard Business School Press, Boston.

Denning, P.J. and Medina-Mora, R. (1995). *Completing the loops*, Interfaces.

Denzin, N.K. (1978). *The Research Act: A theoretical Introduction to Sociological Methods*, second edition, McGraw-Hill Publishing Company.

Denzin, N.K. and Lincoln, Y.S. (1994). Entering the Field of Qualitative Research, In:

Denzin, N.K. and Lincoln, Y.S (eds.) *Handbook of Qualitative Research*, Sage Publications, Inc.

Goldkuhl, G. (1996). Generic business frameworks and action modelling, In Proceedings of *Conference Language/Action Perspective '96*, Springer Verlag.

Goldkuhl, G. (1998). The six phases of business processes - business communication and the exchange of value, accepted to *The twelfth biennial ITS conference (ITS '98)*, Stockholm.

Gummesson, E. (1996). Toward a theoretical framework of relationship marketing, in Sheth, J.N. and Söller, A. (eds.) (1996). *Development, management and governance of relationships*, Humboldt Universität, Berlin.

Habermas, J. (1984). *The theory of communicative action 1. Reason and the rationalization of society*, Beacon Press

Hammer, M. (1990). Reengineering Work: Don't Automate, Obliterate, *Harvard Business Review*, July-August, Vol. 59, No. 4, pp. 104-112.

Hammer, M. and Champy, J. (1993). *Reengineering the Corporation*, Harper Collins.

Hirschheim, R. (1985). *Office Automation: A Social and Organisational Perspective*, Wiley.

Iden, J. (1994). *Six Essays on Business Process Reengineering*, PhD thesis, Department of Information Science, University of Bergen, Norway.

Imai, M. (1986). *Kaizen, the Key to Japan's Competitive Success*, McGraw-Hill Publishing Company.

Keen, P.G.W. (1991). *Shaping the Future: Business Design through Information Technology*, Harvard Business School Press.

Keen, P.G.W. (1997). *The Process Edge*, Harvard Business School Press.

Kling, R. (1987). Defining the Boundaries of Computing across Complex Organisations. In Boland, R. and R. Hirschheim (eds.) *Critical Issues in Information Systems Research*.

Kling, R. and Scacchi, W. (1982). The Web of Computing: Computer Technology as Social Organization, *Advances in Computers*, No. 21, pp. 1-90.

Kotler, P. (1994). *Marketing management. Analysing, Planning, Implementation, and Control*, 8th edition, Prentice-Hall Inc., New Jersey.

Lind, M. and Goldkuhl, G. (1997). Reconstruction of different business processes - a theory and method driven analysis, In *Proceedings of the 2nd International Workshop on Language/action perspective (LAP97)*, Eindhoven University of Technology.

Melin, U. and Ritschel, W. (1998). Information technology's role in process oriented organizations — suitable features of, and strategies for IT, In: Karwowski, W. and Goonetilleke, R. (eds.) *Proceedings of the 6th International Conference on Human aspects of advanced manufacturing: agility and hybrid automation*, Hong Kong University of Science and Technology, IEA Press.

Mintzberg, H. (1991). Five P's for Strategy, second edition; Mintzberg, H. and Quinn, J.B. (eds.), Prentice Hall.

Searle, J.R. (1969). *Speech acts. An essay in the philosophy of language*, Cambridge University Press, London.

Strauss, A. and Corbin, L. (1990). *Basics of Qualitative Research: Grounded Theory Procedures and Techniques*, Sage Publications.

Symons, V. and Walsham, G. (1988). The Evaluation of Information Systems: A Critique, *Journal of Applied Systems Analysis*, Vol. 15, pp. 119-132.

Tapscott, D. and Caston, A. (1993). *Paradigm Shift — The New Promise of Information Technology*, McGraw Hill Publishing Company.

Verharen, E. (1997). *A language-action perspective on the design of co-operative information agents*, PhD thesis, KUB, Tilburg.

Yin, R. K. (1994), *Case Study Research: Design and Methods*, Second Edition, Sage Publications Inc., Thousand Oaks.

TOWARDS THE SHIFT-FREE INTEGRATION OF HARD AND SOFT IS METHODS

Stephen K. Probert[1] and Athena Rogers[2]
[1]*Computing and Information Systems Management Group*
Cranfield University, RMCS, Shrivenham, Swindon SN6 8LA

[2]*School of Computing and Mathematical Sciences,*
Liverpool John Moores University, Liverpool L3 4AF, U. K.

Key words: Methods, Methodologies, Philosophy, Epistemology, Method Integration

Abstract: Shift-free Integration refers to a method by which methodologies (or techniques inherent to them) may be integrated for the analysis of problem domains and the design of subsequent 'solutions', within an acceptable philosophical framework. It is considered viable that this framework could legitimise the integration of any 'soft' technique (such as a conceptual model) with any 'hard' model (such as a data flow diagram). The framework proposed herein supports the Domain of Systems Analysis and Design and harnesses the ancient Chinese symbol of the Diagram of the Supreme Ultimate within which the elements Yin and Yang exist in rotational symmetry; and is itself supported by considerations from contemporary epistemology. Any paradigm or number of paradigms may be represented within the Domain, depending on the academic sympathy of the practitioner concerned. The 'ownership' for the design of optimal information systems within organisations does not necessarily always reside within a Computing domain; it is therefore paramount that this framework is capable of endorsing, in philosophical terms, information processing practices embraced by other domains.

1.   THE DILEMMA

An uncomfortable gulf exists between what practitioners do and what theorists deem as philosophically acceptable; integrating methodologies for systems analysis and design, may well be something analysts 'do', but as yet no agreed framework exists to support this action (Gammack, 1995). Advocates of the soft systems approach have criticised the hard systems following for the propensity of hard methodologies, such as the Structured Systems Analysis and Design Method (SSADM), to commit what may be termed 'errors of the third kind', i.e. proposing solutions to the wrong problems. On the other hand, using a soft approach such as the Soft Systems Methodology (SSM) to elicit accurate requirements necessitates some sort of paradigmatic 'shift', if those requirements are to be translated into a step-wise system design. Utilising different methodologies' 'best bits' within one intervention (a course of action which many practitioners readily undertake), fuels academics' perception that a cogent philosophical framework to support the integration of methodologies for systems analysis and design, is neither clearly discernible or (more to the point) agreed upon. Combining techniques

Systems Development Methods for Databases, Enterprise Modeling, and Workflow Management,
Edited by W. Wojtkowski, *et al.* Kluwer Academic/Plenum Publishing, New York, 1999.

245

(particularly SSM's conceptual model (CM) and SSADM's data flow diagram (DFD)) from differing methodologies is not deemed philosophically acceptable as this essentially equates to a paradigm 'shift'; i.e. moving from an interpretivist SSM 'camp' to a more functionalist SSADM arena or vice versa.

Within SSM, the CM may comprise a set of agreed 'feasible and desirable' activities which could alleviate or broaden a problem domain. However, the CM exists 'below the line' and it resident within what Checkland refers to as the domain in which "Systems Thinking about the Real World" (Checkland and Scholes, 1990: 27) may take place. By its very essence, it is conceptual and perception-laden and if a philosophical framework were to be applied to it, then the CM would fall squarely within the interpretivist paradigm. The DFD however, may be defined as an objective representation of reality. This model is functionalist in nature and ontologically speaking, veers towards realism. Yet it has been noted by researchers in the field that the CM bears a striking resemblance to a DFD and the question was raised as to whether a CM could form the basis of a DFD by, for example, translating activities to processes and by adding inputs and outputs.

At Warwick University (in the U.K.) in 1992, the need to investigate methodology integration was formally recognised (Mingers, 1992). However, advocates of methodology integration were by no means new-comers to this dynamic research field. Miles (1988) had already advocated an "embedded" approach, in preference to soft-front-ending, which he termed "grafting"; Prior (1990) in contrast, had supported the soft-front-ending method as he considered it viable to convert, somewhat mechanistically, the output of the SSM process, stemming from the CM, into a DFD of the required system; Avison and Wood-Harper (1990) had put forward Multiview, a methodology deemed to combine SSM with other information system development tools; Sawyer's approach (1991), had been to see the process of translation from CM to DFD as comprising two epistemological 'shifts'; and finally Stowell and West (1992) had introduced Client-led Design which "...is intended to empower the clients (stakeholders or users) and enable them to exercise full control over the provision of technology to support their information systems." (Stowell, 1995:118). Yet however prolific attempts were to integrate methodologies (and those cited above are but a few), not one has stood exempt from criticism from both researchers and practitioners alike. The challenge was to find an acceptable philosophical framework which would allow for the successful integration of methodologies, without diluting the impact of any particular methodology in use.

2. TOWARDS THE 'SHIFT-FREE' INTEGRATION OF METHODOLOGIES FOR SYSTEMS ANALYSIS AND DESIGN

It is perhaps expedient to start from the premise of what constitutes current thinking in Information Systems development. There is increasing evidence that interpretivist thinking is seeping into Information Systems development in a coherent way. If the Soft Systems Methodology heralded the arrival of an alternative method of addressing (typically) organisation-based complex problems, its more mature form may be identified as being adopted and fashioned into other (commonly referred to as) functionalist methodologies.

In his paper "Empowering the client", Stowell (1995) challenges the previously held dictum that information is the concern of computing (i.e. information technology) as opposed to being the concern of a different distinct area of the domain: information systems. "The operative word here is 'information' as distinct from 'data'. It is the conversion of the data plus the context in which the process takes place that seems to translate what are little more than stimuli into information." (Stowell, 1995:124) In other words, Stowell is distinguishing between data (which is manipulated by technology) and information (which is manipulated by information systems). This may be considered to be a useful observation since it maps neatly with our interpretivist and functionalist dichotomies. The work of researchers such as Liang et al. (1998) demonstrates how interpretivist thinking, in the form of the Appreciative Inquiry Method (AIM), may be used to develop the "...same type of activity models as used in SSM as a way of exploring and describing a user's view of some notional activity." (Liang et al. 1998:168). AIM they claim, may be useful as a 'soft-front end' to Object Oriented Analysis. Further, Gammack (1995) offers an approach which "...attempts to reconcile the 'hard' computing that is possible from objectivised databases representing an organisation's business, or other historically recorded evidence, with the 'soft' or subjective interpretations, judgements, theoretical constructions and appreciation of local situations which characterise human activity systems." (Gammack, 1995:159); and Lewis (1995) posits: "It is suggested that an additional level of data analysis and modelling may be required - one whose concern is the cognitive categories through which the participants in a problem situation make sense of that situation and understand the nature of their own organisation and its environment. The creation of such models would constitute an additional interpretative form of data analysis, enriching and complementing, rather than challenging, existing well-proved techniques." (Lewis, 1995:186). In one sense the dichotomy is being reinforced since interpretivism is being adapted into previously considered functionalist methodologies, but conversely, it is being somewhat disregarded as the implicit need to create a seamless continuum between interpretivism and functionalism emerges.

We are in an arena which is becoming conscious of the dichotomies which exist within it which have hitherto helped shape and define our discipline, but at the same time we are now recognising the need to operate across them. Does this mean that our dichotomies should not exist? Returning to Stowell's differentiation between information and data above, it is evident that in accepting that information is the concern of the information systems specialists, we are somewhat renouncing the computing arena's right to concern itself with analysis (at least what the information systems fraternity mean by analysis). Information is to do with meaning attribution and meaning attribution is to do with perception and perception is what interpretivism concerns itself with. Information systems specialists have the tools and expertise more readily at their disposal to determine system requirements. Once requirements are agreed, then the computing arena is welcome to engage in specific, technical designs. "The methods of information systems definition and design used in the 1970s and 1980s are inadequate for modern needs...the way in which we think about information systems requires a different mind-set to the way we think about computing." (Stowell, 1995:121-122). If analysis was previously the prerogative of computing specialists, this is no longer applicable in our – now more complex - society. The question is now, not which methodology to choose, but which paradigm to embrace. Yet this leads us back to our previous question: Does this mean that our dichotomies should not exist? If we are attempting to shift paradigms in order to complete an analysis and offer a design solution, are the dichotomies inherent to our present philosophy hindering our natural progression from one paradigm to another? Considering the philosophy of our Western world it is indeed unremarkable that such dichotomies exist:

> "The supremacy of thought and reason, of cause and effect, as a guiding star for the perfect rational person is still held as an ideal. Apparently, we have a long-standing fear that rationality will be overwhelmed by chaos and the spiritual by the sensual. The same can be said of dualism or polarity, the traditional Western way to arrange the world and life in mutually exclusive concepts. We think in terms of body/mind, either/or, black and white, good and evil, defining things by their opposites." (Skyttner, 1998:194)

It is worthwhile therefore and essential to this argument that we trace the mechanisms by which we have arrived at the duality of our Western tradition. Western philosophy has its roots in Ancient Greek philosophy; a culture in which science, philosophy and religion were not separated. The split which led to the separation of spirit and matter came with the contrasting philosophies of Heraclitus (who taught that all changes in the world arise from the dynamic and cyclic interplay of opposites) and Parmenides (who considered change to be impossible, but favoured the concept of an indestructible substance as the subject of varying properties). This latterly induced a concentration on the human soul and the question of ethics; and with the domination of the Christian Church, Aristotle's model of the universe which concentrated upon the human soul and a contemplation of God's perfection, endured for the next two thousand years. In the seventeenth century, Descartes' division of nature into two realms, pertaining to mind and matter, led to an extreme formulation of the spirit and matter dualism. The Cartesian division led to Newton's mechanistic view of the world and is directly responsible for the scientific and technological developments of our day. It is ironic now that as we beginning to mistrust science and technology, we are overcoming the fragmentation of Cartesian philosophy and returning to the unity espoused by the Ancient Greeks. Even in philosophical terms it seems we have come full circle:

"Both ordinary people and scientists feel that science - and its offspring, technology - no longer enhances the quality of their lives, but is in fact systematically reducing it...With problems relating to the whole domain of human knowledge, from philosophy to cellular biology, solutions have to be based on something more than the old scientific paradigm. Positivism, lacking in foresight and comprehensive views, now gives a diminishing return in area after area, from social science to quantum physics." (Skyttner, 1998:197-198)

What is interesting is that Eastern philosophy has harnessed the two complementary philosophical traditions, Taoism and Confucianism (or the intuitive and the rational) into one representation of yin and yang which forms the basis of Chinese thought. Yin and Yang exist in a dynamic rotational symmetry, suggesting continuous cyclic movement between the poles which they represent. Each time one of the two forces reaches its extreme it contains in itself already seeds of its opposite.

In remaining locked into positivist thinking, Western science has attempted to apply rational knowledge, via a system of abstract concepts and symbols, characterised by linear, sequential structures, to a natural world of infinite variety and complexity. It is a no wonder that we have failed. Western philosophy, having adopted Descartes' dictum, thinks in terms of opposites; of dichotomies, but Eastern philosophy harnesses these dichotomies into a mutual interplay and dynamic interaction, in which the passage from one 'pole' of a dichotomy to the other can be seen to occur along a smooth continuum. At a deeper level, Eastern philosophers talk about absolute knowledge which is acquired through a non-intellectual experience of reality, arising perhaps from a meditative state; but as somewhat immature beings in a spiritual sense, it is perhaps unreasonable to suggest that we Westerners all begin to meditate to solve our information systems dilemmas! Systems science may well be in its infancy and may well, at this stage, be somewhat ill-defined, but this does mean that systems theorists are prepared to enrich their domain by stretching its boundaries and harnessing other dimensions of thought. Information Systems as a discipline is unstable, largely because it has its roots in the scientific paradigm; but it is, in our present complex world, being forced to address complex information-oriented, human concerns which require a departure from the doctrines of positivistic thinking. The scientific fraternity is resistant to change, we are "...trapped by the omnipotence of the contemporary." (Stowell, 1995:122). "In order to break free from the predominance of positivistic thinking, there must be a significant shift in the way we think about the world..." (Stowell, 1995:122)

3.   THE PROPOSAL

Shift-free Integration is the method proposed. The Domain of Systems Analysis and Design is the arena which represents the totality of our concern. It is maintained that any methodological application will occur within this domain. Superimposed on the Domain is the notional concept of rotational symmetry, represented by the Diagram of the Supreme Ultimate. In ancient Chinese tradition, the dynamic characters of Yin and Yang are illustrated by this Chinese symbol, "Chinese philosophy...has always emphasised the complementary nature of the intuitive and the rational and has represented them by the archetypal pair yin and yang which form the basis of Chinese thought." (Capra, 1991:34). Embracing this philosophy, we may discern a natural tendency for the poles of

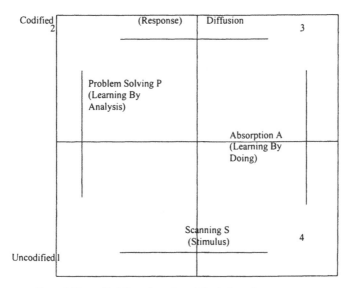

Figure 1. The cyclical flow of new knowledge in the culture-space

The continuums used to demarcate the Domain, may thus be determined by the practitioner, depending on the academic sympathy to which (s)he subscribes. Within the any dichotomy to migrate towards their 'opposites' in an attempt to redefine themselves in terms of that 'opposite'. Each time a pole successfully reaches its extreme, "...it contains in itself already the seed of its opposite." (Capra, 1991:120). Baudrillard (1983) observes, "Everything is metamorphosed into its inverse in order to be perpetuated in its purged form. Every form of power, every situation speaks of itself by denial, in order to attempt to escape, by simulation of death, its real agony." (Baudrillard, 1983:37).

The Domain of Systems Analysis and Design is characterised by harnessing the proliferation of methodological interventions associated with the search for optimal 'solutions'. Methodologies are characterised by their predominant philosophical orientations, but may display varying degrees of 'sympathy' with the opposing philosophy: SSADM is deemed largely functionalist, but veers towards interpretivism in some areas; the inverse is true of SSM. Methodologies contain within them seeds of their 'opposing' methodology. In application, it is paramount to be aware that no methodological intervention is 'absolute' in philosophical terms, but is underpinned by a philosophy which may be defined by its position on a relative continuum. In the Computing field, this continuum may be relevant to the functionalist - interpretivist dichotomy (thus spanning two paradigms), but operating within the Business arena, for example, we may wish to apply some other paradigm, demarcated by different continuums.

Computing field, a number of dichotomies exist which serve to define the domain: interpretivism - functionalism; soft - hard; abstract world[1] - real world; analysis - design.

---

[1] Checkland differentiates between the real world and an area in which systems thinking about the real world may take place. It may be said that this division is not strictly dichotomous, since other abstract thinking (other than systems thinking) could be employed 'below-the-line'.

Had it not been for factors like: the emergence of SSM (Checkland, 1981; Checkland and Scholes, 1990; Checkland and Holwell, 1998), as a philosophically radical methodology (distinct from the hitherto accepted norm); writers like Burrell and Morgan (1979), who constructed a taxonomy of sociological paradigms; and the perceived failure of 'hard' methodologies to provide accurate solutions (Mingers, 1995:18), these dichotomies may not have received the attention that they have, or indeed may not even have been formally identified; and further, the paradigms which exist as a result of these demarcating dichotomies may not therefore have assumed the opposing positions within the field that they have either. If we accept that the design of information systems may be considered to concern the Business sector as well as the Computing field, then it is rational to ask: what other alternative paradigms exist which serve to underpin business practices? Boisot's (1988) model provides us with: "The Cyclical Flow of New Knowledge in the Culture-Space" [Figure 1].

The relevant dichotomies: codified - uncodified and diffused - undiffused, serve to demarcate one paradigm: the Culture-Space. If this is a model that the Business community may well ascribe to, then it is interesting to note that Boisot's model is an interpretation of Shannon and Weaver's (1949) original work on information; a pioneering work for the Computing field, upon which much current theory is based. Perhaps the Business sector has assumed a position that we (as Computing specialists) would rather be in, namely: operating within one paradigm. By reinterpreting the seminal ideas that constructed the current opposing paradigms in the Computing field, we may well have been led to consider information, less as if it were a material object, but rather as the more subjective, perspective-laden entity that it is; data plus meaning. As it stands at present, operating within the functionalist paradigm, we are left somewhat

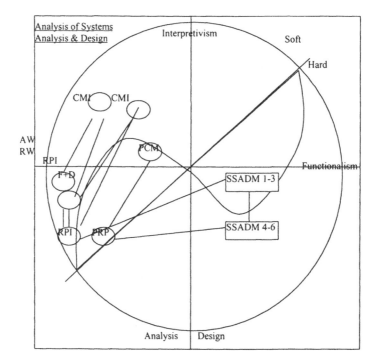

Figure 2. Multi-paradigm shift-free integration

with an 'attempting to describe an amoeba with a clockwork toy'- type syndrome and consequently, we find ourselves plunging into the interpretivist paradigm to master meaning attribution associated with stakeholders' individual interpretations. Has the Business sector 'got it right'? Is this why they are apparently not faced with a philosophical problem? The question is however, superfluous. What is of paramount importance, is to support in philosophical terms, current practice. Systems analysis and design practitioners consistently use methodologies' 'best-bits' in order to arrive at the required 'solution' and within the Computing field this may equate to shifting between paradigms; but if practitioners are largely unaware of their 'crime', theorists recoil at such casual methodological application. Further, it is hardly practical to advocate obliterating the paradigms upon which so much of the discipline is based. Shift-free Integration can thus support technique integration across opposing paradigms as well as within one paradigm.

Focusing on the 'opposing paradigm' case first, the reader is referred to Figure 2: Multi-paradigm Shift-free Integration; a representation of the Domain of Systems Analysis and Design, which supports the integration of methodological techniques (specifically the CM and DFD), within a cogent and philosophically acceptable framework. An initial rich picture (RP[1]) is reformulated via the conceptual model and via the feasible / desirable 'loop' until an agreed primary conceptual model (PCM) helps construct an agreed primary rich picture (PRP). As a real world model, the PRP may then form the basis for a design level DFD. Similarly, if we imagine our model in three dimensions, we may also legitimise the use of an analysis level DFD to help in the reformulation process of the RP. In both cases, there is no illegal 'crime' committed. We are not moving from SSM's interpretivist abstract world to the real world of SSADM's functionalist paradigm (and thus paradigm 'shifting', as we would be if we attempted to move from a CM to a DFD), but simply moving sideways, entirely within the real world, along the same continuum. If we adhere to the belief that continuums define the space between 'poles'; and sympathise with the notion of rotational symmetry, then it follows that a model in application is more or less functionalist, or more or less interpretivist, but never (in reality) absolutely one or the other.

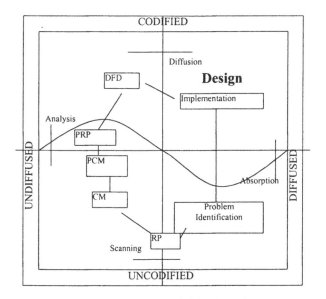

Figure 3. Single-paradigm shift-free integration

Focusing on the single paradigm case, Figure 3: Single-paradigm Shift-free Integration demonstrates a mechanism by which methodological techniques may be linked together within the same philosophical paradigm. If we are working to one paradigm (in this case, Boisot's "Culture-Space"), then superimposing the Cyclical Flow of New Knowledge onto the base of our proposed model, allows us to operate a methodology technique integration without paradigm shifting, or bursting through some binary demarcation between two worlds. Stafford Beer (1985) maintained that a model is neither true nor false, but more or less useful. This is a humbling dictum and one which serves to remind us of the value of flexibility. In figure 3, f we consider *codified* as an interpretation of functionalism and *less codified* as an interpretation of interpretivism (as orientations, not paradigm demarcations), then we are able to use Boisot's Culture-Space as a single-paradigm arena to support the practice of methodology integration. If we adhere to Boisot's framework, then in endorsing the optimisation of information-based systems in a Culture-Space (i.e. an organisational environment), with the notion of cyclic, rotational symmetry (as we do in Shift-free Integration), we can legitimise the passage through the paradigm, via Boisot's four identified phases (scanning, analysis, diffusion and absorption) and attribute individual methodological techniques to those phases, thus verifying Shift-free Integration. Recent (Anglo-Saxon) epistemology may be employed to buttress the arguments propounded so far. This section of the work will be informed by Haack (1993).

In contemporary epistemology, a distinction is usually made between foundationalist and coherentist accounts of epistemology. In essence, the distinction is concerned with the question as to whether any beliefs are privileged over others. It will be argued that both hard and soft approaches adopt predominantly foundationalist strategies. It is doubtful whether coherentist approaches could be sensibly employed within systems analysis, as such approaches are not intended to explain the acquisition of "piecemeal" knowledge, "The characteristic theses of coherentist theories of justification are that justification is exclusively a matter of relations among beliefs, and that it is the coherence of beliefs within a set which justifies the member beliefs." (Haack, 1993:17).

The 'nearest thing' to an example of a coherentist strategy in IS analysis might be the prescriptions to carry out 'cross-referencing' between different products, as should be done when using the U.K.'s Structured Systems Analysis and Design Method (SSADM), for example. If anomalies are generated during the cross-referencing process then this would require some further analysis to be carried out to rectify the situation. However, no methodology (of which the author is presently aware) relies entirely on the procedure of cross-referencing to generate the 'true' requirements. Such faith in this procedure would imply that because cross-referencing between different products can be successfully carried out it follows that the products resulting from the analysis of the information system are entirely correct. What happens in such approaches would seem to be that the 'elementary' elements of analysis (DFDs, entity models, etc) are seen to be epistemologically foundational - but fallible. Cross-referencing exposes some of the errors made during 'elementary' (i.e. foundational) analysis.

Foundationalist epistemologies assume that some beliefs are 'privileged' (fundamental, or more basic) than others:

> "... [A] theory qualifies as foundationalist which subscribes to the theses:

> (FD1) Some justified beliefs are basic; a basic belief is justified independently of the support of any other belief;

> and:

> (FD2) All other justified beliefs are derived; a derived belief is justified via the support, direct or indirect, of a basic belief or beliefs." (Haack, 1993:14).

Both IS approaches (hard and soft) subscribe to this thesis; it is in which type of beliefs that they privilege that their real differences lie. Haack finds that there are many variants of foundationalism. The two which are of most importance in this analysis are (what she characterises as):

1. The experientialist version of empirical foundationalism.

2. The extrinsic version of empirical foundationalism.

Here, 'empiricist' should not be assumed to be 'objectivist', "'Empirical', here, should be understood as roughly equivalent to 'factual', not as necessarily restricted to beliefs about the external world... one style of empirical foundationalism [the experientialist version] takes beliefs about the subject's own, current, conscious states as basic, another [the extrinsic version] takes simple beliefs about the external world as basic..." (Haack, 1993:15).
It can now be concluded that, broadly speaking, soft approaches assume the experientialist version to be the case, whilst the hard approaches assume the extrinsic version to be the case:

> "...[A]ccording to the experientialist version of empirical foundationalism, basic beliefs are justified, not by the support of other beliefs, but by the support of the subject's (sensory and/or introspective) experience; according to the extrinsic version of empirical foundationalism, basic beliefs are justified because of the existence of a causal or law-like connection between the subject's having the belief and the state of affairs which makes it true..." (Haack, 1993:15).

Now, what is one to make of a statement such as this:

> "With regard to epistemology we may identify two extreme positions of positivism and interpretivism. Positivism is characterised by a belief in the existence of causal relationships and general laws that may be identified and investigated through rational action. In contrast, interpretivism allows that no individual account of reality can ever be proven as more correct than another since we are unable to compare them against any objective knowledge of a 'true' reality." (Lewis, 1994:138).

The 'extreme' called 'positivism' may be identified with the extrinsic version of empirical foundationalism, and the 'extreme' called 'interpretivism' may be identified with the experientialist version of empirical foundationalism. However, these positions should not be presented as being in binary opposition to each other; both are variants of foundationalism. Furthermore, the more general distinction between foundationalism and coherentism does not even get a mention in such accounts. Moreover, there are many 'variations on these themes' within contemporary epistemology; and Haack finds that within foundationalism there are four basic variants of both coherentism and foundationalism (some of which are further subdivided). However, these variants need not be discussed further in this work.

4.   CONCLUSION

This reorientation in the way we think about the world is harnessed in Shift-free Integration. The authors have presented a philosophy which: i) helps us define Information Systems as a discipline; ii) harnesses interpretivist and functionalist thinking into a framework which supports the integration of philosophically distinct methodological approaches without paradigm shifting; iii) allows Information Systems developers to make use of interpretative thinking as their need to manage and define information systems would dictate; and which iv) exists in line with our present tendency to disregard the exclusive paradigms that the Western scientific tradition of creating dichotomies has helped to construct, but which rather operates 'across' these paradigms, viewing the dichotomies as extreme ends of continuums along which system developers can 'migrate'. To answer our question: Does this mean that dichotomies should not exist? the author would respond - no. It is not the dichotomies which hinder our progression, it is the way we view them in opposition to each other, rather than considering them in dynamic symmetry which engenders our construction of mutually exclusive paradigms, which we then dare not disregard. All that is required is a shift in thinking. Following Haack (1993), we are arguing that objective judgements always call for the uncritical acceptance of certain standards as to what is - and what is not to count as evidence; correspondingly, subjective judgements about what is the case - whilst lacking standards for their complete acceptance - give indications as to what might be the case (i.e. they are truth-indicative). However, Haack also argues that powerful elements of the coherentist thesis should also be included in a reconstructed epistemology (she calls this 'foundherentism'). Such an approach as Haack suggests would both advocate and enhance the value of any cross-referencing procedures that could be developed between the two approaches to IS analysis (hard and soft). Whilst further discussion lies outside the scope of the present work it can be concluded that the epistemological differences can sensibly be reconciled by admitting:

1. The truth-indicative nature of bona fide subjective judgements about what appears to be the case.

2. The ineliminably subjective nature of objective judgements about what is the case.

Such a view would, at any rate, appear to be endorsed by some of those reflectively engaged in IS practice, such as Gammack - who concludes that, "...[S]ubjectivity and objectivity exist as extremes on a continuum where the difference between them is one of degree rather than of kind." (Gammack, 1995:181). We conclude that such a continuity should be formally recognised, and we have proposed a basis upon which further work can be developed in order to operationalise IS development along such lines.

REFERENCES

Avison, D and Wood-Harper, A (1990): Multiview: An Exploration in Information Systems Development, Blackwell Scientific, Oxford.

Baudrillard, J (1983): Simulations, translated by Paul Foss, Paul Patton and Philip Beitchman, Semiotext(e) Inc., foreign agents series, New York City.

Beer, S (1985): Diagnosing the System for Organisations, John Wiley and Sons Ltd., Chichester.

Burrell, G and Morgan, G (1979): Sociological Paradigms and Organisational Analysis, Gower Publishing Co. Ltd., Aldershot.

Capra, F (1991): The Tao of Physics, 3rd Edition, Flamingo, London.

Checkland, P B (1981): Systems Thinking, Systems Practice, John Wiley and Sons Ltd., Chichester.

Checkland P B and Holwell, S (1998): "Action research: its nature and validity", Systemic Practice and Action Research, Plenum Publishing Corporation, Vol. 11, No. 1.

Checkland, P B and Scholes, J (1990): Soft Systems Methodology in Action, John Wiley and Sons Ltd., Chichester.

Gammack, J (1995): "Modelling subjective requirements objectively", in: Stowell, F (Ed.): Information Systems Provision: The Contribution of Soft Systems Methodology, McGraw Hill, Maidenhead.

Haack, S. (1993), Evidence and Enquiry, Blackwell, Oxford.

Lewis, P. (1994), Information-Systems Development, Pitman, London.

Lewis, P J (1995): "New challenges and directions for data analysis and modelling", in: Stowell, F (Ed.): Information Systems Provision: The Contribution of Soft Systems Methodology, McGraw Hill, Maidenhead.

Liang, Y; West, D and Stowell, F (1998): "An approach to object identification, selection and specification in object-oriented analysis", Information Systems Journal, Blackwell Science Ltd., Vol.8.

Miles, R (1988): "Combining 'hard' and 'soft' systems practice: grafting or embedding?", Journal of Applied Systems Analysis, No. 15.

Mingers, J (1992): "A review of the Soft Systems Methodology and Information Systems seminar", Systemist, Publication of the United Kingdom Systems Society, Vol. 14, No. 3.

Mingers, J (1995): "Using Soft Systems Methodology in the design of Information Systems", in: Stowell, F (Ed.): Information Systems Provision: The Contribution of Soft Systems Methodology, McGraw Hill, Maidenhead. Boisot, M (1988): Information and Organisations (The Manager as Anthropologist), Fontana/ Collins, London.

Prior, R (1990): "Deriving data flow diagrams from a soft systems conceptual model", Systemist, Publication of the United Kingdom Systems Society, Vol. 12, No. 2.

Sawyer, K (1991): "Linking SSM to DFDs: The two epistemological differences", Systemist, Publication of the United Kingdom Systems Society, Vol. 13, No. 2.

Shannon, C E and Weaver, W (1949): The Mathematical Theory of Communication, The University of Illinois Press, Urbana.

Skyttner, L (1998): "The future of systems thinking", Systemic Practice and Action Research, Plenum Publishing Corporation, Vol. 11, No. 2.

Stowell, F A (1995): "Empowering the client: the relevance of SSM and interpretivism to client-led design", in: Stowell, F A (Ed.): Information Systems Provision: The Contribution of Soft Systems Methodology, McGraw Hill, Maidenhead.

Stowell, F A and West, D (1992): "SSM as a vehicle for client-led design of information systems utilising ideal type 'mode 2'", Systemist, Publication of the United Kingdom Systems Society, Vol. 14, No. 3.

IMPROVED SOFTWARE QUALITY THROUGH REVIEWS AND DEFECT PREVENTION:
AN INCREMENTAL IMPROVEMENT PATH

Edwin M. Gary[1], Oddur Benediktsson[2], and Warren Smith[1]
[1]*Dept. of Computer Studies, Glasgow Caledonian University Glasglow G4 OBA, U. K.*

[2]*Visiting Professor at Dept. of Computer Studies, Glasgow Caledonian University, Glasglow G4 OBA, U. K. from University of Iceland, Reykjavik, Iceland*
E-mail: egra@gcal.ac.uk

ABSTRACT

Fixing defects is an integral part of all software development and maintenance effort. This activity is often performed in an informal way . the defects uncovered are fixed on an ongoing basis. Numerous measurements show that this is not an effective approach. It has on the other hand been shown that reviews are the most cost effective way presently known of unearthing defects.

The processes of Review (R) and Defect Prevention (P) are normally intertwined (R&P). If a software organisational unit (OU) is running R&P that is more or less on SPICE level 0-1. The question then is what improvement path should be chosen?

This paper considers a number of reviewing/defect prevention techniques including self-review, buddy review and peer review; and with strong reference to the ISO/IEC 15504 (SPICE) process capability levels suggests a novel framework containing a possible sequence of improvement steps towards the Established Level 3. The paper is supported by interesting data and case studies on reviews from the University of Iceland.

Keywords: review, defect prevention, software process improvement, software engineering.

1. Introduction

Experience over the past two decades has shown that up to 30 to 50 percent of typical software development effort is devoted to correcting defects (rework.) Formal peer reviews (Fagan inspections [4]) have been shown to be an effective way of uncovering defects, finding up to 60-90 percent of all defects and bringing the rework down significantly "as well as providing feedback that enables programmers to avoid injecting defects in future work" [5]. Yet the acceptance of peer review by the software industry in general has been surprisingly rather slow.

Systems Development Methods for Databases, Enterprise Modeling, and Workflow Management,
Edited by W. Wojtkowski, *et al.* Kluwer Academic/Plenum Publishing, New York, 1999.

Other types than peer reviews have also been shown to be highly effective namely self reviews [6] and buddy reviews [1]. These reviews are much lighter in use than peer reviews.

The main conclusion in this paper is that an incremental improvement path can be defined for the review process that commences with self reviews, then goes on to buddy reviews and finally reaches peer reviews. Furthermore, it seems quite plausible that all three types of reviews should be employed on an ongoing basis in a staged fashion.

2. Review technique considerations

The reviews under consideration here are design reviews and code reviews conducted as peer reviews, buddy reviews or self reviews. These reviews are primarily geared to monitoring the product as opposed to the process.

Peer review: Structured process for team review in order to locate defects in author's work item that typically could be a program module or some sheets of design. The process is planned, measured and results are documented. There are different roles to be performed: author, moderator, recorder (scribe), reviewer (peer tester). The essential process steps are *planning* (the moderator assembles the team and plans the schedules the work), *overview* (the work item and associated documents are presented to the reviewers), *preparation* (the reviewers thoroughly read the work item and use checklists to locate potential defects), *review meeting* (all potential defects found are examined and classified and actual defects recorded), *rework* (the author mends the defects that were uncovered) and finally *follow-up* (the moderator verifies that mending has taken place successfully).

Buddy review: A review activity performed by a co-worker in order to locate defects in author's work. Buddy review (also termed two-person review [1]) is known to be practised on an ongoing basis in software development where co-workers pair up and review each others work items as soon as they are completed.

Self review: Thorough examination of own work in order to locate defects. Self reviews are one of the cornerstones of the recently defined Personal Software Processes [6]. When an author is finished writing a work item (e.g. implementation of a module) he or she systematically reviews the work, logs all defects found and classifies the defects according to phase of origin and type and records the time taken to mend the defect.

We will take buddy reviews and self reviews as "private" reviews in that the defects found and summary reports are information that is only seen by the developer(s) involved and is intended for defect prevention on a personal basis. If the data generated is processed further it must be done so on an impersonal basis.

Peer reviews are on the other hand considered to be "public" here and the data that they generate may be used to improve the software processes on a group basis.

The data that is gathered from design or code reviews is of two types:
1. Process data typically includes preparation time, lines of code per hour of preparation time, errors identified during preparation (by category), hours per error found in preparation, review time, lines of code (or design statements) inspected per hour, and errors found per review person-hour (by category).
2. Product data from the review typically includes errors found per line of code (or design statement), action items identified from each review, action items closed for each review, items needing review, and reviews conducted.

Table 1 Percentage of time spent on defect removal (rework)

Project/Student	1	2	3	4	5
1	57	26	22	11	12
2	44	9	28	20	5
3	40	13	12	10	15
4	20	14	16	13	13
5	8	2	8	6	20
6	4	32	30	20	13
7	38	28	21	24	28
8	54	34	16	9	27
9	9	11	3	9	27
Average %	30	19	17	14	18
Stdev	20.4	11.3	8.9	6.2	8.1

The preparation plays a critical role in the review process as it is the study of the review material by each person doing the review. Each does this on his or her own to understand its logic and intent. It is recommended that the review should study: (1) the ranked distribution of error types found by recent reviews, and
(2) checklists of error detection guidelines [8]
There is ample evidence in the literature for the effectiveness of the reviews under consideration here (see for example the compendium by Wheeler [9]).

Table 1 below demonstrates the effectiveness of self review. The results are from the course Software Science at the University of Iceland, Fall 1996. Employment of self review was used to analyse defects generated and measure time spent on rework in C++ programming projects. The projects were single student projects.

The average time spent on defect removal decreased as the work on the projects progressed from 30% in the first project to 14%-18% in the last projects.

3. Review as a process

Review:
1. In systems/software engineering, a formal meeting at which a product or document is presented to the user, customer, or other interested parties for comment and approval. It can be a review of the management and technical progress of the hardware/software development project.
2. In hardware/software engineering, the formal review of an existing or proposed design for the purpose of detection and remedy of design deficiencies that could affect fitness for use. Also reviewed here are the environmental aspects of the product, process, or service, and potential improvements of performance, safety, or economical aspects (ANSI/IEEE Standard 729-1983).
3. In the spiral software development model, the review covers all products developed during the previous cycle, including the plans for the next cycle and the resources required to carry them out. The major objective of the review is to ensure that all concerned parties are mutually committed to the same approach [2].

The emerging international standard ISO/IEC 15504 *Software process assessment* [7] defines a process as "a set of interrelated activities, which transform inputs into outputs". (During the development of this suite of standards it had the acronym SPICE (Software Process Improvement and Capability dEtermination) and we will use that name in this article.)

The process that is termed review process here is concerned both with uncovering defects and mending them. A more appropriate name for the review process is "Review and defect correction process".

The purpose of the review process is to efficiently find and remove defects in work items. The essential process activities are:
- **Plan the review**. Select the work items to be reviewed, assign roles and schedule the work and identify check lists and standards to be employed.
- **Distribute review material**. Meeting. Distribute the material to reviewers and give overview over the work item.
- **Prepare for review meeting**. The reviewers thoroughly examine the work item employing checklists to locate potential defects.
- **Conduct review meeting**. All potential defects found are examined in common and classified and actual defects uncovered are logged.
- **Mend defects**. The author mends the defects that were uncovered recording the time needed to mend each defect and classifies the defects according to type and phase of origin.
- **Follow-up**. The moderator verifies that mending has taken place successfully.

The input to a review is the work item to be scrutinised (pieces of design or code) and the relevant check lists enumerating the types of defects to be looked for. Other documents such as coding standards may also be required as input. The output of a review is firstly the log of the defects uncovered and secondly the mended work item after rework has taken place and thirdly a summary report showing the defect frequencies and fix times by defect classification is needed if the information gathered is to be used for defect prevention.

4. Incremental development of the review process

A novel approach is taken to process improvement in SPICE in that a separate dimension is created for process improvement; the so called capability dimension. There are six capability levels in the SPICE framework and process attributes (PA) that indicate whether a process has reached a given capability:
- **Level 0 - Incomplete**
- **Level 1 - Performed**
  PA 1.1 Process performance attribute
- **Level 2 - Managed**
  PA 2.1 Performance management attribute
  PA 2.2 Work product management attribute
- **Level 3 - Established**
  PA 3.1 Process definition attribute
  PA 3.2 Process resource attribute
- **Level 4 - Predictable**
  PA 4.1 Measurement attribute
  PA 4.2 Process control attribute
- **Level 5 - Optimising**
  PA 5.1 Process change attribute
  PA 5.2 Continuous improvement attribute

Table 2 below gives an indication of the attributes of a review process according to the SPICE capability levels. (Level 0 and Level 5 are omitted for brevity.)

The activities on the different levels are additive in that on Level 2 all the activities on Level 1 must be carried out and so on.

Self reviews and buddy reviews can be seen to operate on any of the four stated levels in addition for that matter Level 0 where the reviews are carried out without any logging or analysis. But by definition a peer review must operate at least on Level 3 more or less.

Table 2 Review process attributes according to SPICE capability levels

1 - Performed	2 - Managed	3 - Established	4 - Predictable
Distribute material. Give overview. Examine work item. Uncover defects. Log defects. Mend defects. Log review and mend effort.	Estimate work. Plan and schedule. Summarise defects. Employ checklists. Verify mends.	Keep process script. Follow defined process. Assign roles. Commit resources. Process historical data. Improve checklists.	Predict defect density. Predict review yield. Predict residual defect density.

Self reviews and buddy reviews have been found effective in reducing rework on an individual basis and improve the quality of the work of individuals [1], [6]. It is also relatively easy to introduce these reviews, for example, in a university student context.

An incremental improvement path may be suggested in the following way: Start with self review for a group of developers and take it up to Level 3-4 on individual basis. Collate check lists and experience gained from the self review and defect prevention results. Use the experience to define buddy review process that can start at Level 2 and continue to use that until Level 3-4 has been reached. At this Level enough material has been gathered and experience gained to define an effective peer review process and start deploying it on Level 3.

5. Conclusions

Peer reviews have been slow in gaining acceptance in the software industry in general even though it has been shown that scarce resources are being wasted due to excessive rework. It is conjectured that the reason for the slow acceptance is that most organisational units are working with their processes on SPICE Level 1-2 and find it hard to start up with a full blown peer review that is on Level 3 to be effective. It is also to be noted that in the CMM (Capability Maturity Model) process framework where the processes are ordered on capability levels that peer review is placed on the third CMM level.

Self reviews and buddy reviews have been shown to be effective in defect detection and prevention on a personal basis and thus serve a definite purpose. These reviews are lightweight in the sense that they can be employed without much discipline and without involving several software developers. It is to be noted, however, that peer reviews have been shown to be highly effective in defect removal and need to be employed where high quality software is being developed.

It is concluded in this paper that an incremental improvement path can be defined in order to get to the stage of deploying peer reviews effectively. That is: start with self reviews, followed by buddy reviews then peer reviews.

It is apparently effective to employ all three types of reviews on an ongoing basis. The question then arises as to defect removal effectiveness of each stage and how much review effort should be put into each stage. The scope (and checklists) of defect detection would vary somewhat with each stage, that is from the author's private scope in a self review to the full blown public scope of a peer review. The authors intend to do further research on theses issues.

References

[1] D. B. Bisant and A. Lyle JR., "A Two-Person Inspection Method to Improve Programming Productivity." IEEE Trans Software Eng Oct. 1989; Vol. 15, No. 10:1294-1304

[2] B. W. Boehm, "A Spiral Model of Software Development and Enhancement." In *Software Engineering Project Management*," R.H.Thayer, ed., Washington, D.C. IEEE Computer Society Press, 1988.

[3] "*Capability Maturity Model Practices*," CMU/SEI-93-TR-25, Software Engineering Institute, Carnegie Mellon University.

[4] M. E. Fagan , "Design and code inspection to reduce errors in program development," IBM System J 1976: Vol. 15, No. 3:182-211.

[5] M. E. Fagan, "Advances in Software Inspection," IEEE Trans Software Eng, July 1986;Vol. 12, No. 7:744-751.

[6] W. S. Humphrey, "*A Discipline for Software Engineering,* " Addison Wesley, 1995.

[7] "ISO/IEC Draft software process assessment standard," parts 1-9, ISO/IEC 15504, (SPICE).

[8] B. A. Kitchenman, B. P. Kitchenman , and J. P. Fellow , "The effects of inspections on software quality and productivity," ICL Technical Journal; May 1986; 3:1:112-122.

[9] D. Wheeler , et al., " Software Inspection - An Industry Best Practice," IEEE; 1996.

IDENTIFYING SUCCESS IN INFORMATION SYSTEMS:
IMPLICATIONS FOR DEVELOPMENT METHODOLOGIES

Steve Page
Bolton Business School, Deane Road, Bolton BL3 5AB, U. K.

INTRODUCTION

Despite having used system development methodologies for nearly 20 years, the development community still manages to build information system failures with alarming regularity. This research considers the most financially successful organisations in the UK., to see if they employ information systems that are liked and used by the workforce (that these systems are helping to meet organisational goals as assumed a priori). The research then identifies a number of indicators behind the workforces' attitudes towards their systems, that give rise to recommendations that systems development methodology builders should incorporate into future versions of development methodologies.

TECHNOLOGY IMPACT & SYSTEM FAILURES

We have been using systems development methodologies (SDM) now for nearly 20 years, yet a cursory glance at almost any daily newspaper, or computing magazine shows that we are still developing incredibly expensive IS failures (e.g. Sauer, 1993; Drummond, 1996; Schneider, 1997). Doherty & King (1997) suggest that it is primarily the organisational issues that lead to computer system failures. Certainly, empirical research shows that the development of IS and the subsequent introduction of computer technology into an organisation does have an impact upon the workforce therein. However, these studies show that the perceived impact may be positive as well as negative. Millman and Hartwick (1987) report that middle managers perceive computer technology positively, reporting increased autonomy, useful delivered systems, and increased control over the results of their work. Conversely, Kaye and Sutton (1985) report that the quality of working life of office workers deteriorates post computerisation. Their findings also indicate that individuals are less able to see the results of their work post computerisation. These factors lead to a negative impact on the workforce. Thus, the suggestion is that management is willing to embrace computer technology (Porter, 1987); whereas, non-managerial staff are likely to be disenchanted by the prospect of computerisation (Turner, 1984).

Systems Development Methods for Databases, Enterprise Modeling, and Workflow Management,
Edited by W. Wojtkowski, *et al.* Kluwer Academic/Plenum Publishing, New York, 1999.

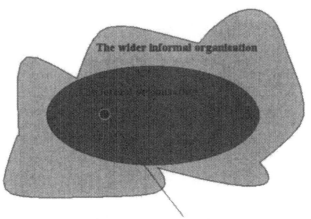

Computerised Information System

Figure 1: Computerised information system residing within its environment.

Avison and Fitzgerald (1995) state that the effective management of information technology (IT) development & implementation requires user involvement at all stages. Wong and Tate (1994, p. 51) concur, stating that:

> The involvement of users in information systems development, whether it is to design new systems or to modify existing ones, is held to be one of the most important factors influencing implementation success or failure.

Poor communication between management and the workforce when a computerised IS is being developed and implemented is a trait of the classic 'structured' approach to systems development still prevalent in many organisations (Warner, 1996). This often leads to under-utilisation of the delivered system, or even system rejection (Sauer, 1993). Computerised IS exist within a multifarious, highly complex informal arena (see figure 1), but Watson & Wood-Harper (1996, p. 63) state that:

> ... analysts have often neglected the way meaning varies with organisational context in favour of developing a general static structure ... analysts have therefore frequently taken only the formal aspects of an information system seriously.

Based on the information concerned in this section, to not produce an IS failure, organisations should adequately address the non-technical, organisational and stakeholder issues surrounding systems development and implementation, and ensure that employees are kept well informed of developments.

RESEARCH AIMS

This research considers how well the issues outlined in the introduction are addressed in the UK's 100 most financially successful (largest) organisations, based on 1994 turnover (Times Books, 1994). The rationale for considering these organisations in this study is that as they are financially successful, it is highly likely that their computerised IS are helping to meet organisational goals.

The research objective is to identify whether these large organisations are employing 'good practice', in terms of ensuring that their information systems have a positive impact upon their workforce. Specifically to:

i) Determine:
   a) The extent to which development needs are addressed (Jayaratna, 1992; Walsham, 1993);
   b) The degree of stakeholder involvement during systems development (Ives & Olson, 1984; Baroudi et al., 1986);
   c) The degree of stakeholder involvement during system implementation (Mason & Mitroff, 1981; Ives & Olson, 1984);
   d) The level of perceived usage (Igbaria et al., 1989; Turner, 1984);
   e) The ability to perform a job post computerisation (Kaye and Sutton, 1985; Millman and Hartwick, 1987);

ii) Determine any implications for new versions of current SDMs, & the development of new SDMs in the future.

METHODOLOGY

This research uses the survey method, and develops a number of hypotheses associated with the study. The associated relationships are tested using the appropriate statistical technique, to test for significance at the 5% level. The research uses a self-administered questionnaire to collect information. The rationale for this choice is twofold:

i) Given the research base, it is not possible to visit the site of every organisation in the survey, this is prohibitive in terms of both time, and finances.

ii) More importantly, as the study requires responses from managers and senior executives, the researcher feels that a better response rate is forthcoming if these people are allowed to complete the questionnaire in their own time.

A condensed version of the research instrument is attached to this paper as an appendix.

RESPONSE TO THE SURVEY

The initial posting results in 54 completed questionnaires being returned - a response rate of 5.8%. With 2 follow-up reminders this increased to a total of 192 completed questionnaires - a final response rate of 20.6%.

As the second reminder only produced a further 37 completed questionnaires, the researcher decided that a third reminder was inappropriate, as the (anticipated) small number of extra responses would not make any statistically significant changes to the results of the study.

All of the industries within which the Top 100 companies operate are represented in this study - this is demonstrated in table 1. Further analysis shows that the total response is *representative* of the whole; that all industries are represented, and that the number of companies participating in the study is sufficient to permit the development and subsequent 'testing' of hypotheses.

Table 1: Analysis of Respondents by Industry. Table compiled from information contained in 'The Times 1000, 1995' (Times Books, 1994, pp. 18-21), all figures expressed to nearest full percent for clarity.

	Total Companies in Survey	Companies Participating	Companies Responding but not Participating	Total Companies Responding
Leisure Industries	16	9 56%	3 19%	12 75%
Food	12	2 17%	6 50%	8 67%
Fuel	8	6 75%	1 13%	7 88%
Packaging & Transport	9	3 33%	2 22%	5 55%
Electronics & Communications	7	2 29%	2 29%	4 58%
Chemicals, Engineering & Textiles	9	4 44%	2 22%	6 66%
Building & Property	6	3 50%	1 17%	4 67%
Stores & Commodities	15	5 33%	6 40%	11 73%
Other	11	3 27%	3 27%	6 54%
Total	93	37 40%	26 28%	63 68%

RESULTS & DISCUSSION

At the personal response level, 134 (equal to 70%) of the respondents are male, thus only 30%, i.e. 58 respondents are female. The reason for the substantial percentage difference of respondents' gender is due to the male dominance of managerial and executive positions generally found throughout organisations in the UK, in keeping with other countries. This is almost exactly the same as Igbaria & Toraskar (1994) report. Similar results are found in other studies, e.g. Parasuraman & Igbaria (1990). Managers account for approximately 43% (83 in number) of respondents, senior executives 21% (41 in number), and the remaining 36% (67 in number) are end-users.

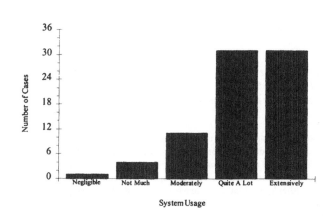

Figure 2: Extent of Actual Usage of System Compared to Potential Number of Users

Turning to the actual use of IS, Mason and Mitroff (1981) state that successful IS development and subsequent implementation is only possible through taking account of the viewpoints of as many stakeholders as possible; a view supported by Bell & Wood-Harper (1998), and Flynn (1998). System usage is reported as being positively related to involvement in the development process by Lucas (1978), or Baroudi et al. (1986).

For this survey, reported system usage is high. Sixty-two out of 78 respondents report their system is used either 'quite a lot', or 'extensively' (see Figure 2). Nevertheless, as is clear from observing Figure 3, there is no consensus of action across the organisations involved in the survey, as to the involvement of users during the systems development process. *(H_o) One* is formulated to test whether a relationship exists, for this study, between system usage and involvement during development.

267

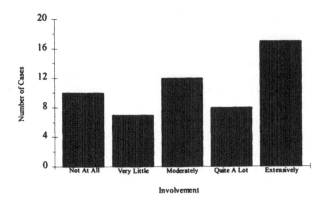

Figure 3: Involvement During Systems Development

*(H_0) One: The perceived level of usage of the delivered computer system
is not related to the degree to which the stakeholders were involved during
the development of the computer system.*

Figure 4 shows the degree to which respondents to the survey perceive that
their needs were addressed when their current computer system was being
developed - only 13 of 86 respondents (11%) reporting that their needs were not
adequately addressed; whereas, 66% of respondents report that their needs were
addressed either 'quite a lot', or 'extensively'. Bell & Wood-Harper (1998) state that
addressing the needs of stakeholders is fundamental if the delivered system is to be
acceptable to stakeholders.

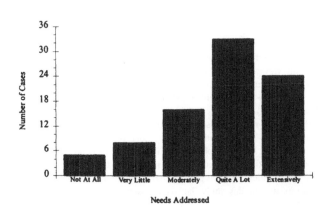

Figure 4: Information Needs Addressed During System Development

Regarding the perceived ability to perform a job post computerisation, Figure 5 demonstrates that of the 54 responses 24 people think that they are 'better' at their job post computerisation, and 16 people think that they are 'much better' at their job. The remaining 13 respondents report 'about the same as before'. Thus, nobody reports being either 'worse', or 'much worse' at their job post computerisation. Kaye & Sutton (1985) report negative perceptions by end-users post computerisation. Similarly, Turner (1984) reports that end-users are likely to be disenchanted by computer systems. The findings from this survey do not concur with either Kaye & Sutton's (1985), or Turner's (1984) findings. (H_0) Two is formulated to test whether, for this study, a relationship exists between the extent to which users' needs are addressed during development and their perceived ability to perform their job post computerisation.

(H₀) Two: *The degree to which stakeholders perceive they are able to perform a job is not related to the degree to which the stakeholders' needs were addressed during the development of the computer system.*

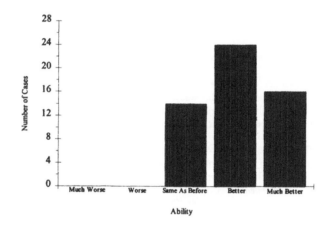

Figure 5: Perceived Ability To Perform A Job Post Computerisation

It is clear from observing Figure 6 that in keeping with involvement during systems development (discussed earlier) there is no consensus of action across the organisations involved in the survey, as to the involvement of end users during system implementation. There are many proponents of involvement during systems implementation, e.g. Baroudi et al. (1986) who state that involvement throughout all stages of development is vital if the delivered system is to be seen as useful by the stakeholders. Similarly to development, Bell & Wood-Harper (1998), and Flynn (1998) advocate that participation (involvement) during the implementation process is necessary if a positive attitude towards the computer system is to follow. Stowell and West (1994a; 1994b) advocate involvement as a prerequisite for system success, stating that the implementation stage should be 'dominated by the clients' who must, *inter alia*, consider the effect(s) that the new system has on the working practises of the staff. However, Kaye and Sutton (1985) report that computerisation has a deleterious effect on the quality of a person's working life, and Ives and Olson's (1984) findings indicate that involvement during development and implementation is not related to system usefulness.

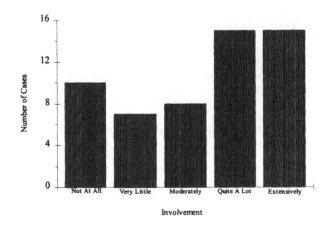

Figure 6: Involvement During Systems Implementation

As already reported, nobody reports they are worse at their job post computerisation (see Figure 6). In fact, 40 out of 54 respondents report that they are either 'better', or 'much better' at their job post computerisation. (H_O) *Three* is formulated to test whether, for this study, a relationship exists between the extent to which users are involved during implementation of their computer system and their perceived ability to perform their job post computerisation.

*(H_O) **Three:** The degree to which a stakeholder perceives that their computer system assists their ability in performing their job is not related to the degree of involvement enjoyed by the stakeholder when the system was being implemented.*

Table 2: Statistical Cross-Tabulation for Stakeholder Involvement During Development against Perceived Level of Usage of the Delivered System

		Involvement		
		Little / Moderate	Considerable	Row Total
Usage	Little / Moderate	10	4	14 23%
	Considerable	32	15	47 77%
	Column Total	42 68.8%	19 31.2%	61 100%
Pearson Chi-Square				
Value	Degrees of Freedom	Minimum Expected Frequency	Number of Cells < Minimum	Significance
0.056	1	4.36	1 = 25%	0.883

OUTCOMES FOR HYPOTHESES

The results of the cross-tabulation for *(H$_O$) One* are represented in table 2, statistically $n=61$, $\alpha=0.05$, and $p=0.816$. Therefore, as $p>\alpha$ (in fact much greater) the null hypothesis *(H$_O$) One* is not rejected, and for this study, there is no significant relationship between the degree of stakeholder involvement during systems development and the perceived level of usage of the delivered computer system, when tested at the 5% level. Unlike Lucas (1978), or Baroudi et al. (1986) who report that system usage is positively related to involvement during the development process, this research found no statistically significant relationship between these two variables. Clearly, there is a need for further study to ascertain whether stakeholder involvement during development is a valid prerequisite for the successful implementation of a computer system, e.g. Leonard-Barton (1988), and Jarvenpaa & Ives (1991); or, whether stakeholder involvement is not particularly important as a determinant of system success, thus confirming the findings of Tait & Vessey (1988), and Flynn & Ade (1996).

(H$_O$) Two cross-tabulates perceived ability to perform a job with the degree to which needs were addressed during systems development. As demonstrated in table 3, statistically, $n=36$, $\alpha=0.05$, and $p=0.020$. Therefore, as $p<\alpha$ the null hypothesis *(H$_O$) Two* is rejected, and for this study, there is a significant relationship between the degree to which a stakeholder's needs were addressed during development of the system and the perceived ability to perform a job post computerisation, when tested at the 5% level.

Table 3: Statistical Cross-Tabulation for Ability to Perform a Job against Needs Addressed During Development

		Needs Addressed		
		Little / None	Considerable	Row Total
Ability	No Improvement	7	2	9 25.%
	Better at Job	9	18	27 75%
	Column Total	16 44.4%	20 55.6%	36 100%
Pearson Chi-Square				
Value	Degrees of Freedom	Minimum Expected Frequency	Number of Cells < Minimum	Significance
5.40	1	4.0	1 = 25%	0.020

As the established relationship is non-directional in the statistical analysis, observation of the actual data is needed to determine whether the relationship is the right way round. By scrutinising this, it is clear that it is the right way round, i.e. the more a stakeholder's needs are addressed during development, the greater the perceived ability to perform a job - post computerisation. Concerning perceived ability to perform a job post computerisation, previous research has proved inconclusive. Kaye & Sutton (1985) report much lower levels post computerisation; whereas, Millman & Hartwick (1987) report much higher levels - as do Porter (1987) and Armstrong (1992), when discussing senior executives. In *this* research, end-users, managers and senior executives alike report increased ability to perform a job post computerisation. In terms of addressing users' needs, the findings from this study corroborate Bell & Wood-Harper's (1998) tenet that addressing stakeholders' needs is important if the delivered system is to assist the stakeholder in performing a job post computerisation. This is a view held by several well-respected authors, and is a fundamental element of Mumford's (1996; 1997) ETHICS methodology.

(H₀) Three cross-tabulates the degree to which a stakeholder perceives that the computer system assists in performing a job with the degree of involvement enjoyed by the stakeholder when the system was being implemented. As table 4 demonstrates, the results of the test after re-coding reported $n=54$, $\alpha=0.05$, and $p=0.018$. Therefore, as $p<\alpha$ the null hypothesis *(H₀) Three* is rejected, and for this study, the relationship demonstrates statistical significance between the stakeholder's perceived level to which the computer system assists in performing a

Table 4: Statistical Cross-Tabulation for Stakeholder Involvement During Implementation against Ability to do a Job

| | | Ability at Job | | |
		No Change	Better	Row Total
Involvement at Implement	Little / None	10	14	24 44.4%
	Moderate / Considerable	4	26	30 55.6%
	Column Total	14 25.9%	40 74.1%	54 100%
Pearson Chi-Square				
Value	Degrees of Freedom	Minimum Expected Frequency	Number of Cells < Minimum	Significance
5.57	1	6.22	1 = 25%	0.018

job, and the degree to which the stakeholder was involved during the system's implementation, when tested at the 5% level. As the established relationship is non-directional in the statistical analysis, observation of the actual data is needed to determine whether the relationship is the right way round. By examining this, it is clear that it is the right way round, i.e. the more a stakeholder is involved during implementation, the greater the perceived ability to perform a job - post computerisation.

The findings support the views of Hirschheim & Klein (1992), Stowell & West (1994a; 1994b), and Mumford (1996) who advocate the involvement of stakeholders, *inter alia*, throughout the implementation process. Lucas (1987) reports that stakeholder attitudes are strongly linked to a system's use and success. Clearly, the ability to perform a job post computerisation is seen as a positive outcome by the respondents to this survey. The findings from this research confirm the findings of Millman & Hartwick (1987), Porter (1987) and Armstrong (1992), who all report an increase in stakeholders' ability to perform a job post computerisation; unlike Tait & Vessey (1988) and Flynn & Ade (1996) who report that stakeholder involvement is not particularly important as a determinant of system success.

CONCLUSIONS

This research has successfully identified two statistically significant, positive relationships between a number of variables pertaining to computer technology development and introduction. These statistically significant, positive relationships lead directly to the following recommendations for methodology developers.

> ➢ Ensure the development methodology fully considers stakeholders' needs during systems development. This increases the perceived usefulness of the delivered system that, in turn, stimulates stakeholders into using the system. Addressing stakeholder needs ultimately leads to a perceived increased ability to perform a job post computerisation (see, for example, Igbaria & Toraskar, 1994).

> ➢ Ensure that the development methodology involves stakeholder participation as much as possible during the implementation of the computer system, as this has a bearing on the perceived usefulness of the delivered system, *viz.* the more involved stakeholders are during implementation, the more likely they are to perceive that the delivered system increases their ability to perform their job (see, for example, Bell & Wood-Harper, 1998; Flynn, 1998).

For this research, no statistical significance is demonstrated in support of the involvement of stakeholders during systems development. Although this is in line with Olson & Ives (1981) and Ives and Olson's (1984) findings, there is a weight of evidence against this finding, e.g. Davis and Olson (1985), Doll & Torkzadeh (1988), Stowell & West (1994a; 1994b), Bell & Wood-Harper (1998), and Flynn (1998). I suggest that this relationship quite possibly does exist, but that the statistical power of the data collected is insufficient to identify it (Page, 1999). This concern is fuelled by that fact that, for this study, involvement during implementation is significantly, positively related to the stakeholders' perceived ability to perform a job post computerisation. Involvement during development and involvement during implementation are obviously very closely related and probably overlap in some organisations. Clearly, further research is needed in this area.

REFERENCES

Armstrong, D. A. (1992), 'The People Factor in EIS Success', in Watson, H. J., Rainer, R. K. & Houdeshel, G., (Eds.), *'Executive Information Systems: Emergence, Development, Impact'*, New York: John Wiley & Sons Ltd, pp. 287-297.

Avison, D. E. & Fitzgerald, G. (1995) *'Information Systems Development: Methodologies, Techniques and Tools',* 2nd Edition, Maidenhead: McGraw-Hill.

Baroudi, J. J., Olson, M. H., & Ives, B. (1986), 'An Empirical Study of the Impact of User Involvement on System Usage and Information Satisfaction', *Communications of the ACM,* Vol. 29, No. 3, pp. 232-238.

Bell, S. & Wood-Harper, A. T. (1998), *'Rapid Information Systems Development: Systems Analysis and Systems Design in an Imperfect World'*, 2nd Edition, Maidenhead: McGraw Hill.

Davis, G. B. & Olson, M. H. (1984), *'Management Information Systems: Conceptual Foundations, Structure and Development'*, 2nd Edition, Maidenhead: McGraw-Hill.

Doherty, N. & King, M. (1997), 'The Treatment of Organisational Issues in IS Development Projects, in *Proceedings of the United Kingdom Academy for Information Systems 2nd Annual Conference, 'Key Issues in Information Systems'*, University of Southampton, School of Management, 2nd-4th April, pp. 363-375.

Doll, W. J. & Torkzadeh, G. (1988), 'The Measurement of End-User Computing Satisfaction', *MIS Quarterly,* Vol. 12, No. 2, June, pp. 259-274.

Drummond, H. (1996), 'The Politics of Risk: Trials and Tribulations of the Taurus Project', *Journal of Information Technology,* Vol 11, pp. 347-357.

Flynn, D. J. (1998), *'Information Systems Requirements: Determination & Analysis'*, 2nd Edition, Maidenhead: McGraw Hill.

Flynn, D. J. & Ade, S. (1996), '*A Survey on the Research into User Participation*', Technical Report, Department of Computation, UMIST, May, 1998.

Hirschheim, R. A. & Klein, H. Z. (1992), 'A Research Agenda for Future Information Systems Development Methodologies', in Cotterman, W. W. & Senn, J. A., (Eds.), '*Challenges and Strategies for Research in Systems Development*', Chichester: John Wiley & Sons Ltd., pp. 235-255.

Igbaria, M., Pavri, F. N. & Huff, S. L. (1989), 'Microcomputer Applications: An Empirical Look at Usage', *Information & Management*, Vol. 16, No. 4, pp. 187-196.

Igbaria, M. & Toraskar, K. (1994), 'Impact of End-User Computing on the Individual: An Integrated Model', *Information Technology and People*, Vol. 6, No. 4, pp. 271-292.

Ives, B. & Olson, M. H. (1984), 'User Involvement and MIS Success: A Review of Research', *Management Science*, Vol. 30, No. 5, pp. 586-603.

Jarvenpaa, S. L. & Ives, B. (1991), 'Executive Involvement and Participation in the Management of Information Technology', *MIS Quarterly*, Vol. 15, No. 2, pp. 205-227.

Jayaratna, N. (1992), 'Should We Link SSM with Information Systems!', *Systemist*, Vol. 13, No. 3, pp. 108-119.

Kaye, A. R. & Sutton, M. J. D. (1985), 'Productivity and Quality of Working Life for Office Principals and the Implications for Office Automation', *Office: Technology and People*, Vol. 2, pp. 257-286.

Leonard-Barton, D. (1988), 'Implementation Characteristics of Organisational Innovations: Limits and Opportunities for Management Strategies, *Communication Research*, Vol. 15, No. 5, pp. 603-631.

Lucas, H. C. (1978), 'Empirical Evidence for a Descriptive Model of Implementation', *MIS Quarterly*, Vol. 2, No. 2, pp. 27-41.

Mason, R. O. & Mitroff, I. I. (1981), '*Challenging Strategic Planning Assumptions*', New York: John Wiley & Sons Ltd.

Millman, Z. & Hartwick, J. (1987), 'The Impact of Automated Office Systems on Middle Managers and Their Work', *MIS Quarterly*, Vol. 11, No. 4, pp. 479-491.

Mumford, E. (1996), '*Systems Design: Ethical Tools for Ethical Change*', Basingstoke: Macmillan Press Ltd.

Mumford, E. (1997), 'Reality of Participative Systems Design: Contributing to Stability in a Rocking Boat', in *Proceedings of the United Kingdom Academy for Information Systems 2nd Annual Conference, 'Key Issues in Information Systems'*, University of Southampton, School of Management, 2nd-4th April, pp. 9-23.

Olson, M. H. & Ives, B. (1981), 'User Involvement in System Design: An Empirical Test of Alternate Approaches', *Information and Management*, Vol. 4, No. 4, pp. 183-196.

Page, S. M. (1996), 'Organisational Culture & Information Systems, in *Proceedings of the United Kingdom Academy for Information Systems 1st Annual Conference, 'The Future of Information Systems'*, Cranfield University, School of Management, 10th-12th April.

Page, S. M. (1999), '*IT Impact: A Survey of Leading UK Companies*', Unpublished MPhil Thesis, School of Management, Leeds Metropolitan University, UK, (Forthcoming).

Parasuraman, S. & Igbaria, M. (1990), 'An Examination of Gender Differences in the Determinants of Computer Anxiety and Attitudes Toward Microcomputers among Managers', *International Journal of Man-Machine Studies*, Vol. 32, pp. 327-340.

Porter, A. L. (1987), 'A Two-Factor Model of the Effects of Office Automation on Employment', *Office: Technology and People*, Vol. 3, pp. 57-76.

Sauer, C. (1993), '*Why Information Systems Fail: A Case Study Approach*', Henley-on-Thames: Alfred Waller.

Schneider, K. (1997), 'Bug Delays £25m Court Case System', *Computer Weekly*, p. 1.

Stowell, F. A. & West, D. (1994a), '*Client-Led Design*', Maidenhead: McGraw-Hill.

Stowell, F. A. & West, D. (1994b), "Soft' Systems Thinking and Information Systems: A Framework for Client Led Design', *Journal of Information Systems*, Vol. 4, pp. 117-127.

Tait, P. & Vessey, I. (1988), 'The Effect of User Involvement on System Success: A Contingency Approach', *MIS Quarterly*, Vol. 12, No. 1, pp. 91-107.

'*The Times 1000, 1995*' (1994), London: Times Books.

Turner, J. A. (1984), 'Computer Mediated Work: The Interplay Between Technology and Structured Jobs', *Communications of the ACM*, Vol. 27, No. 12, pp. 1210-1217.

Walsham, G. (1993), '*Interpreting Information Systems in Organisations*', Chichester: John Wiley & Sons Ltd.

Warner, T. (1996), '*Communication Skills for Information Systems*', London: Pitman Publishing.

Watson, H. & Wood-Harper, A. T. (1996), 'Deconstruction Contexts in Interpreting Methodology', *Journal of Information Technology*, Vol. 11, pp. 59-70.

Wong, E. Y. W. & Tate, G. (1994), 'A Study of User Participation in Information Systems Development', *Journal of Information Technology*, Vol. 9, pp. 51-60.

APPENDIX: RESEARCH QUESTIONNAIRE (Abridged)

ABOUT YOU

1. Which gender are you?

    Male ❑
    Female ❑

2. How many years old were you on the 1st of January 1996?

    20 or less ❑

    21-30 ❑
    31-40 ❑
    41-50 ❑
    51-60 ❑
    61 or more ❑

3. How many years have you worked with computerised systems?

    Never worked with computers ❑
    Less than 1 year ❑
    1-2 years ❑
    3-5 years ❑
    6-10 years ❑
    11 years or more ❑

4. How would you describe your MAIN job role? - tick one box only

    Computer system user ❑
    Manager *(Please go to Q. 16)* ❑
    Senior executive *(Please go to Q. 16)* ❑

END USER COMPUTING

5. To what extent do you understand what the results of your current computer system are used for?

    Not at all ❑
    Very little ❑
    Moderately ❑
    Quite a lot ❑
    Extensively ❑

6. To what extent is the system actually used compared to the number of people who could potentially use it?

    Negligible ❑
    Not much ❑
    Moderately ❑
    Quite a lot ❑
    Extensively ❑

7. When your current computer system was being developed, to what degree were you involved?

    System already developed when I ❑
    started current job role *(Please go to Q.*
    27)

    Not at all ❑
    Very little ❑
    Moderately ❑
    Quite a lot ❑
    Extensively ❑

8. When your current computer system was being implemented, to what degree were you involved?

Not at all	❑
Very little	❑
Moderately	❑
Quite a lot	❑
Extensively	❑

9. How much disruption was there to your working life when your current computer system was being implemented?

None at all	❑
Very little	❑
Moderate	❑
Quite a lot	❑
Extensive	❑

10. How useful is your current computer system compared to your expectations?

Not at all useful	❑
Not very useful	❑
Moderately useful	❑
Quite useful	❑
Very useful	❑

11. To what degree has your ability to perform your job changed since you started using your current computer system?

I'm now much worse at my job	❑
I'm now worse at my job	❑
About the same as before	❑
I'm now better at my job	❑
I'm now much better at my job	❑

12. What level of job satisfaction did you enjoy before you started using your current computer system?

None at all	❑
Very little	❑
Moderate	❑
Quite a lot	❑
Extensive	❑

13. To what degree has your job satisfaction changed since you started using your current computer system?

Decreased a lot	❑
Decreased a little	❑
About the same as before	❑
Increased a little	❑
Increased a lot	❑

14. How motivated to work were you before you started using your current computer system?

Not at all	❑
Very little	❑
Moderately	❑
Quite a lot	❑
Extensively	❑

15. To what degree has your motivation to work changed since you started using your current computer system?

Decreased a lot	❑
Decreased a little	❑
About the same as before	❑
Increased a little	❑
Increased a lot	❑

Please go to Q. 22 next.

276

ABOUT YOUR MANAGEMENT DECISION MAKING

16.   Do you take management decisions based on the results of a computer system?

Yes   □
No - Please explain & go to Q. 27 next.   □

_____

_____

17.   Do you personally use a computer system to assist you in your management decision making?

Yes   □
No *(Please go to Q. 19)*   □

18.   How at ease are you about using your computer system?

Very awkward   □
Slightly awkward   □
Passable   □
Comfortable   □
Very comfortable   □

19.   How effective was your management decision making prior to using your current computer system?

Not effective   □
Not very effective   □
Moderately effective   □
Quite effective   □
Very effective   □

20.   How has the effectiveness of your management decision making changed since you started using your current computer system?

Decreased a lot   □
Decreased a little   □
About the same as before   □
Increased a little   □
Increased a lot   □

21.   Was your computer system already developed when you started your current job role?

No   □
Yes *(Please go to Q. 27)*   □

ABOUT THE DEVELOPMENT OF YOUR CURRENT COMPUTER SYSTEM

22.   When your computer system was being developed, to what degree were your information needs addressed?

Not at all   □
Very little   □
Moderately   □
Quite a lot   □
Extensively   □

23.   How much training were you given prior to the system going live?

None *(Please go to Q. 25)*   □
1 hour or less   □
2-8 hours   □
2-5 days   □
6 days or more   □

277

24. How relevant was the training to your needs?

Not relevant at all	❑
Very little relevance	❑
Moderately relevant	❑
Quite relevant	❑
Very relevant	❑

25. To what extent were you able to make decisions about your job role before you started using your current computer system?

Not at all	❑
Very little	❑
Moderately	❑
Quite a lot	❑
Extensively	❑

26. To what extent has your ability to make decisions about your job role changed since you started using your current computer system?

Decreased a lot	❑
Decreased a little	❑
About the same as before	❑
Increased a little	❑
Increased a lot	❑

ADDITIONAL INFORMATION

27. Are any other comments you wish to make, either about the questionnaire, or your computer system? If so, please give details below:

RESEARCHING ORGANISATIONAL MEMORY

Frada Burstein and Henry Linger
School of Information Management and Systems, Monash University
P. O. Box 197, Caulfield East, 3145, Victoria Australia
Melbourne, AUSTRALIA
E-mail: {frada.burstein; henry.linger}@sims.monash.edu.au

Key words: Organisational Memory, Research Methods, Knowledge Management

Abstract: The modern view of organisations assumes that organisational processes are associated with the creation of new knowledge. Moreover it is recognised that this knowledge needs to be shared and reused. The concept of organisational memory (OM) is indispensable in this respect as a collection of knowledge that can and should be applied in the context of current activities. In this sense, OM is dynamically maintained and facilitates organisational learning. It is these aspects, as components of knowledge management, that enables organisations to evolve and transform in response to their dynamic environment and internal changes. Recognition of organisational knowledge as a resource has prompted an increased interest in research into OM. From the information systems perspective, the focus is on OM systems (OMS) and the technology to support such systems to achieve the desired organisational outcomes. In this paper we constrain our discussion of OMS to its external manifestations (and exclude what people know) because such explicit manifestation of memory can be supported with information technology. However, the danger inherent in focussing on an OMS facilitated by information technology is that organisational knowledge can be (mis)interpreted as simply an information store. This would repeat the history of information systems failure where the mere provision of, or access to, information or technology is deemed to constitute organisational outcomes. We believe that the approach outlined in this paper specifically precludes such misunderstanding

Our objective for this paper is to provide a broad position on the issues related to OMS research and practice. In addition the paper identifies a number of directions which will need to be followed in order to establish a strong theory of OMS with benefits flowing from the application of theory into practice.

1.   INTRODUCTION

Modern views on organisational processes assume these processes to be associated with the creation of new knowledge as well as a significant need for knowledge sharing and reuse. Organisational knowledge becomes a necessary asset for an effective and efficient functioning of the modern

Systems Development Methods for Databases, Enterprise Modeling, and Workflow Management,
Edited by W. Wojtkowski, *et al.* Kluwer Academic/Plenum Publishing, New York, 1999.

279

organisation. A concept of OM is considered relevant in this respect as a collection of informational resources of the organisation that can and should be applied in the context of current activities in order to improve organisational processes. All this has prompted an increased interest in research into the concept of OM in the context of information technology.

OM research has many dimensions and distinct objectives. It cannot be described as an established discipline nor can it be associated with any one particular established discipline. There are researchers who are actively working in this area as well as others who find it relevant to their own research field. As a result of such diversity there are variation in terminology, diversity of the research approaches and philosophies, and several attempts to reflect on existing studies in order to establish a general model of OM system (OMS) applicable to any organisational situation.

Our objective for this paper is to provide a broad review of the issues related to OMS research and practice. In addition the paper identifies a number of directions which will need to be followed in order to establish a strong theory of OMS with benefits flowing from the application of theory into practice.

The paper consists of four sections. The first one provides an introduction to OMS. We overview the current status of OMS research and propose that an information systems perspective is necessary to study OMS. We also present a model of the general structure of the OMS and its role in the organisational transformation process.

The second section presents a contingency model of OMS. The model explores the factors that define the context of an OMS as well as the components that constitute the OMS. The model also outlines what are the expected outcomes from OMS and its contribution to practices of the organisation.

In the third section, the OMS research process is described from the perspective of a research methods taxonomy. It presents a general model of OMS research, which categorises most of the research done to date as theory building, theory application or theory refinement which links theory and practice. This model helps to reveal gaps in the previous studies and to identify areas of interest for future research. From the review of the current status of OMS research we conclude that OMS research is in a 'pre-paradigmatic' stage. This is based on the observation that multiple co-existing paradigms attempt to describe what happens in the current practice of OMS research.

In conclusion we argue for a need for a close link between OMS research and practice and extend the discussion to locate OMS as a core component of knowledge management processes undertaken by organisations.

1.1　From OM to OMS: the IS perspective

From a historical perspective, Organisational Memory (OM) has come moved being purely artefact-oriented towards a concept that covers everything related to management activities, including both physical outcomes, social implications, and any other results and processes associated with the functioning of the organisation.

The early literature on OM dealt with a set of documents (artefacts). Conklin (1992) uses this definition deliberately to differentiate OM from the collective memory; the memory of the people in the organisation who create the OM. From this perspective both OM and collective memory need to coexist for the OM to be useful. Conklin's conclusion is that OM, in his definition, is a by-product of organisational processes. In order to make use of OM, tools are needed which enable the details of the processes to be recorded as part of OM. This is in addition to storing documents. The most obvious problem with this approach is that it reduces the role of OM by treating it as a tool to support workflows.

With information technology being an intrinsic part of modern organisations, an organisational memory system (OMS) is the means of encoding, storing and accessing OM. This view tends to mix OMS with all the other information systems and related processes. This can be a dangerous tendency and is an important issue for OMS research to clarify. The main danger is that such a simplification of the OMS, and an over-emphasis on the technology, can result in a situation where it becomes hard to see the main argument behind the need for the term OMS if it is just a collection of all the information systems in the organisation.

Another deficiency of the technology-oriented view of OMS is that there is limited opportunity for creating and capturing organisational knowledge, which in our view is the main purpose of OMS. The OMS concept has evolved from a static repository (Walsh and Ungson (1991)) through to an information system for organisational effectiveness (Stein and Zwass (1995)). In this paper we extend the OMS concept to focus on organisational transformation which is enabled by OM augmented by information technology (IT). The technology enables the external manifestations of organisational knowledge to be stored and used and provides the facilities for organisational processes that support learning, knowledge creation as well as knowledge reuse. In this perspective, the OMS is viewed as organisational processes that include IT augmented OM. The OMS supports actors to both perform tasks, utilising past experience, as well as to engage in activities that enable them to reflect and act on their experiences with the purpose of transforming the organisation.

1.2 The Structure of OMS

A pragmatic view of OM has undergone a significant evolution. Initial interest in OM arose with the recognition of the usefulness of investigating past events and led to the collection of organisational artefacts; the paper flow. Walsh and Ungson (1990) provided the first detailed review of the sources and storage requirements for this perspective of the OM. They introduced a model of storage of OM as six storage "bins" (individuals, culture, transformations, structures, ecology and external archives). Their model does not refer to the content of the OM but rather concentrates on the means for retaining OM. Stein and Zwass (1995) extend this view from the perspective of the information technology support for OM. The OM model they propose includes a framework that relates organisational effectiveness to OM sub-systems and corresponding mnemonic functions. An overview of the current approaches to OMS is presented, for example, by Morrison (1997). She presented a comprehensive list of OMS characteristics, which are based on previously proposed models.

Our focus is on an analysis of the components of OMS from the point of view of the role it plays in the organisation and its role in organisational transformation. Our model of the structure of an OMS, shown in Figure 1, comprises a hierarchy of components in which more abstract components are constructed from less abstract ones. The interaction between the various components is shown as a process of building the OMS.

The structure of OMS begins from the most concrete level, comprising of 'tangible' entities which we can term collectively *events* and *artefacts*. The collection of artefacts and past events provides the material for the construction of narratives. Narratives are important in that they enable the documentation of history. What is clear is that all organisational actors are engaged in the process of constructing narratives, both explicitly as documentation and implicitly as each actor's mental models of the organisation. The narratives use the artefacts and events as their source material but there is no imperative to use this material as the actual elements of the narrative. Narratives, especially implicit ones, are a means of synthesising diverse material and generalising or abstracting from a large volume of material that is accumulated as a result of the work conducted by the organisation. From this perspective there is potential for a plurality of narratives and stories to be constructed.

Construction of a "story" includes some level of interpretation and requires a level of understanding of the situation by the story-teller. This is not different from the construction of a history of any events. The individual understanding and the background of the author provides a perspective on the content which inevitably (partially) distort an "objective" view on the event (if such a thing exist). Schank and Abelson (1995) state that "storytelling is something we virtually have to do if we want to remember anything at all and it is in the storytelling process that the memory gets informed". The idiosyncratic nature of the story-telling process requires some additional attention on the organisational processes responsible for the development of the OM at this level.

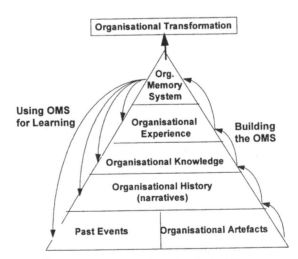

Figure 1. Elements and Roles of Organisational Memory

The potential for conflicting narratives is dissipated by a requirement that narratives share an underlying theme. This requirement is expressed as a consensual, sanctioned and shared understanding that defines the organisation at a point in time. To sustain this understanding narratives need to be continually reconstructed and the stories retold many times. This repetition plays a critical role because it confirms and reinforces the underlying theme. Additionally, such repetition continuously renews and re-affirms the sanctioned, shared understanding that all the actors have of what the organisation is. Importantly, as with all aural traditions, the repetition allows stories to evolve and change to retain their currency. Reconstruction and retelling of narratives provides an opportunity for each actor to incorporate material that they have directly experienced thus ensuring the organisation is inclusive.

Establishing shared understanding is important in defining the organisation but its utility lies in the ability of the narratives to inform and legitimise activities and define the boundaries of sanctioned actions. Narratives encode organisational knowledge when they are used didactically and storytelling becomes a purposeful activity with the intention of achieving organisational learning. The intentional use of narrative in the context of learning allows narratives to be interpreted in a way that reveals the subtext; the intended theme. This process of interpretation encode the knowledge that is carried by the narrative and provides an opportunity for that knowledge (or at least some aspect of it) to be explicitly recorded. This is in line with Sowa's definition of knowledge that it is knowledge itself that prescribes what can be done with it (Sowa, 1994).

However, organisational knowledge does not adequately reflect the skills necessary to perform the work that constitute the "system of purposeful activities" that is the organisation (Spender, 1996). The actor's experience in performing organisational tasks represents tacit knowledge that cannot be "objectified" nor removed from the performance of the task itself. The distinction between organisational knowledge and experience is analogous to declarative and production memory from Anderson's ACT* model (Anderson, 1983). Organisational knowledge can be compared to declarative memory in that it represents "book knowledge" that can be explicitly recorded and is often stored in organisational artefacts. Organisational experience, representing tacit knowledge, can be compared to production memory in that it is the result of understanding, learning and performance; it is a "skill" acquired with time.

Building the OMS relies on the construction and assembly of each element as shown in Fig 1. The construction of each element use the layer below as it source material. But it also has direct access to other elements. For example, to continue the analogy with Anderson's ACT* model (Anderson, 1983), knowledge gained from experience is not directly accessible but can be illustrated with specific reference to artefacts and events.

If the OMS is to conform to the "purposeful activity" view of an organisation then its needs to be used directly to support the processes that enable the organisation to perform its work and to sustain its viability. Organisations are dynamic systems that function in a dynamic environment. Viability requires that organisations evolve and transform in response to internal and external changes. However to ensure organisations remain viable, such transformation needs to be the outcome of organisational learning rather than an unreflective response to change. OMS supports learning processes that enable experience and knowledge to be directly applied to performing current task, or constructing artefacts, as well as growing and creating knowledge (both tacit and explicit). These processes are complex and cannot be hierarchical.

The value of constructing an OMS is that it provides the material for reflection on past activities. Reflection is an activity, which allows actors to learn from past experience with the aim of improving practice. This suggests that a process of 'knowledge discovery' from OMS can be viewed as organisational learning. What is perhaps even more significant is that for such learning to occur requires organisations to acknowledge, sanction and support such activity. Such activity can be considered single loop learning (Argyris and Schon, 1978). For double loop learning, the organisation needs to establish explicit processes to utilise this new knowledge. These processes need to test knowledge resulting from reflection, derive a consensual view of this knowledge and sanction new organisational practice based on that knowledge.

As described above, the purpose of constructing and using an OMS is organisational transformation. An OMS is a system in which all elements in the pyramid interact and are the source material for actors to translate organisational activities into organisational transformation. We see this approach as a natural historical progression of the concept of OM. The view on OMS proposed in this paper represents a holistic perspective, with the evolution of the OMS from explicit, tangible components through the processes of organisational effectiveness to the level of transformation and learning that organisations require in order to achieve competitive advantage. Organisational transformation is also necessary in order for the organisation to remain viable in a dynamic environment in line with Ashby's (1956) concept of 'requisite variety'.

2.   A CONTINGENCY MODEL OF OMS

The concept of OMS as described above requires a complex dynamic construct that is associated with activity that has the potential to support organisational learning, adapt to dynamic environment and allows the organisation to perform effectively in such an environment. We propose and discuss a contingency model, as shown in Figure 2, that defines the requirements to operationalise OMS.

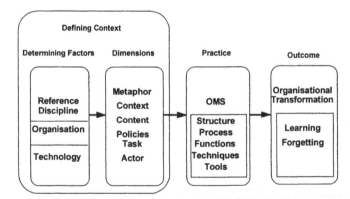

Figure 2. Contingency Model for OMS

2.1   Defining the Context for OMS

The richness of OM literature has led to a situation where studies have been able to selectively choose particular orientation for their OMS which is suited to their purposes. A problem only arises when attempts are made to generalise the outcomes. What is needed is to contextualise the design of the OMS and to explicitly identify the complex interaction of the factors and dimensions addressed by the OMS.

In keeping with an interpretivist approach, we propose that the context of the OMS in the contingency model is defined in terms of fundamental factors and the specific dimensions of the OMS. The factors relate to the specific organisation in which the OMS is embedded, the technology that implements the OMS and the theoretical orientation adopted in terms of the acknowledged reference disciplines. The dimensions identified in the model are derived from the factors but represent the details relevant to a particular OMS. These dimensions are not intended to be exhaustive but rather indicative of the type of detail necessary to contextualise the OMS.

2.2   Determining Factors of OMS

OM needs to be considered situationally to enable the *context* in which memory exists to be defined. The determining factors are the basis for OMS in terms of the intellectual antecedents deriving from the *reference disciplines* and the actual enterprise, the *organisation* which encodes the memory and the pragmatics to realise the OMS, the *technology*. OM has a considerable history from a variety of disciplines and theoretical positions but is essentially an element of organisational theory. OMS is an inclusive term to accommodate the different theoretical positions and accords OM a role as a practical and purposeful aspect of an organisation.

2.2.1   The Organisation as a determinant of OMS

The *organisation*, its people, structures and processes as well as its operating environment provides the concrete framework in which the OMS exists. An OMS cannot be considered an artefact, to be studied in the context of the organisation. It is an integral part of the organisation and evolves as a result of organisational activities but also directly influences those activities. This dialectical nature would indicate that the subject of the OMS is organisational knowledge (or meta knowledge) rather than simply information generated by those activities.

The OMS encodes organisational knowledge in a way that is informed theoretically by the reference disciplines and by the pragmatics of the operation of the organisation in a particular context. The structure of the organisation, its means of production, its deployment of technology and its documentation of its performance simultaneously define the organisation and encode the organisation's knowledge. Organisational evolution provides the temporal aspect of organisational knowledge that is its memory. While evolution is largely driven by the context in which the organisation functions, the direction of that evolution is influenced by prevailing management and organisational theories and practice. Organisational knowledge arises from the adaptation of such theories and practices to the particular situation of the organisation. Information generated by organisational activities can provide the quantitative means to evaluate those theories and practices.

2.2.2   Reference Disciplines for OMS

One of the challenges, and benefit, of the concept of OM is that it is grounded in the works of several *reference disciplines* including, but not limited to, information systems research, management science, economics, systems theory, political theory, organisational theory, decision-making, and communication theory. It is evident that more recent works in OM and its extension into information systems (eg. Sandoe and Olfman, 1992; Stein and Zwass, 1995) require a background in these domains. Our concern here is the influence of these disciplines on the theoretical orientation of the discussion of OM.

One characteristic of the disciplines that deal with OM is that knowledge is the subject of OM. This is important because it is a distinction between OMS and other information systems that also draw on these disciplines. The focus on knowledge brings with it a need for learning. Thus the reference disciplines deal with what is retained in memory, how it is retained and what learning is utilised. There is however a divergence in their orientation about the nature of OM ranging from OM as a consequence of individual experience to considering OM as a socially defined activity. Equally, memory is posited in locations ranging from people (cognitive maps (Weick, 1979), patterns of behaviour (Nelson and Winter, 1982)), culture (Argyris and Schon, 1978), language (Smith, 1982), artefacts (Miller, 1978), procedures (Cyert and March, 1963) and organisational structures (Krippendorff, 1975; Ashby, 1956). This implies different mechanisms by which memory is retained as well as different types of processes for maintaining knowledge.

From the perspective adopted in this paper, the significant distinction is between what Simon (1957) termed natural and artificial memories and Miller (1978) termed individual cognition and artefacts, the latter including all notes, written minutes, financial records etc. In our view, OMS as IT augmented OM goes beyond the notion of a repository of artefacts. In our view artefacts are the materials which represent past experience that an actor can utilise to perform a current task. However, artefacts do not constitute organisational knowledge. What is required is some explicit representation of what the actor understands of the activity so that the artefacts that describe it can be read in context.

It is evident that reference disciplines not only influence how OM is perceived but have a far more fundamental role in determining how the organisation (and its environment) is considered. Thus the current interest in OM by a variety of disciplines can be seen as driven by the obvious changes being brought about by the twin forces of globalisation and the convergence of communications and information technology. Organisations need to be more flexible, responsive but above all engaged in creating knowledge and utilising it. To achieve this organisations need to change and OM is seen as integral to that change.

2.2.3   Organisational technology for OMS

Technology is a determining factor for OMS in that it allows the organisation to construct both the elements of memory and the organisational system which is the OMS. In this sense OMS becomes a dynamic organisational process rather than merely an artefact such as a repository or data store.

OM can include what people know but the study of OMS is facilitated by considering its external manifestations. This constraint is raised because it is possible to support these external aspects of memory with information technology. The use of IT involves information systems, and its associated reference disciplines, in defining the context of OMS. To consider OMS as a dynamic process, facilities are required to support knowledge processes that utilise the organisational knowledge in the OMS. This highlights the contribution of information systems, and intelligent decision support in particular, to the theoretical framework of OMS. Moreover, IT enables artefacts to be constructed that are organisational in nature and allows these artefacts to be widely available throughout the organisation. This accords with a communicative view of OM (Tuomi, 1996) and extends the theoretical framework of OMS by introducing reference disciplines such as cognitive science (Gardner, 1985) and CSCW (DeMichelis, 1994, Kutti, 1991; 1995; Scrivener, 1994; Schmidt and Bannon, 1992; Schmidt and Simone 1996).

Technology allows a constructivist method to be adopted whereby artefacts are built and their use studied. The technology then has a direct affect on the type and nature of the artefact that is constructed which in turn determines to a large extent what aspect of memory is externalised, how it is used and for what purpose. This potentially allows researchers to make more appropriate *intelligent* use of a variety of tools, techniques or methods to encode and distribute organisational knowledge. It also allows them to study existing artefacts from the OMS perspective. This however, has an inherent danger of trivialising OMS; the dialectical aspect of the dynamic evolution organisational knowledge is in danger of being reduced to a one-dimensional information store. This would repeat the history of failure in information systems where the mere provision of, or access to, information (or technology) is deemed to constitute organisational outcomes.

2.3   Multiple Viewpoints: The Dimensions of OMS

To define the context for OMS it is also necessary to consider the determining factors from the perspective of different viewpoints. The dimensions shown in Figure 2 are not intended to be exhaustive but are indicative of the range of issues that need to be considered in OMS. The dimensions are intended to be interdependent.

The nature of memory has been an active topic of discussion since the last century, across a number of disciplines ranging from the humanities to social sciences, biology, psychology and psycho-analysis and more recently artificial intelligence, computer science and information systems. The diversity of views covers a broad spectrum of aspects and features. There is no doubt that in the organisational context, the term *memory* is used extensively as an evocative metaphor "suggesting the promise of infinitely retrievable knowledge and experience." (Ackerman, 1994a). Memory expresses the knowledge used by the organisation to conduct its activities and is stored and retrieved by technical and social methods; the OMS. However all metaphors of memory depend on an idealised anthropomorphic view in which memory is complete in the sense that all material, and by implication its *context* and implied *meaning*, is stored and is immediately retrievable (Ackerman, 1994a; Ackerman and Mandel, 1995). Such an idealised notion can be problematic in that data or information can be readily confused with knowledge and that any information retrieved, as a result of an appropriate search, is considered *ipso facto* useful; the fallacy of data mining.

Implementation of an OMS can vary because, while adhering to the general metaphor, particular disciplinary views of memory allow specific features to be expressed in a particular manner. For example forgetting, an important component of memory, can be considered in a number of ways. In an information system, forgetting is analogous to "... a technical failure in the storage or retrieval of information into or out of (memory)." (Engestrom et al 1991). But forgetting can also be a repression of something unpleasant, from a psychoanalytic perspective, or as a breakdown of the interaction between internal and external inscriptions (Engestrom et al 1991).

In one sense, the organisation is the memory in that its cultural, social and technical systems contain all the necessary elements and artefacts of the OMS. While such a deconstructivist view is appropriate for particular studies, a more pragmatic perspective is usually adopted to identify the content of an OMS. The dimension of *content* is important in that it effectively characterises the OMS in a continuum from 'memory in the small' to a memory of the organisation as a whole.

Ackerman (1994b) proposes a task-based approach, where the content of memory is determined by the actors performing a defined task. In this approach, content combines knowledge and information as effective performance of the task can include documenting the outcome of the task. In task-based memory, the temporal distance between the content and the current activity is relatively small reducing the need to extensively record the context of the information in the memory. Moreover the actors are able to make sense of the information from their tacit knowledge of the task. On the other hand, the OMS requires facilities to support reflection, exploration, experimentation and other knowledge processing activities to enable actors to approach the task as knowledge work rather than just task performance (Iivari and Linger, 1999). The distinction enables the task to be effectively conducted from an organisational perspective in that it provides the means for a process of continuous improvement and ultimately for the task to evolve.

The issue of what tasks are supported is an important dimension of the OMS. At the organisational level, Stein and Zwass's (1995) 4 subsystems define the tasks that are supported. However, the reality of an organisation operating in a rapidly changing environment means that specific tasks will change and require new knowledge while existing ones will need to deal with new and different knowledge. In such an environment being able to define the task and its context is an important aspect of the OMS if it is to remain relevant and contribute to organisational effectiveness.

On the other hand, task-based memory has different imperatives in relation to the task dimension. The major issue is to identify a task that has an organisational dimension (or at least a community dimension) and is relatively stable and invariant. Literature on task-based OMS presents studies of research communities that deal with tasks that are generic to that community (Schatz, 1992, Brown and Duguid, 1991). For example Burstein et al. (1997) describe an OMS to support survey-based epidemiological research. While the methods, tools and techniques of survey-based research can change (and in fact vary between disciplines that use quantitative research methods) the task remains central to epidemiology. The OMS contains knowledge about the methods, tools and techniques, and their evolution, as well as information about individual surveys (and could include survey data). The important aspect is to understand the limitations of the OMS in respect to any variant of the task. The evolution of the OMS (as opposed to its content) will involve the provision of new knowledge processing facilities to enable the actors to 'improve' the task.

The performance of tasks in an organisation is constrained by *policies*, both formal and informal. The importance of policies is that they define what task can be performed by what actors and often in what manner. The extreme example of a this is military doctrine and rules of engagement as applied to an extremely hierarchical organisation. It is evident that the nature of the organisation, its structure and cultural and social norms all contribute to defining how work is distributed and performed. An OMS needs to be consistent and reflect those policies for it to contribute to organisational effectiveness. There is however a contradiction. The more senior the actor using the OMS the more discretionary power the actor has over what work is done and when it is done combined with increased decision latitude about the outcomes.

Implicit in much of the OM literature is that OMS is essentially a managerial facility as it deals with organisational knowledge. This dominant view is challenged when new organisational structures and new forms of work organisation are considered. Flat, network organisations need to adopt post-Fordist work practices so that more people are engaged in what is effectively knowledge work performed in a team environment (Kutti, 1996; Bell, 1976; Blackler, 1995; Blackler et al. 1993; Drucker, 1993; Nonaka and Takeuchi, 1995). In such an environment OMS supports knowledge work and task performance while not necessarily being limited to managerial work. Even in such an environment, policies exist and constrain the OMS. For example in a collegial organisation such as a research unit, policies tend to be externally defined by the research discipline and the research culture of that discipline (Abbott, 1988; Evans and Patel, 1989).

The organisation, as a social and technical system, exists because of the people who are part of it. Yet *actors* are often a forgotten dimension of OMS. Actors need to be considered in an analogous way to the discussion on organisational learning. Organisations do not learn but individuals do; the issue is how individual learning becomes organisational learning. Argyris and Schon (1978) resolve this by proposing actors as agents. In OMS actors define the functionality and content of the OMS.

From the perspective of actors, the most important consideration is whether the OMS is used and whether this use is appropriate and effective. Stein and Zwass (1995) highlight these issues as impediments to OMS adoption and discuss both the individual and organisational aspects of this issue. The OMS represents an additional layer that an actor confronts in performing their work bringing with it additional demands, in terms of input and maintenance, while its benefits are not readily evident. For many actors the use of the OMS is discretionary. This dimension raises significant issue for the development, implementation and use of OMS. OMS at the organisational level involves many actors who would not be directly involved in the development and implementation. The onus is to design the OMS so that it is the OMS that entices the actors to use it in their own interest, as opposed to an organisational interest. Similarly for task-based OMS, significant numbers of actors performing that task need to use the OMS for its full potential to be realised.

2.4   Factors contributing to OMS usage

Contextualising the OMS provides the basis for the design of the OMS and its usage. The content is the significant aspect of an OMS. It is our contention that organisational knowledge is the appropriate content of memory if the OMS is to contribute to organisational transformation. Wijnhoven (1998) points out that the content of memory relates to operational activities, the theoretical and conceptual knowledge and experience that enable those activities to be performed, the knowledge of existing capabilities to perform those activities and importantly information about the content of memory itself so that memory can be used. In this perspective information can illustrate and provide exemplars for knowledge and as such is a legitimate component of memory.

It is apparent that a strict distinction between information and knowledge is not particularly useful as it quickly leads to profound philosophical questions and complex issues of epistemology. While this avenue can provide insight, we are concerned with the practical issues of effective organisational outcomes.

The structure of the memory is a reflection of the tasks and organisational activities which it supports. This is necessary as the actors using the memory need to negotiate the meanings assigned to memory components so that they are able to be deployed to support task performance (Schmidt and Bannon, 1992). This position, which borrows from the CSCW field, is interesting because it implies knowledge construction is both dynamic and temporal where meanings and definitions need to be continually renegotiated. More significantly, it also implies that the structure of the OMS is itself subject to this renegotiation.

Stein and Zwass (1995) and Wijnhoven (1998) both identify memory processes that allow acquisition, retention, maintenance and retrieval of content. Wijnhoven identifies 3 further processes that define memory content; a strategic analysis (top down), current usage (bottom up) and identifying gaps (inside out). Stein and Zwass link these processes to the 4 subsystems that explicitly address organisational effectiveness. Adopting a dynamic view requires that OM is not seen as a static repository, highly structured and fully indexed, but as a dynamic organisational process (Gammack, 1997). This implies that such processes become an integral aspect of task performance and that other organisational activities are introduced that specifically review and reflect on the memory. It is these processes that enable the structure of the OMS to adapt and evolve (Burstein et al. 1997)

This view of OMS provides ample scope for incorporating a wide variety of tools and techniques that can intelligently support the actor in their task performance. For example Sowunmi et al. (1996) demonstrate the use of case based reasoning in an intelligent decision support (IDS) context. The case base stores past experience in terms of situations and outcomes and is used to inform an actor about a current situation. This approach has been extended to suggest a heterogenous architecture combining data, knowledge and case bases with analogical search and inference in an IDS framework (Smith et al 1996). It is important that the nature of the task, in conjunction with the actor, determine the type of tools and techniques that are use incorporated into the OMS for that task. From the design point of view however, the OMS needs to include functions that allow it to adapt and evolve. Such an approach is less computationally tractable than if a static view of memory is adopted.

2.5 Outcome of OMS usage

In our view an OMS is effectively a system that uses IT augmented OM to support post-Fordist work practices (Amin, 1994). This means that pragmatic work, task performance, is integrated with the cognitive aspects of that work, knowledge processing. This approach supports multiple viewpoints and the plurality of private models, as well as the possibility of an organisational perspective which is represented by a consensual model. Importantly, work activity is organised around teams (Kutti, 1996). Such practices are similar to what Schon has discussed as "reflection-in-action" in the context of professional practice (Schon, 1991). He argues a distinction between the knowledge domain and the "epistemology of practice", and his concept of reflection-in-action is an attempt to bridge these two ontologies. In this context, the OMS assumes a communicative perspective. The importance of this approach is that learning is central to integrating these work aspects and the OMS is considered as a tool for learning.

The changing emphasis of memory from a repository to processes of remembering and forgetting presumes that memory is a process not just a structure. Remembering is an interaction between external and internal inscriptions (Engestrom, et al. 1991). An actor engaged in an activity uses materials which record some aspect of the current and/or past events and their own memory. The act of remembering is the process of connecting external material to an internalised image of the activity. The significant issue for OM is the ability to articulate the internal image of the activity that is constructed as a consequence of performing that activity. The implication for OMS is that the content of memory are the explicit, external inscriptions, and these are constrained to those aspects that can be captured electronically. To enhance memory requires some aspect of the internalised image to be explicitly recorded. Inscriptions can be legitimately considered as organisational knowledge, as they represent the collective experience of activities performed by the organisation. Memory then becomes organisational knowledge.

Recording organisational knowledge provides a memory which accommodates individual forgetting in the context of organisational remembering. Forgetting in this context can be considered as a breakdown of the interaction between internal and external inscriptions. The implication of this are significant in that the external records are truly a collective memory in that they are the necessary component for individual remembering. External inscriptions allow an actor to forget much of the detail of the activity, assuming significant aspects of the activity are retained as internal inscriptions. Performance of that activity in the future is facilitated by the actor's understanding of the activity and is enhanced by the memory of previous episodes. Previous episodes are remembered by recalling external inscriptions to sustain the actor's mental images of the activity, the internal inscriptions. It is in this sense that the OMS can be seen as "systems of forgetting"; recording information externally makes it possible for actors to forget the stored information because it is available through the process of collective remembering.

What emerges is that maintenance of memory plays a central role in an OMS. When considering "memory as process", remembering and forgetting, memory maintenance is dynamic in the sense that it is an integral part of those processes and is achieved as a by-product of those processes. An OMS that supports remembering needs to provide facilities for evolving the external inscriptions so that memory reflects organisational patterns in time (Tuomi, 1996). In addition, dynamic memory maintenance requires the OMS to support knowledge processing to enable actors to reflect, explore and experiment. Such knowledge processing constitutes organisational learning that evolves organisational knowledge and can result in knowledge creation.

For the OMS to support the reflective practitioner it needs to be constructed as a tool for organisational learning. The construction of inscriptions that are artefacts of organisational knowledge is the essential element in sustaining double-loop learning which is the mechanism which organisations use to evolve in response to their changing environment in addition to adjusting their strategies (Argyris and Schon, 1978). It is the application and review of this organisational knowledge that enables errors to be "... detected and corrected in ways that involve the modification of an organisation's underlying norms, policies, and objectives" (Argyris and Schon, 1978). The current trends in organisational change are organisational responses that allow "theories [to be] created to understand and predict" behaviour to overcome the constraints of individual and collective "theories of action" thus combining the "discovery of problems with the invention and production of solutions" (Argyris and Schon, 1978). What is of particular interest is how such inventions gain consensual approval in the organisational and are shared between actors. What we suggest is that the process of constructing the OMS is a meta process which Argyris and Schon (1978) identify as "deutero-learning" or learning how to learn.

The contingency model of OMS, in our view, preclude the construction of an OM as an effective strategy in support of organisational restructuring and downsizing. The prevailing view is that, as a consequence of these organisational changes, organisations cannot rely on people and structures to be repositories of internal and external inscriptions. The implication of this view is that this can be compensated with computer based OM. In contrast, while acknowledging the organisational changes that are occurring, we focus on supporting the changing nature of work practices where workers assume more responsibility, are given greater decision latitude, have the authority to change their work and are encouraged to analyse and reflect so as to innovate and improve their practice.

3. IMPLEMENTING OMS: RESEARCH AS PRACTICE

OM has been a subject of research for various disciplines with each discipline determining its own focus and methods of research. In this paper we have argued for an information systems focus; OMS as IT augmented OM. This perspective allows us to adopt an information systems research model (eg., Nunamaker et al. 1990-91; Galliers, 1991) that involves the construction of an OMS as an integral component.

Building a theory involves 'discovery' of new knowledge and is rarely concerned with contributing directly to practice. On the other hand, after the theory is proposed it needs to be tested in the real world to show its validity and to recognise its limitations, as well as to make appropriate refinements according to the new facts and observations made during its application.

Applied research targets 'real world' problems using existing theories or their combinations and employs methods that allow the richness of the problem situation to be fully explored. Often such research requires the development of artefacts necessary for a particular problem at hand and in this way contributes directly to practice. Moreover it has the potential to contribute to the refinement of the theories on which the original work is based. Thus both theory building and theory testing are two stages of research, with theory refinement providing feedback between them. In the case of OMS, where there is no single dominant paradigm, much research is directed towards a definition of the paradigm or to apply paradigms borrowed from established reference disciplines. In this situation the link between the basic and applied research needs to be even tighter. In Figure 3 we propose a model of OMS research that reflects the current situation of OMS research and practice.

3.1   Theory building in OMS

There are several ways to build theory. One is from the ground up based on concept and relation development, integration, and synthesis. This is epitomised in physics by Einstein, whose theories evolved from first principles. Another is by sifting through enormous quantities of data and

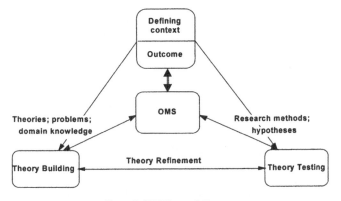

Figure 3. OMS Research Process

identifying unifying patterns. This approach is epitomised by Kepler, who deduced Kepler's Laws of Physics from an analysis of planetary records made a hundred years before by Tyco Brahe. Finally, there is what can be called the "tinkering" approach whereby one builds artefacts and learns iteratively from each design. This approach has been evident in the development, for example, of the transistor and integrated circuits. Building a theory of OMS can proceed along all three lines, with theory development and artefact design first, and empirical studies then building on that foundation. However, in empirical studies of OMS, both the OMS and the organisation in which it resides are the subjects of the study.

Stein's work is based on theory building through the synthesises of a variety of perspectives into a more coherent whole. The work involved deriving a definition of OM and a description both in terms of the contents of memory and the processes associated with it (Stein, 1995). This provided the basis for practical recommendations for manager as well as identifying further research including information systems support for OM. An important outcome of this work is that an OMS is a system by design; OM *per se* has components and processes by design and others that are not consequences of organisational activities. That is the two concepts, OM and OMS are not the same although the literature, unfortunately, tends to gloss over these differences. The importance of the difference lies in the fact that the effects of OM may or may not be positive for the organisation. For example, the "glass ceiling" for women can be considered a negative form of OM. On the other hand, an OMS is meant to be useful in some way by design.

Linger and Burstein (Linger and Burstein, 1998; Burstein et al. 1997), in their work to support researchers, proposed an architecture for a dynamic task-based OMS. The architecture comprises two different levels of representation of the task-related information; pragmatic and conceptual. A process of populating the conceptual level of the OMS requires, and results in, organisational learning. The architecture is also based on communicative perspective on OM, which assumes knowledge sharing and reuse as part of organisational culture. This work involved on the construction of an OMS (Linger et al. 1998, although it was "memory in the small" (Ackerman and Mandel, 1995)

Such studies go some way to establishing a theory of OMS. But they also have implications for the reference disciplines adopted by the studies.

3.2 Theory testing in OMS

Studies which can be classified as a kind of 'applied research' borrow research methods for defining the context of the OMS (reference discipline, types of organisations, etc. as outlined in section 2.2), and test the hypothesis related to the OMS under investigation (see for example, Stein (1995)). The results of such studies contribute to the practice of OMS, refine the theory by providing empirical evidence about the suitability of the chosen approach/paradigm, as well as leading to the advancement of research and practice. However, very few organisation level systems that support memory have been built or studied yet. Most of the work to date has considered systems at the group and individual levels for very targeted tasks or functions.

There is an evident lack of big projects targeted at building OMS for the entire organisation. One of the reasons for this may be the large cost associated with designing and populating the OMS. Another reason, which is even more important, can be resistance to a cultural change which results from the opportunity for better knowledge sharing across the organisation. The third reason can be that to build an OMS is akin to building a complete transaction processing system for an organisation, one that spans the entire organisation and is accessible by all. In order to construct an MIS, the IS community must convince the heads of organisations to build at this level. This results in little theory testing with some exceptions. A number of studies of the application of various technologies to OMS design were reviewed by Stein and Zwass (1995). Another example of theory testing studies is modelling and evaluation of OMS usage at a nuclear power plant performed by Jennex (1997). Other studies include Morisson and Wiser (1996), who conducted an empirical investigation of the usefulness of computer-supported OM as opposed to a traditional paper-based approach. Burstein et al. (1998) conducted a laboratory experiment to test a case-based approach to OMS in a decision support context. Karstein (1996) reported on a longitudinal study performed in order to establish a relationship between the form of OMS depending of the structure of organisation. Johnson and Anderson (1997) conducted a survey to identify the reasons for, and benefits of, introducing OMS into an organisations.

3.3   Methods, Tools and Techniques for OMS

In keeping with the general tradition of IS research, which is well known for the variety of approaches and methods used, there is no one "best" tool and technique used in OMS research. The IS tradition is to follow the contingency approach where the research methodology chosen will depend on the particular circumstances where it will be applied. In OMS, it is the determining factors, as presented in Figure 3, That define which techniques are better suited for that particular system. We believe that most of the research studies on OMS are driven by some kind of reflection on theory or experiences and represent cycles of long-term research projects. From this perspective, action research seems to be a useful research method to approach these kind of studies (Avison, 1993). In action research the role of the investigator is critical as a 'carrier' of the defining factors that determine the research perspective which is in line with the contingency model discussed above.

4.   FROM OMS TO KNOWLEDGE MANAGEMENT

In this paper we present a number of models which reflect our interest in OMS as technologically augmented OM. The discussion shows how a set of dimensions and factors influence OMS, and hence OMS research and practice. We believe the models to be good, operationally defined constructs to establish a paradigm to explain the phenomenon and guide the both the on-going research and continuing construction of operational OMS.

The literature provide enough evidence that organisations are conscious of the need to collect and review their practice and they recognise the benefit of the formal processes and procedures for constructing OMS. However, unless these processes become a part of organisational culture, the success of OMS remains questionable. Empirical studies testing particular interactions between the OMS and the organisation will be of interest but the real key is to generate projects of sufficient magnitude (and funding) that qualify as 'true' organisational memory systems.

We believe the significant outcome of OMS research and practice will be its role as an underlying technology for the emerging field of knowledge management. Moreover, the conceptual framework which we have presented in this paper can play a valuable role in defining knowledge management and prevent the term becoming another marketing fad.

The literature on knowledge management has a remarkable number of definitions of the term, not least because of its currency and because of its perceived importance. Our working definition is:

a broad concept that addresses the full range of processes by which the organisation deploys knowledge

The aim is to add value to material that the organisation already has, support actors performing the activity rather than automating it and apply it on a scale that can be feasibly operationalised. Simultaneously, knowledge management needs to be theoretical based. This requires an understanding of the nature of knowledge work so that processes of knowledge creation and utilisation, as well as organisational learning, can be made explicit. OMS, both theoretically and practically, needs to be consistent with this context to ensure that these processes are effective. From this perspective, memory itself must be a process rather than a repository enhanced with storage and retrieval technology.

OMS, as a technology for knowledge management, has an important role to play within an organisation to ensure that the organisation not only produces its outputs, in terms of products and services, but that it continuously improves these outputs and the means of their production.

5. REFERENCES

Abbott, A., 1988, *The System of Professions: An Essay on the Division of Expert Labor*, The University of Chicago Press, Chicago,

Ackerman, M. S., 1994a, Definitional and Contextual issues in organisational and Group Memories. In *Proceedings of the 27th Hawaii International Conference of System Sciences*, 191-200.

Ackerman, M.S., 1994b, Augmenting the Organisational Memory: A Field Study of Answer Garden, In the *Proceedings of the CSCW 94*, 243-252.

Ackerman, M.S. and Mandel, E., 1995, Memory in the Small: An Application to Provide Task-based Organisational Memory for a Scientific Community. In *the Proceedings of the 28th. Annual Hawaii International Conference on Systems Science.*

Amin, A., 1994, Post-Fordism: a reader; Oxford, Blackwell

Anderson, J. R., 1983, *The Architecture of Cognition.* Cambridge, Mass.: Harvard University Press.

Argyris, C. and Schon, D. A., 1978, *Organizational Learning: A Theory of Action Perspective.* Reading, MA: Addison-Wesley.

Ashby, W. R., 1956, *An Introduction to Cybernetics.* London: Methuen and Co. Ltd.

Avison, D.E., 1993, Research in Information Systems Development and the Discipline of Information Systems, in the *Proceedings of the Fourth Australian Conference on Information Systems*, University of Queensland, Australia, 1-27

Bell, D., 1976, *The Coming of Post-Industrial Society: A Venture of Social Forecasting*, Basic Books, New York, Basic Books, New York, ,First published, 1973.

Blackler, F., 1995, Knowledge, knowledge work and organizations: An overview and interpretation, *Organizations Studies*, Vol. 16, No. 6, pp. 1021-1046

Blackler, F., Reed, M. and Whitaker, A., 1993,, Editorial introduction: Knowledge workers and contemporary organizations, *Journal of Management Studies*, Vol. 30, No. 6, pp. 851-862

Brown, J.S. and Duguid, P., 1991, Organizational learning and communities-of-practice: Toward a unified view of working, learning, and innovation, *Organization Science*, Vol. 2, No. 1, pp. 40-57

Burstein, F., Linger, H., Zaslavsky, A. and Crofts, N., 1997, Towards an Information Systems Framework for Dynamic Organisational Memory, In the *Proceedings of the Hawaii International Conference on Systems Sciences* - HICSS'30, IEEE Press, v 2.

Conklin, E. J.,1992, Capturing Organisational Memory, D. Coleman ,ed, *Groupware '92*, Morgan Kaufman Publishers, 133-137.

Cyert, R.M., and March, J.G., 1963, *A Behavioral Theory of the Firm.* Englewood Cliffs, NJ: Prentice-Hall, Inc.

DeMichelis, G., 1994, A CSCW Environment: Some Requirement" in Scrivener, S.A.R., ed, *Computer-Supported Cooperative Work: The multimedia and networking paradigm.* Aldershot, England, Avebury Technical Report.

Drucker, P.F., 1993, *Post-Capitalist Society,* Harper Business, New York,

Engestrom, Y, Brown, K, Engestrom, R. and Koistinen, K., 1991, Organisational Forgetting: An Activity-Theoretical Perspective. In Middleton, D. and Edwards, D., eds, *Collective Remembering;* Sage Publications Ltd. London, UK

Evans, D.A. and Patel, V.L., eds., 1989, *Cognitive Science in Medicine: Biomedical Modelling,* The MIT Press, Cambridge, MA,

Gardner, H., 1985, *The mind's new science : a history of the cognitive revolution* New York : Basic Books.

Galliers, R. D., 1991, Choosing Appropriate Information Systems Research Approaches: a Revised Taxonomy, in Nissen, Klein, Hirschheim ,eds, *Information System Research: Contemporary Approaches and Emergent Traditions,* Proceedings of the IFIP TC8/WG 8.2 Work.Conf. on the IS Research Arena of the 90's Challenges, Perceptions and Alternative Approaches, Elsevier Science Publishers, 327-345.

Gammack, J. G., 1997, Organisational Memory and Intelligent Decision Support in Shared Information Systems. In the *Proceedings of the Australian Workshop on Intelligent Decision Support* - IDS'97. Monash University, Melbourne Australia.

Iivari, J. and Linger, H., 1999, Knowledge work as collaborative work: a situated activity theory view; In the *Proceedings of the Hawaii International Conference on Systems Sciences HICSS'32,* Hawaii, Collaborative Technology Minitrack,

Jennex, M. E., 1997, *Organizational Memory Effects On Productivity.* Unpublished Doctoral Dissertation, Claremont Graduate School.

Johnson, J and Anderson, J., 1997, Justifying the Information Technology Investment for Organisational Memory, in *the Proceedings of the Hawaii International Conference on Systems Sciences* HICSS'30, Hawaii, Organisational Memory Minitrack, IEEE, Maui, Hawaii.

Karstein, H., 1996, Organisational Memory Profile: Connecting Roles of Organisational Memory to Organisational Form, In *Proceedings of the Annual Hawaii International Conference on Systems Science HICSS'29,* pp 188-196.

Krippendorff, K., 1975, Some Principles of Information Storage and Retrieval in Society. *General Systems,* 20, 15-35.

Kuutti, K., 1991, A concept of activity as a basic unit of analysis for CSCW research, in Bannon, L., Robinson, M. and Schmidt, K., eds, *Proceedings of the Second European Conference on Computer-Supported Cooperative Work,* Kluwer Academic Publishers, Dordrecht, pp.249-264.

Kuutti, K., 1996, Debates in IS and CSCW Research: Anticipating System Design for Post-Fordist Work. In Orlikovsky, Walsham, Jones and DeGross ,Eds, *Information Technology and Change in Organisational Work,* IFIP WG 8.2.

Linger, H. and Burstein, F., 1998, Learning in Organisational Memory Systems: An Intelligent Decision Support Perspective. in the *Proceedings of the Hawaii International Conference on Systems Sciences HICSS'31,* Organisational Memory Minitrack, IEEE, Maui, Hawaii..

Miller, J. G., 1978, *Living Systems.* New York: McGraw-Hill, Inc.

Morrison, J., 1997, Organisational Memory Information Systems Characteristics and Development Strategies, in the *Proceedings of the Hawaii International Conference on Systems Sciences HICSS'30,* Hawaii.

Morrison, J. and Weiser, M, 1996, A Research Framework for Empirical Studies in Organisational Memory, in the *Proceedings of the Hawaii International Conference on Systems Sciences HICSS'29,* III, pp.178-187.

Nelson, R. R., and Winter, S. G., 1982, *An Evolutionary Theory of Economic Change.* Cambridge, MA: Harvard University Press.

Nonaka, I. and Takeuchi, H., 1995, *The Knowledge Creating Company,* Oxford University Press, New York

Nunamaker, J. F. Chen, M. and Purdin, T. D. M., 1990-91,"Systems Development in Information Systems Research, *Journal of Management Information Systems* /Winter 1990-91, Vol.7, No 3, 89-106.

Sandoe, K., and Olfman, L., 1992, Anticipating the Mnemonic Shift: Organizational Remembering and Forgetting in 2001. In *Proceedings of the Thirteenth International Conference on Information Systems,* Dallas: 127-137.

Schank, R. and Abelson, 1995, Knowledge and Memory: the Real Story, ed, R.S.Wyer Jr. *Knowledge and Memory: the Real Story,* Advances in Social Cognition, vol.VIII. Lawrence Erlbaum Associates, 1-85.

Schatz, B.R., 1992, Building an Electronic Community System. *Journal of Management Information Systems,* 8:3, 87-107.

Schon, D.A., 1991, *The Reflective Practitioner: How Professionals Think in Action.* Aldershot, UK, Arena Ashgate Publishing Ltd.

Scrivener, S., 1994, Computer-Supported Cooperative Work, *The Multimedia and Networking Paradigm,* Avebury Technical, England.

Schmidt, K. and Bannon, L., 1992, Taking CSCW seriously, Supporting articulation work, *Computer Supported Cooperative Work,* Vol. 1, pp.7-40

Schmidt, K. and Simone, C., 1996, Coordination mechanisms: Towards a conceptual foundation of CSCW system design, *Computer Supported Cooperative Work,* Vol. 5, pp.155-200

Simon, H., 1957, Administrative Behavior. New York: The Free Press.

Smith, K., 1982, Philosophical Problems in Thinking About Organizational Change. In Paul S. Goodman, Ed. *Change in Organizations San Francisco,* CA: Jossey-Bass Inc., 316-373.

Sowa, J. F., 1994, *Conceptual structures: information processing in mind and machine,* Addison-Wesley.

Sowunmi, A.; Burstein, F.V. and Smith, H.G., 1996, Knowledge Acquisition for an Organisational Memory System, in *the Proceedings of the Hawaii International Conference on Systems Sciences, HICSS*-29 , v 3, 168-177.

Spender, J. C., 1996, Organisational Knowledge, Learning and Memory: three concepts in Search of a theory, *Journal of Organisational Change Management,* 9, 1, 63-78.

Stein, E. W., and Zwass, V., 1995, Actualizing Organizational Memory with Information Systems. *Information Systems Research,* 6,2, June, 85-117.

Stein, E.W., 1995, Organisational Memory: Review Of Concepts And Recommendations For Management, *International Journal of Information Management,* Vol 15, No. 2, 17-32.

Tuomi, I., 1996, The Communicative View on Organisational Memory: Power and Ambiguity in Knowledge Creation Systems. *In Proceedings of the Annual Hawaii International Conference on Systems Science HICSS'29.* 147-155.

Walsh, J.P. and Ungson, G.R., 1991, Organisational Memory. *Academy of Management Review,* 16:1, 57-91.

Weick, K. E., 1979, *The Social Psychology of Organizing* ,2nd ed., Reading, MA: Addison-Wesley.

Wijnhoven, F., 1998, Designing Organisational Memories: Concepts and Methods. *Journal of Organisational Computing and Electronic Commerce,* 8,1, 29-55.

DATA MINING USING NEURAL NETWORKS AND STATISTICAL TECHNIQUES: A COMPARISON

Jozef Zurada and Al F. Salam
Computer Information Systems, College of Business and Public Administration, University of Louisville, Louisville KY 40292, USA

Abstract: This article discusses the similarities and differences between two data mining techniques, neural networks and traditional statistics. It describes how they are used for the tasks of prediction, classification, and clustering, all of which are often performed by financial and marketing managers. The article, which also presents several applications of neural networks, is directed to managers and researchers looking for new artificial intelligence tools for solving business problems.

1.   INTRODUCTION

Current database technology and computer hardware allow researchers to gather, store, access, and manipulate massive volumes of raw data in an efficient and inexpensive manner. In addition, the amount of data collected and warehoused in all industries (such as business, medicine, science, government, and manufacturing) is growing every year at a phenomenal rate. (Data warehouses containing several hundred gigabytes of data are becoming quite common.) Nevertheless, our ability to discover critical, non-obvious nuggets of useful information in the data, nuggets that could influence or help in the decision-making process, is still limited.

Knowledge discovery (KDD) is a new, multidisciplinary field that focuses on the overall process of information discovery from large volumes of warehoused data. The KDD field combines database concepts and theory, machine learning, pattern recognition, statistics, artificial intelligence, uncertainty management, and high-performance computing. Furthermore, the problem of information discovery involves many steps, ranging from data manipulation and retrieval to mathematical and statistical inference, search, and uncertain reasoning. In particular, KDD is an iterative and interactive process that includes the following stages (Brachman *et al*, 1996; Fayyad *et al*, 1996):

(1) learning the application domain (getting to know the data and the task);
(2) data acquisition, preparation, selection, and cleaning;
(3) model and hypothesis development;
(4) data mining;
(5) assessing the discovered knowledge (testing and verification);
(6) interpretation and using discovered knowledge; and finally
(7) visualization of results.

Systems Development Methods for Databases, Enterprise Modeling, and Workflow Management,
Edited by W. Wojtkowski, *et al.* Kluwer Academic/Plenum Publishing, New York, 1999.

299

Although all the phases above seem to be equally important, data mining is the core step in the above process, and it typically occupies about 15-20% of the effort in the overall KDD process. Data mining aims to use existing data to invent new facts and to uncover new relationships previously unknown even to experts thoroughly familiar with the data. The human brain makes such discoveries slowly and sporadically (Grupe and Owrang, 1995).

Typically the following data mining tasks are performed: summarization, clustering, prediction (forecasting), classification, and optimization using the following tools: SQL simple and fuzzy queries, OLAP techniques, statistics (regression-type models, clustering techniques, discriminant analysis), decision trees, case-based reasoning, neural networks, fuzzy logic, and genetic algorithms.

Bigus (1996) claims that data mining is not about complex database queries using SQL or OLAP techniques where the user already has a suspicion about a relationship in the data and wants to pull all of the information together by manually checking and validating a hypothesis. Nor it is about performing statistical tests of hypotheses only using standard statistical techniques known for many years. Data mining should center instead on the automated discovery of new facts and relationships in data looking for the valuable nuggets of business information using new artificial intelligence technologies such as neural networks. Some of the features that make neural networks particularly attractive and useful are their ability to (1) learn from examples, (2) discover relationships/patterns in data, (3) generalize, and (4) adapt to changing conditions. However, Zahavi and Levin (1997) recommend using multiple techniques for data mining simultaneously, including traditional statistical methodology, for two reasons. First, it is difficult to know in advance which particular technique will best uncover the underlying structure of the data. Second, different data mining tools may produce different results when run against the same data. Neural networks are often considered to be statistical methods because they have the ability to identify patterns in data, which was once performed by statistical analysis only. However, both techniques discover patterns in data in a completely different way.

In this paper we discuss the conceptual differences and similarities between neural networks and conventional statistical techniques. We briefly compare and contrast the two techniques and present how they can be applied to certain classes of business problems in the area of marketing and finance. In particular, we discuss the prediction, classification, and clustering tasks (Figure 1). The article is aimed at marketing and financial managers and researchers who are looking for new artificial intelligence tools for solving business problems.

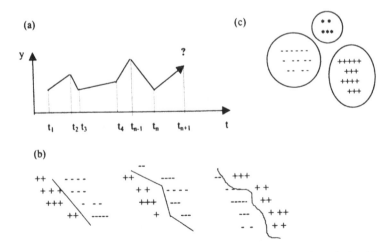

Figure 1. Tasks performed by neural networks and statistical analysis: (a) forecasting, (b) classification, and (c) clustering.

The paper contains six sections. After this first section, the second section contains a brief overview of neural networks. The third section discusses the problem of prediction using the classical regression analysis and the neural network approach. The fourth section and the fifth section concentrate on classification and clustering techniques, respectively, using the two mentioned tools. In particular, in section five we present the self-organizing Kohonen network that enables to group data using unsupervised learning. Also, in sections three through five, we present several marketing and finance applications of neural networks in prediction, classification, and clustering tasks. Finally, section six contains the conclusion.

2.   NEURAL NETWORKS

Artificial neural networks are no longer the subject of computer science alone. They have become highly interdisciplinary and have attracted the attention of researchers from other disciplines as well. Neural networks are nonlinear systems that try to emulate the way the human brain functions and processes information. They are built of highly interconnected nodes, called neurons, that process information. Typically NN models are characterized by their three properties: (1) the computational property, (2) the architecture of the network, and (3) the training/learning property. Neural networks also have specific features that are not present in modern parallel computers.

2.1.   The Computational Property

A typical neuron contains a summation node and an activation function. A neuron accepts an n-element vector $\mathbf{x}$ on input, often called a training pattern. Each element of $\mathbf{x}$ is connected with the summation node by a weight vector $\mathbf{w}$. The summation node produces the value of $net = \sum_{i=1}^{n} x_i w_i$ that is used as the argument in the continuous or binary sigmoidal bipolar activation functions

$$f(net) = \frac{2}{1 + \exp(-\lambda net)} - 1 \quad \text{or} \quad f(net) = \text{sgn}(net) = \begin{cases} +1, net > 0 \\ -1, net < 0 \end{cases},$$

where λ is the steepness of the activation function (Zurada, 1992). Neurons form layers and are highly interconnected within each layer.

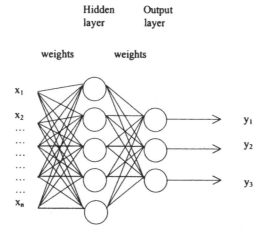

Figure 2. Example of a feed-forward neural network used for prediction and classification.

2.2. Architecture

Several types of neural network architectures have been developed. For example, Figure 2 presents the most common type of a two-layer feed-forward neural network with error back-propagation which is typically used for prediction and classification tasks. The network has two layers, a hidden layer and an output layer. The neurons at the hidden layer receive the signals (values of input vectors) and propagate them concurrently through the network, layer by layer, to the output layer.

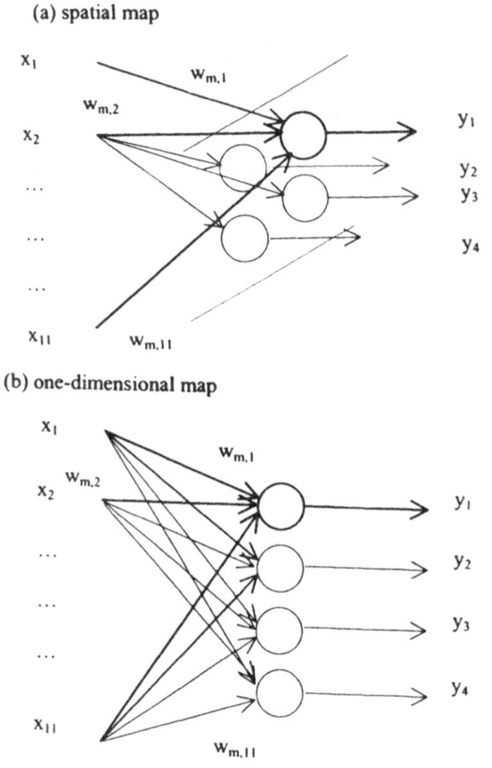

Figure 3. Example of the Kohonen network used for clustering

Figure 3 shows the Kohonen feature maps commonly utilized for clustering or finding similarities in input data. Other architectures, such as the Hopfield network which is used for optimization (to solve the travelling salesman problem), are available, but we will not discuss them in this paper.

2.3. The Learning/Training Property

Neural networks' learning is a process in which a set of input vectors $\mathbf{x}$ is presented sequentially and repeatedly to the input of the network in order to adjust its weights in such a way that similar inputs give the same output. Neural networks learn in two basic modes, the supervised mode and unsupervised mode. In supervised training, the training set consists of (1) the training patterns/examples that appear on input to the neural network (one at a time) and (2) the corresponding desired responses provided by a teacher. The differences between the desired response, vector $\mathbf{d}$, and

302

the network's actual response, vector **y**, for each single training pattern modify the

$$E = \sum_{j=1}^{N} (\mathbf{d}_j - \mathbf{y}_j)^2 / N$$

weights of the network in all layers. The training continues until the mean squared error for the entire training set containing N training vectors is reduced to zero or a sufficiently small value close to zero.

In unsupervised learning, only the input vectors (training patterns) are available, and a teacher is not present. The network is able to find similarities in input data and classify them into clusters. We will discuss this type of learning more thoroughly in section 5.3. In both types of learning, the knowledge (what the network learned) is encoded in weights of the network.

2.4. Features

Jain and Mao (1996) claim that neural networks have many desirable characteristics such as

  (1) no requirement for an *a priori* system model,
  (2) massive parallelism,
  (3) distributed representation and computation,
  (4) learning ability,
  (5) generalization ability,
  (6) adaptivity,
  (7) fault tolerance,
  (8) robustness, and
  (9) low energy consumption

that are not present in von Neumann or modern parallel computers.

These features make them particularly suitable for solving certain classes of business problems in the area of prediction, classification, and clustering, to name a few. In a prediction/forecasting task, we are given a set of n samples $\{y(t_1), y(t_2), y(t_3), \ldots, y(t_n)\}$ in a time sequence $t_1, t_2, t_3, \ldots, t_n$ and we need to predict the value of $y(t_{n+1})$ at some future time t_{n+1} (Figure 1a). The task of pattern classification is to assign an input pattern represented by a feature vector to one of many prespecified classes (Figure 1b). A clustering/categorization algorithm explores the similarity between the patterns and places similar patterns in a cluster (Figure 1c).

3.   PREDICTION/FORECASTING

Marketing managers often need to forecast retail sales over the planning period, while financial auditors and creditors face the difficult task of predicting the future earnings of a firm. These forecasts are necessary to make decisions regarding managing labor, scheduling production, ordering raw materials, or estimating the future stock value of a firm. Using databases containing the information about the market demand for the product, consumer's income, and price of the product, or information regarding the past financial performance of a firm, marketing and financial managers were using multiple linear regression, non-linear regression, or logistic regression for prediction. For the sake of simplicity, our discussion is limited to linear regression models.

3.1.   Linear Regression Model

In matrix algebra notation, a linear regression model is written as

$$\mathbf{y}_p = \mathbf{X}\beta + \varepsilon$$

where **X** is the $n \times k$ design matrix (n is the number observations and k is the number of independent variables or predictors), β is the $k \times 1$ vector of unknown parameters, ε is the $n \times 1$ vector of unknown errors, and $\mathbf{y}_p$ is the $n \times 1$ vector that represents a dependent variable that needs to be predicted. The first column of **X** is usually a vector of 1s used in estimating the intercept term.

The statistical theory of linear models is based on some very strict classical assumptions:

(1) the form of the model is correct
(2) dependent variables are measured without error
(3) the expected value of errors is zero
(4) the variance of the errors is a constant across observations
(5) the errors are uncorrelated across observations and normally distributed.

When these assumptions are not met, the results should be interpreted in a cautious fashion with a discounted certainty in the significance probabilities (SAS© User's Guide: Statistics, 1985).

3.2. Deficiencies of Linear Regression Models

A priori knowledge of the model (the regression equation) represents a tremendous obstacle for a marketing or financial manager. Therefore, he or she typically needs to specify a family of models and select one that best fits the data. The quality of fit is measured by the error term between the predicted variable y_p and the experimental variable y_e. The least square method is normally used to fit the model. In addition, when a series of data exhibits turning points, trend, or non-linearity, linear regression models have difficulties to capture them (Venugopal and Baets, 1994). Also, most traditional regression models are designed to predict one dependent variable at a time. Besides these issues, several other problems such as multicollinearity and heteroscedasticity may be involved. For example, correlation between variables may lead to a multicollinearity problem, or if a single value is missing in an observation, it needs to be dropped for all observations.

3.3. Neural Network Approach

When neural networks are used for forecasting, assumptions 1 through 5 above do not have to be met and the fact that independent variables may be correlated does not create a problem either. A network with one hidden layer is sufficient to approximate any continuous function. The number of neurons in the input and output layers is equal to the number of independent and dependent variables, respectively. The number of neurons in the hidden layer is typically determined experimentally.

Unlike traditional regression models, neural networks can handle multiple dependent variables. However, they do not enable researchers to set the confidence limits for estimated parameters. It is also very difficult, if not impossible, to explain why the neural network generated a particular decision/prediction because the underlying mathematical equation is unknown and the knowledge learned is encoded in the network's weights. In other words, neural network methods are non-parametric in the sense that functional form does not need to be specified *a priori*. Neural networks learn from examples by adapting their connection weights. They can easily handle non-linear data series by mapping an n-dimensional input space in to a k-dimensional output space. Unlike regression analysis, neural networks perform well for missing or incomplete data.

3.4. Applications of Neural Networks in Prediction

Predicting airline passenger volume using traditional statistical methods is a challenging task because it is difficult to build an appropriate regression model and there is no guarantee that statistical techniques will give the best results. As mentioned, these techniques require assumptions about the underlying function in which errors must be independent and normally distributed with a zero mean and a constant variance across observations. Nam and Prybutok (1997) compared the performance of neural networks against the exponential smoothing and time-series regression methods for predicting the passenger volume.

The data sets that they used for training and testing consist of the monthly volume of international airline passengers between U.S. and Mexico for the period of 1982 through 1993. The time plots of the passenger data showed that the number of passengers had an upward trend and depended on the season. The training and testing data sets were used to generate (1) the time-series regression, (2) exponential smoothing, and (3) neural network models to test the predictive capability of the three. In addition, three levels of training sample sizes were selected: -- 3 years (1988-90), 6 years (1985-90), and 9 years (1982-90) -- to forecast the years 1991-93.

A two-layer feed forward neural network with error back propagation was used (Figure 2). The neural network architecture contained 15 neurons in the hidden layer and one neuron in the output layer that represented the volume of passengers. There were 12 variables on input to the neural network that were to capture the seasonality in the data and the time trend. The performance of the neural network was compared to those of the time-series regression analysis and exponential smoothing using the mean absolute percentage error (MAPE), and the F-test was used to find out if the difference in MAPEs of the different models were statistically significant. The study found that in the three scenarios of the 3-, 6-, and 9-year training set sizes, the NN performed better than both above mentioned methods with the average MAPEs of 5.2%, 5.7%, and 4.6%, respectively. In addition, Nam *et al* (1997) noted that the most accurate prediction was achieved for 9-year training size (MAPE of 4.6%) and that the F-test showed that the error difference with the other two models was significant at 0.05 level.

Venugopal and Baets (1994) present a conceptual two-layer feed-forward network. The network accepts on input five variables such as the consumer's disposable income, size of the population, price of the product, price of the substitute, and finally price of complementation products, and predicts two outputs (dependent variables) such as the retail sales and the time it takes to break even.

4. CLASSIFICATION

The decision making often involves classifying a pattern into one of two or more groups. For example, many companies adopt catalog marketing or mail marketing to sell their products to the customers. It is clear that the marketing managers want to achieve a very good response rate. In order to maximize the response rate, marketing managers need to choose customers who would be willing to buy their product(s) and drop those who are unlikely to do so. Thus, customers need to be divided into two groups, prospective and non-prospective customers, based on the data about them stored in the database. Being able to reliably classify the financial health of a firm (as healthy or distressed) is important to bank lending officers, market analysts, investors, creditors, and auditors. For many years, discriminant analysis was used for these types of classification.

4.1. Discriminant Analysis

The purpose of statistical discriminant analysis is to find a linear or nonlinear discriminant function that allows to determine/guess which class a pattern belongs to based on the knowledge of the independent variables only. A linear form of this function written in matrix notation can be:

$\mathbf{z} = \mathbf{X}a$

where $\mathbf{X}$ is the $n \times k$ design matrix (n is the number of observations and k is the number of independent variables), a is the $k \times 1$ vector of unknown parameters, $\mathbf{z}$ is the $n \times 1$ vector of discriminant scores, and the first column of $\mathbf{X}$ is usually a vector of 1s used in estimating the intercept term. The function can be constructed using Fisher's method, which maximizes the ratio of between-groups to within-groups variances. The vector $\mathbf{z}$ of discriminant scores combines the information contained in the matrix $\mathbf{X}$ containing k independent variables. Because discriminant scores themselves are not generally equal to the values assigned to classes, a classification rule is used to translate these scores into class membership predictions. For example, if $C \leq z_i$, then a pattern i is assigned to class 1; otherwise, if $C > z_i$, a pattern i is assigned to class 2. A value C is called a cut-off value that can be determined in a number of ways (Venugopal and Baets, 1994).

4.2. Deficiencies of Discriminant Analysis

For discriminant analysis to be reliable, the following assumptions have to be met. Independent variables have to be approximately normally distributed and have equal covariances within classes/groups. In practice, data that describe the characteristics of customers or financial performance of companies do violate these conditions. It all affects the classification accuracy of the analysis and the appropriate forms of the classification rule. In addition, a discriminant function cannot fully capture the subtle relationships between the complex factors that cause bankruptcy or represent a customer.

4.3. Neural Network Approach

In recent years, many businesses started to use neural networks for classification. It is easy to construct a two- or three-layer feed-forward network with error back propagation in which the number of independent variables and the number of classes represent the number of neurons in the input layer and the output layer, respectively (Figure 2). Neural networks make no assumptions about the underlying statistical distribution in the data. Unlike discriminant analysis, neural networks adapt very well to the situations of missing or incomplete data, or to the addition of new data to the sample. The classification accuracy of discriminant analysis may be negatively affected, especially when new cases are drawn from a different distribution than the old ones. In other words, one cannot apply the discriminant function which was developed from the independent variables representing old cases to the classification of new cases. On the other hand, neural networks such as be easily retrained using new examples to improve their classification capability. They will have the knowledge about both new and old cases embedded in their weights.

4.4. Applications of Neural Networks in Classification

Bigus (1996) describes an interesting application in which he uses a two layer feed-forward neural network to rank the customers according to their "goodness." In addition, he performed a sensitivity analysis to understand which customer attributes are the best predictors of profitability. He selected the following company data for training the neural network: (1) years in business, (2) number of employees, (3) type of business, (4) revenue, (5) average number of orders, (6) average revenue per order, and (7) average profit per order. His "goodness" measure was a score that was obtained by multiplying the average number of orders by the average profit of those orders. He used a threshold function and converted a continuous goodness score into four somewhat discrete categories: A, B, C, and D, with A representing the best current (and prospective) customers and D representing the worst. Consequently, Bigus (1996) has four output variables (neurons) representing how "'good" the customers are. He divided training and test data into two equal sets having 200 patterns each, and established the acceptance criteria for the classifier to be considered trained. His goal was to correctly classify 90% of the customers into classes A, B, C, and D both during training and testing. When training was finished, Bigus (1996) used 200 patterns to test the performance of the neural network. The neural network correctly classified 191 out of 200 patterns (90.5%) of the customers into appropriate classes.

Being able to classify the financial health of a firm reliably is very essential to bank lending officers, market analysts, investors, creditors, and auditors. For many years, these individuals were using conventional statistical methods such as financial ratios, discriminant analysis, and logistic regression to classify firms as healthy or distressed. However, they cannot easily understand how all the complex factors that cause bankruptcy interplay, or consequently, build a discriminant analysis model. Therefore, many financial experts have started to use neural network techniques for classification.

Most of the studies previously performed involved bankruptcy prediction for two states only: healthy versus bankrupt (Coats and Fant, 1993; Fletcher and Goss, 1993). These studies found that the neural network approach is better than traditional regression analysis techniques because the former could classify healthy versus bankrupt firms better than the latter. Because firms experience different levels of distress before becoming bankrupt, Zurada et al (1997) compared the classification accuracy rates of the neural networks to those from logistic regression models for a four-level response variable.

Zurada et al (1997) examined four levels of distress: healthy (state 0), dividend cuts (state 1), loan defaults (state 2), and bankruptcy (state 3). They selected a two-layer feed-forward neural network with error back-propagation that is typically used in the majority of financial applications. The input to the neural network contained 9 variables that represented historical financial ratios collected over the period of three years for small and large firms traded on NASDAQ. There were four different models of data to examine. As a result, there were 12 different scenarios tested.

To train the network, they used an unbalanced, but realistic, training sample of 204 firms. The number of neurons in the hidden layer and the output layer was 12 and 4, respectively. In the output layer, each neuron was representing one of the four states. For testing, they again used an unbalanced sample containing 141 of the holdout firms.

Due to an excellent classification of healthy firms (state 0), the overall network's classification capability was better then that obtained from the regression models. However, the regression analysis appeared to give more accurate predictions for the three remaining states: dividend cuts (state 1), loan defaults (state 2), and bankruptcy (state 3). Zurada et al (1997) concluded that the neural network was not superior to traditional regression models.

In another paper, Zurada et al (1999) collapsed the distressed states 1 through 3 into a single unhealthy state. As a result, there were only two states analyzed: healthy versus bankrupt. Zurada et al (1999) again compared the classification accuracy of the neural network to that of the logistic regression. It turned out that the regression technique classified firms marginally better than its neural network counterpart. The authors concluded that when classification accuracy is the most important goal, a neural network tool should be used for a multi-state financial distress response variable, and logistic regression and neural networks produce comparable results for a two-state financial distress response variable. When testing the usefulness of specific information is the most important goal, researchers should use logistic regression analysis.

5.   CLUSTERING/GROUPING

Business applications of clustering are mainly in the marketing arena (Peacock, 1998). One of the many challenges that face any marketing business is to understand its customers at many levels. First, what services and products interest the customer the most? Second, what does the average customer look like? Is he or she single or married with children? What is the customer's average age and income? With information on (1) customers' demographics, interests, habits, economic status, and geographic information, (2) product information, and (3) customers' sales transactions available in databases, the marketing business can target sales promotions directly at the current and new customers who are most likely to buy a product from the company. This technique is called target marketing, and it relies principally on the ability to segment or cluster the total market into several specialized groups that can be served with a higher degree of satisfaction for the customer and profit for the marketing company. In other words, by knowing what a customer is interested in, companies can lower marketing costs through more effective mailings, and customer satisfaction is improved because they are not receiving what they perceive as "junk mail."

5.1.   Statistical Cluster Analysis

There is a conceptual difference between cluster analysis and discriminant analysis. The former creates groups, but the latter assigns patterns to existing groups. Clustering is understood to be grouping similar objects and separating dissimilar ones. The most common similarity rule is the Euclidean distance between two patterns $\mathbf{x}$, for $\mathbf{x}_i$ defined as:

$$\|\mathbf{x} - \mathbf{x}_i\| = \sqrt{(\mathbf{x} - \mathbf{x}_i)'(\mathbf{x} - \mathbf{x}_i)}$$

The smaller the distance, the closer the patterns. Using the above formula, the distances between all pairs of points are computed, and then a distance can be chosen arbitrarily to discriminate clusters (Zurada, 1992). The formula above can be generalized to compute the distance between any two of the n patterns. There is a number of other statistical cluster search algorithms such as maximum-distance or k-means. If the population of patterns is sufficiently well separated, almost any clustering method will perform well (SAS© User's Guide: Statistics, 1985).

5.2.   Difficulties in Cluster Analysis

Most cluster analysis methods are heuristics and are not supported by an extensive body of statistical reasoning (Venugopal and Baets, 1994). Also, the cluster analysis methods (1) require again some strict assumptions about underlying distribution of data patterns, (2) cannot satisfactorily determine the number of clusters necessary to separate patterns, and (3) are sensitive to a large sample size. In addition, it is always difficult for researchers to select an appropriate similarity measure (SAS© User's Guide: Statistics, 1985). For these reasons, market analysts began to use neural networks that can perform clustering and do not suffer from the above limitations.

5.3.   Neural Network Approach

The Kohonen network is one of the neural network architectures commonly used for clustering. It classifies N input vectors $\mathbf{x}$ into one of the specified number of p categories (Zurada, 1992). Kohonen feature maps are feed-forward networks that use an unsupervised training algorithm. During the process called self-organization, they configure the output units into a topological or spatial map (Figure 3). The network consists of a single competitive output layer only, and the input vector is fully connected to this layer. When an input pattern, vector $\mathbf{x}$, is presented to the network, the units in the output layer compete with each other for the right to be declared a winner. The winning neuron is the one that has the maximum response due to input $\mathbf{x}$. In other words, the winner is typically the neuron whose incoming connection weights $\mathbf{w}_m$ are the closest (in terms of the Euclidean distance) to the input pattern (Figure 3, in bold). The winner's connection weights $\mathbf{w}_m$ are then adjusted by the term $\Delta\mathbf{w}_m = \alpha(\mathbf{x}\text{-}\mathbf{w}_m)$ in the direction of the input pattern by a factor determined by a small learning constant $\alpha > 0$ selected heuristically.

Often, in this network, not only do the winner's weights get adjusted, but also the weights of the neighboring neurons get moved closer to the input pattern. As training progresses, the size of the neighborhood fanning out from the winning neuron is decreased. In other words, initially large numbers of neuron weights' will be updated, and later on smaller and smaller numbers are updated. Similarly, the learning constant α typically decreases as learning progresses. Finally, at the end of training only the weights of the winning neuron are adjusted. A simple example of the Kohonen network for $p=4$ clusters (output neurons) and vectors $\mathbf{x}$, containing 11 elements each, is shown in Figure 3.

5.4.   Applications of Neural Networks in Clustering

Bigus (1996) used the Kohonen network for target marketing. He constructed the network that can cluster customers with respect to their potential interest in five

Table 1. Composition of four clusters by age and by percentage of the population

Segment #	Segment size in %	Average age of customer
1	42.8	43
2	24.8	42
3	20.4	52
4	11.9	26

Table 2. Spending habits of the analyzed customers

Segment #	Maximum $ spent on	Minimum $ spent on
1	Home appliances	Sports equipment
2	Exercise equipment	Appliances
3	Furniture	Appliances, sports equipment, & entertainment
4	Entertainment, exercise & sports equipment	Appliances & furniture

product categories. Using SQL queries, he selected the following customers' attributes from the database: (1) age, (2) sex, (3) marital status, (4) income, and (5) homeowner status, combined with customer interest information that included five product categories: (6) sporting goods, (7) home exercise equipment, (8) home appliances, (9) entertainment, (10) home furnishings, and (11) total amount spent. All numerical fields such as the customer age or income have been mapped to the range [0.0,1.0], whereas all alphanumeric fields such as sex ('M', 'F', 'U' - unknown) or homeownership ('Y', 'N', 'U' - unknown) have been scaled to (1.0, 0.0, and 0.5), respectively. These 11 variables were applied to the input of the Kohonen network. Because it was decided arbitrarily to divide customers into four groups, there were four neurons (2-by-2 layout) in the output layer. The training started with a learning constant $\alpha=1.0$ that decreased 0.05 per training cycle. Thus, it took 20 training cycles to train the network.

The Kohonen map split the customer into four segments (clusters). The descriptive statistics concerning these segments is shown in Table 1. Table 2 presents the spending habits of the customers with respect to five product categories (sports equipment, exercise equipment, appliance, entertainment, and furniture) in each of the four segments. It is clear from Table 2 that the customers who fell into segment 1 should be mailed promotions on home appliances, while the customers who belong to segment 4 should be mailed entertainment and sports and exercise equipment catalogs. The customers in segments 2 and 3 should be targeted with exercise equipment and furniture advertisements, respectively.

It might be interesting to know that in clustering there is no "right" answer, and the customers could be partitioned into 6, 8, or even 10 distinct groups. Also, because the initial weights of the Kohonen network are initialized to random values each time the training starts, it is very probable that that when marketers run the same network on the same 4 clusters, the composition of the segments may be slightly different.

Kohonen networks usually need much less training time than feed-forward networks with error back-propagation and have fewer training parameters to set. However, these types of networks suffer from some limitations as well. Because of a single-layer architecture, this network cannot efficiently handle linearly nonseparable patterns. Moreover, weights may get stuck in isolated regions without forming the expected clusters. In other words, some neurons only are declared the winners and get their weights adjusted. In such a case, the training needs to start from scratch with a new set of randomly selected weights.

6. CONCLUSION

The paper is written for managers and researchers looking for new artificial intelligence tools for solving business problems. In the paper, we addressed the issues concerning the similarities and differences between two data mining techniques, neural networks and traditional statistics. Both techniques are successfully used for solving certain classes of business problems involving prediction, classification, and clustering tasks. However, they solve these tasks in completely different ways. The two methods can also complement each other by using them in parallel to produce a very powerful data mining tool.

Both techniques have some drawbacks. For example, unlike traditional statistical techniques, neural networks can neither adequately explain the outcomes they produce, nor enable researchers to set the confidence limits for estimated parameters. In addition, some of their architectures may suffer from a long training time. Also, it may be difficult to select the appropriate network configuration for a given task, and it is often necessary to preprocess raw data and select variables for training. Preprocessing data and selecting variables may require a great deal of time and skill.

On the positive side, however, neural networks are much more robust than statistical techniques in discovering patterns because they do not require any assumptions about underlying distribution of data patterns. They can tolerate faults and missing values in observations, can be trained from examples, adapt easily to changing conditions, and process data in parallel.

REFERENCES

Bigus, J.P, 1996, *Data Mining with Neural Networks*, McGraw-Hill, New York.

Brachman, R.J., Khabaza, T., Kloesgen, W., Piatetsky-Shapiro, G., and Simoudis, E., 1996, Mining Business Databases. *Communications of the ACM*, New York, **39**, No. 11, 42-48.

Coats, P.K., and Fant, F.L., 1993, Recognizing Financial Distress Patterns Using a Neural Network Tool. *Financial Management*, **22**, September, 142-150.

Fayyad, U., Piatetsky-Shapiro, G., and Smyth, P., 1996, The KDD Process for Extracting Useful Knowledge from Volumes of Data. *Communications of the ACM*, New York, **39**, No. 11, 27-34.

Fletcher, D., and Goss, E., 1993, Forecasting with Neural Networks: An Application Using Bankruptcy Data. *Information & Management*, **24**, No. 3, 159-167.

Grupe, F.H., and Owrang, M., 1995, Data Base Mining: Discovering New Knowledge and Competitive Advantage. *Information Systems Management*, Fall issue.

Jain, A.K., Mao, J., and Mohiuddin, K.M., 1996, Artificial Neural Networks: A Tutorial. *Computer*, **29**, No. 3, 31-44.

Nam, K., Yi, J., and Prybutok, V.R., 1997, Predicting Airline Passenger Volume. *The Journal Business of Forecasting Methods & Systems*, **16**, No. 1, 14-16.

Peacock, P.R., 1998, Data Mining in Marketing:Part I. *Marketing Management*, **6**, No. 4, 8-18.

Peacock, P.R., 1998, Data Mining in Marketing:Part II, *Marketing Management*, **7**, No. 1, 14-25.

SAS© User's Guide: Statistics, Version 5 Edition. Cary, NC: SAS Institute Inc., 1985.

Venugopal, V., and Baets W., 1994, Neural Networks and Statistical Techniques in Marketing Research: A Conceptual Comparison. *Marketing Intelligence & Planning*, **12**, No. 7, 30-38.

Zahavi, J., and Levin N., 1997, Issues and Problems in Applying Neural Computing to Target Marketing", *Journal of Direct Marketing*, Vol. 11, No. 4, pp. 63-75.

Zurada, J., Foster, B.P., Ward, T.J, and Barker, R.M., 1997, A Comparison of the Ability of Neural Networks and Logit Regression Models to Predict Levels of Financial Distress. *Systems Development Methods for the 21st Century*, (G. Wojtkowski, W. Wojtkowski, S. Wrycza, and J. Zupancic , eds.), Plenum Press, New York, pp. 291-295.

Zurada, J., Foster, B.P., Ward, T.J., and Barker, R.M., 1999, Neural Networks Versus Logit Regression Models for Predicting Financial Distress Response Variables. *Journal of Applied Business Research*, **15**, No. 1, 21-29.

Zurada, Jacek M., 1992, *Introduction to Artificial Neural Systems*, West Publishing Company, St. Paul, MN)

THE YEAR 2000 PROBLEM IN SMALL COMPANIES IN SLOVENIA

Borut Verber, Uroš Jere, and Jože Zupančič
University of Maribor, Faculty of Organisational Sciences,
400 Kranj, Kidriceva 55a Slovenia
Ph.: +386 64 37 42 82, Fax: +386 64 37 42 99
E-mail: {Borut.verber; uros.jere; jose.zupancic}@fov.univ-mb.si

ABSTRACT: Results of an investigation conducted in Slovenia in 1998, focusing on the Year 2000 problem in small Slovenian companies are presented. The level of awareness of the problem among managers of these companies is assessed, and actions, plans expectations and approaches to the solution of the problem is surveyed. Based on the survey, some recommendations for managers of small companies in connection with the Y2K issue are given.

1. INTRODUCTION

The Year 2000 problem (Y2K Problem, Millennium Problem) stems from application software that represents a year using two digits and thereby creating identical data values for the years 1900 and 2000. The six-digit date format (YYMMDD or MMDDYY) has arisen from the historical need to spare space on early mass storage devices. Many early programmers believed that future programming techniques would surpass their original work and therefore saw no need to indicate a four-digit year. Technology has advanced rapidly, yet, hardware and operating systems progress has supported backward compatibility. Therefore, many original programs have never become functionally obsolete (Schulz, 1998).

Recently the Y2K problem has been popularized in mass media. Books have been published aiming at computer professionals (e.g. Feiler and Butler, 1998; Koller 1997) and at millions of people who use computers or in some way depend on computers (e.g. Yourdon and Yourdon, 1998; Lord and Sapp 1998). Due to its specific characteristics, the Y2K problem has been described as a crisis without precedence in human history (DeJesus, 1998), and called »time bomb 2000« or »millennium bug« in the mass media. Large amounts of money have been spent on solving the problem; an estimate by Gardner Group which is frequently cited in the literature, is over 600 billion US$ worldwide.

Systems Development Methods for Databases, Enterprise Modeling, and Workflow Management.
Edited by W. Wojtkowski, *et al.* Kluwer Academic/Plenum Publishing, New York, 1999.

Investigations have shown that the Y2K problem can affect all computer based systems, including mainframes, PCs, and embedded systems. Embedded chips exist in a wide range of engineering tools, process control and security systems, even in cars, office equipment and household appliances with internal clocking mechanisms. One such example are fax machines which often play an important role in business communications, particularly in small companies. Internet represents a specific case: Internet based applications may fail, because servers, operating systems and switches may fail. There might be minor service blackouts all over the Internet, seriously affecting Internet based business. Yet, the Internet as a whole may stay up because it still has enough redundancy.

2. YEAR 2000 PROBLEM IN SMALL COMPANIES

Small companies are a major component of the business environment throughout the world. Further, it is expected that the importance of small and medium sized companies will increase - particularly in the manufacturing sector - due to a shift in the strategic policy of many major organizations - the movement towards a »hollow« factory where most assembly work is sub-contracted rather than done in-house (Levy and Powell, 1998). In Slovenia, much like in other former communist countries, many new small companies have been founded after the change of the economic and political system in 1991. In 1996, the 33,356 small firms represented 93.2% of all companies in Slovenia, and their number is still increasing.

Many articles, papers and books dealing with the Y2K problem and its solutions have been published, and a large amount of the millennium bug related information is available at various Internet sites. Most of them aimed at large companies, hence neglecting the issue in small organizations. Previous studies conducted in Slovenia (Mesarič et al., 1996; Zupančič, 1997; Leskovar et al., 1998) addressing large companies showed that the awareness of the Y2K problem among information system (IS) and general managers had increased considerably in the past few years. Quite many organizations have made some sort of an impact analysis although they apparently underestimate the cost and effort of solving the millennium problem.

Recent investigations by Gartner Group indicate that the Year 2000 issue has been neglected particularly in small companies and in government institutions. Small companies often represent an important part of the supply and/or value chain of large companies. Information systems in some small companies may fail due to the millennium bug which may strongly impact the performance of the whole chain.

In general, small companies lag behind large businesses in the use of information technology (IT) due to a special condition commonly referred to as resource poverty. This condition is characterized by severe constraints on financial resources, a lack of in-house expertise, and a short term management perspective imposed by a volatile competitive environment. On the other hand, small businesses can be more innovative than large organizations because they are less bound by bureaucracy and cumbersome

314

organizational systems, are more flexible and able to respond to changing customer needs. Because of scarce financial resources, most small businesses tend to choose the lowest cost information system (IS) and often underestimate the amount of time and effort required for IS implementation. Hence, small businesses face a greater risk in IS implementation than larger businesses (Thong et al. 1997).

Nowadays, most small companies use computer technology. With introduction of micro computers, file servers and networks small firms have the potential to take advantage of the same technology that large companies have access to. At the same time, they are becoming more and more dependent on information technology (IT). Investigations showed that IT can have a strong positive effect on business performance and improve the ability of small enterprises to compete, both nationally and internationally (Doukidis et. al., 1994; Serwer, 1995). A small business, by its nature, must commit resources from a limited pool and therefore must implement to its full IT potential (Winston and Dologite 1999).

Small companies mostly use PCs and PC networks. Many have Y2K non-compliant versions of PC software tools, or antiquated DOS software. Solving the Y2K bug on PCs and fixing PC developed applications are considered less sophisticated and less critical than solving the problem on mainframe computers. Never the less, investigations in large organization showed that the Y2K problem can strongly affect PCs and PC based information systems too.

For example, an impact analysis in a large organization in Slovenia (Tič et al., 1998) revealed that only 22.1% of its nearly 3000 PCs were fully year 2000 compliant. More than 71.4% of the PCs required an upgrade (most frequently new BIOS), while 6.5% had to be replaced. About one third of application software (32%) was assessed as Year 2000 compliant - assuming that it was used according to the vendor's documentation. Another third (34%) required minor corrections (installation of Fix Pack), and one third (33%) was found incapable to deal with dates from the next century and should be replaced or rewritten.

The above example which can probably not be straightway generalized to small companies, indicates that many PCs which support some vital business operations in some firms may fail at the beginning of the next century. The risk for small businesses is even bigger because they usually use older and more obsolete hardware and software than large companies. Further, they almost never have their own IS professionals who are able to analyze the impact of the Year 2000 bug on their information systems. They therefore strongly depend on IS related expertise from outside. Such outside expertise and assistance are usually inadequate to deal with the large number of PCs in small organizations. This is also one of the reasons that even some large companies have realized that the best approach to resolving the PC related year 2000 issue is to make the PC users responsible. To achieve this, they have to educate them, and give them tools and procedures for coping with the problem (Rollier, 1999).

Our investigation shows that managers and owners of small businesses have learned about the Y2K problem only from the mass media (newspapers, TV, radio, ...), friends and business partners, and that they underestimate its impact on the company's business. Due to the Slovenia's still unstable economic situation and its lack of a national long-term strategy, small firms receive but a limited support from the state. So far, no action has been taken by any government institution to increase the awareness of managers in small organizations of the millennium problem and/or to provide a framework for assistance.

Our paper presents results of a study carried out in October 1998 among small companies in Slovenia. It was focused on the awareness of managers in small companies of the millennium problem, approaches and efforts aimed at the solving of the Y2K problem.

Table 1: Job position of the respondents

Position	Number of respondents
Top executive / owner	38 (42.2%)
IS professional	14 (15.5%)
Accountant	13 (14.4%)
Sales manager	9 (10.0%)
Other	12 (13.3%)
No response	4 (4.4%)
Total	90 (100.0%)

3. RESEARCH APPROACH

To collect the survey data, a questionnaire with 20 demographic questions and questions related to Year 2000 problem was developed. It was tested in 5 selected organizations. Several contacts with respondents helped to improve accuracy and relevance of the questionnaire. In September 1998, the questionnaire was mailed to unidentified managers of 500 manufacturing and service firms with not more than 100 employees. Businesses were randomly selected from a commercially available directory (Kompas Xnet). A letter with a short description of the Y2K problem was added to the questionnaire. It was addressed to the managerial stuff of the company.

Ninety-two questionnaires were returned. Two of them were not usable. This yields a 18% response rate which is common for this type of studies. Respondents who supplied data came from a wide variety of business backgrounds and sizes. The average number of total employees was 36.3. The position of the respondents in the organization is presented in Table 1. In the sample most organizations (57) were service oriented, 32 were manufacture oriented, and one was a non-profit organization.

4. RESULTS

The majority of organizations from our sample (80%) used only PCs. The remaining 20% used minicomputers or a mainframe in addition to PCs. Most of organizations which used other computers than PCs companies came from data processing businesses. The average number of PCs was 10.5 per organization (one PC per 3.5 employees).

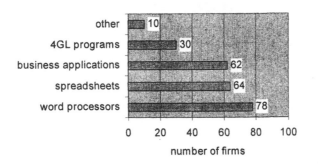

Figure 1: Type of software used by sampled firms

The oldest PC still used in one of the sampled firms was from 1990. The average purchase year of the oldest PC or PCs in the company which are still in use was 1993,3. This means that they used motherboard 80286, 80287 or 80386, and are definitely year 2000 incompatible. We can expect that software which was running on these computer was not Year 2000 compliant.

The major part of the surveyed firms used more than one operating system on their PCs. Most of them (75 - 83.3%) used Windows 95. Then DOS (52 - 57.7%), Windows 3.x (26 - 28.9%), Windows 98 (19 - 21.1%) and Windows NT (17 - 18.9%). Eight firms used other PC or mainframe based operation systems such as *VMS, UNIX, Novell* and *OS/2*.

Most companies used word processors, spreadsheets and some business applications such as accounting, inventory management etc. (Figure 1). One third (30) firms ran programs developed using a 4GL or a tool such as Delphi, MS Access or Paradox.

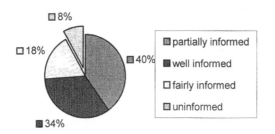

Figure 2. Level of respondents' familiarity with the Year 2000 problem

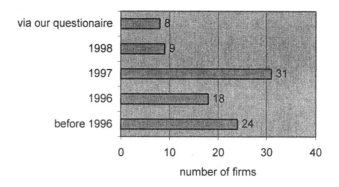

Figure 3. When did respondents hear about the Year 2000 problem for the first time

Figure 2 shows to what level respondents were familiar with the year 2000 issue. Relatively few (8%) responded that our questionnaire was their first information about the Y2K issue. This probably can not be generalized to all small companies in Slovenia, as the response rate to our questionnaire could be lower among companies that knew nothing about the Y2K problem. Figure 3 shows when respondents heard about the Y2K issue for the first time.

Respondents were asked to indicate their major sources of information about the Y2K problem. Responses are summarized in Figure 4. The fact that the prevailing source of information were newspapers and TV indicates that knowledge of the Y2K problem in small companies could be very superficial. Most (62) of the organizations out of the 84 who responded this question were not informed about the Year 2000 issue by the Chamber of Commerce or any other government institution, 12 organizations received some kind of basic information, and 10 had no idea whatsoever of the problem.

The question »Have you made an impact analysis or assessment of the Y2K problem in your organization?« was responded by 81 out of 90 participating organizations. Forty-four organizations (54%) have not analyzed the issue and 35 (46%) have made an assessment or are working on it. Most of them (28) did it in 1998, and only 3 in 1996 or before. Contacts with some organizations who have not responded this question indicated that they have not analyzed the problem yet. Sixteen out of 23 companies whose primary business is computing have made the analysis.

Figure 5 summarizes responses to the question »When will you start (or have you started) solving the problem?«. Twelve out of 90 respondent gave no answers to this question. Nine out of 18 companies who responded that they had solved the problem were informed about it by mass media.

Respondents were asked to give a qualitative evaluation of the impact on their organizations and a qualitative assessment of the cost of solving the Y2K problem. The majority (82 out of 90) responded this question. We didn't ask them to give us a quantitative assessment because previous investigations in Slovenia (Mesarič et.al. 1996, Zupančič, 1997, Leskovar et al. 1999) had failed to collect any usable data about the cost. Even in this case only 51 out of 90 respondents provided an answer to this question. Surprisingly enough, many of them found the cost critical, but not the problem itself. Figures 6 and 7 present the responses.

Most companies from the computing business (19 out of 23) responded that Year 2000 showed no or only a minor problem to them. The remaining four companies considered it a serious problem.

Figures 8 and 9 show expectations of respondents regarding hardware and PC based application packages.

More than a half (52%) of the surveyed companies used or intended to use external support (consulting) to solve the Y2K problem, 7% of respondents thought that adequate support was not available in Slovenia, and only a few (4%) planned to solve it without external support. The remaining 37% responded that they were not sure about external support.

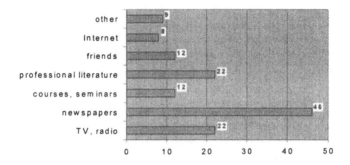

Figure 4: The primary source of information about the Year 2000 problem.

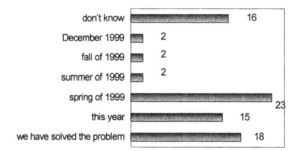

Figure 5: Time when organizations intend to start solving the Y2K problem

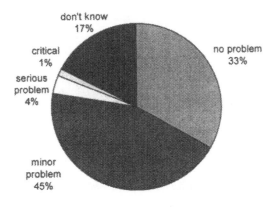

Figure 6: Estimated size of the problem

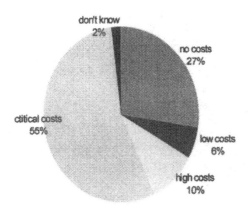

Figure 7. Estimated cost of solving the Year 2000 problem

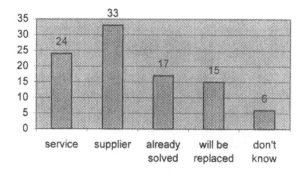

Figure 8: Respondents expectations as to by whom the hardware related Y2K problems will be solved.

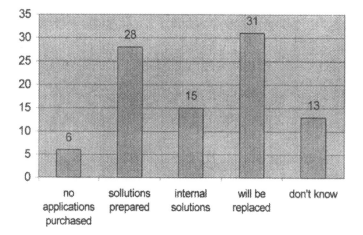

Figure 9: : Respondents expectations as to by whom the purchased applications related Y2K problems will be solved.

5   DISCUSSION

Our survey shows that most of the small companies are aware of the Y2K problem about which they have been but superficially informed, primarily from mass media. Only about one fourth have learned about the Y2K problem from the professional literature. Therefore, the understanding of the problem in most small companies may be inadequate and the attitudes unconstructive. The majority of the organizations have not started solving the problem yet and have no plans about it either. They will most likely take no preventive actions, and will simply waiting to see what will happen at the turn of the century rather than get foreseeable problem under control before they develop into uncontrollable ones.

Most of the surveyed mangers think that there is not enough expertise in their company to deal with the Y2K problem and that they will need external support. They put high expectations in their hardware and software suppliers and organizations who maintain their computing systems. We can expect that these suppliers and organizations will be overwhelmed by urgent calls for support and help around the turn of the century and will not be able to provide the required service. Therefore, we recommend to mangers of small companies to get better acquainted with the Y2K issue, related risks and ways of solving the problem with minimum external help. A similar approach has been taken by some large companies who have a high level inhouse expertise. They have realized that they have too many PCs and that their IS staff simply can not manage the problem all by themselves. They train their PC users, help them to identify the problems, and give them tools and procedures for taking care of their computers (Rollier 1999).

Software and hardware suppliers in Slovenia strive (quite successfully, so far) to avoid any responsibility for delivering the Y2K noncompliant products. Slovenian legislation governing the liability of suppliers for their products is much less rigorous and followed less strictly as for example in the USA. On the web site of one of the domestic software vendors we can read the following statement (translated by the authors of the paper) which can provide a good example of their attitude to the issue:

»Although the application *AAA* uses a two digit year, the program will work in year 00 (2000) and later. Regarding the general and specific warranty conditions for software, our company *XXX* is **not responsible** for eventual loss of data, regardless of the reason. The user is always responsible for data protection, therefore we recommend to back up the data regularly (daily or weekly). The user is responsible for the judgement if the software is suitable for use or not. The **responsibility for its proper functioning is also on the side of the user**. This applies also to the **functioning of software and hardware after 1999**«.

Testing hardware and software, including PCs, for the year 2000 compliance can be very difficult. Numerous software tools to test your PC are available from different vendors, and even free from the Internet. Yet nobody guarantees that they will detect all the Year 2000 related problems. Though vendors claim that new versions of their hardware and software are compliant, people who tested them found the Y2K problems even with the most recent versions of products. For example, even the new Pentium II should have some Y2K problems (Rollier, 1999) and may have a tendency to die when tested. Microsoft is testing its products extensively and maintains a »living« web site - Product Guide - (http://www.microsoft.com.year2000) with updated results of these tests and other related information. Microsoft further claims that all its 32 bit products are year 2000 compliant, and gives diverse recommendations for other products. Nevertheless, Microsoft recommends to use packaged software wherever possible since »Y2K problems with products from well known vendors are relatively rare«.

A considerable number of surveyed managers believe that their information systems are Year 2000 compliant. Many of them have learned about the year 2000 issue mainly from media and believe their hardware and software suppliers for the company's IS is ready for the year 2000. On the other hand, many experts (e. g. DeJessus, 1998), based on the external outcome of numerous tests, recommend not to trust Year 2000 compliance claims by external consultants and hardware and software suppliers. Therefore, companies should consider the possibility that some applications and/or computers will fail and those which may run fine might produce incorrect information (which is often very difficult to detect).For the bather case they should prepare a contingency plan.

Interestingly, Microsoft recommends businesses to simplify their work processes so as to make them automation and control easier, and to develop an »emergency« system using a small computer and »manual« work procedures, to ensure that at least the core business functions will work properly. Similarly, (DeJesus, 1998) recommends that companies should develop Plan B, a contingency plan just for the case that their computer systems go south as they come to the year 2000. This means to keep paper records of the last period of 1999, such as the November and December payroll, and generate and approximate the January payroll, so that they will have something to work from when their computerized information system fails. Further, they should have a printed list of contacts for every supplier and customer, and have procedures mapped out for billing, invoices, payroll, accounting, sales administration, and anything else their computers routinely handle.

Companies should also involve their suppliers and customers into the preparation of their contingency plan. Due to the worldwide competition, customers require shorter and shorter lead times for products and »just in time« deliveries, and organizations are becoming more and more interdependent. A supplier with a computing problem can represent a risk for the whole business. Increased cooperation among business partners can help to avoid such pitfalls.

Our investigation focuses on small companies in Slovenia. Its findings can probably not be generalized to newly created small businesses in other East and Central European countries, or even in emerging markets in Asia because these countries are quite diverse in terms of the levels of economic activity, application of IT, growth rate in also culture. Further research conducted in several countries of East and Central Europe would be necessary to assess on the side the extent to which organizations in this part of the world will be affected by the Y2K problem and on the other response to the issue.

REFERENCES

DeJesus EX. Year 2000 survival guide, Byte 1998 July: 52-62.

Doukidis G, Lybereas P, Galliers RD. Information system planning in small companies, Proceedings of the Fourth International Conference on Information Systems Development - ISD'94; 1994 September 20-22; Bled; Maribor: University of Maribor, 1994.

Faulkner A. The Year 2000 problem and it's implication for the information profession, Journal of Information Science 1998; 24:255-65

Feiler J, Butler B. *Finding and Fixing Year 2000 Problem - A Guide for Small Businesses and Organizations.* Academic Press, 1998

Mesarič T, Jere U,. Horvat G., Informacijski sistemi na prehodu v novo tisočletje (Information Systems at the Millennium Transition). Organizacija 1996; 29:365-73

Koller K.E. *The Year 2000 Survival Guide*, In Line Publishing Inc., 1998

Leskovar R, Šinkovc M, Zupančič J.: "The Year 2000 problem in Slovenia." In *Evolution and Challenges in System Development*, eds. J. Zupančič et al., Plenum Press, 567-74, 1999

Levy M, Powell P. SME flexibility and the role of information systems, Small Business Economics, 1998; 11:183-96

Lord J. Sapp E. *Survival Guide for the Year 2000 Problem: How to Save Yourself from the Most Deadly Screw-Up in History*, J.Marion Publishers, 1997

Rollier B. Responses to Y2Kquestion posed to ISWORLD, http://ubmail.ubalt.edu/? brollier/y2k.html, 1999

Schulz JE. Managing a Y2K Project - starting now. IEEE Software 1998; May/June: 65-71

Serwer A.E., 1995, Paths to wealth: the new economy, Fortune 1998; 20 February: 57-62

Thong JYL, Yap C-S. Raman K. S., Environments for information systems implementation in small businesses, Journal of Organizational Computing and Electronic Commerce, 1997; 7: 253-78

Tič A, Ružič Z, Simšič S. Reševanje problema leta 2000 na osebnih računalnikih (Solving the Year 200 Problems on PCs), unpublished report, 1998

Winston E.R, Dologite D G., Achieving IT infusion: a conceptual model for small businesses, Information Resources Management Journal 1999; 12: 26-38

Zupančič J. "Addressing the Year 2000 issue - the case of Slovenia." In *System Development Methods For the Next Century,* Wojtkowski WG et al., Eds. Plenum Press, 1997

Yourdon E, , Yourdon J. *Time Bomb 2000.* Prentice Hall PTR, 1998

USING DISCOURSE ANALYSIS TO UNDERSTAND IS PLANNING AND DEVELOPMENT IN ORGANIZATION:
A CASE STUDY

John A. A. Sillince and G. Harindranath
School of Management, Royal Holloway, University of London,
Egham, Surrey, TW20 0EX, U. K.
Ph.: +44 1784 443780, Fax: +44 1784 439854
E-mail: Jsllince@rhbnc.ac.uk; g.harindranath@rhbnc.ac.uk

ABSTRACT

Discourse analysis can take understanding about information systems planning and development further than is possible with current approaches. The two added value items in this paper are those of 'talk-as-action' and social constructionism. Both are used to elaborate a structurationist view that values and ideas (those of the actor or agent) which appear in discourse early on in the case study become accepted as rules or as roles (the organizational structure) and which become resources for individuals (the actor or agent again) to use by the end of the case study.

KEYWORDS

Information systems (IS) planning and development, discourse analysis, 'talk-as-action', social constructionism.

INTRODUCTION

By itself, talk is viewed conventionally as ineffectual ("only talk") and separate from action ("talk first, act later"). But in recent years there has been an increasing emphasis in the management, organization behavior, and communication literatures on processes which combine talk and action into one, more complex unity. Such processes include single, double and triple-loop learning (Argyris, 1993, Argyris & Schon, 1978), reflective practice (Schon, 1983), sense-making as reflection on action (Weick, 1995), enactment (Starbuck, 1985) and organizational learning as acted procedure as well as internalized interpretation of those actions (Senge, 1990).

This has led to the notion, principally put forward by discourse analysts of "talk-as-action", which sets out to establish the intimately connected nature of talk and action, by positing a model which shows the multiple relationships which exist between conversational activity and content, identity, skills and emotions, and action (Grant *et al.,* 1998). Reality is subjectively understood and interpreted (Weber, 1947), and social interaction leads to a process of social construction of reality (Berger & Luckmann, 1966). Conversation is a means of constructing that social reality. Two people must put themselves in the other's position, in order to establish a "reciprocity of perspectives" (Schutz, 1967) through which

Systems Development Methods for Databases, Enterprise Modeling, and Workflow Management.
Edited by W. Wojtkowski, *et al.* Kluwer Academic/Plenum Publishing, New York, 1999.

each can accept the other's reality. Through this process things seen initially as problematic become things seen without deliberative interpretation (Merleau-Ponty, 1962) and thus become "taken-for-granted", in the "natural attitude" (Schutz, 1967).

Talk constructs reality. An example of this within the information systems field is the social construction of the "technical". Anthropologists have identified means by which societies create boundaries in order to motivate action. Often an action must occur because of God, money, or time (Douglas, 1973). In the information systems community, these imperatives are "safety", "budget" and "deadline". But an important distinction exists between what is "technical" (and therefore outside of the bargaining arena) and the "social" (which can be negotiated and argued about). To make something seem "technical" removes it from discussion and renders it taken-for-granted. The effect of conflict and political processes within IS planning and development is to reveal to participants that what they have regarded as undeniably technical aspects of IS development are in fact socially interpreted and socially constructed (Sillince & Mouakket, 1997).

Our case study described below examines two such concepts – commitment and risk - both of which are treated technically in many contexts. For instance, commitment is often supported by technical props such as deadlines, reports, budgets, rules and procedures; risk is often supported by mathematical theory and economic theory. But, as the case study shows, when disagreements emerged between the various project participants, these very concepts were revealed to be socially constructed.

Talk performs as well as informs. Also, analysis of utterances has led to their characterization as "speech acts" such as promising, requesting, commanding and so on (Searle, 1969). Some have suggested that these speech acts are to be understood not only as providing information (a request informs us that the requestor wishes us to do something) but also as performing an action (such as when a request warns that anger or retaliation may follow) (Austin, 1962). In the case study below, we will demonstrate how talk acted as a 'symbolic action' (Johnson, 1990) during the course of the project.

Talk as 'structurational' action. Discourses are also actions in the sense that they create influential social rules and roles, which influence later social actors. This process is 'structurational' (Giddens, 1984) because the individual actor or agent is viewed as initiating actions which effect changes in the social structure of the organization, a structure which later constrains her actions.

This iterative, structurational process will be demonstrated in our case study below by means of showing that values and ideas, which appear in discourse, later become accepted as constraining rules and roles.

Talk-as-cognition. Just as talk and action are not separate, Edwards (1997) has argued that talk and cognition are not separate. People use talk as a means of perceiving, enacting and interpreting the world. And so the way people talk illuminates and constrain the way they view the world. Wallmacq & Sims (1998: 125-6) discuss the case of a woman who had previously started a software training company, which grew very successfully. Her account of that process is illuminating: "The firm was becoming too big: there were too many of us, and so we decided to split: I asked my second in command if he would manage the new firm in Luxembourg; he accepted and so that is what we did". The metaphor conjured up was one of cellular division. But it was not only talk. The metaphor channeled her feeling of being marginalised and uncomfortable at losing her power into an ambition to, in the words of one of her employees, "reproduce her firm by photocopy". The cellular division metaphor was thus both talk and action.

The case study will show how conversations between the various project participants actually went on to influence and even constrain the manner in which the project unfolded.

Performance. Another intellectual seed for the talk-as-action approach has been ethnomethodology's concern for the pursuit of participants' own categories, the use of participants' resources (everyday rhetorical discourse, rules of thumb) as analytical data, the focus of attention on the dramaturgical notion of "performance" (Garfinkel, 1967) – which combines talk and action as essential elements - and therefore the interest in breaches, slips, and other presentation problems.

One way to represent the talk-as-action approach is in terms of the context in which action takes place. The talk-as-action approach regards "the situation of action as transformable. It is identifiable as the reflexive product of the organized activities of the participants. As such, it is on-goingly "discovered", maintained, and altered as a project and product of ordinary actions. Situational constitution is essentially a 'local' and immanent product of methodic procedure rather than the result of 'pre-existing' agreement on 'matters of fact'" (Heritage, 1984: 132). The actions of "discovered, maintained, and altered" are deliberately chosen by Heritage (1984) to indicate that they are involved in the making of a performance.

In the case study presented below, the issue of how well the performance was carried out hinged on conflicting discourses generated by the various project participants.

THE CASE STUDY: AMBULATORY CARE AT CMT

This research focussed on a major healthcare project at CMT, an acute hospital in the UK National Health Service (NHS) employing 1300 people and with an annual turnover of £50m. £16m of funding had become available through land sales and central government funding through the Private Finance Initiative (PFI), for the CMT Hospital NHS Trust (henceforth, referred to as the Trust) to establish an Ambulatory Care and Diagnostic Center (ACAD) adjacent to the main hospital. Rapid service delivery and medical process redesign of this kind was adventurous but politically controversial and was still at an early stage of political acceptance or diffusion into acute care. Only a few pioneer examples existed in Australia, Switzerland and the USA.

The new ambulatory care center required both changes to business and medical processes and social structures, because it focussed on rapid throughput and computerized scheduling. Also it required a careful relating of new systems to existing hospital functions and information systems. Although the new unit was planned to be in a new building, it was to be sited within the much larger main hospital complex and draw staff largely from there. The new information system had to link with General Practitioners (GPs), the main hospital at CMT and other NHS care units, and had to solve complex scheduling problems, in order to radically shorten patient care. The approach not only required restructuring of working practices but it also required new technology - for example, MRI and CT scanning which enable rapid interactional diagnosis. The ACAD design was intended to facilitate the performance of imaging-guided interventions, involving, for example, clinical procedures requiring imaging and endoscopy, or imaging and surgery.

CMT joined together with a Consortium consisting of a software developer, a hardware manufacturer, and a facilities management supplier in order to develop ACAD. One of CMT's directors, the Director of Contracts & Clinical Activities, was also project manager, acting as chairman of the ACAD Steering Committee, which included the Chief Executive, Chairman and six other directors of CMT, together with representatives from the other Consortium partners, the ACAD architect, and the software requirements team.

The authors recorded meetings of the Steering Group and the meetings of the Groups within ACAD between June 1996 and June 1998. The Clinical Group involved the clinicians in reengineering their working practices. The Design Group commissioned architects who began design work on the new building between June 1996 and January 1997 based on a master plan, intending to commence construction by April 1997. The IS Group commissioned a software requirements team to produce a requirements document by April 1997, with the intention of software development by the Consortium partner responsible after that date. The Negotiation Group dealt with contracts and agreements between the Consortium partners.

A number of meetings and discussions had been held at CMT with regard to the development and implementation of ambulatory care. We particularly focus on those meetings and discussions that dealt with the issue of partnership between the hospital and the Consortium partners. Typically, these meetings included the ACAD Steering Group meetings, as well as meetings of one or more of the above mentioned ACAD Groups. One of the main problems that the hospital and the Consortium partners encountered was that of uncertainty in relation to the information systems within ACAD. This uncertainty was holding back the Consortium partners and CMT from coming to a legally binding agreement for the development of ambulatory care facilities and infrastructure for ACAD. There were several processes taking place. One was building design, the other was the requirements analysis for the ACAD information systems, and yet another was the redesign of clinical processes to accommodate ambulatory care principles.

The hospital wanted to slow things down because the information systems design was still uncertain, and this was having a knock-on effect on the building design:

"The building itself will depend upon what kind of activities information systems can do, such as scheduling, links to GPs, imaging storage and retrieval, tele-medicine, electronic medical records, and whether there is paperless information systems" (Systems Analysis Consultant, Steering Group Meeting, 31.7.96)

However, the other members of the Consortium wanted to obtain commitment (via a contract) as early as possible in order to maximize the chances of the project being undertaken and their getting paid for it. The facilities management supplier partner in the Consortium, in particular, was supposed to play a crucial integrative role and thus was the most accountable (and hence vulnerable to risk) because it depended on the other partners for its own successful project completion:

"Let's look at the risks. There is the cost of the building overrun, and the risk of [the Consortium]. They are both accountable to [the facilities management supplier]" (Director of Contracts, Steering Group Meeting, 30.7.96).

The greatest source of risk for the facilities management supplier was that they had responsibility for an activity, which was controlled by others:

"[The facilities management supplier] are concerned that they would have to manage and operate a building that was not designed by them. They have reservations about taking the risks involved with the design defects which may arise later" (Director of Contracts, Negotiation Group Meeting, 16.8.96).

This led to a delicate negotiation, which put commitment at the heart of a rhetorical process.

COMMITMENT AS A RHETORICAL CONSTRUCT

Commitment is such a broad concept that it requires specific occasions, settings and images to render it a convincing value. Several linguistic devices were used to convince audiences about the need for commitment, such as the use of a sense of urgency:

"But outpatients activity cannot be ignored. So we have to start work on the Main Hospital issues pretty soon" (Director of Contracts, Steering Group Meeting, 3.9.96)

Also there was the use of a town planning blight metaphor:

"Now we have a plan. But we need to act on it now. We should not get caught in a planning blight" (Director of Contracts, Information Systems Group Meeting, 2.10.96).

DELAY AND COMMITMENT RHETORICALLY PRESENTED AS OPPOSITES

Although the hospital wanted to delay the project, its managers privately admitted that delay and commitment were oppositions. In general, contracts were viewed favorably as means of reducing risk. So it was not the principle of contracts that was at issue but the questions of when to sign one. There was no doubt that a contract reduces risk:

"In order to avoid unnecessary risk, all attendees felt that the master contract date should be set for March/April 1997" (Minutes of ACAD Timeout Conference 25.7.96).

However, various individuals at the hospital privately believed delay was in the Trust's interest. They had several arguments in favor of delay and against commitment. One was that internally generated uncertainty (because of novel and untried information systems) meant that commitment should be delayed:

"There is uncertainty about information systems, which may lead to not signing the commercial agreement contract. Signing as late as possible would be better as the technology is changing and we could get the latest. In January 1999 ACAD opens and so between March and November 1997 would be a better period to sign up" (Director of Contracts, Steering Group Meeting, 30.7.96)".

Another reason the hospital favored delay was that delay meant better technology:

"We should postpone the master contract of [the Consortium] to March/April '97. Till then all facilities like information systems will have time to decide what they want and so will be able to plan" (Hospital General Manager, Steering Group Meeting, 30.7.96).

Another was that commitment too early would be penalized financially:

"If you sign too early and want to change then they'll want more money (Chairman, Steering Group Meeting, 30.7.96)".

This body of opinion viewed commitment in pejorative terms, for example, depicting it using a prison metaphor:

"Should the contract be legally binding or non-legally binding? – if legally binding then the hospital will be locked into the contract" (Director of Contracts, Steering Group Meeting, 12.8.96).

Also, the software supplier wanted delay:

"[The software supplier] want some money up front. X [the supplier's representative] asked after the Negotiation Group Meeting on the 16th of August. Y [the hospital's Director of Contracts] reassured X that they will get the contract. X seems more like a salesman and doesn't seem to be interested in a 'partnership' approach. [The software supplier] may not sign a contract until the information system uncertainties are sorted out. It might take at least 12 months. This was after the [software supplier] person came down from the USA and felt unhappy about the information system uncertainties" (Director of Contracts, Steering Group Meeting, 19.8.96).

However, other Consortium partners had a different view. For them, any delay in signing the final contract or coming to a final and legally binding contract was a lack of commitment from the Trust towards the role to be played by the Consortium. Although, the Consortium agreed to a 'letter of comfort' before signing the final contract, it was not entirely convinced of the Trust's intentions. For instance, the facilities management supplier, one of the Consortium partners, insisted on a so-called "letter of comfort":

"The company's culture makes it difficult for me to convince [the company] that [CMT] is committed firmly to the deal and to [the Consortium]. A "memo of comfort" with (the FD's) sign on it should do the trick" (Facilities Management Supplier's representative, Steering Group Meeting, 16.8.96).

DELAY AND COMMITMENT RHETORICALLY PRESENTED AS COMPLEMENTS

The hospital wanted delay but knew that the Consortium wanted commitment. Its way of getting over this presentation problem was to rhetorically present the two values of delay and commitment as mutually reinforcing complements. The hospital's way of promising commitment and yet achieving a delay was to offer an outline agreement, called a "Heads of Terms agreement" (HoT) and to speed up the building part of the project. The arguments were that the Consortium's need for commitment was answered by a formal and binding agreement:

"[The hospital Chairman] recognized the Consortium's need for a firm and timely partnership agreement and it was agreed that binding 'Heads of Terms' would be completed by December 1996 at the latest" (Minutes of ACAD Timeout Conference 25.7.96).

"A legally binding Heads of Terms agreement by December will sort things out" (Director of Contracts, Negotiation Group Meeting, 16.8.96).

One way of having delay and commitment was to delay the main project (of work and information system redesign) but to go ahead on the building. The argument was that a need for commitment by the Consortium should be answered by speeding up the building and construction element of the project:

"3 million out of the 18 million goes to [the Consortium]. So let's concentrate on the building and engineering of ACAD and give [the Consortium] a simple reassurance of their involvement (Director of Contracts, Steering Group Meeting, 30.7.96)".

Another way of getting delay and commitment was to pay for commitment in incremental portions, avoiding any need for trust:

"These options are going to be difficult as now [the Consortium company] want to be paid upfront for putting in any further resources" (Director of Contracts, Information Systems Group Meeting, 2.10.96).

Thus, the Trust wanted long-term collaboration but short-term delay, whereas its Consortium partner wanted short-term commitment on a contractual rather than collaborative basis. The Trust therefore rhetorically presented the notion of delay and commitment as complementary concepts (both reinforced collaboration) and this constrained the way it dealt with its partners. Commitment was thus viewed in several divergent ways, depending on which point of view was expressed. The Consortium defined it in decision terms as the opposite of delay. The Consortium also saw commitment as a divisible entity, to be bought in installments through up front payments from the Trust. On the other hand, the Trust also saw commitment as a divisible entity, but for an entirely different purpose – to achieve incremental increases in later (and therefore better) technology and information systems. At the same time, the above examples also show how talk can act as 'symbolic action'. For instance, the talk concerning progress on the new building was seen to reassure the builders and the architects.

LEGAL AND COLLABORATIVE COMMITMENT RHETORICALLY PRESENTED AS OPPOSITES

The hospital saw commitment in collaborative terms of shared aims and a commitment of shared resources in working together toward what they saw as a novel scheme. ACAD represented valuable collective intellectual property and this justified careful design, which necessitated a collaborative view of commitment. The hospital's collaborative definition of commitment, and the Consortium's legal view, are starkly contrasted in the view that the work can be appropriated for each party's commercial benefit only if first there is a sharing of knowledge:

"There's tremendous intellectual value for [the Consortium company] from this project. But they're being front-ended by 'salesmen', led by people who just want all the answers now. [The hospital] needs to think through all the logistics. Solutions will take time. [The Consortium company's] representatives cannot just sell. We are prepared to educate them" (Director of Contracts, Facilities Management Group Meeting, 2.9.96).

Collaborative commitment to shared design work was viewed as reducing risk.

"By being involved with the designers, [the facilities management supplier] can be certain that the building will be designed well and that they can then manage it well" (Director of Finance, Negotiation Group Meeting, 16.8.96).

This view of commitment was antithetical to a legal, contract-based view of commitment:

"[The hospital] feels that the agreements at this stage cannot be made legally binding as it would cause damage to both parties" (Director of Finance, Negotiation Group Meeting, 16.8.96).

The suppliers took the same view, about the opposition between legal and collaborative commitment. The difference was that they wanted legal commitment. Collaborative commitment put them in the position where they depended on parties over whom they had no control; such as on the systems analyst who was developing requirements:

"[The facilities management supplier] is concerned that [the Systems Analysis Consultant] should get full support from [the hospital] as trust people have to decide on procedures and this will have implications for how [the Consultant] performs" (Facilities Management Supplier's representative, Negotiation Group Meeting, 16.8.96).

Similarly the facilities management supplier was perceived as taking a legal view of commitment:

"[The facilities management supplier] will not give 15 million unless the contract's written" (Director of Contracts, Steering Group Meeting, 30.7.96)".

Even the systems analysis consultant, so supportive of the hospital's views in other ways, considered a lack of clarity increased the risk of non-delivery

"Where we have no clarity, we should leave it out of the IT contract. This is because [the software supplier] will get paid for it, but they will not deliver! We know they cannot deliver when there is no clarity" (Systems Analysis Consultant, Information Systems Group, 2.10.96).

The following argument between the hospital and the Consortium during a Negotiation Group Meeting show the differences in perception between the two groups:

"We feel that the Heads of Terms agreement is transferring the risk off the Building Design Group to us (Consortium Partner, Negotiation Group Meeting, 27.9.96).

"We are not. But we are all agreed that the building design will affect the way you work. So you are involved in the building design. The HoT in fact only provides an extra channel for you to use, i.e. apart from approaching us, you can approach the builders to express disapproval if you are not entirely happy with the design" (Director of Finance, Negotiation Group Meeting, 27.9.96).

"But we don't want to do this ... We do not want to be responsible for this and we do not want to go to the architects" (Consortium partner, Negotiation Group Meeting, 27.9.96).

"If you are not interested in this, then how different is this [CMT-Consortium] deal from a normal tender? [The hospital] could actually tender facilities management, equipment and IT in the normal way. Why are we here then? Why are we negotiating with one Consortium - one party? The idea of the deal was that you will share in the risk involved" (Director of Finance, Negotiation Group Meeting, 27.9.96).

"We cannot take any responsibility for building design. This is a significant shift in the deal since April 1996" (Consortium Partner, Negotiation Group Meeting, 27.9.96).

Similarly, at the Steering Group Meeting, the Director of Contracts made the following statement:

"Careful analysis of the [Consortium] submission indicates that ... there is insufficient risk transfer from the Trust to [the Consortium]; and there is little evidence of partnership and understanding which will be of prime importance in any on-going relationship over a 5 year period and, potentially, beyond. [The Steering Group] is asked to make the following recommendations to the Trust Board: that [the Consortium's] final proposal does not satisfy either the Trust's requirements for service and value or PFI requirements for risk transfer; that, accordingly, there is no further merit in further negotiations with [the Consortium]; that the Trust go back to the market and seek a new PFI partnership" (Director of Contracts, Steering Group Meeting, 5.8.97).

Thus, the way the Consortium partners talked to the hospital Trust led the Trust to perceive its partners as "'salesmen'... people who want all the answers now". This further constrained what the Trust said and lowered its expectation of a long term and open collaboration and a sharing of intellectual capital. Indeed, this led to the Trust's decision to cease further negotiation with the Consortium partners and to the eventual collapse of the Consortium.

TRUST AS A RHETORICAL CONSTRUCT

We have already seen how commitment was viewed in several divergent ways. It was seen in decision terms as the opposite of delay. It was also seen as a divisible entity, to be bought in installments. It was also seen in symbolic terms, to be created by means of reassurance. It was also seen in psychological terms as the effect of persuasion. Because of these divergent views of commitment, it was important that the parties should trust each other. Indeed, the size of the risks seemed to be a justification for the value of trusting each other. Trust was given the mantle of a savior in difficult circumstances. Here the argument is that the more the risk, the more that trust is needed:

"The issue is trust and the management of risks. There are risks for both sides. But with trust the parties can sort it out" (Chairman, Negotiation Group Meeting, 16.8.96).

The significance of this as rhetoric for a public audience is shown by considering the private skepticism sometimes demonstrated, such as anticipation of the effects of withdrawal of parties from the Consortium:

"We don't know the reaction of [the hospital] if [the software supplier] drops out...[The systems analysis consultant] needs to prepare for risk management by changing the information systems plan which will be tendered out to several solution providers (if [the software supplier] jumps off). The plan as of now is not strong enough to meet a tender" (Director of Contracts, Information Systems Group Meeting, 2.10.96).

Being in a company whose culture did not value trust was even used as a rhetorical weapon. For example, the facilities management supplier representative argued that it needed full-scale legal commitment (i.e. a contract) because it had no past history as a company of entering into the kind of vague collaborative arrangements the hospital was favoring. A past history of suspicion was by a sleight of hand transformed from being a problem to being a negotiating strength:

"The company's culture makes it difficult for me to convince [the company] that [CMT] is committed firmly to the deal and to [the Consortium]" (Facilities Management Supplier's representative, Negotiation Group Meeting, 16.8.96).

The hospital's rather self-centered view was that it could use co-option, and that it could absorb outsiders (such as the systems analyst consultant) into its own concerns without having to commit itself to a formal and binding agreement. Here is the "success story" of the "outsider transformed into insider". The implication is that all Consortium partners should allow themselves to become similarly co-opted:

"[The Systems Analyst Consultant] works as though he is a 'Trust director' in charge of IT. That way he gets support from everyone at [the hospital]. His voice is the Trust's voice and he understands [the Trust] and NHS [National Health Service] management" (Director of Contracts, Negotiation Group Meeting, 16.8.96).

PERFORMANCE AS OPPOSITION BETWEEN DISCOURSES

The following conversation expresses the two opposed discourses – the hospital's discourse of trust, and the Consortium company's discourse of risk. The Consortium company asked for a "letter of comfort" which the hospital provided. The hospital Chairman sent a letter to the software supplier who wanted to be compensated for the hours they spent at the hospital. To the Consortium's chagrin, however, this was not possible because the hospital wanted all payments to be for deliverables.

"Our company feel that the Heads of Terms ought to be "Heads" of terms and not details. We're spending far too much time negotiating these HoTs rather than having a simple memo of understanding and negotiating the contract itself."" (Consortium company representative, Negotiations Group Meeting, 27.9.96).

to which the hospital replied that the details were merely informatively and not punitively intended:

"The details are simply to provide clarity to the Consortium company. Who's going to be responsible for what etc." (Director of Finance, Negotiations Group Meeting, 27.9.96).

A document full of assignments of responsibilities affected the Consortium the most because it was the most vulnerable to risks created by its constituent partners being dependent on each other without being able to control each other's work:

"HoTs are not what we expected. It shows a normal "design and build", which is not what we're here for. We simply need a "strategic-level" document. If HoTs are to be legally binding, then let's sign it but we need more flesh around all the clauses. We need to know exactly what we're in for. We feel that the HoT is transferring the risk off the Building Design Group to the us" (Consortium Partner, Negotiations Group Meeting, 27.9.96).

The hospital wanted the Consortium to negotiate solutions bilaterally with individual partners:

"Apart from approaching, us you can approach the builders to express disapproval if you're not entirely happy with the design" (Director of Finance, Negotiations Group Meeting, 27.9.96).

This was strongly rejected by the Consortium:

"But we don't want to do this. If the performance of any equipment is hampered by building design error, we only want to come to you. We do not want to be responsible for this and we do not want to go to the architects" (Consortium Partner, Negotiations Group Meeting, 27.9.96).

The disagreement underlined the impossibility of writing a contract to legislate for a situation where partners do not trust each other to share risks:

"If you're not interested in this, then how different is this [CMT-Consortium] deal from a normal tender? [The hospital] could actually tender the facilities management, equipment and IT in the normal way. Why are we here then? Why are we negotiating with one Consortium – one party? The idea of the deal was that you would share in the risk involved" (Director of Finance, Negotiations Group Meeting, 27.9.96).

Even more dysfunctional was the software supplier's unidirectional "user states requirements to supplier" model:

"Our understanding was that you [the hospital] will come to us with what you need – requirements and we would provide those "(Software supplier representative, Negotiations Group Meeting, 27.9.96).

Again the hospital referred to its vision of shared risk by means of collaborative learning and by jointly appropriated discoveries:

"We thought we're getting the intellectual input from a world class organization!" (Director of Finance, Negotiations Group Meeting, 27.9.96).

The answer was combative, but in a disputatious and negative way:

"We're a strong Consortium! And we are in fact going to see another hospital! Can we have some interim consultancy payments – because we are working entirely on speculation" (Consortium Partner, Negotiations Group Meeting, 27.9.96).

As we have shown above, the issue of how well the performance was carried out depended on conflict between two types of discourse – the Trust talked about trust, whereas its partners talked about risk. The same events and facts evoked these two opposed reactions. The necessary arguments to span the gulf, which existed, were not provided. The performance (of agreement, of reconciliation of opposed sets of priorities) was not convincingly carried out. Thus contrary to the hospital Trust's confident expectation, there was no pre-existing agreement. Agreement can only come about as the result of negotiation in which bargaining strength is influenced by the persuasiveness of individuals' performances.

CONCLUSIONS

In this paper, we have used a case study to show that discourse analysis can enhance our understanding of the processes involved in information systems planning and development, and especially so in situations involving partnerships and negotiations. The case study has shown that the predominant 'model' of IS development, which proposes a sequence consisting of deliberation, followed by decision and action, is unrealistic, and that instead there was a process whereby two discourses, one of trust/commitment, and one of risk, challenged one another. These two discourses were resolved by one voice, (the Consortium's), being silenced by the Consortium's collapse.

The case also shows that talk is not "just" talk, but that it is also action: by the end of the study period, the collaborative view of commitment (rather than the legal view of commitment) had become the "dominant logic". Indeed, values and ideas (those of the project participants) which appear in discourse early on in the case study become accepted as rules or as roles (the organizational structure) and which become resources for individuals (again, the project participants) to use by the end of the case study. The linear model of systems development, which is predicated on the idea of talking (to users first), followed by action (implementation and testing) later, is thus inappropriate and even untenable in this case.

We have also seen how talk can actually construct reality, and even illuminate and constrain the way we view the world during the course of a project. In the IS domain, concepts that are seen as undeniably technical aspects of systems design, development or implementation take on entirely different meanings during a project's lifetime as they become socially interpreted and socially constructed. In our case study, concepts such as commitment and risk that are often seen as 'technical' were revealed to be socially constructed by various project participants to support their particular positions, as disagreements and conflicts emerged. Thus, the use of discourse analysis to examine 'talk' (those of the agents or actors involved) can improve our understanding of 'actions' (again, those of agents or actors) during systems development and provide valuable feedback to move projects forward. Exposing hidden conflicts and the reasons for disagreements could be the first step towards their speedy resolution.

REFERENCES

Argyris C., 1993, *Knowledge for action*, Jossey-Bass, San Francisco.

Argyris C., and Schon D.A., 1978, *Organizational learning: a theory of action perspective*, Addison Wesley, Reading, MA.

Austin J.L., 1962, *How to do things with words*, Clarendon, Oxford.

Berger P., and Luckmann T.L., 1966, *The social construction of knowledge: a treatise on the sociology of knowledge*, Doubleday, Garden City, New York.

Douglas M., 1973, *Natural symbols*, Pantheon Books, London.

Edwards D., 1997, *Discourse and cognition*, Sage, London.

Garfinkel H., 1967, *Studies in ethnomethodology*, Prentice Hall, Englewood Cliffs.

Giddens A., 1984, *The constitution of society*, University of California Press, Berkeley.

Grant D., Keenoy T., and Oswick C., (eds), 1998, *Discourse and organization*, Sage, London.

Heritage J.C., 1984, *Garfinkel and ethnomethodology*, Polity, Cambridge.

Johnson G., 1990, 'Managing strategic change: the role of symbolic action', *British Journal of Management*, 1, 183-200.

Merleau-Ponty M., 1962, *Phenomenology of perception*, Routledge and Kegan Paul, London.

Schon D.A., 1983, *The reflective practitioner*, Basic Books, New York.

Schutz A., 1967, *Collected papers*, Nijhoff, Den Haag.

Searle J.R., 1969, *Speech acts: an essay in the philosophy of language*, Cambridge University Press, London.

Senge P.M., 1990, *The fifth discipline: the art and practice of the learning organization*, Doubleday, New York.

Sillince J.A.A. and Mouakket S., 1997, 'Varieties of political process during systems development', *Information Systems Research*, 8, (3), 1-30.

Starbuck W.H., 1985, 'Acting first and thinking later', in Pennings J.M., (ed) *Organizational strategy and change*, Jossey-Bass, San Francisco.

Wallmacq A., and Sims D., 1998, 'The struggle with sense', 119-134 in Grant D., Keenoy T., and Oswick C., (eds), 1998, *Discourse and organization*, Sage, London.

Weber M., 1947, *The theory of social and economic organization* (Henderson A.H. and Parsons T., eds and trans) Free Press, Glencoe, IL.

Weick, K., 1995, *Sense-making in organizations*, Sage, London.

Wieder D.L., 1974, *Telling the code*, in Turner R., (ed), *Ethnomethodology*, Penguin, Harmondsworth.

INFORMATION SYSTEM DEVELOPMENT METHODOLOGY:
THE BPR CHALLENGE

Vaclav Repa
Dept. of Information Technologies, Prague University
of Economics, W. Churchill sq. 4, Prague, Czech Republic

Key words: Business Processes Reengineering, Information Systems Development

Abstract: One of the most significant phenomena the nineties has brought is the
Business Processes Reengineering (BPR). Information technology (IT) with all
its coherent aspects (including the information systems development
methodologies) plays a very important role here. It works as the trigger as well
as the enabler of BPR. On the other hand, the evolution of the theory of
business process reengineering works as a very strong feedback for
information systems development methodologies themselves. It sets new tasks,
which have to be anticipated by ISD methodologies. It also discovers new
facts and consequences, which unexpectedly illuminate some current problems
of ISD methodologies. This paper aims to outline the influence of this BPR
challenge on the theory of IS analysis methodology.

1.   EVOLUTION OF THE INFORMATION SYSTEMS ANALYSIS METHODOLOGY

At present, the situation in the field of Information Systems Analysis
seems to be relatively stabilised. The virtual battle between "structured" and
"object" paradigms is finished and there comes the time for transition to the
higher level of quality. Today the negation is not the proper way. Current
objective is to make the methodology complete, consistent and prepared to
anticipate significant trends in business and technology. After the past
evolution the unifying feature of the Information Systems Analysis
Methodology, in which all main principles manifest themselves, is the
separation of different points of view. The following kinds of separation are
substantial:
- Data and processes as well as objects and functions are taken into
  consideration separately
- abstract (high level) concepts, and their relationships, are separated from
  the subordinated (more detailed) ones using hierarchical abstractions
- Modelling as the key principle gives the developer an abstract view of the
  general characteristics of the information system deprived of their
  particular shape, which is complicated by various non-essential aspects

Systems Development Methods for Databases, Enterprise Modeling, and Workflow Management,
Edited by W. Wojtkowski, *et al.* Kluwer Academic/Plenum Publishing, New York, 1999.

337

– Building an information system gradually on three different levels - conceptual, technological and implementational (the "Three Architectures" principle) reflects the effort to separate three different, relatively independent problem areas: the essential concept based on the model of reality, technological design and implementation shape of the system

The main reason for the separation is the effort to simplify problems, which have to be solved. Even if the system is relatively small, there are too many aspects in too complex relationships to describe them simply (i.e. clearly). In addition, there are too many relatively simple problems in too complex relationships to solve them at the same time. *Separation* seems to be *the vital condition for the mental control of a problem.*

In the following text general characteristics of the main forms of separation mentioned above are described as well as the evolution of IS analysis methods from this point of view.

1.1 Data and processes / objects and functions separation

In the structured approach to IS development, there is a significant difference between two basic components of IS - data and processes. Of course, this difference is partly caused by the evolution of IS development methods. The early 70's brought the first revolution in IS development - data analysis and data modelling. The first significant paradigm was formulated: "information system is based on the model of the real world". Unfortunately, this paradigm was formulated just in the field of the "data component" of IS and have led to the well-known statement of James Martin that data are more stable than functions, therefore the model of the real world is just the database. Although this idea has started a long discussion on the differences between data and processes, it has brought to life the main principle of IS analysis - *the principle of modelling*. This principle is discussed in the following paragraph. In addition, the evolution of technology had a significant influence on the evolution of IS development methods. Database systems supporting limitation of non-creative activities in the IS development process even more emphasised the "separation paradigm". Good design of the database was considered as the most important and creative activity. Processes in the information system have been regarded as not so important and thus easily implementable (with the use of database systems support - report generators, user interface generators, query languages etc.) according to rapidly changing user requirements. Later evolution of data analysis and modelling has taken into account also the procedural dimension of the real world model (in the form of procedural database languages (4GL), mechanisms of integrity constraints, triggers etc.). However, the "separation paradigm" is in place - all these real world functionality characteristics are regarded as characteristics of the data model. On the other hand the evolution of function-focused analysis methods also accepts the separation of data and procedural dimensions of IS. For example Yourdon method (Yourdon, E. (1989)) talks about three separate parts of the conceptual model of the IS - data model, behaviour (function) model and control model. The last two models form the procedural model of the IS. Its separation into two models reflects the inability to arrive at a settlement with the complexity of relationships between the top-down hierarchical structure of system functions and not top-down structure of data entities. From this

338

contradiction between two antagonistic concepts of hierarchical structuring follows all well-known problems with the control flows in functional structure. This contradiction will be discussed in detail in the following paragraphs. So the structured approach to IS analysis looks into the information system from two different points of view:

– data
– functions

Each of them follows a specific logic and requires a specific language (i.e. tools as DFD or ERD) for description of the IS structure.

Object-oriented (OO) methods try to overcome these problems and contradictions of the structured approach by encapsulating data together with functions ("methods" of objects). There are techniques and rules forcing us to think about data in the context of their processing by functions only. This approach helps the analyst to formulate control operations as well as data processing operations with respect of the meaning of the data processed (i.e. meaning of the object). Thus, also the problem with placing control flows in functional structure is solved. The existence of the object as the collection of data (attributes) and functions (methods) is to be the right reason for data processing operations control (strictly speaking: the object life cycle).

Unfortunately, the object-oriented approach is suitable to model just one dimension of reality. The conceptual object model is the net of really existing objects jointed by their common actions. For the conceptual object model to be clear there is no place for such operations, which model the using of the information system itself. These operations are not technological (in the sense of the Principle of Three Architectures - see below) - their place is also on the conceptual level of modelling.

1.2   Modelling

The principle of modelling was first formulated from the data point of view: contents and structure of database objects reflect the contents and structure of real world objects. Correctness of the data model is measured via its similarity to the real world. For such measuring there must be exactly defined the term "similarity". Therefore the special tool - Entity Relationship Diagram (ERD) has been developed (Chen P.P.S. (1976)). ERD describes the essential characteristics of the real world: objects and their mutual relationships. It is constructed to be able to describe exactly objects and their relationships in the same way as we see them in the real world. At the same time this model describes the essential requirements for the database - it must contain the information about the same objects and their relationships. The form in which a particular database describes these facts always depends on technological and implementation characteristics of the environment in which the database is realised. However, the essence of the model remains the same. Because of the need to describe the same database in its various shapes (essential, technological, implementation) the *principle of different architectures* has been formulated. This principle, generalised to the scope of the whole system (not only its database) is discussed below. The modelling principle proves to be general too - some parts of system processes have to be regarded as the model of the real world also. However, the main problem of the so-called structured approach in IS development is that it is not able to recognise which system processes form the model of the real world and which do not. Such recognition requires separation of the modelling operations from the other ones and organising them into the special algorithms according to real world objects and their relationships. This point of view is not reachable under the "structured paradigm" without accepting the natural unity of the modelling system processes and the data in a database. Acceptance of the natural unity of the modelling processes and the data entities, formulated as the main OO principle, enables us to solve Yourdon's problems with control processes - the essential controlling algorithms follow from the entity life histories.

As shown above the Modelling Principle seems to be general and independent of existing paradigms. Each new paradigm can only specify its place in IS development but cannot eliminate or limit it.

1.3   Hierarchical abstraction

Hierarchical abstractions are the means for decomposing the elements of designed information system to the level of detail. Higher level concepts consist of the lower level ones. On each level of detail the elements of developed IS and their relationships are described. The elements on each higher (i.e. non-elementary) level of detail are abstract concepts. Only the lowest (i.e. most detailed, elementary) level contains definite elements. There is the "tree structure" of dependencies between the concepts of the higher and lower levels, so that each element has only one parent element on the higher level (with the exception of the highest element - root of the tree) and can have several child elements on the lower level (with the exception of the lowest elements - leaves of the tree). Hierarchical abstractions are of two basic types:
- *Aggregation.* Subordinated elements are parts of the superior concept.
- *Generalisation.* Subordinated elements are particular types of the superior concept.

The aggregation type of abstraction is typically used for decomposing the functions into sub-functions (using Top-Down procedure) while the generalisation type of abstraction is typically used for decomposing the entities of the conceptual data model into sub-entities. Incompatibility of these two basic approaches to the concept decomposition inside the structured methods has often played the role of a source of vital problems of the "structured paradigm".

1.4   The "Three Architectures" principle

The principle of "Three Architectures" was mentioned in the paragraph in which the "modelling principle" was discussed. These two principles have very much to do with one another. Separation of the implementation and technology-dependent aspects of developed information system from the conceptual ones is the vital condition for putting the Modelling Principle into practice. Without such separation, the developer would not be able to see (and to discuss it with the user) the model of the real world as the functional and database structure of developed IS. Three levels of the model of the IS seem to be essential:
- Conceptual model represents the clear model of the real world, which is not "contaminated" by the non-essential aspects given by assumed technology and implementation environment of the system
- Technological model is based on the conceptual model enriched by the aspects given by assumed technology. Technological aspects often significantly change the original (conceptual) shape of the system. For

example, 3GL technology using sequential files leads to the data structures considerably distant from the conceptual entities and their relationships. On the other hand, relational database technology preserves a maximum of the original shape of the conceptual data model

– Implementation model depends on the used technology taken into consideration in the technological model respecting also implementation details given by the used particular environment. Thus, the implementation model is even more distant from the real world than the technological one.

Such a model of the three different views on the same thing has some general characteristics:

– each view has specific logic and requires specific methods of examining and specific language for description, which match this logic

– to retain the consistency between particular views it is necessary to have some means (i.e. methods and techniques) for the transition of the contents of one view into the next view

The following figure illustrates the essential relationships between the three architectures:

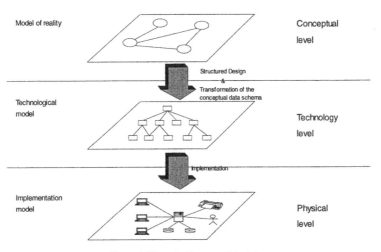

Figure 1. Three Architectures Principle

So each of these three levels of IS development represents a specific goal, a specific type of developer's activity, specific techniques and tools to use. The transition of the design from one to the next level also requires specific activities, techniques and tools. The methods, techniques, activities and tools used on three levels of IS development differ from the functional, data and object points of view. The following tables set out an example of these differences on the conceptual and technological levels:

The tables also show some important characteristics of the object-oriented approach:

- OO approach combines activities and tools of both - data as well as function-oriented approaches
- OO approach drives for the unified language for both - conceptual as well as logical levels of modelling

Table 1. Functional point of view

Level	Activity	Tools	Techniques
CONCEPTUAL	Functional Analysis	Data Flow Diagram, Structure Diagram, State Transition Diagram	Event Partitioning
LOGICAL	Program Modules Design	Structure Chart	Modular Programming, Composite Design, Information Hiding etc.

Table 2. Data point of view

Level	Activity	Tools	Techniques
CONCEPTUAL	Data Analysis	Entity Relationship Diagram (Chen)	Normalisation, Integration (i.e. Canonical Procedure)
LOGICAL	Logical Database Design	Entity Relationship Diagram (Martin)	Transformation of data model into the logical data structures

Table 3. Object point of view

Level	Activity	Tools	Techniques
CONCEPTUAL	Object Analysis	Objects Diagram, Objects Communication Diagram, State Transition Diagram	Normalisation, Integration
LOGICAL	Object Design	Objects Diagram, Objects Communication Diagram	Transformation into the logical data structures, Information Hiding etc.

- OO approach covers all aspects of the data point of view and some (not the all) aspects of the functional point of view. Especially the Event Partitioning Technique for the conceptual modelling of functions is not acceptable here.

1.5   Integrity rules as the tool for eliminating negative consequences of the separations

Integrity rules are the rules, use of which is necessary for keeping the consistency of different views on the same thing. As strong is the separation into the different views, so strong and bulky have to be the consistency rules. Therefore the role of these rules in the structured methods of IS development is very significant. Edward Yourdon (Yourdon, E. (1989)) named integrity rules as the "rules for balancing the different diagrams". This name truthfully expresses the source of main the problems in the structured approach. The most important areas of consistency are balancing the data model (ERD) against the function model (DFD) and balancing the function model (DFD) against the control model (STD). Integrity rules used in the structured methods of IS development define potential problems but do not offer the way to prevent them. As mentioned above the object-oriented methods try to overcome the problem of separation by encapsulating data together with functions. This approach solves some aspects of separating the data from the function models. However as there is the need of several different kinds of separation in the conceptual analysis methods and techniques (see above) the integrity rules still remain very important even in OO methods (for example see: Rumbaugh J., Blaha M., Premerlani W., Eddy F., Lorensen W. (1991)).

2. POINTING PROBLEMS

2.1 The Conceptual Model Problem

One of the basic principles of the IS analysis is *the principle of modelling*. According to this principle the model of IS has to be based upon the model of the real world. Under the term „real world" we understand the *objective substance of the activities to be supported by the IS and of the facts to be stored in the IS*. This demand is only met in the „static" parts of the traditional conceptual model (i.e. in the data or object model of the reality). In the model of system's behaviour (functional model, Use Cases etc.) we model the information system's dynamics rather than the dynamics of the real world. We model there not only the objects, but also the users of the IS, not only the information sources, but also its targets. On the other hand, it is obvious that also the way in which the IS should behave (and should be used) is substantial. It arises from the rules of the real world - from the business activities which define the sense of the IS in the form of the business need for information. So the crucial question is as follows: *which from the real world actions and processes are so substantial that they are to be modelled*? Some solution is offered by the object-oriented methods. A model of the real world as a system of objects encapsulating the data with appropriate actions speaks not only about the data which the IS stores but also about the actions with the data and their sequences (processes). The system of conceptual objects and their interaction models that part of the real world dynamics that follows from the nature of the objects (their life cycles) and their relationships. However, it does not model that part of the real world dynamics which follows from the substance of the information need - from the nature of the business.

So there are at least two kinds of „dynamics" of the real world to be analysed within the process of IS development:

– Dynamics of the real world objects and their relationships given by their conceptual nature (real world conditions and constraints)

– Dynamics of the business activities given by the conceptual nature of the business processes (business nature).

2.2 The Techniques Contradiction Problem

Modelling of the dynamics of the real world objects and their relationships is the main subject of OO Analysis Methodologies (Rumbaugh J., Blaha M., Premerlani W., Eddy F., Lorensen W. (1991), Coad P., Yourdon E. (1990)). Object model represents "the static description" of the real world. It describes what objects and their mutual relationships the real world consists of. So the proper technique for analysing the objects should be based on the well known "*normalisation technique*", which defines the objects via grouping their attributes caring of their exact assignment to just one entity. One of the strongest features of the OO Analysis Methodologies is the necessity of consistency of the views on data and on operations. According to this fact, the *real world actions have to be submitted to the objects* as the elements of their methods in the conceptual object model.

In the functional approach to the real world modelling, the *Event Partitioning Approach* proposed by Yourdon (Yourdon (1989)) is used. This technique is based on grouping the real world actions according to their time dependencies: the *function structure of the system is submitted to the events*, which are mutually time-independent. It is derived from the substantial need of the combinations of the events. Combining the time-independent events, anyway, means storing the information on them. However, the stored information is matter of the data model. As the data-stores in this approach work as the crossroads of the events (and of the function communication), this approach represents the orthogonal view on data and operations in comparison with the "normalisation" technique. On the other hand, the time-dependencies of the actions have to be rightful part of the conceptual model.

It seems that there is a contradiction between the two approaches existing, while each of them is obviously right. Such a situation cannot mean nothing other than there is some "third-party view" missing. This view should bring the explanation of the situation and define the meaning of the contradiction of current views.

There has been some work made about the convergence of these two orthogonal approaches, which is also to be a subject of interest (see Jackson, M.A. (1982), Repa V. (1995), Repa V. (1996)).

3. CONCLUSIONS - BUSINESS PROCESSES ANALYSIS IN THE CONTEXT OF INFORMATION SYSTEMS DEVELOPMENT

Let us draw some conclusions from the previous paragraphs. It seems that there are two basic orthogonal views of the "real world":
– Object view which emphasises the structure of the real world
– Process view which emphasises the real world behaviour
The first view represents the *objects and their mutual relationships* while the second one represents the *business processes*.

Of course, the object model also speaks about the behaviour - in the form of entity life algorithms (methods ordering). Such behaviour is seen from the point of view of objects and their relationships. It says nothing about the superior reasons for it. So, the *behaviour of the objects should be regarded as the structural aspect of the real world.*

The significant aspect of the real world behaviour, seen from the process point of view, which is not present in the object point of view, is that there has to be the *superior reason for the real world behaviour, independent of the object life rules*. In practice, it means that for each business process some reason in the form of the goal, objective, and/or external input event (customer requirement) must exist. Business process as the collection of the actions, ordered by time and influencing the objects (their internal states and their mutual behaviour), is something more than just a random heap of the actions.

Because of the consideration made above, we can regard the *Event Partitioning Approach* as a suitable *technique for the conceptual modelling of the business processes.*

It is obvious that the two basic views described above are different views of the same thing. As mentioned on the first chapter this fact always cause the need for consistency rules. The following table outlines the basic facts, which should be used in the consistency rules:

Table 4. Outline of the consistency rules requirements (different meaning of the same facts)

Fact	Object Model	Business Process Model
Event	Stimulus for: object internal state change possible communication with other objects (send the message) in the case of the "common action"	Stimulus for: operation execution process state change output production possible communication with other processes (processes co-ordination)
Data Change	Consequence of the internal state change	Consequence of: operation execution (product) process state change
Exception	Exceptional object state	Abnormal process termination

While in the theory of management the "business process orientation" is quite a new phenomenon, in the information systems development methodologies the activities, which the process of analysing business processes consists of, are not completely new. There are various approaches to model the real world dynamics in the ISD methodologies. Some of them are focused just on business processes modelling (Lundeberg M., Goldkuhl G., Nilsson A. (1981), BSP (1984), Turner, W.S., Langerhorst, R.P., Hice G.F., Eilers, H.B., Uijttenbroek, A.A. (1987)) or on the processes modelling at least (Yourdon, E. (1989)). Business processes modelling activities in these methodologies are usually disseminated among other modelling activities in the form of current state analysis, information needs analysis, time dependencies analysis etc. As the most "business processes oriented" methodology see ISAC in Lundeberg M., Goldkuhl G., Nilsson A. (1981). The really new aspect in this field is the need for *detachment of the business processes modelling activities from other modelling activities* (i.e. modelling the static real world structure as well as modelling the internal dynamics of the objects) in the ISD methodologies. Analysis and modelling of business processes seems to be a separate activity, which should precede other ISD activities. The main reason for such detachment is the universality of the conceptual Business Processes Model. It works as the basis not only for information system development, but also for workflow implementation as well as for business process reengineering (see *Figure 2*).

Figure 3 shows Conceptual Business Process Model (BPM) and Conceptual Business Objects Model (BOM) as the base for the conceptual model of the information system. BPM gives to the process of conceptual IS

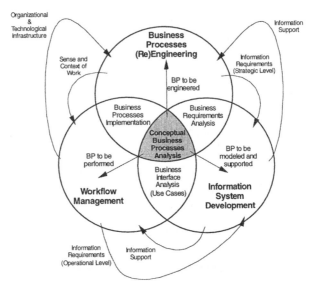

Figure 2. BPR vs. ISD vs. Workflow Management

modelling the information about the necessary structure of the interface functions in the form of business products, actors and activities identified. BOM gives to the process of conceptual IS modelling the information about the necessary real world structure in the form of business products, actors and objects identified. In fact, most of the traditional ISD analysis activities should be performed in the form (and with the purpose) of business processes analysis (that naturally includes also the analysis of business objects). The conceptual model of the information system consists of two main parts. Central part - object model of the IS - represents the clear model of the real world structure from the IS point of view. It describes those real world objects and their relationships, which are to be supported (and thus modelled) by the IS. The dynamics of the real world, contained in this model, is seen here from the object's point of view - in the form of life cycles of the objects. The peripheral part - process model of the IS - represents the model of the real world behaviour from the IS point of view. It describes those real world processes and their relationships, which are to be supported (and thus modelled) by the IS. In terms of the IS it represents input/output interface function structure.

Let us take cognizance of two basic kinds of dynamics, described by the conceptual IS model. The dynamics, contained in the object model, follows from the substance of the objects (their life cycles which have to be respected by the IS). On the other hand, the dynamics of the real world, described in the process model, follows from the business processes, which are to be supported by the developed information system. The substantial role of the information system is to support the consistent interconnection of these two points of view. An information system has to support business processes, respecting the inviolable nature of the objects, expressed by their

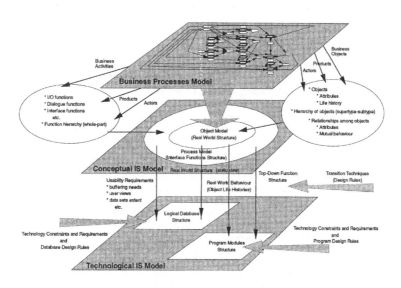

Figure 3. Business Processes Analysis as the Starting Point for Information System Development

life cycles, in the sense of the integrity rules, roughly outlined in the *Table 4*.

The role of the ISD in this conception, in contrast to its traditional role, is reduced to the modelling and designing the information system itself, abstracting the aspects, which are not connected directly with it. Most of the analytical effort and experience now should be moved into the "information technology independent" and more general conceptual business processes modelling as a separate activity, which should precede other ISD activities. Therefore the methodology of business processes modelling, based on the Event Partitioning Approach (Yourdon, E. (1989)) should be developed. Such BPM methodology should satisfy the general needs of business processes management as well as the needs of IS development process, respecting the convergence of these fields as described above (see *Figure 2*).

For the first outline of such BPM methodology, see Repa V., Bergner M., Chlapek D. (1997), Repa V. (1998), and Repa V. (1999).

REFERENCES

BSP (1984) "Business System Planning: Information Systems Planning Guide", IBM, GE20-0527-4.

Coad P., Yourdon E. (1990) „Object-Oriented Analysis", Prentice-Hall Inc., NJ.

Chen P.P.S. (1976) "The Entity Relationship Model - Towards a Unified View of Data", ACM TODS, Vol. 1 No.1.

Donovan J.J. (1994) „Business Re-engineering with Information Technology", Prentice-Hall Inc., Englewood Cliffs, NJ.

Goodland M., Mc. Lean J. (1995) „From BPR Vision to IS Nightmare in Business", in Proceedings of 5th. Conference on Business Information Technology BIT '95, Department of Business Information Technology, Manchester Metropolitan University.

Greenwood R.M., Robertson I., Snowdon R.A., Warboys B.C. (1995) „Active Models in Business", in Proceedings of 5th. Conference on Business Information Technology BIT '95, Department of Business Information Technology, Manchester Metropolitan University.

Hammer M., Champy J. (1994) „Reengineering the Corporation: A Manifesto for Business Evolution", Harper Business, New York.

Jackson, M.A. (1982) „System Development", Prentice-Hall Inc., Englewood Cliffs, NJ.

Lundeberg M., Goldkuhl G., Nilsson A. (1981) "Information Systems Development - A Systematic Approach"", Prentice-Hall Inc., Englewood Cliffs, NJ.

Repa V., Bergner M., Chlapek D. (1997) „Modelling the Enterprise Activities", research paper, University of Economics, Prague.

Repa V. (1999) „Business Processes Based Information Systems Development", Proceedings of the BIS 99 International Conference, Springer Verlag, London.

Repa V. (1998) „Methodology for Business Processes Analysis", Proceedings of the ISD 98 International Conference, Bled.

Repa V. (1996) „Object Life Cycle Modelling in the Client-Server Applications Development Using Structured Methodology", Proceedings of the ISD 96 International Conference, Sopot.

Repa V. (1995) „Hybrid development methodology", in Proceedings of 5th. Conference on Business Information Technology BIT '95, Department of Business Information Technology, Manchester Metropolitan University.

Repa V. (1994) „Seeking the Actual Reasons for the „New Paradigm" in the Area of IS Analysis", Proceedings of the ISD 94 International Conference, Bled.

Rumbaugh J., Blaha M., Premerlani W., Eddy F., Lorensen W. (1991) „Object-Oriented Modeling and Design", Prentice-Hall Inc., Englewood Cliffs, NJ.

Scheer, A. -W. (1992) „Architecture of Integrated Information Systems -Foundations of Enterprise-Modelling", Berlin.

Scheer, A. -W. (1994) „Business Process Engineering - Reference Models for Industrial Enterprises", Berlin.

Turner, W.S., Langerhorst, R.P., Hice G.F., Eilers, H.B., Uijttenbroek, A.A. (1987) „SDM, system development methodology, North-Holland.

Yourdon, E. (1989) „Modern Structured Analysis", Prentice-Hall Inc., Englewood Cliffs, NJ.

PERVASIVE IT SYSTEM:
ENSURING A POSITIVE CONTRIBUTION TO THE BOTTOM LINE

Robert Moreton
Professor of Information Systems, School of Computing and IT
University of Wolverhampton, 35/49 Lichfield Street,
Wolverhampton, WV1 1EL U. K.
Ph.: +44 (0) 1902 321 462, Fax: +44 (0) 1902 321491
E-mail: moreton@wlv.ac.uk

ABSTRACT

This paper examines the implications for organisations of the widespread/pervasive use of IT to underpin business processes. First it addresses the critical question of how businesses can evaluate the output of the IT function to ensure that it makes a positive contribution towards the organisation's objectives. Then the paper deals with the key issue of IT architecture, and concludes by discussing the need for organisations to 'learn' about IT to ensure that the technology can be used to support continuous adaptation of the business.

KEYWORDS

Strategy, investment, architecture, infrastructure, technology evaluation

This paper is derived from a series of commercial research projects which were undertaken by the author, as part of the CSC Index Foundation research programme. The paper examines the implications for organisations of the widespread/pervasive use of IT to underpin business processes. First it addresses the critical question of how businesses can evaluate the output of the IT function to ensure that it makes a positive contribution towards the organisation's objectives. Then the paper deals with the key issue of IT architecture, and concludes by discussing the need for organisations to 'learn' about IT to ensure that the technology can be used to support continuous adaptation of the business.

1   Pervasive IT requires a fundamental review of IS/IT investment

The power of IT, as its use becomes more widespread/pervasive, lies in its integrating potential, both internal and external to an organisation (Moreton and Chester, 1997). This calls for a fundamental review of IS/IT strategy in which the emphasis is on building an infrastructure for connectivity and information sharing, and separating infrastructure issues from business process/team support issues is crucial.

Systems Development Methods for Databases, Enterprise Modeling, and Workflow Management,
Edited by W. Wojtkowski, *et al.* Kluwer Academic/Plenum Publishing, New York, 1999.

Moreton and Chester argue that, like any other investment decision, IT investment must support the needs of the organisation. IT is a tool which can significantly improve the way an organisation exploits its information resources. If IT is to be used as a strategic/competitive weapon, then it should be considered along with all other related investment opportunities (rather than merely among other IS/IT opportunities). Decisions about IT investments will then be made in the same way as any other business investments, with senior managers agreeing priorities by debating the issues and reaching a consensus (Strassmann, 1997). It is the management process that is important, rather than the absolute level of expenditure. The significance of management processes in determining successful exploitation of IT in small to medium enterprises (SMEs) is addressed by Levy et al, 1998.

The criteria that are appropriate for justifying different types of IT investment differ according to the purpose of the investment. In particular, they differ according to the nature of the benefits that are to be achieved by the proposed systems. Consequently, it is important to distinguish between the different types of IT investment if appropriate evaluation criteria are to be applied when justifying systems. Five different types of investment can be identified (mandatory, business performance/productivity, competitive advantage, infrastructure, research), and each may relate to different business aims, such as satisfying regulatory requirements, maintaining market share or enhancing IS/IT capability (Willcocks and Lester, 1995).

To be properly assessed, some types of IT investment can require extensive knowledge and experience of the business. Management judgement is therefore an essential element of investment appraisal. The judgmental aspects of investment evaluation become more important both as the need to understand the demands of the market and as the organisational resources to meet those demands increase. For example, factors such as customers' or suppliers' reactions and competitors' likely responses need to be taken into account. Activities such as market research and product or service promotion may be vital ingredients in the evaluation and planning process. These may well change both the structure and the total cost of project budgets.

Markus and Keil (1994) propose that senior management should expect design reviews to cover projections of the penetration of the system and the extent and quality of its use. They write that there should also be periodic reviews of all key operational information systems to see if they are still contributing to the business's goals.

These messages were affirmed by Michael J. Earl in his book *Management Strategies For Information Technology* (1989). From his study of UK organisations, he reported that, in those companies which successfully took advantage of IT, the exploitation of IT (through business processes) was clearly differentiated from the development of the IT strategy and infrastructure (providing business support). He also observed that IT architecture frameworks are required to balance control with flexibility in infrastructure development. It is to this issue that we now turn.

2   A major investment in infrastructure and applications

The pervasive use of IT systems inevitably results in a large scale investment that can only be justified by the business policies and plans relating to long term competitiveness (CSC Index, 1990). One of the key requirements is that the technology enables business and application connectivity and integration, which will often lead to a major investment in infrastructure and applications. Thus, the ability of the organisation to manage the technology in relation to the business and its information needs is critical to successful investment in IT. Investing in IT to maximise the business benefits is a business issue first and a technology issue second.

Davenport et al. (1989) report that the companies they studied that were using IT more effectively seemed to have internal management principles which are used to evaluate any proposal for an IT system and to guide management decisions in this area. These researchers suggest that a set of these principles would be an advantage for any company. The examples they give include:

- Computing systems should facilitate company-wide (global) information consistency.
- All product data should be accessible through a common systems such as order-processing.
- All product groups should be self-sufficient in their information systems capabilities.

While the use of such principles as guidelines is commendable, care should be taken that they do not become management by an unchangeable, inflexible constitution. A particular principle may turn out to be unsuitable in the light of technological or methodological progress and should not be used to hold the business back from utilising modern IT hardware, tools and techniques.

2.1 Developing systems connectivity and integration

The ability to integrate information from different areas of the business, and to enable people in different parts of the organisation to work together using the same information, is an essential requirement of organisations where IT is in widespread/pervasive use. In most organisations, there is an enormous amount of existing computer-based information that could be used to advantage by the business. However, the opportunities may not be realised, partly because few people are aware of what is available, and partly because of the technical difficulties in assembling and integrating the information. A coherent technical infrastructure helps to solve both these problems.

Kaplan (1990) has argued, that the demand for connectivity derives directly from the general shift away from technology-focused management to data-resource management. He suggests that such a shift requires a greater degree of connectivity between applications, and their associated databases. In turn, this leads to distributed computing and the development of 'intranets'.

Keen (1991) has provided a framework for discussing connectivity and interoperatability in terms of what he calls 'reach' and 'range'. *Reach* refers to the locations and organisations with which systems can, or need to, interwork. *Range* describes the nature of the interaction that is available or needed. An electronic data interchange system, which, for instance, automatically initiates payments on receipt of goods, requires a high degree of connectivity and integration between the supplier's and the buyer's systems - displaying 'intermediate reach, high range'.

In general, the reach and range of systems demanded by organisations are growing. Decisions about the required levels of connectivity and integration, both internal and external to the organisation, are essentially business decisions, although IT management will provide advice on feasibility, costs and implications. For example, rules to govern the way in which data items, such as customer codes, are created and maintained will also be required (along with data interchange standards) to safeguard future integration paths. However, a well-defined technical infrastructure will enable the integration of information from different functional systems - a prerequisite for the process working required by transformed organisations.

2.2 Investing in infrastructure and applications

The ability to exploit opportunities swiftly and effectively (or to respond to competitive threats) requires an adequate and flexible IS/IT infrastructure. There are three important components of this infrastructure:

- Telecommunications systems. These need to be adequate to meet the foreseeable needs of applications that extend out to customers, intermediaries, or suppliers (Roberts and Flight, 1995). The technical infrastructure must also provide a sufficiently reliable service so that business managers have the confidence to use them for linking with the organisation's trading partners.

- Databases and their associated access systems. These need to be organised so that relevant data can be accessed effectively and so that this can be done without compromising confidentiality or other security issues (particularly when access is by third parties such as customers).

- Systems development capability. Given that systems development resources are scarce, it is essential to use these resources on applications that add value and increase competitiveness.

Keen (1991) noted the extent to which the cash flow of US companies is based on electronic transactions. More than half the revenue of banks, for instance, in most major money centres now derives from automatic teller machine (ATM) transactions, foreign exchange trading and electronic funds transfer. He contends that this will reach 90 per cent or more by the year 2000. He argues that technical infrastructure is therefore of fundamental importance to most major organisations, and he has derived a set of policy-level requirements that could be used as an agenda for defining a technical architecture.

The process of defining a technical infrastructure must not only produce a plan for systems implementation, but it must also build commitment to the plan among the users who will have to utilise it in operation. Without this commitment, it is impossible to ensure that the applications and infrastructure will be consistently used and maintained. This point is strongly made in relation to Business Process Re-engineering (BPR) by Mumford and Hendricks (1997). By closely involving business managers and trade unionists in their planning, Sweden Post (cited in Moreton and Chester, 1997), for instance, has explicitly acknowledged that investment in IT will only be of value if it enables rather than constrains organisational development in its widest context.

2.3 Justifying infrastructure requires business judgement

The ability of an organisation to manage the technology in relation to the business and its information needs is critical to successful investment in, and exploitation of,

IT. Justifying this investment in the technical infrastructure can create a major problem for two reasons: the subjective value of the investments, and the scale of the requirement.

The issue of the subjective value of the investments arises because the value of infrastructure investment derives from its role in facilitating the development and successful operation of appropriate applications. The benefits of the investment stem from these applications, rather than the infrastructure itself. While the size of the investment may be judged with some accuracy, the value of the applications' benefits may be a matter of judgement.

The scale of the requirement arises because the infrastructure is rarely limited to one application or to one business manager's area of responsibility. The ultimate benefits of the infrastructure investment accrue to the organisation as a whole. Such benefits are not always as readily demonstrable as improvements in the short-term performance of individual business units. Managers may therefore find it difficult to take an organisation-wide view of the value of the infrastructure, and so may be reluctant to underwrite the investment.

This, again, emphasises the need for senior business managers to be involved in the implementation decision-making for organisational transformation. It is necessary for corporate management to make a judgement on whether the expected benefits of the proposed infrastructure justify the investment. In this way, also, the two extremes of investment on technical criteria alone, and of loading the whole investment onto a single application can be avoided. Corporate management should make a judgement based on the known cost of developing the infrastructure, compared with an estimate of the likely benefits of the applications portfolio that the infrastructure is required to support. As far as is possible these benefits should be quantified, and, ideally, expressed in financial terms. If the investment is required to improve the performance of the existing infrastructure, rather than to support new applications, the judgement should be based on the strategic or operational value of the business benefits that will arise from that improvement.

In preparing the formal business case for investing in the technical infrastructure, the emphasis should therefore be on the future capabilities that the new infrastructure will provide. This situation is analogous to the justification of factory-automation systems in the automobile industry. The point is illustrated by Clemons (1991), who states that: "The problems of evaluating investments in factory-automation are illustrative [of the problems of evaluating IT investments]. This should not be viewed as new applications for making today's automobiles, but rather as new technology for making tomorrow's automobiles. This requires assumptions about (the future direction of the business)."

A long term view is required because a technical infrastructure which conforms with a planned technical architecture, usually costs more in the short to medium term than one constructed in an ad hoc fashion. Individual investments must be undertaken only if their impact on other parts of the infrastructure is recognised. The technical architecture may also require 'enabling' investments to be made, which provide no direct payback. These might include interim solutions to link systems together, for example wide-area network gateways and EDI services (Roberts and Flight, 1995).

Many formal methods and techniques are available for evaluating IT investments to ensure that they match business priorities (Norris, 1996, Willcocks, 1995). Methods are typically based on ranking the business contributions, both quantitative and qualitative, and risks according to criteria that are relevant to the organisation. The criteria can vary according to the priorities of each business. Thus the methods support the decision-making process by helping to assess the relative contribution of proposed investments in a disciplined and systematic way.

A comprehensive method of setting priorities for IT investment were developed by Parker and Benson (1988). Their method is based on an approach which they call 'information economics', and involves scoring potential projects on the basis of ten system features. The features are used to assess both the business justification of the project and its technical viability. Weights for the different features can be set by an organisation to reflect its own priorities for IT investment and the features of its technical architecture. Projects can then be ranked in terms of their weighted scores.

Although this approach implies a mechanistic appraisal, and contains a subjective basis for many of the scores, it does provide a useful checklist for assessing the wider impact of introducing systems, rather than focusing on limited financial criteria. Indeed, its greatest value may well come from the explicit evaluation of issues during the assessment process, rather than from its statistical outcome (Willcocks, 1993).

3   The need for research and development

To successfully exploit the relevant technologies there needs to be a continuous programme of research and development. During the research for this paper we visited organisations that had very different views on the importance of IT to their businesses. We believe that whilst some recognise the contribution that IT can make to organisational success others underrate its value.

In our view, the best way to examine the potential of IT is to set up a joint programme by managers and the information systems function. The aims of the programme are:

> - to understand the potential for emerging (and existing) technologies within the organisation, and to assess when they will be usable - both technically and economically;
>
> - to identify customer-service needs and threats;
>
> - to underpin the informating process, by acquiring sufficient knowledge about the technologies so that the organisation can understand their potential application, seize opportunities with limited risk of technical failure, and respond swiftly to innovative uses of IT by competitors.

One organisation with whom the author has worked, CSC-Index (1985), has a well established technology tracking method that enables IT planners to judge which developments are now, or are likely to be, relevant to their business. In this way an organisation can determine its investment priorities for IT. The four steps are:

> - review the whole technological field and select all the developments that might have an application within the organisation;

- estimate the current level of maturity of each of the technological developments. Four broad divisions of technology (embryonic, pacing, key and base) can be defined. Base technologies: an organisation must master to be an effective competitor. Key technologies: provide competitive advantage. Pacing technologies: could become tomorrow's key technologies. Embryonic technologies: are at the laboratory stage with some some limited commercial experimentation;

- forecast how each development will mature over the next five years and plot the likely change of maturity;

- pick out those developments that seem likely to become relevant during the planning time-frame and track these carefully. Follow the rate of development and also their early uses, especially by competitors.

An effective R&D programme must include some investment to build on core competence in pacing technologies and some effort to gain intelligence, from sources such as customers, industry-watchers and universities, to help identify and evaluate these techniques. At the same time, disciplined judgements about commitments to key technologies are necessary; enthusiastic overspending on advanced IT can undercut essential support for more mature technologies.

A further factor to consider when selecting from the technological options, is the ability of the IS function to manage the chosen technology. There is a relationship between technological maturity, as described above, and the IS management capability required to harness that technology. Management maturity can be defined in terms of the role of the IS function within the organisation. At its least mature and influential, the role is purely responsive. Its stages of maturity go from being industry-led, to being determined by the organisation's business strategy, and at its most mature is strategy influencing.

4   Summary

In this paper the author has examined the implications for IT investment and development of the wide scale use of IT to underpin business processes. The paper identified the contribution that IT can make to organisational success, and emphasised that this may not be a trivial task. The remainder of the paper dealt with the key issue of IT architecture, and discussed the need for organisations to 'learn' about IT to ensure that the technology can be used to support continuous adaptation of the business.

REFERENCES

Clemons E K. Evaluation of Strategic Investments in Information Technology. Communications of the ACM, Volume 34 number 1, January 1991.

CSC-Index. Developing and Implementing a Systems Strategy. CSC-Index Foundation Research Report, October 1985.

CSC-Index. Getting Value from Information Technology. CSC-Index Foundation Research Report, June 1990.

Davenport TH, Hammer M, Metsisto T J. How Executives Can Shape Their Company's Information Systems. Harvard Business Review, March-April, 1989, pp. 130-134.

Earl M J. Management Strategies for Information Technology. Prentice Hall, 1989.

Kaplan R. Trading in a Tired Technology. Datamation, Volume 36, number 16, 15 August 1990.

Keen P G W. Shaping the Future: Business Design Through IT. Harvard Business School Press, 1991.

Levy M, Powell P and Yetton P. "SMEs and the Gains from IS: From Cost Reduction to Value Added". Proceedings IFIP WG 8.2 & 8.6, Information Systems: Current Issues and Future Changes, Helsinki, December 1998 pp 377-392.

Lindsey A. H. and Hoffman P. R. "Bridging traditional and object technologies: Creating transitional applications", IBM Systems Journal Vol 36, No. 1 - Application Development, 1997, pp 32-48.

Markus M L and Keil M. If We Build It, They Will Come: Designing Information Systems That People Want To Use. Sloan Management Review, Summer 1994, pp. 11-25.

Moreton R. and Chester M., "Transforming the Business: the IT contribution". McGraw Hill, London, 1997.   .

Mumford E. and Hendricks R. Reengineering Rhetoric and Reality: the rise and fall of a management fashion. http://bprc.warwick.ac.uk/rc-repb-6.html, 1997. Accessed January 1999.

Norris   G."Post-Investment Appraisal". In Willcocks L (ed) Investing in Information Systems: Evaluation and Management". Chapman Hall London, 1996.

Parker M M and Benson R J. Information Economics: Linking Business Performance to IT. Prentice Hall, 1988.

Roberts R and Flight G. "The Enabling Role Of EDI In Business Process Re-Engineering", http://infosys.kingston.ac.uk/ISSchool/Staff/Papers/Roberts/EDI_BPR.html, March 1995, accessed January 1999.

Strassmann P A. "The Squandered Computer: Evaluating the Business Alignment of Information Technologies", Information Economics Press, 1997.

Willcocks L and Lester S. "Evaluating the Feasibility of Information Technology Investments". Oxford Institute of Information Management, RDP93/1, 1993.

Willcocks L and Lester S. "In Search of IT Productivity: Assessment Issues". Oxford Institute of Information Management, RDP95/7, 1995.

Willcocks L, Currie W and Mason D. "Information systems at work: people, politics and technology", London, McGraw-Hill, 1998.

THE CHALLENGE:
PROVIDING AN INFORMATION SYSTEM FOR A CLINICAL LABORATORY

H. M. Aus, M. Haucke, and A. Steimer
Institute for Virology, University of Wuerzburg
Versbacherstrasse 7, 97078 Wuerzburg, Germany
Ph.: +49 931 201 3963, Fax: +49 931 201 3934
E-mail: ause@vim.uni-wuerzburg.de

ABSTRACT.

This paper chronicles database redesign projects at two separate laboratory departments that test human specimens with many similar methods and tasks. Both departments had earlier purchased the same software package to support the daily laboratory testing and administrative operation. Our objective was to replace the flat file data heap with a modern database while preserving the original software's strengths. This paper describes briefly the laboratory routine, our Structured Query Language (SQL) redesign concept, the outcome of the redesign project, and some lessons learned.

1. INTRODUCTION - THE LABORATORY CHALLENGE.

The daily laboratory routine is summarized schematically in Figure 1. The process is not a simple top to bottom operation because intermediate test results may indicate that additional tests or revised test procedures are needed.

The laboratory routine starts with entering the information about the patient specimens from the physicians and hospitals. Many specimens are from the same patient received in irregular intervals, which can be as short as one day or as long as several years. None-the-less, the database software must correctly identify all known medical diagnostic information associated with each patient and specimen.

After receiving a specimen, the database is first queried for any previous information about the patient. Based on the archived data and any new additional information, the staff determines the testing procedures and schedule. The detection of viral infections includes indirect methods to test for antibodies, and direct methods to test for antigens and viral genomes (DNA, RNA). The test method and the human "virus" (antibody, antigen or viral genome) define each laboratory test. Some viruses require several tests.

Testing for a human virus may require a time history of viral activity in a patient's specimens collected over a specific interval. To complicate the logistics, specimen testing can require many different procedures in several laboratories and the various procedures seldom finish on the same day. Intermediate results may necessitate repeated or additional testing under different conditions. The test results receive a test code based either on measured values, calculated values or qualitative codes (positive, negative, etc.). Lastly, the laboratory findings are reported to

Systems Development Methods for Databases, Enterprise Modeling, and Workflow Management,
Edited by W. Wojtkowski, *et al.* Kluwer Academic/Plenum Publishing, New York, 1999.

357

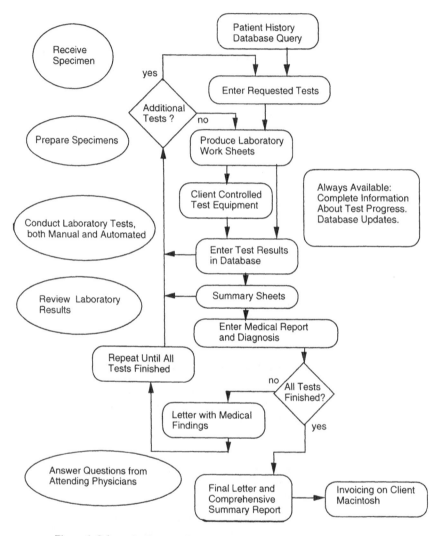

Figure 1. Schematic diagram of the workflow in our daily laboratory routine.

Table 1. Overview of the data stored in our IS-SQL application environment.

1,400,000	On-line test results since 1.10.1987.
208,000	On-line specimens tested since 1.10.87.
117,000	Patients On-line.
28	Programs.
63	SQL Tables
48	SQL Indices.
79	Storage areas with table and index data distributed over 4 disks

Table 2. Some inherent challenges that confront a hospital and clinical information system.

- Each patient's specimen undergoes many laboratory tests in many different laboratories.
- A patient's birth date is often unknown.
- Duplicate and chronological data are inherent in a clinical laboratory information system.
- Many patients have either the same name or birthday, or both.
- Each patient can have, in his or her lifetime, many:
    - Name changes through marriage, divorce, adoption, etc.
    - Addresses.
    - Doctors.
    - Visits to physicians and hospitals.
    - Specimens taken and tested.
    - Treatments.
    - X-rays.
    - Surgical operations.
    - Transplants.
- New and mutated human viruses are continuously being discovered.
- Continuous evolution in instrumentation, testing methods, regulations, invoicing requirements and reporting formalities.
- Summarize data for epidemiological queries, studies and reports.
- Communicate with a large variety of other systems in the hospital.

the attending physician, and the clinic or the patient or his health insurance is invoiced for the services rendered.

The production database stores all the diagnostic and virological information for future reference (Table 1.). In addition to the individual specimens and patient data, the database also stores a complete description of the laboratory activities. This pool of SQL tables contains the parameters that describe the daily laboratory routine. These parameters include: hospital and physician addresses, types of human specimen material tested in our labs, virus types, test methods, test and virus matrix, result and diagnostic codes, fixed texts, parameters for other laboratory tasks, and invoicing. The pool tables change only rarely. The information system (IS), however, mutates permanently because the laboratory methods, equipment and staff evolve endlessly. None-the-less, the database must be totally transparent especially when the laboratory physician advises the attending physician about his patient's medical history and condition.

Some additional challenges are summarized in Table 2. Continuous evolution, customized end-user interfaces, revised testing methods, adaptation to change and reliable operation are primary information system challenges.

2. THE SQL IMPLEMENTATION.

The initial commercial laboratory package was installed in our Institute in 1985. The software had previously been installed at several European universities in the late 1970s and 1980s. The software had evolved over the years and contained many features that supported the various laboratory tasks. Shortly after our software was installed, we began to cooperate with another institute that had installed the same software several years earlier. Both laboratory departments had many common software needs. In 1990 we decided to switch from the legacy flat file application to a SQL based implementation while preserving the many strengths of the initial application. The other institute decided to switch also.

In our department, we decided to utilize 3GL compilers because we wanted absolute control of the software and it was not obvious to us that several mission critical laboratory operations could readily be implemented in a 4GL tool. Third Generation Languages (3GL) include the classical compilers such as Pascal, Basic, C, and FORTRAN. In this paper, Fourth Generation Language (4GL) tools include a wide variety of visual programming, World Wide Web (WWW) development packages, and objected-oriented tools and other component based technology. 4GL provide standardized boilerplate input and output formats which can lead to unwanted and incomplete displays, and may require unnecessarily complex input sequences. Furthermore, the input routines in many packages are too time-consuming and awkward for a proficient data entry clerk who enters information without looking at or waiting for the window display screen. For obvious financial reasons, we chose the compilers and Rdb-SQL database that were included in our University's campus software license. This turned out to be an excellent choice. Substantial differences exist among the variety of commercial SQL database engines. During the development, we have learned to appreciate Rdb-SQL's procedural language, crash recovery safeguards, extensive monitor and administrative tools, and reliability. The Oracle Corporation, Red Wood Shores, California, USA currently owns Rdb. Both departments decided to use the Rdb-SQL database engine.

In the second laboratory department, the project objective was also to replace the flat file data heap with a SQL database while preserving the legacy software's mode of operation, function and user interface. These goals were quickly forgotten soon after the project began. There the programmers decided to implement the new software with 4GL tools and ignored the existing software's function in the laboratory operation. Seven years and several groups of 2 – 3 programmers later, the legacy system is still in operation. The programmers have failed to produce a replacement system that the users can accept. The first group of software engineers left after they failed to convince the laboratory staff to change the laboratory procedures to fit the software. The department has contracted another software house to develop an IS with a different set of 4GL tools. Whether or not the current attempts produce a successful information system remains to be seen. The critical question is whether or not the department can adapt their organization to the 4GL limitations. The programmers still ignore the existing software and the department's internal know-how.

An essential part of the project was to study, understand and appreciate the strength of the existing 3GL software, which the users appreciated and make a large, complex system easier to use. One of our objectives was to port the laboratory information from the old system to the new SQL database. Porting the data to the new system enabled us to test the new IS design with more than five years worth of data, but required an extensive programming effort. Table 3 lists some reasons for porting any old data to a new IS. Tests with the users showed that the information system was too slow and that the new SQL table design was incomplete. An acceptable IS response time required a complete redesign of the SQL tables and reprogramming the whole application. The port was well worth the effort but was often tedious and time consuming. The redesigned SQL tables formed the bases of the SQL tables in our IS at the present time, Table 1.

Many changes and additions have occurred since the laboratories began using our new information system on January 1, 1994. New laboratory equipment, testing procedures and regulations have required new programs and new or modified SQL tables. As an example, one of our previous PC client machines required eight SQL tables and six programs. Four years later, the newer instrumentation is easier to use, provides more reliable results, and requires only three SQL tables and two programs. The Euro currency and new hospital and national health insurance regulations in Germany also necessitate a continuous revision of the invoicing software. The new rules require additional columns in existing tables, access to the hospital administration's computer for the work order numbers and general billing and accounting procedures. Billing procedures differ for the University hospital, external hospitals, insurance companies, private patients and scientific studies, etc. To date, our 3GL SQL Information System has met all these challenges.

The users were and still are continuously involved in the IS development, especially during the interface design, testing, and debugging phases. IS development is a continuum of concurrent processes and not a sequence of disjoint tasks.

3. DISCUSSION.

Many books and articles have been written on clinical informatics and computers in health care. Degoulet et al. (1994) is a concise introduction to Clinical Informatics and is a translation from the original French text. Dragan et al. (1994) and Shortliffe et al. (1990) are two pragmatic books, which document design and implementation of health care information systems. The articles (chapters) in both books contain many interesting case studies written by professionals in both academia and industry. The general design prerequisites (Table 4) for a patient care information management system is relevant to all information systems (Metzger and Teich 1994; Fagan and Perreault 1990).

The predominate difference between patient health care systems and other information intensive applications is the concern for the patient's welfare (Shortliffe et al. 1990; Drazen 1994). The contributors to the ISD'97, Sixth International Conference on Information Systems Development- Methods and Tools (Wojtkowski et al. 1997), held August 11-14, 1997 in Boise, Idaho raise many general IS design and implementation issues (Table 5) that apply directly to the clinical environment (Pereira 1997; Paetau 1997; Kirikova & Grundspenkis 1997; Kirveennummi et al.). Our experience has been that these software development processes are not separate, disjoint, sequential processes, but rather concurrent, continuous interaction between all users and the software engineers (Nagasundaram 1997; Vogel 1997).

All these factors, and more, contribute to successful Information Systems. However, one highly significant but rarely mentioned topic is the software engineer's attitude, qualification, reliability and willingness to collaborate effectively within the framework of the project objectives and mission. All too often software engineers are requested to build "the software equivalent of 'aircraft carriers' with little more experience in building more than glorified 'row boats'" (Vogel et al 1997). Dynamic programmers eagerly introduce visual programming and 4GL methodologies, and demand expensive 21-inch workstation. Only the latest version of glossy shrink-wrapped software is good enough. High gloss software tools and the newest big screen workstations do not, however, guarantee high quality applications. Many projects fail because

Table 3. Reasons for porting legacy data to a new IS development.

- Train the user early in the development process with known data.
- Decrease risk of missing important requirements.
- Expose the users to the advantages of the new ideas in the new system during development.
- Re-design the new software until it fits the organization and satisfies the specifications.
- Require the programmers to understand both the legacy system and the organization.
- Test how the new system fits into the organization and workflow early in development.

Table 4. Design prerequisites for an information system. *

- The IS must be easy to use and require little or no training.
- The IS must minimize manual data input time and maximize usage incentives.
- The IS must be available whenever needed, especially during business decisions.
- The IS must provide quick and value-added access to information.
- The IS must be designed to fit actual workflow situations.
- The IS must integrate a large variety of hardware and software support tools.
- The IS must manage change.

* Adapted from Metzger and Teich (1994).

Table 5. Potpourri of fundamental issues faced when developing information systems. *

- Requirements for complete, concise and consistent definitions.
- User participation in the specification, development and testing processes.
- User attitudes, actual usage and user satisfaction.
- Creative group support systems.
- Reliable and uniform development methods, tools and support systems.
- Software development methodologies, tools, reuse and cost reduction.
- Organizational obstacles, boundaries, defenses and traditions.

*From Wojtkowski et al. (1997).

Table 6. Our list of factors for a successful Information System Design.

- Take the time to understand the organization, the tasks and any legacy software.
- Involve all the users.
- IS development is a continuous, concurrent set of user and developer tasks, especially during testing.
- Modern programming tools are not a solution for every application.
- Match the IS technology to the dynamics of the organization, tasks and workflow.
- Consistent, complete and concise design definitions are a pipedream.
- Plan for change.
- The missing IS functions are always the most important.

---

the staff disregard the people agenda and exaggerate the technology (Pereira 1997). With today's computer and network hardware, software development is often more a people issue than a technical problem.

As in any software project, we would welcome reliable, fast development methods, faster debugging methods, reusable software, complete and easy to use class libraries, team development tools, and shorter development cycles (Frizgerald 1997; Paetau 1997; Nagasundaram 1997). Our observation has, however, been that the 4GL tools require a substantial financial and manpower investment (Vogel et al. 1997). Furthermore, we are reluctant to base IS design on the exciting but rapidly changing development tools (Willard et al. 1999). Any incomplete functions and libraries in the software tool introduce additional problems in an already complex project. Sometimes the libraries in the newest version of a popular visual programming tool are not backward compatible with the older version, which means reprogramming the whole application. Similarly, modifying an object-oriented class requires re-testing all the sub-classes (Paetau 1997). More critically, one cannot assume that these methods will solve all future challenges (Cheung et al. 1997; Paetau 1997). Nothing is more important to the user than the unfulfilled requirements and wishes. Furthermore, a 4GL implementation can be more expensive and time consuming than a 3GL implementation. Small groups like ours can ill-afford to invest in such methods only to experience that the method can not provide the required information in a timely manner and in a useful format. Too often we have heard programmers try to convince users to adapt the new software and change organization's mode of operation (Kirikova & Grundspenkis 1997; Kirveennummi 1997; Pereira 1997). Usually, there is little or no obvious benefit for the users, only for the programmers. In some cases it is impossible to change the organization's function. Our doubts about being able to implement certain critical IS features with fixed boilerplate solutions proved to be correct (Pereira 1997; Post & van der Rijst 1997). Programs that confront users with unnatural procedures and constraints for accessing and entering information are unlikely to be used and accepted (Fagan and Perreault 1990). Not surprisingly, many software companies and groups do not use object-oriented methods (Paetau 1997).

Although many information systems have been developed over the last 20 – 40 years, the critical factors for successful IS development remain elusive (Shaw & Pereira 1997). Table 6 shows a short list of the lessons learned in our experience. Above all, the IS must support and provide substantial added value to a organization's core business (Drazen 1994; Metzger and Marwaha)

4. CONCLUSION.

The contributors at ISD'97 in Boise express their concern about the need for better, reliable, cost-effective software tools and they describe many interesting solutions and suggestions (Wojtkowski et al. 1997). We share the authors' concerns but we do not share their enthusiasm for the available methodologies. Many hurdles and shortcomings, including people issues, too long design and implementation cycles, cumbersome input and output, and transferring successful research results to the pragmatic real world, need to be overcome before developers can use the new tools with confidence (Pereira 1997; Metzger 1994).

The question is not whether computers play a critical, value-added role in information management, but rather how can we ensure that the technology supports the organization's business effectively (Fagan and Perreault 1990). The outcome depends on the planners, the users, policy makers, system developers and the informatics professionals.

The laboratory information system presented in this paper enjoys a high degree of user acceptance primarily because the users are continuously involved in the design and implementation work. Furthermore, the 3GL-user interface is adaptable to all new requirements and doesn't hinder the laboratory function and schedules. The newer IS preserves many of legacy system's advantages and includes all the laboratory information from the older system. Our laboratory IS has also been integrated into the larger hospital IS on our campus.

All the design, redesign, testing, repeated effort and attention to detail was often quite tedious and time consuming, but well worth the effort. The present software concept, which has solved all challenges since 1994, is based on that preparatory work. We are confident that the LIS can deal effectively with future challenges. Remember, IS development is not a disjoint, sequential set of tasks but rather a continuum of many interesting and challenging processes.

ACKNOWLEDGMENTS.

The authors are grateful to the following people for many motivating project discussions: F. Albert, A. Pohl-Koppe, J. Schubert, L. Steimer, and B. Weissbrich, at the Institute of Virology, University of Wuerzburg.

REFERENCES.

Degoulet, P., Phister, B., Fieschi, M. , 1996, „Introduction to Clinical Informatics (Computers in Health Care)". Springer-Verlag, Berlin.
Drazen, E.L., Metzger, J.B., Ritter, J.L., Schneider, M.K., 1994. „Patient Care Information Systems Successful Design and Implementation." Springer-Verlag, Berlin.
Fagan, M., Perreault, L.E., 1990, „The Future of Computer Applications in Health Care." In (Medical Informatics: Computer Applications in Health Care.)" In Shortliffe, E.H., et al. 1990, Addison-Wesley..
Shortliffe, E.H., Perreault, L.E., Wierhold, G., Fagan, L.M., eds., 1990, „Medical Informatics: Computer Applications in Health Care." Addison-Wesley, New York.
Willard, K.E., Hallgren, J.H., Connelly, D.P., 1999, „W3 Based Medical Information Systems vs Custom Client Server Applications." University of Minnesota Hospital and Clinic Proceedings Web Site: http://www.ncsa.uiuc.edu/SDG/IT94/Proceedings/ MedTrack/willard/UMHC_www/UMHC_www_paper.html
Wojtkowski, W.G., Wojtkowski, W., Wrycza, S., Zupancic, J., eds., 1997, „Systems Development Methods for the Next Century." Plenum Press, New York.

The following articles all appear in „Patient Care Information Systems Successful Design and Implementation." Springer-Verlag, Berlin, (Drazen, E.L., et al. 1994).

Drazen, E.L., 1994 „Physician's and Nurses' Acceptance of Computers." InDrazen, E.L., et al., Springer-Verlag, pp. 31-50.
Metzger, J.B., 1994, „The Potential Contributions of Patient Information Systems." InDrazen, E.L., et al.. Springer-Verlag, pp. 1-30.
Metzger, J.B., Teich, J.M., 1994, „Designing acceptable Patient Care Systems." InDrazen, E.L., et al., Springer-Verlag, pp.83-132.
Metzger, J.B., Marwaha, S., 1994, „Redefining the Patient Care Information System." InDrazen, E.L., et al., Springer-Verlag, pp. 187-212.

The following articles all appear in „Systems Development Methods for the Next Century." Plenum Press, New York, (Wojtkowski et al. 1997):

Cheung, K.S., Chow, K.O., Cheung, T.Y., 1997, „A Feature-Based Approach for Consistent Object-Oriented Requirements Specification." In Wojtkowski et al., Plenum Press, pp. 31-38.

Eriksson, I., Nissen, H.-E., 1997, „Information Systems for Responsible Actors: Basic Concepts for Understanding and Action." In Wojtkowski et al., Plenum Press, pp. 391-408.

Fritzgerald, B., 1997, „System Development Methodlogies: Toime to Advance the Clock," ." In Wojtkowski et al., Plenum Press, pp. 127-140

Kirikova, M., Grundspenkis, J., 1997, „Organisational Defenses in Requirements Engineering." In Wojtkowski et al., Plenum Press, pp. 227-240.

Kirveennummi, M., 1997, „Objectives and Reality: User Participation in Information System Development." In Wojtkowski et al., Plenum Press, pp. 115-126.

Krogh, B., 1997, „Modelling the Dynamics of Cooperative Work Arrangements." In Wojtkowski et al., Plenum Press, pp. 61-74.

Nagasundaram, M., 1997. „Creativity, Group Support Systems, and Systems Development." In Wojtkowski et al., Plenum Press, pp. 39-48.

Paetau, P., 1997, „Use of Object-Orientation for Information Systems Development in Software Companies in Finland: A Proposal for Study." In Wojtkowski et al., Plenum Press, pp. 23-30.

Post, H.A., van der Rijst, N.B.J., 1997, „A Dynamic Approach in Information System Development." In Wojtkowski et al., Plenum Press, pp. 545-558.

Pereira, R.E., 1997, „Organizational Impact of Component-Based Systems Development". In Wojtkowski et al. 1997, Plenum Press, pp. 241- 257.

Shaw, T., Pereira, R.E., 1997, „The Sources of Power between Information System Developers and End Users: A Resource Dependence Perspective." In Wojtkowski et al. 1997, Plenum Press, pp. 463-471.

Vogel, D., van Genuchten, M., Rodgers, T., 1997, „Team Support for Software Development." In Wojtkowski et al., Plenum Press, pp. 49-60.

THE CONCEPT OF IMPROVING AND DEVELOPING AN INFORMATION SYSTEM FOR THE NEEDS OF MANAGING A POLISH ENTERPRISE

Adam Nowicki

Faculty of Management and Computer Science, Wroclaw University of Economics, ul. Komandorska 118/120, 53-345 Wroclaw, Poland
E-mail: nowicki@han ae.wroc.pl

ABSTRACT

The article presents a concept of improving and developing an information system (IS) of an enterprise. The concept was implemented in many companies in Poland. The considerations are divided into four parts. In the first part the important role of an IS for managing an enterprise is stressed. The second part explains the essence of improvement of the IS where improving and developing activities were identified. In the third part the assumptions of the concept of improving and developing of an IS are stated. The fourth part presents a model of improving strategy of an IS. The model includes the phase of formulating the strategy and the phase of its execution.

1. INTRODUCTION

In theoretical and practical economy Information Systems play a very important role in management of an enterprise. IS perform a variety of tasks and functions. Among other things Lucey (1995) and O'Brien point that IS:

- are the main source of information that makes it possible to perform activities that form the current situation and future perspectives for an enterprise,

- provide interactions between Management System and Performing System (manufacturing and services),

- have an influence on the cost level of a business, making a feedback which allow to perform decision correcting activities or allow communication between senders and recipients of information,

- help to develop competitive products and services which can provide strategic advantage for an enterprise,

- are a „treasure" of every enterprise because they help to find innovative physical resources (row and processed materials and energy).

Actually, every Information System is the image of the enterprise. Information System running smoothly and efficiently ought to provide the recipient with necessary information in a proper form about past, current and future condition of the enterprise. The information is necessary to make proper decisions. Applying

Systems Development Methods for Databases, Enterprise Modeling, and Workflow Management,
Edited by W. Wojtkowski, *et al.* Kluwer Academic/Plenum Publishing, New York, 1999.

367

information for the purpose of managing depends on execution of information processes which takes places in a particular system. Correctly organized Information System can function only when there are clearly settled rules of information processing. Different Information Systems have been created that allow increasing of abilities of Information System in the area of resolving complex problems of management. Various computer technologies, hardware and software solutions and particular managing strategies made the Information System continuously improving and developing.

Problems of improving and developing IS are fairly new in Polish economy.

Political and social changes that began in our country in 1989 caused that centralized management system forced by socialistic economy and lasting for 45 years was replaced with free-market economy. From the moment of introduction of new principles of free-market economy, traditional Information System had to be changed. Access to modern technologies, barrier avoidance between different countries and large demand for new, innovative products forced companies to introduce deep structural, proprietary and organizational changes. Realization of the changes pointed onto important role of IS in achieving goals of an enterprise.

The improvement of IS begins to be steered into introducing new generations of IS that support making decisions. New classes of systems are created: Management Information Systems called Managerial Team Supporting Systems and Decision Support System. Applications running under network operating system Novell NetWare and DOS operating system. They are created mainly in dBase standard and written in Clipper, C and Pascal programming languages. Systems run under Windows NT and OS/2 appears on the market. Applications client-server become more and more popular. Many computer companies start to produce Management Information Systems based on database management systems such as Oracle, Informix or Progress. More information on that topic can be found in the article Kosiak and Nowicki, (1995).

Recently, the market of hardware and software has got a stable shape. A lot of important worldwide or domestic known computer companies start to collaborate with many economical sectors. Economical growth causes companies to think about comprehensive solutions instead of temporary solutions. Some examples could be Integrated Information Systems for supporting management, manufacturing and distribution which were put into practice, in MRP (Material Requirement Planning) and MARP II (Manufacturing Resource Planning) standards. Examples of modules servicing managing and manufacturing are shown in Table I.

The aim of the article is to present a concept of improving and developing an IS, which was put into practice in different enterprises in the 90s (Nowicki 1997). The result of the concept is the model of improving and developing an IS. The considerations regarding the model will be preceded by explanations of the essence of improving and developing an IS. At the end of the article the features of the presented strategy will be specified.

2. THE ESSENCE OF IMPROVING AND DEVELOPING INFORMATION SYSTEMS

Looking at experiences of western countries it can be noticed that improving and developing IS is considered mainly in aspects of designing IS, strategic planning, solving business problems and making organizational changes in an enterprise. The following authors present such a view in their publications: Awad (1988), Boar (1997), Curtis (1989), Lucey (1995), O'Brein (1993), Ward and Griffiths (1996) and Wang (1995).

The practical experiences reveal that **improving and developing IS** is a process made up of deliberate functional and structural changes. As a result of these

Table I. Examples software modules in different classes of Information Systems existing in Poland in the 90's

CLASSES OF SYSTEMS	EXAMPLES SOFTWARE MODULES
- DATA PROCESSING SYSTEMS (DPS)	Material Economy (ME), Finances and Accountancy (FA), Staff and Salary (SS), Costs (C), Trade Turnover (TT), Transport (T)
- MANAGEMENT INFORMATION SYSTEMS (MIS)	Firm CSBI-PRO/MIS system, Firm TETA-MIS module, FIRM SIMPL-MIS module
- DECISION SUPPORT SYSTEMS (DSS)	Standard MRP (Material Requirement Planning) e.g. firm JBA-21 system, CSBI-MFG/PRO, SAP AG-R/3 system, Tetra-CHAMEleon, IBM-MAPICS system, ICL-MAX, Siemens Nixdorf - COMET sytem
- OFFICE AUTOMATION SYSTEMS (OAS)	Microsoft Office, LOTUS NOTES

changes the system achieves higher levels of parameters that provide achieving goals of an enterprise. The parameters are typical quality and quantity characteristics such as system memory, reliability, efficiency of generating information, data processing productivity, efficiency and costs of running the system.

The process of improving and developing IS consists of activities that can be divided into two categories:

- modernizing activities,
- developing activities.

The modernizing activities include continuous analysis of the system, finding and removing faults and weak points of the system and substituting some parts of the system by others, so it could manage to perform certain tasks.

Current state of computing technology forces two ways of improving the system:

1) organizational and functional activities,
2) technological activities.

The main directions of organizational and functional activities include:

- making the system more flexible and open to changes in management system and organizational structure of an enterprise,
- processes of integration of function, modules and databases of information systems,
- creating efficient communication between spread organizational units of an enterprise,
- making processes of searching and finding information for different end-users of managerial team more efficient,

- increasing safety of the data protection stored, processed and transmitted in the system.

- Technological activities have a specialized character because they deal with hardware and software solutions. They include:

- current upgrading of hardware and networks,

- introducing new versions of system software, software tools and application programs,

- modifying telecommunication among different organizational units,

- applying widespread area network technologies for outside communication inside a particular organization,

- creating standards of formats of stored data and procedures in communication system,

- choosing effective tools for data protection by applying proper protection systems.

The second type of activities plays a very different role in the process of improving IS. As a result it causes a development of the system.

Developing activities are connected with introducing structural and functional solutions which were not applied in the past. The solutions are the reaction for a very fast progress in science and technology.

The result of the developing activities is introduction of new methods of their aggregation and disaggregation and introduction of new ways of processing and transmitting information. New generations of systems are created which have better functional parameters then systems previously used. For instance, apart from Data Processing Systems (DPS), Office Automation Systems (OAS), Computer Aided Manufacturing (CAM), Computer Aided Design (CAD) or Decision Support Systems (DSS), Executive Information Systems (EIS) and Expert Systems (ES) are introduced. Introduction of developing activities prepares an enterprise to achieve its strategic goals in new technological circumstances supported by advanced information technology.

3. CONCEPT ASSUMPTIONS

Improvement of an Information System means creation of new concepts that can be achieved as a result of updating and developing activities. These goals were defined in point 2 in this article. Their specification is the result of applied research formulation. Consequently the progress of the research determines methods, principles and activities of the process of improving the system. Therefore we can say **that the improvement is a creative process which happens according to particular strategy**.

The choice of proper strategy of improvement of an Information System is a result of a process, which consists of four levels. The levels are:

- knowledge and vision,

- strategic thinking,

- strategic research,

- strategy.

Figure 1. Levels of strategic depiction of improving of a system.

The mentioned levels are shown in Figure 1.

The presented levels occur in Boar's model of strategy (1997, p.111). In the model the **strategy** is inspired by visionary strategic thinking and implemented through strategic planning built on a solid foundation of knowledge. This approach makes up a basic assumption of the presented concept of the improvement of an IS.

Knowledge and vision are basic to formulate the strategy of the system improvement. Knowledge reflects intellectual abilities of the research team. The knowledge allows the research team to create the vision of the system, i.e. the image of the system from the perspective of the time. For such a system we have to initially determine the values of its utilitarian parameters that will bring its required state nearer to its ideal state - without faults and failures. The vision of the future system should be the result of carefully planned processes of rising the efficiency of strategic management and active adaptation of the enterprise to conditions signalized by weak signals from the environment. The imaged system should be adequate to the vision of the development of the whole enterprise.

Strategic thinking is the ability of the research team to achieve planned future solutions of the system. First of all the system ought to support the enterprise in keeping up the certain position on the market. This requires solutions from the range of information technology. In the process of strategic thinking we should plan modernizing and developing undertakings for the whole enterprise, going consequently through all functional domains, manufacturing, marketing and so on. The planning sets for members of the team clearly settled goals and methods for their realization. Moreover, the planning makes easier prediction of future problems and allows their solving in the correct time.

Strategic research aims at the recognition of problems of improving and specifying of predicted structural and functional changes in the Information System. The usefulness of the research expresses in the way of identification of information needs and the area of communication between particular levels of the management. According to target of the research, financial and perceptual possibilities of the team we can choose the set of methods and technologies specifying in the best way the current and required state of the system. An important element of the research is preparation of multivariant solutions of changes in the system. Information needed for these goals is supplied by the situation analysis of the system. This kind of analysis considers failures of the system, assessment of applied technology and the chances of the technology to change economic processes.

Execution of the research and analyses mentioned above shows the start of creative processes that heads for specifying of the strategy of improving an Information System.

The strategy of improving an Information System (IS) is the specified set of means of making structural and functional changes in the system for the purpose of management.

The contents of the improving strategy of the system are activities including the formulating of the strategy with specifying its goals, types and factors of their selection, and its research and organizational execution together with fixing costs and effects of these activities. In that aspect the strategy points on organizational-research procedures allowing to achieve planned goals. The goal of the strategy is to lay directions of modernizing and developing activities. The effects of strategic activities is the updating and development of the system. The changes can be carried out in different ways, so we can say about different strategies of improvement.

4.   STRATEGY MODEL

To treat the strategy as a concept of improvement we have to stress that it is connected with activities that change the structure and functions of the IS. The activities have a functional character and describe dependencies stemming from a chosen research procedure. The matter of description of dependencies are economical events and phenomena which reflect the IS. The common way to find different dependencies is modeling.

The models implemented in the improving of an IS can concern the description of the subject of improvement and variants of modernizing solutions created during the improving process (see point 2 of this article).

Improvement strategy modeling comprises all procedural activities that are applied by the research team. On the other hand, the model of a **strategy of improvement** consists of a set of activities including the phase of formulating strategy and the phase of its execution. Graphical representation of the model of a strategy is shown in Figure 2.

From the phases shown in the diagram stems the following observations:

Phase I. **Formulating the strategy** concentrates on recognition of the problems of improving the Information System. Their solution is passed to the executive team, which fixes the diagnosis of functional position of the enterprise and carries out situational analysis of the system. This phase aims at specifying the goals of the strategy, its types and to making the choice the final strategy. This part of strategy includes the recognition, analytical, assessive and creative processes.

Phase II. **Realization of the strategy** determines modernizing and developing activities that consists of designing and preparation of executive documentation. The final steps of that phase is a preparation of detailed plan of designed activities, analysis of accounting of costs

and effects and assessment of the risk of selected strategy of improving the Information System for managing of the enterprise. The end of putting into practice the modernizing and developing activities means   solution of the problems of improvement and achieving the goal of the strategy, i.e. introducing new changes and improvements both functional and structural.

The basic moment of transition from the phase of formulating strategy to the phase of its execution is the assessment of selected strategy by the managerial team of the enterprise. If achieved results of the decisive choice are satisfying, the following step is the acceptance of specified arrangements. In the opposite case all elements in the area of formulating the strategy of the improvement of the Information System have to be verified. New information needs appear in the meantime on particular levels of the management create renewed strategic approach. The necessity of their solving is a beginning of the next strategic approach.

5.   CONCLUSIONS

Social and political changes taking place in Poland require modern management of a company.

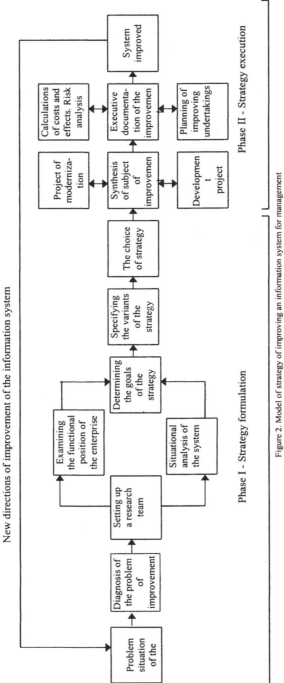

New directions of improvement of the information system

Phase I - Strategy formulation

Phase II - Strategy execution

Figure 2. Model of strategy of improving an information system for management

Source: Own research

373

Acquired research experiences show that functions and goals of management system can be achieved by a continuous process of improving and developing the IS in a company. This process has to be carried out according to a certain strategy.

The strategy of improving and developing an IS presented in this article is characterized by the following features:

1) Strategy expresses certain attitude that has different methodological origins and different practical applications.

2) Model of strategy has a procedural shape and includes activities from the area of its formulation and realization.

3) Realization of the strategy means implementation of new directions of structural and functional changes of the Information System.

4) Strategy is connected with achieving goals including inner elements of the system that are designated to modern management, to react to outside signals appearing in the enterprise environment.

5) Strategy requires resources that are necessary to its implementation. The resources are result of decisions making by the enterprise management and the team executing undertakings from the area of improving the Information System. Formulating the strategy and its realization requires constant engaging of the managerial team in the process of effective and efficient managing of the enterprise.

REFERENCES

Awad, E.M.(1988): Management Information Systems. Concept. Structure and Applications. The Benjamin Inc., California.

Boar, B.H. (1997): Strategic Thinking for Information Technology. Wiley, Inc., New York.

Curtis, G. (1989): Business Information Systems. Analysis, Design and Practice. Addison – Wesley Inc., Wokingham, England.

Kosiak, M. and Nowicki, A. (1997): Development of Management Information Systems in Poland. [W:] Systems Development Methods for the Next Century. Proceedings of The Sixth International Conference ISD'97, Boise, Idaho, USA. Plenum Press, New York.

Nowicki, A. (1997): Research program of developing and improving an information system in an enterprise. [In:] Scientific works of Academy of Economy in Wroclaw. Informatyka ekonomiczna nr 1/ 1997, Wrocław.

Lucey, T. (1995): Management Information Systems. The Gurnsey Press Co. LTD., London.

O'Brein, J.A. (1993): Management Information Systems: A Managerial End User Perspective. IRWIN, Inc., Illinois.

Ward, J. and Griffiths, P. (1996): Strategic Planning for Information Systems. J.Wiley Inc., New York.

Wang, R.Y. (1995): Information Technology in Action. Trends and Perspectives. Prentice Hall, Englewood Cliffs, New Jersey.

THE IMPLICATIONS OF CHANGING ORGANISATIONAL STRUCTURES FOR THE FUTURE OF THE INFORMATION SYSTEM (IS) FUNCTION

Robert Moreton and Myrvin Chester
School of Computing and IT, University of Wolverhampton,
35/49 Lichfield Street, Wolverhampton, WV1 1EL, U. K.
Ph.: +44 (0) 1902 321462, Fax: +44 (0) 1902 321491
E-mail: r.moreton@wlv.ac.uk; M.F.Chester@wlv.ac.uk

Abstract
There are a number of aspects of recent changes in organisational structures that have profound implications for the future of the Information Systems (IS) function. This paper considers the issues which must be addressed if IS staff are to play a full part in the future development of organisations. Issues addressed include: a planning framework for the IS function itself, aligning the organisation of the IS function with the business processes, and the additional knowledge and skills required by IS management and staff. The paper contends that, by applying general management principles, IS management can maximise the contribution of teams and individuals, and that by this means, the IS function, as a whole, will be able to make an effective contribution to the process of organisational change.

Keywords
business process orientation, organisation of the IS function, change management, motivating potential of IS jobs, organisational transformation

Introduction

There are a number of aspects of recent changes in organisational structures that have profound implications for the future of the Information Systems (IS) function. These include:
- The change in structure from a functional orientation to a business process orientation.
  - The changed focus in system design so that the IS/IT requirements are treated as part of a larger system that has behavioural as well as economic aims and objectives. Such projects are multi-disciplinary and business led.
  - The requirement to see changing business needs as an opportunity for market advantage, with its implication for the full support of systems through their entire life cycle.
  - The tailoring of jobs both to meet the preferences of individuals, and to exploit their strengths, which implies flexibility in system design and operation, as well as additional attention to the individual needs of users.

For IS staff to play a full part in the future development of organisations, the function needs to participate in the following:
  - the formulation of the business vision and alignment with it of IS/IT strategy and plans;
- the design, provision, and support of business processes and their constituent systems in order to enable higher levels of customer satisfaction;
  - the design of jobs so that the employee contribution to the business is maximised in both the short and longer term;

Systems Development Methods for Databases, Enterprise Modeling, and Workflow Management,
Edited by W. Wojtkowski, *et al.* Kluwer Academic/Plenum Publishing, New York, 1999.

375

- the process of managing change.

The IS function also has the main responsibility for determining and providing the information technology infrastructure needed to enable the applications and services that are required. Against this background, this paper discusses a framework for the IS function to use in planning their contribution to future IS development.

1 The planning framework for the IS function itself

In their book, Moreton and Chester (1997) argue that the three key components for success in IS development are alignment of plans with the business vision, commitment and support to the vision and the plans that spell out the steps to its realisation, and competence/mastery of all of the tasks associated both with change and with the exploitation of the new capabilities.

They argue that from the IS department perspective, there are four important communities within the business:

-        There is an important impact on the relationship between the IS function and the board, since this is the management body that will lead the integration of business, organisation and information technology strategies in the formulation of the business vision.

-        The IS department will have new and changed relationships with the 'owners' of business processes and with functional managers respectively.

-        'Users' are another interest group. By users, we mean any individual for whom information technology enables a significant part of their job.

-        Within the IS function itself, which needs to change in a number of respects, if it is to play an effective part in the change processes.

Rockart and Hofman (1992) have attempted to provide a framework for addressing the issues involved in improving the development environment, which is particularly relevant to a transforming organisation. They argue that the *key business process* of the IS function is the systems development process, and that changing the way systems are developed, to support the organisation, is itself a major *business process redesign*. Rockart and Scott Morton (1984), adapting from Leavitt (1965), developed a conceptual model of five organisational components which must be kept in balance when introducing change to an organisation: technology, corporate strategy, organisational structure, managerial processes, and individuals and their roles. The basic premise is that changes in one aspect must be balanced by changes in other aspects for the organisation to remain effective. The model can be used to discuss the organisational changes that are necessary to introduce IT-related changes into an organisation successfully.

In this model, Rockart and Hofman acknowledge that *culture* pervades all the other components. Schein (1985) defines culture as the 'basic assumptions and beliefs shared by members of an organisation'. It relies on, and is built from, the policies, procedures, and methods that are associated with 'the way things have been done' in all five elements of the model. The traditional IS culture, with its strong technical orientation, is not appropriate for the types of development processes and applications required in organisations for the future. Butler and Fitzgerald (1998) argue that for the IS function to play a significant role in facilitating organisational change, it has, itself, to undergo considerable transformation in terms of IT strategy, IS infrastructure and processes. Interestingly, Schein believes that major cultural changes can only be accomplished through extremely effective, almost charismatic, leadership or by changing the key personnel involved.

2 The aims of the IS function and the steps necessary to achieve them

Although IS departments will each need to consider carefully the 'target' position for which it is aiming, taking into account the situation in its own business, it is possible to provide some generalised comments. We have based these comments on the lessons derived from a series of interviews with around 30 European companies, and have related them to the different user groups.

2.1 Working with the Board

We have found that different organisations can have very different relationships with their Board, and these relationships are partly a consequence of the systems track record, partly related to Board members' understanding of the potential contribution of information technology, and partly related to cultural and personal empathy.

The main aim is to achieve a situation in which the information technology strategy is aligned with a vision that takes appropriate account of information technology, and then to maintain that position. To achieve this the following steps are necessary:

- IS directors need to determine the relative importance of information technology as a factor for their businesses.
- The Board needs to understand the potential contribution of information technology to the business and the nature and scale of the investment required.
- Assuming that information technology is either 'very important' or 'crucial' to a business, the IS director *must* be closely and continuously involved in the process of generating the vision. This will depend on both the standing of the IS function with the Board and on the quality of the educational programme.
- Although the creation of the vision is a process that the chief executive will lead, most businesses also find it helpful to have a senior manager in the role of facilitator. This is a role that some IS directors are well equipped for, particularly if they have business knowledge, and skills and experience in the management of change, in addition to information technology expertise.
- In addition to this deep initial involvement in the creation of a vision, and the progressive alignment of information technology strategy with it, the IS director must strive continuously to keep the information technology strategy co-ordinated with it.

2.2 Working with middle management

The organisation structures, jobs, and systems that enable the aims of improved organisational effectiveness to be achieved will be devised and implemented by teams that are led and facilitated by the middle managers who are responsible for the business processes. Functional managers will also have an important part to play since they will be responsible for professional standards and practices and possibly for resourcing.

The main aim is to provide systems that enable the business to achieve the desired outcomes - in both behavioural and economic terms - over the entire life cycle. To achieve this, the following steps are necessary:

- Projects need to identify, at the outset, both the behavioural and economic objectives of the system, and these need to be in line with those defined by the vision.
- Systems teams need to work with other specialists, middle managers and users to identify (and possibly prototype) organisations and systems that can meet these aims.
- Systems development and implementation plans need to take account of the probable steep learning curve for users, systems staff and other specialists and ensure that these plans are regarded as credible and achievable by the business.
- The application systems and supporting infrastructure have to be designed and built such that they can enable the business of exploiting change rather than being inhibited by it.
- Development projects and their teams need to be scheduled to support the business beyond the activities of initial implementation, and include support as the users become competent in the use of the system, and see opportunities to exploit the new capability.
- Support teams need to be organised to provide continuous assistance, as and when this is needed, to cope with and take advantage of change.

2.3 Working with users

Virtually all office staff, of all levels are now, or soon will be, 'users'. The aims of organisations generally include a desire for positive employee benefits and in particular increased job satisfaction. As well as any merit this may have in its own right, it is also essential if the business is to achieve the adaptability and economic benefits associated business survival.

The main aim is to design jobs and systems in a coherent way. They must be able to be carried out effectively by the *individual*; and they must contribute to the achievement of the business vision (Zuboff, 1988). To achieve this, the following steps are necessary:

-       Job design and information system design (along with structures and reward systems) need to be regarded as different aspects of a larger system.

-       Systems teams must listen to users' information and job requirements, and not assume them from indirect fact finding - usually from supervisors or managers. Again, prototyping of jobs and systems can be helpful.

-       Systems need to be sufficiently flexible that they enable the job holders to adapt them their own style of working.

-       To provide job satisfaction, systems need to enable the job holder to work effectively.

3 Aligning the organisation of the IS function with the business processes

Increasingly, organisations are restructuring to serve their main business processes. In all of the businesses that we interviewed, functional departments remain and have responsibility for resource provision and allocation, and for the relevant professional and operational standards. Typically a small, highly expert staff group represents and develops the functional viewpoint, and provides relevant planning, and perhaps operation, of those elements of the business infrastructure that are in its domain. We believe that this is an organisational model that applies to IS as much as any other functional department.

3.1 Clarify the IS organisation and responsibilities

Two factors are of particular importance in determining the organisation of the IS function. First is the need to take account of the management style of the business as a whole. Second is the extent to which it is appropriate for some IS responsibilities to be formally devolved to users. To ensure appropriate alignment between the IS function and the business, it is necessary to match the IS management style with the management style of the business. Thus, in a heavily centralised organisation, in which head office has control over functional divisions, IS should be centralised so that the systems strategy is developed close to the business decision makers. At the other extreme, in highly diversified groups, such as conglomerates, where there is little common business between operating divisions, each business unit should be responsible for its own IT. In such an organisation, there is little scope for developing common systems. In each of these styles of business, it is important that IS staff can have contact with the real users of their systems. This contact can be more difficult to achieve in a heavily centralised company.

In practice, there are several reasons for adopting a more centralised approach to IS management than is suggested by group management style. One is the need to provide flexibility for the group's future organisational choices. Groups with divisions which replicate common business processes, such as multiple retailers, can benefit from co-operation and shared approaches to business ventures, for example, the use of shared store-card systems. This type of organisation might restructure so that, for instance, certain common functions such as personnel or marketing are put under one line manager. Autonomous development of IT systems might lead to the utilisation of mutually incompatible hardware or software, thereby constraining the potential for realigning the business units.

A second issue constraining the organisation of the IS function, is the trend to integrate systems, which involves the design of corporate databases to support a variety of business and decision-support applications. A further reason for adopting a more centralised organisational style is the scarcity of skilled resources. While some traditional technical skills, such as operational knowledge and third-generation language capability, are less in demand, there is a demand for new technical skills, for database design, fourth-generation language programming and network communication design. There is also a growing need for 'hybrid' staff with equal ability in the business and the technological environments. A more centralised organisational style will make it easier for organisations to provide suitable career structures for IS staff. Career planning needs to be centrally co-ordinated so that staff can move freely around the organisation, between business units, or between business units and the centre.

The final organisation issue arises from the need to support systems throughout their life cycle. Situating the systems staff in the business does much to foster such support, because the staff and application or maintenance teams work on those tasks that the business regards as important. One potential problem with this is an excessively short term planning horizon that limits the scope and resources for strategic developments. This is one reason why IS management need to ensure that they understand the way in which the business vision is developing, and are working to keep information technology plans in alignment.

3.2 Assess the current structure of the IS function

IS management should review the current structure to determine whether it is contributing to, or hampering, performance. To evaluate whether the IS function is achieving its goals, three criteria can be applied:

- strategic direction: does the IS function help the business to achieve its aims and objectives (both now and for the future)?

- effective contribution: is the IS function well regarded and does it support user needs?

- efficient delivery: does the IS function meet its operational and quality targets?

Positive answers to each of these questions will indicate that the IS function is well placed to support the organisation. Where the IS department is functioning in a strategic or turnaround environment McFarlan and McKenney (1983), and where it is perceived by line managers to be ineffective, the organisation must take urgent action to review the alignment of the IS function with the management structure of the business. As a minimum, good communication channels, at a senior level, must be established between the IS function and the line/business management.

A department that is operating in a factory or support environment must support its users and ensure that it is well regarded by them. In these environments, the IS workload is likely to contain a high proportion of system enhancements (that is, perfective or adaptive maintenance), and user-developed applications. The management of IS/user relationships is critical and user-support personnel should be located close to those users to whom they are providing support.

Where an organisation is satisfied that its IS function is operating as effectively as possible in its particular environment, it can then undertake a review of IS efficiency. (There is little value to be gained from enhancing productivity if the contribution of the department is seen by line managers to be 'peripheral' in helping them to achieve their goals.) This, of course, is not a one-off exercise, IS effectiveness and efficiency should be subject to frequent review.

4 Clarifying the user responsibilities for IS activities

The devolution of control for IS activities to users will take place as the IS function re-aligns its responsibilities in line with the business organisation. The devolution of control was given significant impetus with the proliferation of personal computers and client-server systems through the 1980s and 1990s. This devolution/distribution needs to be managed in such a way that order is maintained without initiative being stifled.

4.1 Responsibilities can be formalised by reference to three levels of applications

Responsibilities can be formalised by reference to the three levels of typical applications (Moreton, 1996). These systems can be classified as core (corporate), departmental (local), and personal.

Core, or corporate, applications are essential to the day-to-day operation of the business. In general they maintain and update the common corporate databases, and often provide a base for subsequent applications to use. These systems, which typically exploit database technology and process high volumes of transactions, require specialist skills if they are to be developed efficiently. These systems must conform with central policy guidelines to ensure that a coherent software infrastructure is maintained. Senior managers will take the lead in deciding what systems should be developed, and in managing their development and implementation. Responsibility for IS innovation is therefore shared between business managers and the IS department.

Departmental, or local, applications are used by a business unit or a department within a business unit. Their purpose is to achieve the particular objectives of that business unit, and they do not normally affect the day-to-day operations of other business units. It is appropriate, therefore, that business unit managers should have control over what systems are implemented. However, because the data and programs may be shared by other departments in future, the systems should conform to the corporate policies and guidelines established by central IS management.

Development of these applications is likely to be undertaken by the users themselves, utilising newer technologies such as end-user computing and office automation. Successful end-user computing requires a high level of control to be exercised by users. This is particularly significant for business-critical systems, for which speed of development and a close fit to requirements, rather than technical efficiency, are critical. The role of systems specialists with these applications is to provide education, support and guidance.

Personal systems are those developed for their own use by end-users on microcomputers, using a variety of tools and techniques such as word processors, spreadsheets and database management packages. These systems are entirely in the control of the users, and the role of the IS function is limited to providing them with approved packages, training in their use, and essential technical assistance. These systems, if successful, in time may well become business unit systems requiring access to corporate databases to enable users to manipulate the data locally. It is important, therefore, that end-users should conform to conventions, established by the IS function, for the company as a whole. Organisations, for example, might allow users to choose whichever personal computer they require but may only connect standard products to the corporate network.

With the division of responsibilities described above, most business unit projects can be developed by users, while the IS function concentrates on those that are shared by several business-units. Responsibility and accountability for performance are shifted to the users for all types of system. This places the onus on business management for the planning, use and control of IS systems within their organisation.

A matrix, developed by the Oxford Institute of Management (Feeney *et al*, 1987) can be used to define the respective responsibilities for system development of users and the IS function. The matrix takes into account the strategic importance of future systems developments and the maturity of the technology (not the stage of assimilation reached by the particular organisation) required for these applications. If the strategic impact of applications is assessed as *low*, and the technology required as *mature*, operational efficiency is the prime objective. IS specialists should be given responsibility for their development, with the user manager having ultimate responsibility for their implementation and exploitation. Where the strategic impact is *low*, but *new* technology is required, the technical risk is high and the potential benefit to the business very limited, and the application should probably not be developed. The combination of *high* perceived strategic impact and relatively *mature* technology means that the users need to be in real control of the systems strategy (the 'what'), while IS specialists control the methods (the 'how') of systems development. These could be corporate or business-unit systems. Applications that have a *high* strategic impact and use *new, immature* technology should be entirely within the users' control, with a singular focus on systems effectiveness. These are the departmental and personal systems.

5 IS management and staff need additional knowledge and skills

In section 2, we identified a number of key aims for the IS function. There is an implication that IS management must maintain effective communication with senior management, and that all aspects of the business vision - and the ways in which its components fit together and interact - need to be fully understood. If this appreciation of the way in which the business works is confined only to IS management there will not be any ability to deliver and support the systems that are needed for changes to take place. It follows that education and training are needed throughout the IS function if staff are to become competent in - and then master - their changed role and the demands it will make of them. They are likely to need education and training in four skills areas (Moreton, 1995):

- *collaborative working*, which requires the integration of IS staff with business staff for applications development. Hence systems staff must possess appropriate interpersonal skills, ability to work in teams, and skills in the use of tools for prototyping/Rapid Application Development. Joint development teams work best when they consist of a small number of staff who are dedicated to a task or project until it is complete;

- *the application of organisation and job design principles*, which involve specification of the human computer interface and socio-technical systems, development of job enrichment programmes, and opportunities for adaptive learning through job experience. Very often centralised technical departments exert their influence in a conservative direction, not just in choice of technology, but also in organisational and job design (Taylor and Katambwe, 1988);

- *change management*, which requires an understanding of social processes in organisations. As business environments become less stable, a key requirement is to achieve forms of organising which permit rapidity and flexibility of response within a stable culture. Consequently systems staff need to be adept at handling the 'political' climate in which these changes will be introduced. The ability of any organisation to exploit IT is determined by the pace at which change can be absorbed, this is often dictated by the speed and effect with which staff can be educated in the proper use and potential of available technology;

- *building adaptable systems*, which permit a continuous evaluation of needs. The utilisation of appropriate methods is one aspect, but the adoption of business drivers rather than technology drivers is another. At the same time, systems staff need technical skills to utilise appropriate software tools, and a co-operative and progressive attitude towards users is required.

Much of the process of change is complex, and as we explained earlier, it takes considerable time to agree and gain commitment to change on this scale. From an IS perspective it is likely also to require investment in new methods, result in changed priorities, and lead to a new form of organisation structure. Whatever the underlying attitudes, some of the additional knowledge outlined above, such as improved communication and a more collaborative approach, does not need to wait for a momentous business decision to proceed. The necessary learning can be rapidly implemented at a functional, team, and individual level.

6 Maximising the contribution of the individual

The importance of highly motivated staff in a successful organisation is reflected in the staff contribution to the IS function. Motivating staff means equipping them for the new roles that are emerging from the re-alignment of the IS function to the business, and taking positive steps to maximise the contribution of each individual.

6.1 Apply the general principles of staff management

There are four specific actions that can be undertaken by IS management to ensure individuals make the appropriate contributions to organisational objectives:

- *broaden the scope of jobs*. As the role of the IS professional becomes more diverse, it is necessary to broaden the scope of IS jobs. To develop the types of systems that are required for successful organisations, it will be critical for IS staff to have a good knowledge of the business. Collaborative working requires IS staff to have people-oriented skills. These requirements mean that IS staff must be able to operate much more flexibly than in the past. In response to these trends, there is a need to move towards a more 'hybrid' role, such as analyst/programmer, using modern development tools.

However, care must be taken in the unthinking utilisation of hybrid roles, as there may well be advantages in specialisation that should not be lost without careful consideration. As suggested by Chester (1992) and Chester and O'Brien (1995), the quality of the software product must not be adversely affected by the use of any tools, methods, or roles.

- *introduce job rotation.* Job rotation is a useful way of broadening the skills of the individual, increasing job interest, and improving motivation and productivity. Philips, the multinational electronics company based in the Netherlands, for example, provides positive encouragement for job rotation. The philosophy is one of encouraging change, fresh insight, and creativity, while trying to minimise 'ownership' or 'dedication' of systems. Turnover of IS staff at Philips is very low, at about two per cent per year.

- *provide a flexible career structure.* With companies moving towards a more distributed organisation for IS, career advancement for dispersed staff becomes one of the critical factors in achieving a successful devolution of responsibilities. To provide a flexible career path, systems managers must recognise the wider roles that are required of the IS and provide more scope for 'lateral' development. This would allow IS staff to seek career advancement within other areas of the business as well as in the IS department. The key is to provide a structured framework of suitable career opportunities recognising the potential value of both technical and non-technical skills. In many businesses, this will also require a change in recruitment patterns to ensure recruits have the personality characteristics that will allow them to operate successfully in broader business-oriented roles.

- *fit jobs to people.* In times of increasing staff shortages, greater flexibility can be obtained by fitting jobs to people rather than vice versa. This can be achieved by undertaking an audit of staff skills and expertise, identifying individual strengths and weaknesses, and to restructure jobs accordingly.

6.2 Wide variations exist in the motivating potential of IS jobs

The motivating potential of jobs within the IS function vary widely. These variations are due to the different nature of the work, the degree of scope and work variety and the level of work feedback. For example, studies by Couger and Zawacki (1980) and Couger and Colter (1985) indicate maintenance work has a significantly lower motivating potential than the others. The implication is clear: either tasks with low motivating potential should be minimised as much as possible at the individual level; or staff engaged full time in such tasks need to be carefully selected.

6.3 Personality differences affect job performance

There are significant differences in personality characteristics between IS staff and the population in general. This has implications for IS management, in terms both of matching individuals to jobs and of mixing personalities within teams.

Personality, which can be defined as the characteristics that determine how a person thinks and behaves, can be measured in a number of different ways. The Myers-Briggs Type Indicator (MBTI) has been used to represent four interrelated dimensions of personality (Lyons, 1985). It shows that, compared to the population at large, IS staff tend to be more introverted, intuitive, thinking, and judgmental. As a result, compared to the average, they are insensitive, short of communication skills, and 'loners', preferring to work by themselves rather than as part of a team.

Couger and Zawacki (1980) have developed two measures which provide further insight into the personality types of IS staff. These two measures are: *social-need strength* (SNS), which is a measure of the need of staff for social contact; and *growth-need strength* (GNS), which is a measure of an individual's need for accomplishment, learning and developing, and for being stimulated and challenged. These studies point to the need for IS staff to be trained in communication skills if they are to participate fully in collaborative working. At the same time, it is important to match individuals with a high GNS with a job that provides high motivating potential.

6.4 Provide a supportive work environment

There is evidence that the productivity of IS staff is considerably affected by the work environment. In an original study, De Marco and Lister (1987) reviewed the performance of over 600 system development staff. They compared the results of the top-performing and bottom-performing quarter of staff. Performance was measured as the time taken to perform a standard programming task. The top quarter, who reported much more favourable working environments than those in the bottom quarter, performed 2.6 times better than the bottom quarter. Although this study does not prove that a better working environment will help people to perform better, because other factors may account for performance difference, it does suggest that work environment is a factor which should be taken into account for effective staff performance and management.

7 Organising teams to ensure effective performance

Although it is normal for IS work to take place in teams, very often little true team work takes place in practice. Much of the work, particularly in the main system-build phase, can be made up from units of complete, self-contained tasks. Each task can be undertaken by an individual, with short, though critical, periods of communication between the individuals performing the tasks. It is possible, for instance, to identify the interactions that are essential between a programmer and the program specifier (Chester (1992) and Chester and O'Brien (1995)). This principle is analogous to those used for high quality system design work, which is based on low coupling (little dialogue between modules) combined with high cohesion (grouping together highly interrelated tasks). The main purpose of grouping individuals into teams is to ensure that everyone is committed to, and working towards, achieving the overall objective of developing a successful system. The key task for the team leader is to ensure that an individual's goals are aligned with those of the team.

7.1 Team composition is critical to success

Although much systems deveopment work can be accomplished by individuals, there are times when active team working is needed in every project, such as during the specification and design phases. Team composition and ensuring that the roles of the individual are clearly defined then become critical (after Belbin, 1981). White and Leifer (1986) noted that systems development work tends to become more routine in the later project phases. Hence, team composition requirements differ by development phase. Teams consisting of people with similar personalities work together best on simple routine tasks, such teams encourage co-operation and communication. Thus teams made up of people with similar personalities will be more appropriate during the later project stages, when the extent to which work is routine is the greatest. Conversely, teams made up of unlike individuals work better during the earlier stages of a project when the amount of routine work is smaller. Such teams are good for problem-solving tasks, and for tasks involving complex decision making, because the team members stimulate each other, producing a higher level of performance and quality. However, teams of unlike individuals can create a great deal of conflict, while teams of similar people encourage conformity, which can lead to non-productive activity if the team norms for, say, work output, quality and working practices, are not consistent with team goals. If individual and team goals are coincident, increasing cohesion leads to improved productivity. Cohesion reduces with increasing team size (CSC-Index, 1993).

7.2 Team leaders can strongly influence attitudes and performance

The role of team leaders in IS development is a facilitating one; it is primarily oriented to helping individual team members increase their personal reward and satisfaction by aligning individual goals with team goals and by ensuring that this relationship is well understood. Another important characteristic of team leadership is the flexibility to adapt leadership style to suit the circumstances of the moment. This is required as the project and team requirements change from phase to phase of systems development, and in order to handle individual team members and different types of conflict.

The implications for the IS function are that team leaders should be selected on the basis not only of their technical expertise, but also on their ability to co-ordinate and to motivate team members.

Summary

In this paper, we have considered the way in which future organisation changes will change the role of the IS function at the departmental, team and individual level. These changes place the same demanding requirements on the IS function as they do on the organisation as a whole. In order to succeed, the relationship of IS/IT staff and the 'users' must be positive, and the responsibilities of each must be clearly defined. In many instances, this will require additional knowledge and skills, and changed attitudes on the part of IS staff at all levels. However, as we discussed, by applying general management principles, IS management can maximise the contribution of teams and individuals. By this means, the IS function, as a whole, will be able to make an effective contribution to the process of organisational change.

References

Belbin, R.M. (1981) *Management Teams - Why They Succeed or Fail*. Heinemann.

Butler, T. and Fitzgerald, B. (1998) Enterprise Transformation and the Alignment of Business and Information Technology Startegies: Lessons from Practice. *Proceedings IFIP WG8.2 & 8.6, Information Systems: Current Issues and Future Changes*, pp393-416 Helsinki December.

Chester, M.F. (1992) Analyst-programmers and their effect upon software engineering practices. Dissertation for MSc. in Information Technology, University of Nottingham, Nottingham.

Chester, M.F. and O'Brien, M. (1995) Analyst-programmers seen as harmful to software quality. *Third BCS-ISM Conference on Information Systems Methodologies*, 1995.

CSC Index (1992) Rapid Application Development. PEP Paper 25, August, CSC Index, London.

Couger, J.D. and Zawacki, R.A. (1980) *Motivating and Managing Computer Personnel*. John Wiley.

Couger, J.D. and Colter, M.A. (1985) *Improved Productivity Through Motivation*. Prentice Hall, Englewood Cliffs, NJ.

De Marco, T. and Lister, T. (1987) *Peopleware: Productive Projects and Teams*. Dorset House.

Feeney, D.F., Edwards, B.R. and Earl, M.J. (1987) *Complex Organisations and the Information Systems Function - a Research Study*. Oxford Institute of Information Management, Oxford.

Leavitt, H.J. (1965) Applying organisational change in industry: structural, technological and humanistic approaches. In *Handbook of Organisations* ed. J.G. March Rand McNally.

Lyons, M.L. (1985) The DP psyche. *Datamation*, 15 August.

Moreton R and Chester M F. (1997) *Transforming the Organisation: the IT contribution*. Addison Wesley, London.

Moreton R. (1996) Managing information in a distributed systems environment, *Proceedings Second South China International Symposium on Business and IT*, pp981-993, Macau, November.

Moreton R (1995) Transforming the Organisation: the contribution of the IS function. *Journal of Strategic Information Systems*, pps 49-163 V4 N2.

McFarlan, F.W. and McKenney, J.L. (1983) *Corporate Information Systems Management*. Irwin.

Rockart, J.F. and Hofman, J.D. (1992) Systems delivery: evolving new strategies. *Sloan Management Review*, Summer.

Rockart, J.F. and Scott Morton, M.S. (1984) Implications of changes in information technology for corporate strategy. *Interfaces*, January-February.

Schein, E.H. (1985) *Organisational Culture and Leadership*. Jersey-Bass.

Taylor, J.R. and Mulamba Katambwe, J. (1988) Are new technologies really reshaping our organisations? *Computer Communications* V11, N5.

White, K.B. and Leifer, R. (1986) IS development success: perspectives from project team participants. *MIS Quarterly*, September.

Zuboff, S. (1988) *The Age of the Smart Machine*. Heinemann, Oxford.

MIDDLEWARE ORIENTATION[1]
INVERSE SOFTWARE DEVELOPMENT STRATEGY

Jaroslav Kral
Dept. of Software Engineering,
Faculty of Mathematics and Physics, Charles University
Malostranské nám. 25, 11000 Prague, Czech Republic
E-mail: kral@ksi.ms.mff.cuni.cz

Key words: Open information systems, large-scale software architectures, autonomous components, middleware orientation, federated information systems

Abstract: Large software systems tend now to be assembled from large autonomous units (autonomous components, AC) having many features of almost independent applications. The collaboration AC's is supported by a sophisticated middleware. The situation, when the system is assembled from large (black box) co-operating AC's is already quite common. Autonomous components are often complete or almost complete applications, e.g. legacy systems. The components can be developed and/or supplied by different vendors. Typical examples are an OLAP system or a workflow system integrated into an existing information system. AC's are used like services in human society. They work autonomously in parallel way and collaborate asynchronously like post offices and/or transportation firms in human society. It is if an activity requires a service it need not wait till the service is accomplished. The properties of such systems depend crucially on the infrastructure (middleware) supporting the collaboration of the components (data transmission, encryption, firewalls, addressing modes etc) and requires a new way of thinking - middleware orientation. This fact is not reflected enough by standard CASE systems. The construction of systems from autonomous components requires techniques different from the ones used in the development of individual components. Some examples of the application of middleware orientation are given. The examples deal with cases when the specification and design of middleware and the use of it is most important part of the software system development.

1. AUTONOMOUS COMPONENTS

The experience gained from the development of large systems supports the conclusion that class is a too small unit (see Finch 1998) to be used an universal building brick of the software architecture in the large. Class libraries are too huge to be used easily. The reuse of classes of object oriented programs is then limited and substantially less common than in the case of COBOL based systems which were usually assembled from large autonomous programs Finch (1998). It is very difficult to find a proper class

[1] Supported by the Grant Agency of the Czech republic, grant No. 201/98/0532

Systems Development Methods for Databases, Enterprise Modeling, and Workflow Management,
Edited by W. Wojtkowski, *et al.* Kluwer Academic/Plenum Publishing, New York, 1999.

385

in the ocean of classes and to use it properly. It could be often easier to develop a new implementation of the class than to find and understand an existing one. Moreover the use of small object methods is to some degree similar to the antipattern code cutting, Brown (1998). As such it has some well-known drawbacks.

The solution proposed by e.g. UML (1998) is based on the grouping of objects into packages (PA) with a narrow interface. There are techniques allowing dynamic replacements of the objects encapsulated in PA (see e.g. Plášil et al, 1998). PA should implement the 'complete' activities modeled by the technology of use cases, see Jacobson 1995 or UML. 1999.

Object technology tends to prefer synchronous collaboration of quite small units (classes, packages) based mainly on the (remote) method interception (RMI). This is a technique conceptually near to the technique of remote procedure call (RPC). System based on RMI tends to be one sequential program. Parallelism is treated as an added, not an inherent, feature. The design of the system views the system as a monolith, which can be later decomposed into parts running eventually on different computers. This attitude is especially apparent in the Rumbaugh (1991) methodology, the main predecessor of UML.

PA is a good concept. It reduces the interface size to be considered during the system design and therefore enhances reusability. It follows from the previous discussion that it is, however, not a good tool for the integration of large autonomous components (AC) delivered by various vendors. This is a consequence of the fact that the collaboration of packages in UML is still based on RMI.

Assembling of information systems from autonomous components is conceptually different from the object-oriented attitude based on classes and/or packages of classes. It is best visible in the case, when we need to build a truly open system allowing the integration and dynamic replacement of third party products and/or legacy systems. All the known successful solutions are based on autonomous components. The interface of AC's is defined via a specialized formal language. A good example is SQL used in data layer.

In the case of AC oriented development the properties of link and-or transport layers of the communication subsystem (CS) supporting the collaboration of AC's must be taken into consideration during the design and even during requirements specification phases (see the examples below). CS influences substantially the architecture of the system as well as their properties like system accessibility, infrastructure services like encryption, message logging and routing, and client services (e.g. browsing tools). The design is therefore middleware oriented.

The middleware oriented design (MOD) is not properly supported by present CASE systems. The visibility of middleware (MW) during specification and design phases changes substantially the set of activities in which the customer should take part as well as the priorities of the activities. The modeling power of structured models (like data flow diagrams, DFD) as well as object oriented techniques like UML (1999) are not sufficient. DFD assume static links between components, UML excludes DFD almost completely although some useful features of DFD (data stores, data passing) are not replaced properly by other concepts. Method call is not always a good interface for complex components requiring communication in the form of complex languages. UML and other object oriented methodologies tend to view the system as a static monolith that can be eventually later decomposed into components able to work in a distributed way. It is good for the design of some client-server architectures, but it is not appropriate for the integration of large components provided by third parties. It not suited well to the design of the client part almost independently on the server part (see e.g. the browser clients in Internet environment).

The specification of the system must therefore often include the design and/or the customization of the middleware. In this case we speak about inverse development strategy (IDS) as the emphasis is not at the development of components. Let us give some examples of inverse strategy, its necessity, and promises.

2. EXAMPLES OF INVERSE DEVELOPMENT STRATEGY

We shall show the advantages of the inverse (middleware oriented) development strategy We show that the use of it is often necessary and that it offers new engineering paradigms and/or opportunities.

2.1 Information system of the Czech Republic

At present there is a lot of incompatible information systems (IS) developed and used by various state organizations like tax offices, declare offices, local authorities, etc. This system was criticized, as many data has to be entered repeatedly and are not accessible enough. Data are not generally accessible for citizens as well as for the state authorities. It was planned to design a new state information system as a monolith from scratch. The monolith was intended to be decomposed later into components supporting particular offices. This attempt has failed and it was OK that it happened. Let us give some reasons for it.

1. The monolithic system is very expensive. Its development would last of several years due to the known dependencies between the system size, development effort, and development duration, see e.g. Boehm (1981). During the development time the existing information system (IS) would have to be used. It was very likely, that many parts of the developed system would become obsolete during the development time.
2. There would be problems to convert data and activities of existing systems (the danger of big jump).
3. The responsibility for the correctness of data and/or operability of the particular systems should be delegated to individuals and organizations producing and using the data. Their primary interest is to have correct and actual data and an operable system. The interests of the others are not so strong. The use of almost autonomous subsystems must be therefore preferred, as particular offices having the primary interests could 'own' their data and information (sub)systems.
4. The individual offices can and must guarantee security and correctness of the data. It is easily accessible if the subsystems are autonomous as much as possible.
5. The learning curve during the system introduction can be less steep (it causes less frustration of the users).
6. It can be shown that the decomposed solution could be cheaper, Král (1981), even in the case when no third party components are integrated.
7. The heads of offices need not worry about their positions and/or power. We cannot neglect the influence of lobbies and or long term experience

of officials with their particular system. So it is good to integrate the existing subsystems if it makes sense (and it is often the case, Warren, 1999).

8. Business process reingeneering can be processed gradually. The whole system is scalable and modifiable.

It is clear now that the following requirements must be met.

- The existing information (sub)systems must be integrated into the resulting system as such with the exception of gates and/or services allowing to make certain services and/or data generally accessible for authorized people.
- The system must be designed in such a way that almost independent development of information subsystems is possible.
- As the requirements of outsiders on IS can vary substantially the subsystem interface should be easily modifiable.

The problem of state information system is therefore to a high degree the problem of the design of communication infrastructure between components offering:

1. Modifiable access of all citizens to the system.
2. Collaboration between autonomous subsystems under the condition that the individual components can be added and/or modifies quite freely.

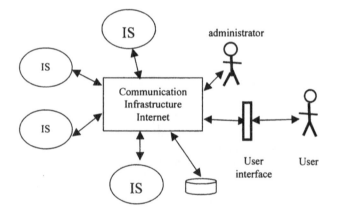

Fig. 1. The structure of state IS.

3. Security services (access rights, cryptographic services, rules of data replications, etc).
4. General and powerful message formats and standards for (meta)data communicated between information subsystems.

It is not difficult to see that the communication subsystem must fulfil further requirements like the data replication, general firewall tool, data warehouses etc.

Further analysis implies the use of Internet with some tools allowing flexible definitions of message formats, presentation tools designed by the system administrator. The possible structure is given in the fig. 1. The architecture allows to negotiate cooperation in a decentralized way, see Král (1999) for details.

2.2 Information system of a municipal office.

A large Czech town wants to modernize its information system (IS). Several IS's are in use. A tender was declared. It was intended to modernize and/or automate some activities (e.g. workflow control) not automated satisfactory yet. The system should use an existing communication infrastructure.

The properties of the infrastructure were not specified enough in the tender requirements. The integration of the existing information subsystems was not specified enough in the responses to the tender, as the responses were oriented towards newly installed components only. The possibility to coordinate the design of the communication subsystem to be compatible with the future solution of state (governmental) information system was not considered as well as the properties of the interface to town (and to the state at the same time) system for citizens. Traditional way of projection and development seems to bring substantial losses and/or dissatisfaction with the future system.

This was caused by the fact that the properties of the middleware were on included in the system requirements. It not too difficult to see that software houses need not be too interested in the openness of systems. It could increase the freedom of customers, they need not use products of the same supplier.

2.3 Development and testing of a mission critical system

We show in this example that the collaboration of autonomous components supports interesting paradigms. Assume that a control system of a mission critical system (e.g. atom plant control software) is to be developed. The conditions are:
- The control software must be developed without experiments with the controlled system,
- The software must be prepared to react on the disasters and accidents which can be tested virtually only.
- The reactions of the control software must be not only correct but also in time as the system has real-time features (see fig. 2).

A well known solution (see e.g. Král (1998), chapter 11, see also Král (1986,1991) is based on the following turns.
1. The drivers implementing the communication with the controlled system are encapsulated in a separate program called *I/O Drivers* (a process in the sense of operation systems).

The program *Control Logic* implements control logic. It communicates with *I/O Drivers* via an asynchronous message passing system (Fig. 2a, Fig 2b). It is obvious that the control logic can be decomposed into several programs communicating via PP as well. The development of the system can then be incremental. Communication with the components not implemented yet can be replaced by a dialog with the operator or with a prototype of the missing component. This is a very effective prototyping technique. The operator interface must be, however, configurable. The architecture of the system is shown in fig. 2.

To test the system off line we can redirect the messages from *I/O Drivers* to a program *Simulator* simulating the controlled system. This is not difficult, but it has some drawbacks.
- If we want to redirect the messages, some places of *Control Logic* must be changed. It can cause errors

– The message routing cannot be changed dynamically while the system is running. It disables some interesting techniques.

The passed messages should be logged in a file for future analysis. It is better not include this service into Control Logic. The solution in the case that the response times are not too short can be the following. All messages between *Control Logic* and *I/O Drivers* are sent via *Post Program* acting like a post office. The identifier of the addressee augments every message. The identification of the addressee is based on its logical number and a table maintained by the system administrator. *Post Program* (PP) logs the messages together with the time when the message is sent.

Technically the PP sends the message to the subsystem determined by addressee identifier using a table T establishing the correspondence between addressee identifier and the target subsystem. Routing of the messages is performed by the change of the contents of T. It can be done dynamically. Prototyping of AC's can be based on the "replacement" of the communication with the component by communication with operator supported by a simple syntax directed tool transforming binary data into character form.

The technique can be used as a infrastructure supporting the decomposition of the application logic into several AC's. This turn is shown in the fig. 2 d)

The system we are discussing was successfully implemented in a flexible manufacturing system. The communication between components was based on certain services of an operating system, so the system was not completely open. On the other hand it was very stable and it is still in use. The lifetime of it is more than ten years.

It clear from this example how important is the existence of free powerful globally used middleware. Such a tool allows all the above mentioned techniques to be used but adds substantially new features, especially teh possibility to design open modifiable systems with dynamiacally changeable functions.

2.4   Architecture of a commercial system

The Lawson Software system is an example how is the design of CS influences basic properties of the generated system. Several years ago Lawson Software was faced with the problem to add to its system the connection via Internet, workflow systems facilities, and an OLAP analytic tool. LS itself developed Internet based flexible communication system LOGAN. It was one of the first satisfactory systems of such a type. It was quite successful, Howlett (1996).

Workflow system FileNet was bought and attached via a newly developed messaging system called WOMA (Open Workflow Messaging System). The link is so flexible that it allows a tight collaboration with the LS financial modules. It allows e.g. activity based costing. The system File Net can be easily replaced by another workflow system. The analytical engine performing online analytical processing (OLAP) was bought from the firm EssBase. It can be replaced by a system from an another firm. The structure of the system is given in Fig. 2. Note that it is similar to the final structure from fig. 1. The internal LS message passing bus is conjectured on the basis of the system behavior. It was not fully disclosed. It probably enabled an easy implementation of an another technique called drill down - drill around offering an integrated graphical user interface to several databases and systems.

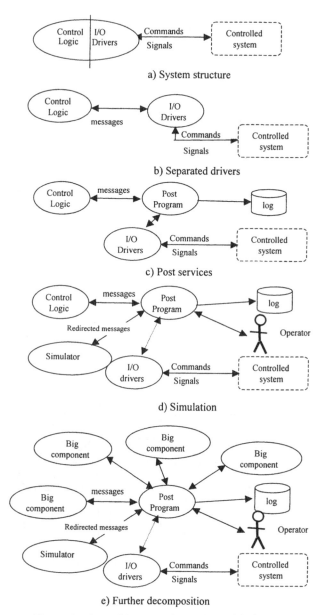

a) System structure

b) Separated drivers

c) Post services

d) Simulation

e) Further decomposition

Figure 2. The architecture of a mission critical system.

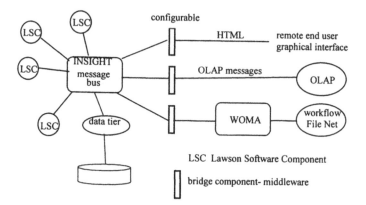

Fig. 3. The structure of Lawson INSIGHT.

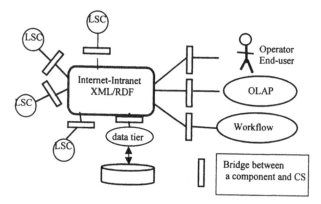

Fig. 4. The possible Internet solution.

One of the pleasant effects of the discussed solution is the possibility of simulation of future changes starting from current data and performing all actions corresponding to expected or planned development. Note that the properties of Lawson INSIGHT depend heavily on its architecture. The architecture of the system is formed by a group of modules or almost independent applications connected via an infrastructure using the technology of message passing systems communicating via some message passing bridges.

The port to Internet is configurable by a tool that can be used by end users. The solution from fig. 3. has the drawback that it contains several different CS and many different ports. Generally $m(m-1)/2$ ports are needed to interconnect m components. If we use one CS with common message format we need m ports/gates only. The format can be defined in metadata expressed in XML or RDF languages. In this case the system could have the architecture from fig. 4[2]. This philosophy is used in the new system Lawson INSIGHT II.

3.   ATTRIBUTES OF MIDDLEWARE ORIENTATION

The above given examples show the communication system cannot be treated as a service which need not be a topic of the requirements specification and/or design. The above given examples indicate that the required software architectures have a lot of common features. Let us summarize the substantial features of communication subsystems in our examples.

1. Asynchronous peer-to-peer collaboration of autonomous components acting in parallel like the services in human society.
2. Dynamic change of addressees. System should enable an easy replacement of components. Very important is the possibility to find the addressee according to the format or contents of the message.
3. Flexible user interface supported as a part of communication system in our sense. The basic properties of the interface should be defined in the system design.
4. Many functions of the communication system administration should be specified in system requirements. This is e.g. the case of message routing.
5. The online administration tools

It is important from the methodological point of view that many problems are passed from individual components to communication systems (message formats, data security, universal user interface tools etc.). In other words, if we use the terminology of OSI standards, CS must cover not only the line and transport layers but also the presentation layer and some functions of application layer.

There is yet another problem. On of the most difficult problem of the requirements specification is the decision what is to automate and what not Král (1998) Equally difficult seems to be the decision what to solve in components and what to require from middleware.

---

[2] Many other new versions of commercial customisable information systems use a similar philosophy.

4.   IMPLEMENTATION ISSUES

4.1   Middleware, components, and metadata.

The freedom in the choice of addressees implies the necessity to standardize formats of communicated messages and/or data as all the potential addressees must accept message formats a to some degree understand their semantics. The attempt to standardize data formats (e.g. for invoices), i.e. to fix the formats, is not too promising. To explain the problem in details let us discuss the experience with EDI/EDIFACT standards defining formats of (economic) documents like invoices etc. There are substantial problems Moudrý (1998) if such formats are to be used in a worldwide company. The contents of documents with the same purpose, e.g. invoice, differ in various states due to the local laws and/or habits. There are even differences between different enterprises. The attempt to unify the documents is then a hopeless task.

More promising is the solution not to standardize data built to standardize metadata (data on data) and to transport metadata together with data. Moreover it is good to offer a flexibility of metadata by providing a metadata definition tools, i.e. to provide metametadata facilities. The advantage is not the high flexibility only. There are no commonly used different de facto standards of metadata system. The only widely used system having some features near to desired metadata solution is the Internet HTML protocol. It is good that it is a de facto standard. We are therefore not limited by obsolete or too limiting existing solutions.

This fact was taken into account by the W3C consortium. The successor SGML of HTML is proposed in such a way, that it can be used as metadata system. There are two subsets of SGML proposed as metadata systems. XML, Extensible Markup Language XML (1999) and RDF (1999), resource definition frameworks.

A closer look on these two standards shows that they are something like programming languages. Instead sending the data only the data are sent inside a program in the form of constants. Execution of such a program implies the intended interpretation of the transported data. RDF is intended to be a metadata specification tool. The XML solution offers something more than merely a metadata. The consequences of it are not clear yet.

The advantage of XML and RDF is the possibility to use Internet as a transport and a presentation layer. This observation has a very important consequence for the identification of the possible addressees of the message. There are two promising ways how to find an addressee. Both suppose that each autonomous component offers a syntax parser of acceptable messages. A component is identified as an addressee of a message if the message has syntax structure acceptable by the component.

A more refined identification of addressees can be based on the execution of XML/RDF programs, i.e. analyzing not only the syntax but to some degree the values of transported data.

The discussed addressing schema can be used to support the dynamic reconfiguration of the system. The addressee can be found on WWW and the used in plug and play style. The identified component can be used as it is (in situ) or copied somewhere according some further criteria. This is a problem to be solved in software agent systems like Aglets (1998) by IBM. Similar problems are solved in multiagent systems Fisher et al. (1999).

4.2   Dynamic configuration

The identification or looking for the addressee based on the syntax analysis of the messages can be implemented as a method/service offered by

the components. The two main strategies are possible:
- The potential addressees of a message of certain type are at first identified using some form of the broadcasting, the subsequent messages are send to the detected components. The administrator of the system can support the choice of the components. This is a substantial requirement, as we cannot always suppose that the identification of the message destination can be completely automated.
- The addressee is searched dynamically for every message. This solution is general but expensive and insecure. The addressees can be mobile and the addressing schema can involve the analysis of the values of the transported data. We can use some turns proposed in CORBA.

Middleware oriented system integration is very simplified and can be to a high degree automated and/or performed by the users. The system integration schema in fact replaces configuration management.

5. CONCLUSIONS

The inverse development strategy is not supported by the by standard CASE tools. DFD is too static to model the dynamic changes of the configuration of the system. It to some degree holds for OO models. The binding based on method interception tends to prefer synchronous collaboration. If we analyze the methods proposed by e.g. Rumbaugh we see that the system is designed as a monolith and then decomposed. The packages in UML are used as white boxes. So the OO models are not suited well to design open systems.

Inverse design of IS is based on the concept of big autonomous components integrated as black boxes. The components collaborate asynchronously as a peer to peer network of parallel activities. The collaboration is based on a rich communication language. The components can by easily replaced by other ones. Missing components can be easily simulated. This is a very effective way of prototyping. New components can be easily integrated. It supports incremental development and/or enhancements with its well-known advantages.

The system is scalable. The most important is that it is possible to integrate, even dynamically, the third party products and legacy systems. Inherent limits of customized systems led to the renaissance of legacy systems, see Warren (1999).

The design of communication subsystems must be part of the system design. New paradigms and techniques are needed for it. These paradigms are substantially different from the paradigms known for the development individual autonomous components. New modeling tools are needed.

On the other hand many questions to be solved in middleware oriented systems have counterparts in many software engineering problems. Examples are software agents, e.g. Aglets (1998), the protocols and design of communication systems, distributed databases, data replication, see Král (1999a) for details.

REFERENCES

Aglets, 1998, http://www.trl.ibm.co.jp/aglets/index.html. *URL on Aglets*, the IBM variant of SW agents

Booch, G., Rumbaugh, J., Jacobson, I., 1995, *The Unified Method for Object-Oriented Development , Version 0.8*. Metamodel and Notation. Relational Software Corporation.

Brown, W.J., 1998, Antipatterns. Refactoring Software, Architectures, and Projects in Crisis, John Wiley & Sons.

BaaN, 1999, http://www.baan.com. Home page of BaaN company.

Boehm, B. W., 1981, Software Engineering Economics, Prentice Hall, 1981.

CACM, 1995,The Promise and the Cost of Object Technology. A five years Forecast, Comm. Of ACM, Vol. 38. Oct. 1995.

Finch, L., 1998, http://www.ddj.com/oped/1998/finc.htm So much OO, So Little Reuse, Dr. Dobb's Web Site, May 1998.

Fisher, K., Oliveria, E., Štěpánková, O., Multiagent Systems: Which Research for which Applications. To appear.

Howlett, D., 1996, Lawson Adds Web and Workflow. PCUser, 16 Oct. 1996.

Jacobson, I., et al, 1995, Object Oriented Software Engineering: A Use Case Driven Approach, second printing, Addison Wesley, 1995.

Král, J., Demner, J., 1991, Software Engineering, (in Czech), Academia Praha, 1991.

Král, J., 1986,Software Physics and Software Paradigms, 10th IFIP Information Processing Congress, Dublin, North Holland, 1986.

Král, J., 1996, Object Orientation in Information Systems. A useful but not Universal Tool, Proceedings of ISD96 Conference, Gdansk, 1996.

Král, J., 1998, Information Systems. Design, Development, Use. Science Veletiny, 357pp. In Czech.

Král, J., 1998a, Architecture of Open Information Systems. In *Evolution and Challenges in Systems Developments*, (Wojtkowski, W.G., Wrycza, S., Županič, J., eds.), 7th int. Conf. on Information Systems, Bled, Slovenia, Sept. 21-23., 1998. Plenum Press

Král, J., 1999, Middleware Orientation Can Simplify the Development and the Future Use of the Czech State Information System. To be presented at the conference *System Integration'99*, Prague.

Král, J., 1999a, *Open Problems of Middleware Oriented Software Architectures*, to appear.

Lawson Software, 1997, *Lawson Software Documents*, 1997.

Lawson Software, 1999, http://www.lawson.com.

Moudry, J., 1998, Private Communication, 1998.

Plášil, F., Bálek, D., Janeček, R., 1998, SOFA-DCUP: An Architecture for Component Trading and Dynamic Updating. *Proceedings of ICCDS'98*, May 4-6, 1998.

RDF, 1999, http://www.rdf.com/.*Resource Description Framework*. A Proposal of the W3C consorcium.

Rumbaugh, J., Blaha, M., Premerlani, W., Eddy, F., Lorensen, W.,1991, *Object Oriented Modeling and Design*, Prentice Hall, 1991.

SAP, 1999, http>//www.sap.com. *Home page of SAP company*.

Serain, D., 1999, *Middleware*, Springer V.

UML, 1999, http://uml.systemhouse.mci.com/artifacts. *Home page containing links to the proposed standard of UML*.

Warren, I., 1995, *The Renaissance of Legacy Systems. Method Support for software system Evolution*, Springer V., (1999).

W3C, 1999, http://www.w3.org. *Home page of W3C consortium*.

XML, 1999,.http://www.xml.com/xml/pub/ *Extensible Markup Language*. A proposal of W3C consortium.

SELECTED DETERMINANTS OF APPLYING MULTIMEDIA TECHNIQUES FOR DESIGNING DECISION SUPPORT SYSTEMS

Celina M. Olszak
Faculty of Management
Katowice University of Economics
Bogucicka 3, 40-266 Katowice, POLAND

1. INTRODUCTION

The creation of computer decision support systems (DSS) is characterised by constant efforts to adopt procedures that will help decision makers with more effective use of their memory and certain abilities including, inter alia, perception or creative problem solving, and with transition to more extensive and differently organised knowledge. The designing of such procedures constitutes one of the most challenging areas as far as the development of decision making tools is concerned [17].

Dynamic development of multimedia technology evokes reflection on its potential for designing DSS. Multimedia play a significant role in controlling cognitive processes of a human being. They refer to a mental model and a natural way of human perception. They facilitate faster communication, learning, expansion of knowledge, exercises in imagination, and as a result more effective process of the decision making [2, 7, 18]. Therefore, the designing of a multimedia interface that corresponds closely with a mental model of a user seems to be particularly important. Some interesting results have been accomplished within the framework of cognitive psychology. Such knowledge has also provided the author with some inspiration for presenting a conceptual interface by means of multimedia techniques.

2. COGNITIVE PSYCHOLOGY IN THE THEORY OF DECISION SUPPORT SYSTEMS

Psychological issues have frequently been raised in relation to creating computerised decision support systems. The greatest achievements in this field belong to H. Simon and A. Newell who, on the basis of empirical research, formulated perceptive structure of a brain. Their theory resulted in some methodological consequences in both research on the processes of problem solving and computer decision support systems [12, 16]. The fascinating results of neurophysiological research are worth mentioning here. They enabled to ascertain that the left cerebral hemisphere is so-called "rational" or "cold" hemisphere, responsible for logical thinking and language speaking. It is facts and sharp contrasts orientated and it works in an analytical way. It is based on the sequential ("step by step") and short-term ("now and here") processing. It is focused on fragmentary images of reality. However, the right cerebral hemisphere may be described as an expressive or "hot" one and it allows us to see images, feel emotions, etc. It works in a more intuitive and holistic way. The left cerebral hemisphere provides decision makers with abilities to make proper decisions in the situation of the lack or inadequacy of

Systems Development Methods for Databases, Enterprise Modeling, and Workflow Management,
Edited by W. Wojtkowski, *et al.* Kluwer Academic/Plenum Publishing, New York, 1999.

397

information. It is responsible for creating comprehensive structures evaluated in respect of their internal cohesion and completeness [9].

Former DSSs were, in many cases, orientated towards supporting a decision maker's left cerebral hemisphere. As a consequence they favoured a formulation of decision makers' empirical and logical attitudes, i.e. attitudes preferring facts, measurements, detailed evaluations – analysis by means of clear, hierarchical and centralised structure.

In cases of necessity for fast orientation in complex situations that are difficult to outline in terms of a quantitative form, a decision maker – a visionary - makes their appearance. They are able to evaluate situations globally, and create new ideas and strategies. This type of mind prefers graphic and analogue forms of presenting information and events. Decision makers who incline towards global understanding of a problem believe that every organisation is a "battlefield" characterised by decentralised (functional) structure that is to provide its all members with new opportunities for participation.

It is the research on human memory that casts some light on human being's cognitive processes. The discovery of two kinds of memory, long and short-term, is of fundamental importance for science. It is highly probable that long-term memory, being practically unlimited, constitutes the main store of human knowledge. Long-term memory is associative in its character. Processes of association originate among speaking, problem solving, etc. Information stored in this type of memory is not directly accessible. It has to be sought, recalled and activated. Images or real names are perfect "reminders". On the other hand, short-term memory is dramatically limited. However, its content is directly accessible. It is the major system governing processing of information and, as a result, decisional processes [16].

Analysing functions of former DSSs in the context of decision makers' capacity for the expansion of cognitive processes, it is possible to draw the following conclusions:

1. DSSs currently in use prefer decision maker's long-term memory. They do not support short-term memory and cause a decision maker to learn various names, procedures, etc.
2. DSSs frequently provide a decision maker with numerical information. They rarely utilise analogue code that would be familiar to a decision maker and could facilitate the implementation of a conceptualisation.
3. DSSs are orientated towards data files and models having nothing in common with the way of thinking and formulating decisions.

For the reasons mentioned above, it is necessary to consider how to design, in the context of complex problems of decision making, modern DSSs in order to ensure their better support for decision makers' cognitive and perceptive processes. It seems that this gap may be filled by a multimedia interface. This element is constantly underestimated by designers, although it constitutes an important bridge between a user and the rest of DSS.

3. MULTIMEDIA TECHNIQUES IN THE REGULATION OF DECISION MAKERS' COGNITIVE PROCESSES.

Multimedia facilitate activating human memory, evoking associations, remembering information, and finally generating new informational and decisional needs. They participate in the regulation of cognitive processes more actively than any other technique because they are orientated towards two levels of receiving information by a human being: conscious and subconscious. The former refers to the transfer of concrete information, data, facts, etc. The latter concerns the evocation of certain emotions, feelings, conditions and readiness to take particular actions. The above fact is proved by empirical research that emphasises a high correlation between the reception of information and human being's understanding of a problem, their behaviour and actions [11].

The analysis of research being conducted seems to suggest that multimedia participation in the regulation of cognitive processes may be resolved into three significant stages:

- cognitive – when a decision maker receives information,
- emotional – when new needs and ideas arise, and
- volitional (causative) – when problems are solved and decisions are made.
  It is expressed in, inter alia, the fact that multimedia:
- enables to compare and reorganise a huge amount of information in a clear and "friendly" form,
- enables to interpret information complexly – multimedia is orientated towards the synthesis of information, as opposed to former techniques that were focused on the data analysis (e.g. results interpretation in a multidimensional space),
- facilitates decoding, storing and retrieving huge amounts of information from a decision maker's long-term memory; the information is essential for the analysis of a decisional situation (the limiting of human short-term memory results in some difficulties in the interpretation of huge amounts of information),
- helps identify relations between positions of little importance (at a particular moment) to the analysed problem, and
- encourages to see a problem from different perspectives and in different ways.

Creative regulation of human cognitive processes by means of multimedia is expressed, above all, in numerous forms of communication offered. Great emphasis put on the multitude of information and intuitive access to resources of information distinguish multimedia from other forms [2, 14]. As a consequence, multimedia allows information to be quickly converted into knowledge that is later sent, processed and utilised for decision making. Specialists in interpersonal communication point out that the transfer of information in form of text influences the outcome of communication in 7%. Properly modulated voice reaches 38%. Visual communication achieves 55% [10].

4. AN OUTLINE OF MULTIMEDIA INTERFACE CONCEPTION

The designing of the user interface has always been treated with little attention, although it proves decisive for the utility of a generated system.

Some interesting results have been achieved within the framework of cognitive psychology. The defining, inter alia, of a user's mental model offers firm bases for working out solutions within software engineering. The definition of a modern user interface closely corresponds with the role that a user's mental model takes in the interaction with a computer system. The user sees the system (its structure and dynamics) via their mental model. That is why the most important objective is to design such user interface that would be as consistent with its individual mental model as possible [1].

A good user interface expresses information (stored in the system) in terms that are familiar to the user. Such result may only be accomplished by the careful design of various layers of the user interface. The physical interface is the "obvious" part of an interface, and it is probably for this reason that it tends to receive relatively more attention in professional literature. However, the conceptual interface is "less obvious". It is more dependent upon tasks of the system, user's style of work and their psycho-sociological qualities. The conceptual interface exists, as the mental model of a user does, although it is frequently difficult to reveal what it looks like and how it works. In many cases it is also hardly possible to maintain a clear distinction between the physical and conceptual interfaces. Icons, which are major elements of physical interfaces, are simultaneously major components of the conceptual interface, and they represent conceptual entities and relations. The appropriate interface contains objects that the user needs in order to express their intentions in an intuitive way. Such objects may be expressed by means of signs, symbols, sound, slide, etc. Thus the broad context of application and the way it appears to the user should be taken into consideration while designing user interfaces. The main task of the user interface is to take the application functionality and make it visible to and manipulable by the user. Visual languages, multimedia technology, 3D, associative style of hypertext and object technology facilitate the designing of such mental communication agent (interface) [6, 13, 15].

On the basis of presented knowledge on the user's mental model it is possible to distinguish the following levels of the user interface designing -fig. 1 [1]:

- level of tasks and targets of the system analysing (the analysis of users' tasks, their requirements and needs of perceptive processes support),
- lexical level (the analysis of user's language and applied metaphors),
- semantic level (the user's analysis of the model of the system perception),
- syntactic level (selection of the GUI standard and environment of the interface implementation), and
- level of presentation (creating application, selecting tools for input and output of information).

The first level of the interface designing refers to the stage of garnering information on potential users, i.e. their tasks, needs of perceptive processes support,

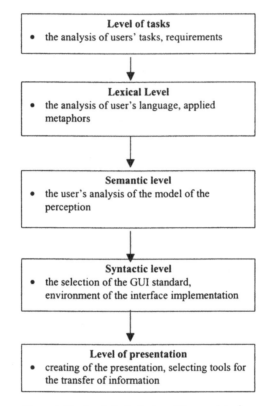

Figure 1. The levels in designing of the user interface

anticipation of needs, context of system usage, etc. This is the suitable moment to undertake e.g. preliminary analysis of types of mind on the basis of Jung's typology. A potential interface should always provide alternative techniques of supporting perceptive processes and communication with the system.

Within the framework of the lexical level it is necessary to focus on the description of the language and notions applied by the user. One should be aware that the user's vocabulary determines, to some extent, their thinking, categorisation of phenomena and objects. Certain groups of nouns, verbs and adjectives, that are most frequently utilised for the user's description of reality, may be distinguished and then associated with some symbolic references. At this stage it seems important to analyse metaphors and capabilities offered by the effect that evokes potential users' associations by means of applying appropriate symbols, sounds , colours, etc. Those elements constitute specific influence on the course and character of the whole process of communication with a computer system. Perception of colour, for instance, is one of the most subjective qualities of human perception and appears to be a relatively complex issue. The same colour or presentation background may evoke positive or negative reactions depending on a situation.

The semantic level should be responsible for modelling the interface conceptual structure, and as a result the required way of the user's perception of the system. The basic task is to create so-called user's views that determine their perception of the system structure and its dynamics – and simultaneously the interaction with the system. The user's view does not necessarily have to be identical to the model of a system but it has to be compatible with it. Since the objects are not exactly identical to those creating the system, the unquestionable relation of implementability projecting the states of objects of user's views onto the states of objects of the system must be established [10]. User's views should be created with particular regard for the results of the analysis of the methods of reality perception, and for metaphors and notions used by the user.

Created user's views become the basis for the technical design. Within the framework of syntactic level the standard of GUI and the environment of interface implementation should be selected in the first place. User's views are then transferred to designs of views of selected methodology of designing (e.g. object methodology). It works on the principle that the user's views are supplemented with some technical details associated with a selected environment, GUI and a programming language [10].

On the other hand, selecting media for particular objects presentation one should express their characteristics and stay as close to the original metaphors as possible. The implementation of computer techniques means that the transfer of information is limited, e.g. by the screen space. Hence, it is extremely important to create a strong, consistent layout and achieve a balance of the various elements on the screen. Well designed screen should be clear, it cannot provide too much information at a time, and nonessential elements should not take a more interesting form than central objects. Being able to utilise sound, animation, colours, gradients, contrasts, etc. properly is of great importance [5, 11]. Therefore, the designing of those elements should be supported by the analysis carried out at lexical and semantic levels.

5. CONCLUSIONS

Although the above reflections on the multimedia participation in the regulation of decision makers' cognitive processes are certainly far from being complete and perfect, they draw attention to the necessity of a more frequent usage of such techniques in designing DSSs. It seems particularly important to apply them while creating the interface for a decision maker. This area, so often neglected by scientists, deserves, according to the author, more attention and requires detailed and thorough research.

REFERENCES

1. Barfield, L., 1993, *User interface. Concepts & Design*. Addison –Wesley, New York.
2. Buford, J., 1994, *Multimedia Systems*. Addison-Wesley, New York.
3. Burger, J., 1995, *Multimedia for Decision Makers. A Business Primer*. Addison-Wesley, New York.
4. Cox, N., Manley, C., and Chea, F., 1995, *LAN Times Guide to Multimdia Networking*. Osborne McGraw-Hill, Berkeley, California.
5. England, E., Finney, A., 1996. *Managing multimedia*. Addison-Wesley, New York.
6. Galitz, W.O., 1996, *Essential Guide to User Interface Design*. Wiley Computer Pub.
7. Jeffcoate, J., 1996, *Multimedia in Practice*. Prentice Hall, New York.
8. Kaplan, R.M., 1997, *Inteligent Multimedia Systems*. Wiley &Sons, Inc. New York.
9. Kozielecki, J., 1977, *Psychologiczna teoria decyzji*. PWN, Warszawa.
10. Kuźniarz, L., Piasecki, M., 1998, "Zarys obiektowej metodologii analizy i projektowania multimedialnego interfejsu użytkownika." *In Multimedialne i sieciowe systemy informacyjne*, Cz. Daniłowicz, Wrocław.

11. Lindstrom, R., 1994, *Business Week Guide to Multimedia Presentations.* Osborne McGraw-Hill, New York.
12. Newell, A., Simon, H.A., 1972, *The Human Problem Solving.* Englewood Cliffs N.J.
13. Newman, W.M., Lamming, M.G., 1995, *Interactive System Design.* Addison-Wesley.
14. Nielsen, J., 1994, *Multimedia and Hypertext. The Internet and beyond.* Academic Press, New York.
15. Parsaye, K., Chignel, M., 1993, *Inteligent Database Tools & Applications. Hyperinformation Access, Data Quality, Visualization, Automatic Discovery.* Wiley & Sons, New York.
16. Simon, H., 1982, *Podejmowanie decyzji kierowniczych. Nowe nurty.* PWE, Warszawa.
17. Sprague, R. H., 1994, A Framework for the Development of Decision Support Systems. In *Management of Information Systems,* P Gray, W. King, E. McLean , H. Watson, ed. The Dryden Press, London.
18. Szuprowicz, B., 1997, *Multimedia Tools for Managers.* Amacom, New York.

TOWARD EFFECTIVE RISK MANAGEMENT IN IS DEVELOPMENT: INTRODUCTION TO RISK BASED SCORECARD

Piotr Krawczyk

Dept. of Computer Science and Information Systems
University of Jyvaskyla P. O. Box 35 SF-40351 Jyvaskyla, Finland
Ph.: +358 (14) 60 24 77, Fax: +358 (14) 60 25 44
E-mail: piotr@jytko.jyu.fi

KEYWORDS

IS Development; Risk Management; Balanced Scorecard

ABSTRACT

Though, both academics and practitioners, recognise the importance of risks related to information systems development (ISD) process, relatively little attention has been paid to this problem. In this paper we distinguish three types of ISD activity, representing three levels of IT concerns:

- the development of an organisation's IT strategy (the strategy level);
- the selection of a portfolio of IS projects (the planning level);
- the management of a system development (the management level);

Senior IT executives need to be aware of all these risks and their interrelationships, though they cannot be heavily involved with all of them. Responsibility for the various risks is probably best handled if the threats are evaluated by the stakeholders directly involved with them. We need guidance for how such a division could be developed and managed.This paper builds on premises that organisations should be planned and managed from multiple perspectives and it introduces the concept of Risk Based Balanced Scorecard.

INTRODUCTION

The chapter incorporates the description of three out of four levels defined in a holistic approach towards IT risk management by Ginzberg and Lyytinen (1995). The fourth level, called operational, is beyond the scope of this paper and represents the day-to-day management of an IT infrastructure.

Strategy Level

The business impact of a poor fit between business strategy and technology strategy can be assessed in terms of opportunity costs (options lost) to the organisation due to its inability to act (see, e.g., Clemons, 1995 and 1998). Henderson and Venkatraman (1993)

Systems Development Methods for Databases, Enterprise Modeling, and Workflow Management,
Edited by W. Wojtkowski, *et al.* Kluwer Academic/Plenum Publishing, New York, 1999.

403

have proposed that investments in IS development can be analysed using the theory of real options.

Another type of strategy level risk deals with the increased vulnerability of the organisation due to a high level of technological dependency. Perrow (1984) provides an insightful analyses of the ways that system complexity and tight couplings can result in unanticipated modes of operation and consequent performance failures.
However, little attention has typically been paid to risks at this level in either the practitioner or research literature.

Planning and Portfolio Level

At this level the major organisational risk deals with the difficulty of reconfiguring the hardware and software infrastructure in order to implement the organisation's technology strategy (Henderson and Venkatraman, 1992). Failure at this level may result in insufficient infrastructure capacity to meet growing demands for IT services, inability to deliver specific services, lack of integration among IS applications, poor data quality, high training costs.
There has been considerable attention paid to pieces of this level in IT management literature (e.g. Ward and Griffiths 1996; Ciborra and Jelassi 1994), including: mechanisms of analysing and managing risks of the overall systems portfolio and technology platform development (e.g., Elam et al. 1988, Chapter 5); developing and appropriate balance of risk in the portfolio of IS development projects (e.g. Mc Farlan, 1981); and, adjusting the architecture of choices to comply with these strategic planning contingencies (Earl,1989).

Management Level

The major risks at this level deal with the failure to design a system with the needed capabilities or to implement the selected system in a timely and cost effective manner. Strategic importance of critical technology projects, hence the problem of ability to manage them, is well described in (Willcocks et al. 1997). This level has been the focus of a large body of research, including work on software risk management and the substantial literature on systems implementation (Boehm 1988). Textbooks (e.g., Charette, 1989) and tutorials (e.g., Boehm, 1989) have been devoted to clarifying major risks in software development and to establishing mechanism for effectively dealing with them. A good summary and analysis of the major studies in this area has recently been completed by (Lyytinen et al., 1994). Overall, this level of IT risks has been much more thoroughly investigated and is better understood than two higher levels.

Interaction between levels- Holistic Approach

An important characteristic of the IT risk framework is the interactions between risk levels across the company structure (see ITRM model Figure 1). For example, a choice of a technology architecture at the Planning Level affects directly the formulation of the systems development projects and project level risks at the Project Level. These interactions, however, run in both directions and may affect multiple levels at the same time. Though not depicted in Figure 1, each level in the hierarchy should provide feedback to each higher level. Consequently, a failure at the management level may impact an organisation's thinking at the Portfolio or Strategy levels. These interactions have not been well documented, and are not completely understood. Yet, it is a key aspect of the CIO's responsibility to manage these interactions and to understand the associated risk impacts.

METHODS

A more detailed description of instrument validation can be found in Krawczyk's (1999a) article. We developed a generic model of risk related to ISD process and its management operationalized by 23 latent variables (see Figure 1).

ISDRM Model

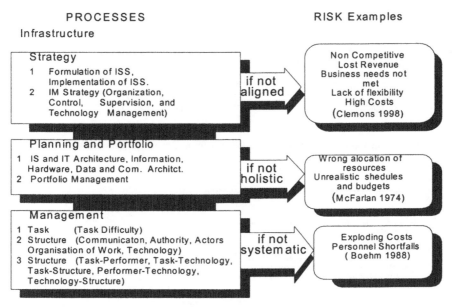

Figure 1 Risk Management in Information Systems Development Process

Measurement

Based on (Heikkinen 1996) thesis, a list of risk factors was elaborated by referring to over 20 publications on IT management threats (see Table 1).

Table 1 Risk Factors Itemised Across Four Levels of IT Management

Identified Risk Factors	Source by Author (for full description see list of references)
STRATEGY LEVEL	
Lack or Imperfection of IS Strategy	Earl 1989, p 67
IS Rigidity	Clemons 1995, p 65-66
Difficult Implementation of IS Strategy	Henderson & Venkatraman 1992, p 7
Faulty Organising in Information Management	Lacity & Hirschheim 1992, p 5
Inadequate Control /Supervision	Earl 1989, p 119, 159, 166 Ginzberg & Moulton 1990, p8
Inadequate Technology Management	Lyytinen 1994, p 11
PLANNING & PORTFOLIO LEVEL	
Lack or Imperfection of Information Architecture	Leppanen 1996
Lack or Imperfection of IS Architecture	Halttunen 1990a

	Ginzberg & Lyytinen 1995
Lack or Imperfection of Hardware Architecture	Ginzberg Moulton 1990, p 5, 6
Lack or Imperfection of Data Comm. Architecture	Turban, McLean &Wetherbe 1996
Lack or Not Systematic Portfolio Management	Earl 1989, p73, 74 Mc Farlan, 1981
MANAGEMENT LEVEL	
Task Difficulty	Beath 1983, p138 Lyytinen 1994
Communication	Mantei 1981
Authority	Fiegener & Coakley 1995 p 58
Organisation of Work	Margetts & Willcocks 1994 p 8-17
Actors Involved	Boehm & Ross 1989, p 902 Lyytinen 1994, p10
Technology	Lyytinen 1994, p11
Task-Performer	Lyytinen 1994, p11
Task-Technology	Lyytinen 1994, p11
Task-Structure	Mantei 1981
Performer-Technology	Lyytinen 1994, p11
Performer-Structure	Lyytinen 1994, p11
Technology-Structure	Lyytinen 1994, p11

Likert scale was applied upon each risk factor in form of two major questions (see Table 2).

Table 2

2.5 Management of The Information System Portfolio

a) Is there a systematic method in use for examining
the coverage of the systems portfolio in the organisation? y/n
b) What method ?_____
i.e. BSP (Business Systems planning) by IBM
Other: _____
c) Is there a systematic method for the evaluation of current system portfolio in terms of business
value added and technical quality improvement ? y/n
d) What method are used in evaluating/examining the systems portfolio?
i.e. (Systems audit grid)
Other: _____
e) What risks, might be triggered by lack of systematic IS portfolio management ?
_____

In your opinion, how important is the risk factor (Lack of systematic IS portfolio management) to your organisation?

1	2	3	4	5
meaningless	minor	mediocre	important	very important

How well do you think, in your organisation, the factor (Lack of systematic IS portfolio management), and risks it causes, are managed?

1	2	3	4	5
poorly	tolerably	mediocre	well	very well

Sample

In our pilot study we conducted 15 structured interviews with CIO's and senior IT managers in the years 1997/98 . From the database of 2080 Finnish CIOs and IT managers (over 90 percent of the population), addresses were chosen randomly, and an interview based questionnaire was sent to 500 of them (over 20 percent).

Design

In order to triangulate our research with independent views 23 software engineers were interviewed from companies subcontracting to the aforementioned organisations.

Content Validity

Based on Heikkinen's (1996) thesis, a list of risk factors was elaborated by referring to over 20 publications on IT management uncertainty (see Table 1). In order to check if the instrument measures cover all vital properties under investigation (Burt, 1976, p.4).

406

Construct Validity

Independent Samples Test was applied (see table 4 in the appendix) to check stability of measures across two different methodologies: interview vs. questionnaire data gathering (Campbell and Fiske, 1959; Cronbach, 1971).

Reliability

All measures but the first, related to the importance of identified risk factors at the strategic level (.52 Cronobach alpha), evidenced strong reliability and validity. Internal consistency reliability estimates using Cronobach's alpha ranged from .72 for evaluation of IT risk management performance at strategy level to .84 for evaluation of IT risk management performance at planning and portfolio level. However, it was an overall Cronbach alpha .91 for the 30 items across the four levels that showed strong reliability of the applied measures (Cronbach, 1951). The high Cronobach alpha gives empirical support for advisability of the holistic 4-level approach.

Sampling Bias

We obtained 5 percent respond rate (24 out of 500 questionnaires were returned with valid answers) comparing to 10 percent achieved by Ernst & Young (1998) Global Annual Information Security Survey based on the same CIO/IT Managers database. The higher respond rate of E&Y survey could be explained by the maturity of the instrument with regard to its content and structure, the portfolio of E&Y clients included in the sample, and the professional editing of the questionnaire form. We investigated carefully reasons for not responding. Direct e-mails were sent to 128 managers who failed to fill the IT Risk Management enquiry form .

RESULTS   (for descriptive statistics see Table 3 in the Appendix)

This is a short version of result analysis from research in progress described in Krawczyk's (1999b).

Contradictions that are inherent in the making of meaning, knowledge and decision allow the organisation to learn and adapt. Consensus about shared meanings enables co-ordinated activity, but divergent interpretations ensure robustness (Choo 1998).

Shared Understanding (CIOs vs. Software Engineers)

We present cognitive map of IT management competence based on relation between the level of risk importance per factor and CIO's ability to manage it (see Figure 2).

The survey results showed much agreement between CIOs and SE- Software Engineers on many areas of IT risk management. Both parties share understanding, and acknowledge independently, big risks related to the rigidity of IS strategy, strategic information management, and technology management (Ward J, Grffiths p. 1998 pp 359-441; Corder 1989.); the portfolio of IS and data-communication architecture; communication, authority, performer and structure at project level; as well as information security and practices and procedures at operational level.

Differences (CIOs vs. Software Engineers)

Some differences in IT risks assessment among the two groups may be explained by the heterogeneity of the actors involved (different status in the organisation, technical vs. business background, different levels and areas of job responsibility and distinctive to every individual "world of thoughts"). Characteristic pattern can be found in ITRM performance assessment. Software Engineers as an independent control group claimed that areas like strategy formulation (see e.g. Ward and Griffiths 1998 pp 490-539), strategic technology management, portfolio of hardware and data communication architecture (see e.g. Ross and Beath 1996), organisation of work (with underlying issues of communication within a team, use of authority and skills of project members) as well as general rules and procedures for

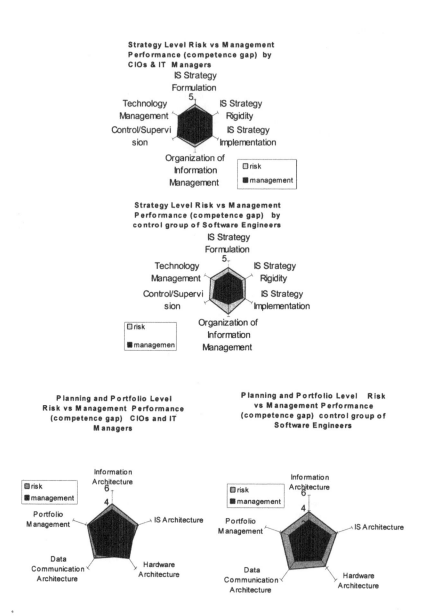

Strategy Level Risk vs Management Performance (competence gap) by CIOs & IT Managers

IS Strategy Formulation

Technology Management

Control/Supervision

Organization of Information Management

IS Strategy Rigidity

IS Strategy Implementation

□ risk
■ management

Strategy Level Risk vs Management Performance (competence gap) by control group of Software Engineers

IS Strategy Formulation

Technology Management

Control/Supervision

Organization of Information Management

IS Strategy Rigidity

IS Strategy Implementation

□ risk
■ managemen

Planning and Portfolio Level Risk vs Management Performance (competence gap) CIOs and IT Managers

Information Architecture

Portfolio Management

Data Communication Architecture

IS Architecture

Hardware Architecture

□ risk
■ management

Planning and Portfolio Level Risk vs Management Performance (competence gap) control group of Software Engineers

Information Architecture

Portfolio Management

Data Communication Architecture

IS Architecture

Hardware Architecture

□ risk
■ management

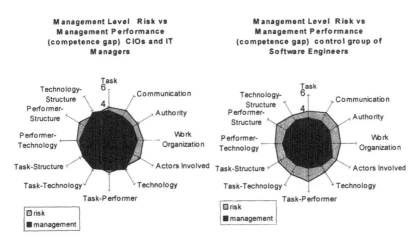

Figure 2 ISD Risk and Management Performance Across Three Levels (based on scales depicted in Table 2)

Table 3 Risk Based Balanced Scorecard (example)

Management Level

General:

Identifier:	1
Name:	Ambiguous quality level
Description:	The requested quality level is so ambiguous that it is impossible to understand what is to be reached.
Owner:	IT Manager
Classification:	Technology
Nature:	Specification
Domain:	Software

Causes:

Identifier:	Name:
1	Undecided customer.
2	Quality plan missing.
3	Uncomplete quality plan.

Impact:

Description:	Product quality is not adapted to user's needs (sub or over-quality).

Strategy:
Risk management strategy:
Actions:

Identifier:	Name:
1	Write a quality plan
2	Re-formulate quality requirements.
3	Give users a test version.

Table 4 Check List (sample)

Management Level

<u>Risk Definitions:</u>

☐ The requested quality level is so ambiguous that it is impossible to understand what is to be reached.

☐ Customer asks for changes which are not documented, treated and planned, do not lead to contract negotiation. Management is not kept informed.

☐ Data are numerous and complex, leading to performance problems in storage, treatment and exchanges.

<u>Cause Definitions:</u>

☐ Undecided customer.

☐ Quality plan missing.

☐ Uncomplete quality plan.

☐ Users' requests badly managed.

☐ Users remarks not known.

<u>Impact Descriptions:</u>

☐ Product quality is not adapted to user's needs (sub or over-quality).

☐ Every user's remark is considered as a request for change (project drifts in cost and time scale) or no request is taken into account (client is unsatisfied).

☐ Product performance during execution does not address requirements (problems due to inputs/outputs).

☐ Efficiency of resources is insufficient.

personnel, are seriously under-managed. From the risk perspective, the formulation and the implementation of IS strategy is of paramount importance to CIOs, whereas SEs were concern with complexity of structure-task-technology relations at management level (see Cash, McFarlan and McKenney 1992) and personnel related risks at operational level. Software engineers in contrast to senior IT managers seem to underestimate importance of IS strategy formulation . This may lead to serious breaches in IS strategy implementation process due to lack of commitment, shared understanding and mutual support for IT and Business strategy alignment. In fact 50 percent of senior IT managers reported lack of IT strategy as the major threat. 45 percent of CIOs reported lack of alignment between IT and business strategy as high risk , yet common problem.

CONCLUSIONS

The differences in risk and related management performance should be analysed and appropriate course of action should be assigned. Different types of cognitive gaps lead to the activation of different information strategies to bridge those gaps.

Risk Based Balanced Scorecard

A Balanced Scorecard process as defined by Kaplan and Norton 1996 provides a framework for thinking about performance planning and management from the multiple perspectives. We present simple mechanism for building and sharing organisational knowledge on risks related to ISD and ways to manage them   (see management level example Table 3 and Table 4). More precise measures of risks identified in ISD process can be elaborated along time and cost dimensions.

REFERENCES

Beath C.M. " Strategies for Managing MIS Projects: A Transaction Cost Approach. Computers and Information
Systems Graduate School of Management, University of California, Los Angeles.

Boehm, B.W. "A Spiral Model of Software Development and Enhancement", Computer IEEE, 1988.

Boehm B.W., "Software Risk Management" IEEE Computer Society Press 1989.

Boehm B.W., Ross R. "Theory of Software Project Management: Principles and Examples" IEEE Transaction on
Software Engineering 15 (7), 902-916, 1989.

Cash, J., McFarlan, W. And McKenney, J. 1992 "Corporate Information Systems Management", Irwin, Boston..

Charette R.N., " Software Engineering Risk Analysis and Management", McGraw-Hill, 1989.

Choo W.C. "The Knowing Organization: How Organizations Use Information To Construct Meaning, Create
Knowledge, and Make Decisions", Oxford University Press, 1998.

Ciborra C., Jelassi T. "Strategic Information Systems " Wiley Series in Information Systems 1994.

Clark D.D., Wilson D.W. " A Comparison of Commercial and Military Computer
Security Policies", Proceedings of the 1987 IEEE Symposium on Security .

Clemons E. "Using Scenario Analysis to Manage the Strategic Risk of Reengineering",
Sloan Management Review, 1995 pp.61-71.

Clemons E.K., Thatcher M. E., Materna R. " Strategic Implications of Infrastructure as
a Consumable Resource (Taking the "Real" Out of Corporate Real Estate), 1998.

Corder, C. "Taming your Company Computer", McGraw-Hill 1989.

Earl M. "Management Strategies for Information Tecnology", Prentice Hall, 1989.

Elam J., Ginzberg P., Zmud R. " Transforming the IS Organization, Washington , ICIT Press, 1988.

Fiegener M., Coakley J. " CIO "Impression Management- Problems and Practices", Journal of Systems
Management, Nov/Dec 1995.

Finne T. "A Knowledge-Based DSS for Information Security Analysis. Licentiate Thesis. Abo Akademi ,
Ekonomisk-Statsvetenskapliga Fakulteten. 1995

Ginzberg M., Moulton R. "Information Services Risk Management" ,
Proceedings of the 5th Jerusalem Conference on Information Technology, 1990.

Ginzberg M., Lyytinen K. "Information Technology Risk at the Top: The Definition and Management of Risk By
Senior Executives. A Proposal to the SIM International Advanced Practices Council. September 12 1995.

Heikkinen K. "Tietohallinnon Riskitekijat ja Niiden Arviointi" Master Degree Thesis, Department of Information
Systems, Unversity of Jyvaskyla 1996.

Henderson J. C. and Venkatraman N. 1992, Strategic Alignment: A Model for Organisational Transformation
through Information Technology in Kochan and Keen P.G.W 1991, Shaping the Future : Business
Design Through Information Technology, Harvard Business School Press, Boston MA.

Kaplan R.S and Norton D.P. "The Balanced Scorecard" , HBS Press, ISBN 0-87584-651-3,1996.

Kendall R.A.H " Risk Management for Executives (A practical Approach to Controlling Business Risks " Pitman
Publishing 1998.

Krawczyk P. (a) Towards Effective IT Risk Management Structures Across the Organization. Proceedings of 1999
Information Resources Management Association International Conference, Hershey, PA, USA, May 16-
19, 1999, Idea Group Publishing 1999.

Krawczyk P. (b) IT Risk Management as a Knowledge Domain. Proceedings of Workshop on Futures in
Information Systems and Software Engineering Research Stockholm, Sweden, April 23-24 1999.

Lacity M., Hirschheim R. "The Information Systems Outsourcing Bandwagon: Look Before You Leap". Working
paper. 1992

Leppanen M. "Contextual framework for the analysis and design of information systems development
methodologies", Manuscript, University of Jyväskylä, 1996.

Lyytinen K. Mathiassen L. Ropponen "An Organizational Analysis of Software Risk
Management Approaches", University of Jyvaskyla, working paper, 1994.

Mantei M. The Effect of Programming Team Structures on Programming Tasks". Communication of the ACM,
March , vol. 24, 3.

Margetts H., Willcocks L. "Informatization in public sector organizations: Distinctive or common risks?"
Information and the Public Sector , Vol. 3, No. 1 p 8-17, 1992.

McFarlan W.F. " Portfolio Approach to Information Systems" Harvard Business Review, 1974.

McFarlan W.F. " Portfolio Approach to Information Systems" Harvard Business Review, vol.59(4), pp 42-150,
1981.

Neumann P. "Computer Related Risks" ACM Press, Addison Wesley 1995

Perrow C. Normal Accidents: Living with High Risk Technologies, New York: Basic Books, 1984.

Ross J. W., Beath C.M. Goodhue D.L. " Develop Long-Term Competitiveness Through IT Assets", Sloan
Management Review 1996.

Turban E., McLean E.and Wetherbe J. " Information Technology for Management", chapter 9, by New York: John
Wiley & Sons, 1996.

Ward J., Griffiths P. " Strategic Planning for Information Systems" Wiley Series for Information Systems 1996.

Willcocks L., Feeny D., Islei G. " Managing IT as a Strategic Resource" McGraw-Hill , p 203-233, 1997.

APPENDIX   Table 3 Descriptive Statistics

Strategy Level   **Descriptive Statistics**

	N	Mean	Std. Deviation
IS Strategy Formulation -Risk	58	4.043	.813
IS Strategy Formulation Mngmt	58	3.259	.909
IS Strategy Rigidity -Risk	58	3.698	.873
IS Strategy Rigidity Mngmt	58	3.379	1.006
IS Strategy Implementation - Risk	58	3.724	.937
Management of IS Strategy Implementation	58	3.448	.877
Organization of Information Management - Risk	58	4.147	.806
Information Management	58	3.224	.961
Control/Supervision- Risk	58	3.862	.883
Control and Supervision Mngmt	58	3.534	.863
Technology Management - Risk	58	3.5862	.9135
Management of Technology	58	3.422	1.025
Valid N (listwise)	58		

Planning & Portfolio Level
Descriptive Statistics

	N	Mean	Std. Deviation
Information Architecture - Risk	58	3.9914	.7863
Management of Information Architecture	58	3.103	.877
IS Architecture - Risk	58	3.9483	.8619
Management of IS Architecture	58	3.276	.812
Hardware Architecture - Risk	58	3.7500	1.0055
Management of Hardware Architecture	58	3.741	1.069
Data Communication Architecture -Risk	58	4.2414	.6834
Management of Data-Communication Architecture	58	3.466	.898
Portfolio Mngmt - Risk	55	3.7273	.9370
Project Portfolio Mngmt	55	3.191	.895
Valid N (listwise)	55		

413

Management Level   Descriptive Statistics

	N	Mean	Std. Deviation
Task -Risk	58	3.9224	.8100
Task Mngmt	58	3.405	.861
Communication - Risk	58	4.1207	.7741
Communication Mngmt	58	3.267	.756
Authority Risk	58	3.9914	.8909
Authority Mngmt	58	3.241	.885
Work Organization -Risk	58	4.0259	.8294
Work Organization Mngmt	58	3.207	.982
Actors Involved -Risk	58	4.1810	.8721
Management of Actors Involved	58	3.362	.888
Technology Risk	58	3.7845	.9324
Technology Management	58	3.552	.793
Task-Performer Risk	57	4.0439	1.0189
Task-Performer Mngmt	57	3.360	.972
Task-Technology Risk	57	3.7018	1.0345
Task-Technology Mngmt	57	3.474	.928
Task-Structure Risk	57	3.8246	.9472
Task-Structure Mngmt	57	3.421	.963
Performer-Technology -Risk	57	3.9035	.7874
Performer-Technology Mngmt	57	3.447	.754
Perofrmer-Structure Risk	57	4.2895	.8070
Performer-Structure Mngmt	57	3.211	.861
Technology-Structure Risk	57	3.6579	.9872
Technology-Structure Mngmt	57	3.553	.742
Valid N (listwise)	57		

Table 4 Independent Samples Test for Construct Validity at Strategy Level

		Levene's Test for Equality of Variances		t-test for Equality of					95% Confidence Interval of the	
		F	Sig.	t	df	Sig. (2-tailed)	Mean Difference	Std. Error Difference	Lower	Upper
IS Strategy Formulation -Risk	Equal variances assumed	.071	.791	-.130	37	.897	-3.750E-02	.288	-.622	.547
	Equal variances not assumed			-.131	30.423	.897	-3.750E-02	.287	-.623	.548
IS Strategy Formulation Mngmt	Equal variances assumed	.089	.767	.000	37	1.000	.000	.309	-.627	.627
	Equal variances not assumed			.000	28.438	1.000	.000	.314	-.642	.642
IS Strategy -Risk	Equal variances assumed	6.732	.013	-.126	37	.901	-3.750E-02	.298	-.642	.567
	Equal variances not assumed			-.109	18.537	.914	-3.750E-02	.344	-.759	.684
IS Strategy Rigidity Mngmt	Equal variances assumed	1.766	.192	-.352	37	.727	-.117	.332	-.789	.555
	Equal variances not assumed			-.328	23.584	.746	-.117	.355	-.851	.617
IS Strategy Implementation - Risk	Equal variances assumed	.480	.493	-1.283	37	.208	-.379	.296	-.978	.220
	Equal variances not assumed			-1.239	26.544	.226	-.379	.306	-1.008	.249
Management of Strategy Implementation	Equal variances assumed	1.717	.198	-.056	37	.956	-1.667E-02	.298	-.620	.587
	Equal variances not assumed			-.061	36.790	.951	-1.667E-02	.272	-.567	.534
Organization of Information Management - Risk	Equal variances assumed	1.712	.199	-.070	37	.945	-2.083E-02	.298	-.624	.582
	Equal variances not assumed			-.074	35.263	.941	-2.083E-02	.280	-.589	.547
Information Management	Equal variances assumed	2.176	.149	.190	37	.850	5.833E-02	.306	-.562	.679
	Equal variances not assumed			.203	35.322	.841	5.833E-02	.288	-.526	.643
Control/Supervision- Risk	Equal variances assumed	.443	.510	.301	37	.765	9.583E-02	.319	-.550	.742
	Equal variances not assumed			.290	26.418	.774	9.583E-02	.331	-.583	.775
Control and Supervision	Equal variances assumed	3.570	.067	2.054	37	.047	.625	.304	8.439E-03	1.242
	Equal variances not assumed			2.197	35.700	.035	.625	.284	4.800E-02	1.202
Technology Management - Risk	Equal variances assumed	.375	.544	.839	37	.407	.2333	.2782	-.3304	.7970
	Equal variances not assumed			.795	24.933	.434	.2333	.2933	-.3709	.8375
Management of Technology	Equal variances assumed	.588	.448	.116	37	.908	3.333E-02	.288	-.549	.616

HYPERMEDIA BASED DISTANCE LEARNING SYSTEMS AND ITS APPLICATIONS

Christian Andreas Schumann[1], Karsten Wagner[2], Andrey Oleg Luntovskiy[3], Eugene K. Ovsyannikov[4], and Elena M. Osipova[5]

[1]*Institute for Regional and Local Development Inc., West Saxony University at Zwickau, the Centre for New Studies Forms, Germany*

[2]*Institute for Regional and Local Development Inc., Chemnitz, Germany*

[3]*National Technical University of Ukraine <<KPI>>, CAD Dept., Kiev, Ukraine*

[4]*St. Petersburg Institute for Informatics and Automation, The Russian Academy of Science, St. Petersburg, Russia*

[5]*Bureau for International and Interdisciplinary Projects, St. Petersburg State University, Interdisciplinary Centre for Advanced Professional Education, St. Petersburg, Russia*

Key words: Courseware, Distance Learning (DL) via Internet, E-commerce, Information Systems (IS), Internet and Internet-II, Hypermedia, Multimedia Network Communication, Multimedia educational software, Telework (TW), Web Technologies and Design.

Abstract: Some approaches to Distance Learning and problems of courseware for Distance Learning IS elaboration using modern Hypermedia tools and technologies are considered. One of the most important of possible applications is so called Telework. There are represented some criteria for effectiveness evaluation of DL and Multimedia educational software (courseware). In conclusion some directions for development of infrastructure and media in NIS are proposed, having the final goal an integration of national DL Information Systems into global one.

Foreword

This paper was elaborated in the framework of the research project .Access to Large-Scale Facilities", was initiated by the European Union in order to promote the knowledge transfer and co-operation of scientists from different European countries and later since 1995 from New Independent States from Eastern Europe. The co-ordinator is the Institute for Regional and Local Development (ITKE e.V., WWW: http://www.tu- chemnitz.de/mbv/FabrAutom/Access-e.htm) in closer collaboration with TU Chemnitz and the Institute of Industrial Sciences and Factory Systems (IBF, WWW: http://www.tu- chemnitz.de/mbv/).

Main research and project themes are [6,7]:

- *CAD and Numeric Calculations Technologies;*
- *Plant Planning and Simulation;*
- *Production Planning and Controlling;*
- *Manufacturing Control, Cell Controlling Systems, Data Acquisition;*
- *Manufacturing and Material Flow Systems;*
- *Computer and Communication Systems.*

Among the topics are investigated are Computer and Multimedia Communications (so called *Hypermedia*) and its issues, such as

- *Distance Learning via Internet;*
- *Telework;*
- *Multimedia training courses development*

just this paper has been dedicated to.

Introduction. The impact of Distance Learning on Telework activities

Development of Hypermedia Communication tools gives a good opportunity to adapt socially and involve in labour activities people who, according to the psycho-physical, medical, mental and other reasons, cannot actively participate in normal life. In the first place Telework [15,20] should be addressed to the following target groups:

- self-employed persons;
- disabled persons;
- house wives;
- far rural regions living persons;
- external and remote universities students (for obtaining a profession in the most quickly and cheap way);
- as well as improvement and renewal qualification requiring persons .

Elaboration of Hypermedia Based Distance Learning Systems permits those people not only to reach a necessary education and professional level but to be involved immediately in different job activities. Due to these objectives additional work places could be created, so to say .at home., and society could obtain simultaneously the additional labour resources and increase its intellectual potential. That's why numerous European institutions pay special attention to the problems of information system development for TW and DL via Internet:

Telework, a key component of the Information Society, has the potential to affect and benefit the whole range of economic activity - large organisations, small- and medium-sized enterprises, micro-enterprises, and the self-employed - as well as the operation and delivery of public services and the effectiveness of the political process. (see at Home Page of EUROPEAN CHARTER FOR TELEWORK, Diplomat. Project, WWW: http://www.wise-forum.org/).

Here we would like to state some ideas about TW and DL. On the way to solving this problem already existing approaches and tools of Multimedia, Distance Learning, as well as Internet services should be widely used. Europe, New Independent States are a single market with a thousand boundaries. There is an urgent need for world-wide guidelines on the infrastructure creation, technological development and technical implementation of TW.

418

Table 1. Distance Learning basic media

Print-based	Audio-visual or technology-based	Human interaction
• Books, journals	• Audio-cassettes/disks/ CDs	• Learner-tutor telephone conversation
• Study guides	• Radio broadcasts	• Learner-learner telephone conversation
• Self-teaching texts: ▪ .tutorials-in-print. ▪ workbooks with audio/videotape ▪ Self-tests, project guides	• Films	• Videoconferencing
•	• Video-cassettes	• Computer conferencing
•	• Television	• "Conversation" via E-mail
•	• Multimedia	• WWW-based conversation
• Maps, charts, schemes	• Interactive video	
• Hand-written materials		

1. Distance Learning

Here is given a brief overview of the current state of the problem [1,2,3,5,11,16,17].

1.1. Distance Learning approaches

What do different DL media and methods offer? Some of the media available in DL are as follows (Table 1):

1.2. Multimedia educational software

Multimedia educational software is literally an integral part of a Distance Learning information system. Multimedia contents can be used both via Internet and as client-side applications, working from the CD-ROM or hard disk depending on the hardware and networking facilities available. Comparing Multimedia courses with other ways of presenting educational information it may be stated that high level of interactivity provided by such courses as well as usage of additional attractive features like animations, pictures, voice explanations, videos and animated cartoon characters in the role of virtual teachers, etc. increase the level of perception of information and the level of interest in studies also.

The structure of Multimedia program should have hierarchical nature and contain parts similar to those described for the Web-course. Multimedia application can also use hyperlinks and provide non-linear navigation in the course.

Development of Multimedia applications has its own peculiarities. Mainly it depends on the pedagogical goals set for this course, i.e. whether it is a course for self-studies or group discussions should be previewed; if feedback is necessary or not, if yes - immediate or delayed, whether videoconferencing is possible and desirable or not, etc. The answers on these questions determine the structure of a learning unit, information contents per unit, duration of a lesson.

Among the most popular tools for developing Multimedia applications can be mentioned the following languages and environments:

1. C ++ *
2. Delphi *
3. Macromedia Director **
4. Toolbook **
5. Java ***

---

* - these languages are normally used by teams of professional programmers in commercial enterprises
** - these environments are used by professional programmers as well as specialists in universities mainly for development of Multimedia educational software and making presentations and Encyclopaedia
*** - Java can be used for development of both Internet programs (based on applets) and applications for client-side usage. The main disadvantage - low speed of work on the level of application in comparison with the other 4 tools mentioned.

Of course this is not the full list of possible means for development of Multimedia educational software. Many specialised programs also should be used for preparation of the media components, i.e. sound files, animations, videos, pictures, etc.

Although Multimedia applications can look and function in a number of different ways, some criteria for evaluation of the programs should be worked out. The criteria must fit both technical and pedagogical requirements resulting from the nature of DL systems.

They could be as follows (Table 2):

Table 2. Multimedia courseware features

• Easiness of using a course	Simple interface, minimal set of commands and options to choose from
• Navigation complexity	Acyclic nature of hyperlinks, hierarchical structure of the course
• Hardware requirements	The less the better
• Compreh ensiveness of information contents	Easy to understand sentences without using of special terms not related to the subject of the course, correctly formulated questions in quizzes and other forms of tests
• Possibility of course adaptation	Options, previewed for adaptation of a courseware to the level of learner's knowledge and speed of perception of new information.

Table 3. Properties of E-mail as DL media

• Location of the learners	Different place and different time
• Access ability	Easy
• Way to present information	Linear
• Elements of the information	Text, hypermedia attachments
• Learning strategy	Learner oriented
• Model of interaction	One-to-one, one-to-many
• Type of interaction	Off-line
• Way of interaction	Text
• Strategy of interaction	Teacher oriented
• Object of interaction in learner's point of view	Tutor, other learners, study material
• Teacher's possibility to control the interaction	Low
• Costs of the production	Low software costs, low labour costs
• Hardware requirements on the client side	Low
• Costs of the updating	Cheap to update

421

• Costs dependent on the number of learners	Fixed
• Speed	Quick material production, quick material delivery
• Updating	Easy, quick redelivery

1.3. Distance Learning via Internet

Here we have the following possibilities [4,8,9,10,12,13]. **Using E-mail technology for delivery of distance course, teaching and learning.** Historically, the method of sending printed lesson materials through the post mails to learners, who work in a predominately self-study manner to complete activities based on the printed materials is used. Typically after that then the learner mails the materials back to a tutor, who reviews them and provides some form of feedback, again often through the post.

The basic points characterising this media for teaching and learning are as follows (Table 3):

E-mail is convenient to use both in DL as well as in traditional educational methods, especially in large groups, or when the tutor is hard to reach. Tutors are more easier to answer a message electronically because they can decide themselves when is it to answer. In this case e-mail has an additional value. It does not replace other communication channels. A drawback is, that there is more time spent on communication, so a course will take more time than originally. In distance education settings e-mail could replace originally media, like letters, telephone and radio. The drawbacks of these 'old' media are obvious.

Let's examine the benefits of simple E-mail for a tutor. As a further improvement over face-to-face meetings during office hours at least for certain sorts of learner questions, e-mail messages can be handled in various ways:

- A tutor can print the series of learner messages out, and consider a joint response rather then individual ones, if the same question appears from a number of learners. If the point at issue seems sufficiently important, the instructor can send the response to the whole class, and thus prevent further repetitions of the same question. Electronic copies of both questions and responses can be saved in a directory for the course or kept available in a folder in the e-mail environment, in case it is handy to re-use an earlier response, by re-addressing and revising it and then forwarding it.

- A tutor can answer directly on the message, or on an edited version of the message, and return it. This is handy when a learner ask a complicated set of questions.

- A tutor can keep copies, electronically or in printout form, of what he receives and what he sends. This is easier and more efficient than making notes after a face-to-face meeting and can serve as a remainder of points that need attention later in the course, or in the next round of the course.

- A style of E-mail messages encourages friendly and informal contact between tutor and learners, a type of contact sometimes lacking in post-secondary situations.

Messages can become extra lectures or tutorials, if the tutor reflects in a lengthily fashion about an issue. They allow reflective follow-up to a lecture by both tutor and learners.

Using conferencing technology. Computer conferencing allows a group of people to hold a discussion via computer. Members of the group can use the system to post messages to the whole group, and discussions can thus take place over a period of time. It is similar in many ways to email, except that the way it organises messages makes it much easier to follow the threads of discussions between group members.

CHAT and Talks services have already became a habitually phenomena. The videoconferences are a powerful tool, but expensive too. But one of the most convenient in sense of factor .cost-effectiveness. is combined WWW-MPEG-technology where15-minutes videos are sent and played in quasi - .on-line. mode as preamble to the lesson and then tutor and learners communicate with each other via WWW-on-line.

Using discussion groups and electronic journals. There are literally thousands of special interest discussion groups, each individually controlled by the tools known as a LISTSERVs, Usenet News, BBS etc. Discussions are often moderated by a list owner (tutor), but this is not always the case. Most lists can be provided to the learner either in a digest form or on a post-by-post basis. Any member of the list may take part in a conversation or begin a new topic. Also there are electronic journals (commonly known as E-journals or E-texts) distributed, moderated on-line magazines usually dedicated to a specific area of research. Lists and e-journals could be joined by posting E-mail.

Using WWW technology for delivery of distance course, teaching and learning. Now we are observing increasing number of courses that are offered completely over the Internet and more specifically, completely through a WWW-site. WWW is a collection of protocols and standards to access the information via Internet. These standards and protocols define a particular kind of information. Using WWW servers as delivery platform allows to:
- present learning material as Hypertext material;
- use embedded Multimedia (Hypermedia materials and courseware);
- provide interactivity during on-line learning process.

Therefore, WWW (and also Hypermedia) is a platform for collaborative communication (Table 4).

Table 4. Properties of WWW as DL media

• Location of learners	Different place and different time
• Access ability	Complicate
• Way to present information	Non-linear
• Elements of the information	Text, colour, audio, graphic, animation
• Learning strategy	Learner oriented
• Model of interaction	Branching
• Type of interaction	On-line

• Way of interaction	Text, visual
• Strategy of interaction	Material oriented
• Object of interaction in learner's point of view	Study material
• Teacher's possibility to control the interaction	On-line lessons, videoconferencing . high, medium
• Costs of the production	High software costs, high labour costs
• Hardware requirements	Medium to high
• Costs of the updating	Cheap to update
• Speed	Slow material production, quick material delivery
• Updating	Complex, quick redelivery

There are many benefits according to the decision of maintenance a WWW-site as a central integrative environment for a courseware. Some of these are generic to any central electronic site but are made particularly flexible through the WWW-hyperlink organisation. Learners can access the environment at any time, from a variety of locations (this is useful not only for learners already enrolled in a course but also those considering taking the course, to help them in decision making about course). For example, there is no need for either the tutor or learners to be inconvenienced if learners did not get a copy of a particular handout; they can simply download it or print it out from the WWW-site. Additionally, one can use hypertext to organise enormous amounts of data in a relatively lucid fashion, using menus, key word searches, even .click-able. graphics, sounds, animations, videos as a means to link a learner to more and more information.

2. Concepts of Hypermedia courseware implementation

2.1. Structure requirements to courseware

To our opinion it's worth to recommend for Multimedia- and Web- courses to have the following structure units:
- About the training course (home page, index-page, generics);
- Bar;
- Schedule of the course;
- Outline of the course;
- Chapter page or directory;
- Lessons/overview page;
- History page;
- Learner application page;
- References;
- Glossary;
- Discussion;
- Registration/authorising page and utilities.

424

Let's describe them in details. The item .About the course. has the Institution Logo, Course Logo and the following topics, such as "Duration and Structure", "Certificate", "Who can benefit from the course", "Acknowledgements", subscription information, .Copyright information., authorising tools.

At the beginning of all chapters and lessons the logo of the course should be shown. At the end of all pages of the course corresponding copyright information should be shown.

The item .Bar. will suit to purpose of using frames. Every time two frames should be presented on the screen:

- the bar (constant);
- information page (variable).

The bar should help a learner to be orientated in the course material. The bar contains icons with following titles: "About", "Schedule", "Outline", "History", "Education", "References", "Glossary", "Discussion", "Registration". All pages except "About" contain at the beginning and at the end a line with the following links: "Previous", "Next", "Up", "What's New", "Help", "Tutor", "Bar". These links help to navigate in the course. They serve the same goal as the bar, but not alike the bar they are page-dependent. Links "Previous", "Next", and "Up" let a learner go in the course one page back, forward or up. Link "What's New" follows to an HTML document with news, relevant to the page. Link "Help" follows to a document with instructions that help learners to learn more effectively. Link "Tutor" lets the learner easily send an E-mail message to a tutor responsible for the chapter (and page). Link "Bar" helps to show the bar when it is not present (it could happen for example when a learner reaches an information page directly by its URL).

The item .Schedule of the course. contains a list of chapters in all modules with corresponding periods of instructions, and links to all chapters.

The item .Outline of the course. contains a hierarchy of chapters and lessons in all modules without links to all chapters. The outline page is formatted primarily for printing it out.

The item .Chapter page. or .Chapter directory. for any chapter should contain:

-      a list of titles of all lessons with corresponding links,
-      short (1 paragraph size) descriptions of all lessons of the chapter,
-      description "After this chapter you should be able to understand and explain briefly".

The item .Lesson/overview page. is a page, divided in a number of smaller parts - lessons and overviews. Any overview or lesson starts with the course logo and ends with the copyright information. Both overviews and lessons can contain links to supporting documents that compose the last fourth level of hierarchical structure of the course. Any overview or lesson has its own simple (not hierarchical) outline at the beginning that defines main parts of an overview or a lesson and their order. At the end there is a section: "Additional Material", where there are references to additional (not obligatory) learning material. The last two sections in an outline of a lesson are: "Activity", and "Tests". There are no such sections in an outline of an overview. "Activity" section contains description of activities the learners should accomplish in the lesson. "Tests" section contains a link to the test page.

2.2. Educational Web-server and its information implementation

Educational Web-server tools use Internet/Intranet-services such as Web, E-mail, News, FTP etc. to organise the co-operative learning process and curricula supporting in the framework of educational Department at an University or other institution.

Information content should be as follows:

- Multimedia processed data and courseware as
  - Texts with and/or without layout;
  - Pictures, drawings and charts in 2D- and 3D-graphics in GIF, JPEG, 3DS, PNP;
  - Sounds and Speech in WAV, MIDI;
  - Animations, Videos in AVI, MPEG, MOV;
- Training courses and presentations in PPT;
- Hypermedia documents using HTML, Java Scripts, CGI-Scenarios;
- Distributed applications, Java applets;
- 3D-Animations and Virtual Worlds in WRL (WWW: http://vrml.org/)using VRML such as
  - on results of some physical or technical experiments;
  - for modelling of physical phenomena or designed objects;
- Network libraries with Web, VRML and Multimedia resources such as
  - shareware tools;
  - electronic books;
  - demo examples;
- Questions to self-control and learning programmes.

Access to courseware. Obviously there could be a commercial and free-of-charge courseware. Therefore the important aspects of Hypermedia based DL systems are development and implementation of technologies of learner access to courseware data:

- Learner authentication;
- Learner authorisation;
- Separation of access levels depending on knowledge level and psycho-physical characteristics of learner. One of the most important tasks here is adequate modeling of learner's person and elaboration of approaches to knowledge level evaluation.

Here could be used the standard solutions such as:

- Learner/tutor/administrator Access Matrix using;
- HTTPS-protocol (WWW: http://whatis.com/https.htm);
- Public key cryptographic systems, respectively Digital Signature and data encryption, e.g. PGP (WWW: http://www.pgpi.com/);
- Learner/tutor/administrator signature OCR-systems (i.e. handwriting pattern recognition) and so on.

Learners knowledge evaluation. To test the knowledge level should be used these kinds of test questions:

- **Open questions.** Fill in the correct answer in the Right answer field. Fill in expected wrong answers in the Wrong answers field.

- **Multiple choice questions**. Questions with alternatives. We recommend to use the maximum number of alternatives - 6.
- **Matching questions**. In matching question a learner has to link all items of one row to the items of another row to do comparing.
- **Observation**. Describe the task to be observed. We recommend to clearly define the actions to be performed.
- **Assignment**. Describe the concrete tasks a learner should perform. The tests could be organised so that learners are assigned tasks and answer tasks by e-mail or automatically. This kind of task should be quite intricate so that learners must show what they understand about the topic. They should not be able to find answers directly from text books or other materials.

The evaluation procedures can be performed in such modes:

- **On-line evaluation**. One computer conference item may be set up as a discussion area about the course requirements, activities, and learning. Learners can express their concerns and raise issues about course activities in this item for instructor response. Private E-mail exchange may be used too. These provide useful formative data for fixing the current course, future offerings of the same course, and the current design of the similar on-line courses.
- **In-person evaluation**. It may be in-person session for learners to meet with the tutor. They are encouraged to convey constructive (or critical) views of the learning experience so it becomes possible to make important adjustments for the second half of the course.
- **Written evaluation**. To sum up overall impression of the course, written questionnaire for the learner to complete may be used. This instrument includes questions about course effectiveness as well as some dealing specifically with learning on-line at a distance and so on.

Examples of educational Web-server implementation. Some examples of fragmentary implementation (most in Russian and Ukrainian) have been given on the pages of Web-site exists at the CAD Department of National Technical University of Ukraine .Kiev Polytechnic Institute. NTUU-KPI [4,13] since 1996 (WWW: http://cad.ntu-kpi.kiev.ua/). The server has a lot of network resources and learning materials tied to curricula of the Department and educational targets. The courses on the subjects of *Web-Technologies and Design, Networks Design, Banking Computer Technologies, Electronic Simulation and Data Security* have found its support on the pages of this server, i.e. under the following addresses:

- http://cad.ntu-kpi.kiev.ua/resources/index.html
- http://cad.ntu-kpi.kiev.ua/˜demch/academic/index.html
- http://cad.ntu-kpi.kiev.ua/people/lunt/my_courses.html
- http://cad.ntu-kpi.kiev.ua/copernicus/index.html

The scientists of the CAD Dept. were admitted to participate as partners in joint European projects, such as COPERNICUS.

2.3. Multimedia courseware implementation

Obviously, the highest efficiency of education processes could be obtained when Multimedia education is supported with modern communication tools [14,18].

Multimedia courseware. Multimedia training courses are mostly implemented on CD-ROMs. This is the current state of the art. The reasons are evident:

- Low rates of information transmission over existing communication channels;
- Limited Internet access for a lot of learner categories;
- High hardware requirements.

In fact, Web-courses offer more flexibility for learners while Multimedia courseware on CDs give more ways of material representation and as a result can have more influence on learner's study effectiveness.

If to combine advantages of both, i.e. to be able to transmit Multimedia courses via Internet connections in real time, then we'll receive the new generation of educational courses . providing real-time interaction and Multimedia features at fully extent. This can become possible with the development of Internet-II, the project aimed at implementation of ATM-technologies. It will considerably increase the rate of information transmission and make possible the thins described above. The Interdisciplinary Centre for Advanced Professional Education (ICAPE) of Saint-Petersburg State University (SPbSU, WWW:http://icape.nw.ru) is engaged in this project from the part of Saint-Petersburg State University. As main educational centres will be connected by Internet-II channels, knowledge transfer will become a reality (WWW: http://web.mit.edu/is/nextgen/

Multimedia conferences in real time give to the learner an opportunity not only passively admit the knowledge, but also to participate in discussions via high-speed Internet networks. The best technical solution for such kind of educational centres are Multimedia .Network-Satellite. workstations [19]. Some positive experiences have been collected at the Saint-Petersburg Institute for Informatics and Automation (SPIIRAS) in collaboration with European programs INTAS (WWW: http://www.ujf-grenoble.fr/GE/GENS6/BoursesINTAS.html) and COPERNICUS (WWW: http://www.cordis.lu/inco/src/organiza2.htm), in joint projects under the auspices of European Community and other programs (WWW: http://www.spiiras.nw.ru/text/relations.html)

Examples of educational Multimedia courseware implementation. ICAPE of Saint-Petersburg State University is actively implementing the elements of Web-based Distance Learning. To give an example, it is creating the so-called .ecological ring. in the Regional Network for Education, Science and Culture (ROKSON, WWW: http://www.nw.ru), which it is the managing node of. One of the sites of this ring is represented by the Ecological Department at ICAPE (WWW: http://www.ecosafe.nw.ru) and contains a lot of educational information issues, useful for people, engaged in advancing their knowledge or self-education as well as for students of Faculties for Geographic and Biology of SPbSU. This site contains information:

- Distance Learning training courses *Mathematical Modelling in Geographical Researches*, *Information Technologies of Ecological Safety;*

- Hyperlinks to other WWW-resources on DL in the field of environmental sciences, examples of computer simulators;
- Information of the Centre of Ecological Safety of Russian Academy of Sciences, e.g. .Baltic-21. (WWW: http://www.ecosafe.nw.ru) etc.

Conclusions

Telework, Hypermedia Communications, Distance Learning, New Educational Forms Development, by its nature, are world-wide-spread. TW and DL transactions bring a high performance and effectiveness of knowledge transfer and interchange in business processes. Studies, work, jobs, purchase orders, payments, services and goods move at very high speed around the world. But the World, Europe and New Independent States have for their own great number of nations, multitude of regions, languages and dialects, rules, regulations, registrations, variety of technical and technological platforms and infrastructures.

For the NIS, including Russia and Ukraine, the development of DL and TW has some special features:

- Geographical extent;
- Economical difficulties;
- Low population urbanisation;
- Scientific forces and educational means distribution irregularity, concentration of intellectual potential in large urban centres;
- Labour distribution irregularity;
- Lack of infrastructure and media means development.

Therefore, it is necessary to elaborate two complementary ways for TW and DL development in NIS:

- **For large urban centres** with developed infrastructure and media it's worth to provide joint projects aimed at creation with western partners of common Information Space for implementation of modern TW and DL technologies on the already existing basis.
- **For far rural regions** it's necessary to launch grant-supported and joint projects aimed at acceleration of information transfer system development and adaptation of their environments for modern computer network technologies needs.

The resulting approach could be helpful to further promotion and development of TW, DL, and, therefore, of Information Society in the world.

Acknowledgments

This paper was prepared on the basis of research results of different scientists from Germany, Russia and Ukraine and as a fruit of common works in the framework of the European project PECO/NIS at Technical University (TU) of Chemnitz. New Independent States (NIS) authors are grateful to the co-ordinators of this project.

References

[1] B.Collis. Networking and Distance Learning for Teachers: a classification of possibilities. Journal of Information Technology for Teacher Education, vol.4, No.2, 1995, pp.117-135.

[2]   B.Collis. Tele-learning: Making Connections, Faculty of Educational Science and Technology University of Twente, Enschede, Netherlands, 1995.

[3]   Communication and the On-line Classroom, Ed. Zane L. Berge and Mauri P. Collins, 1995.

[4]   Demchenko, Yu. V. Cooperative and contributive Learning . the real choice to push professional education in Networking Information Technologies in Ukraine and CIS. Proceedings JENC8. . 1997, pp.822-1-822-7.

[5]   ECC COCOS 2.0. User's Guide. 1993. Andersen Consulting - ECC. Arthur Andersen & Co., S.C., Eastmond Dan, Ziegahn Linda, Instructional Design for the On-line Classroom, Computer Mediated Institute of Industrial Sciences and Factory Systems (IBF) Year Report'97, TU Chemnitz-Zwickau, 1997.

[6]   Institute of Industrial Sciences and Factory Systems (IBF) Year Report '97, TU Chemnitz 1997.

[7]   Institute of Industrial Sciences and Factory Systems (IBF) Year Report'98, TU Chemnitz, 1998.

[8]   Kovgar, Vladimir B. The Methodology Aspects of New Course Creation for Flexible Distance Learning. - Dissertation to Award of Degree of MBA. - Kiev, 1997.

[9]   Leshchenko,M.A., Dobronogov, A.V., Luntovskiy, A.O.
Providing information in the Internet about international educational and telecommunication projects //A Conference under the support of TEMPUS-TACIS T-JEP-08570-94 .Computer Networks in Higher Education., 26-28 May 1997 National Technical University of Ukraine «KPI», Kiev, 1997.

[10]   Luntovskiy, Andrey O. .New information technologies and its meaning for higher education management reform.. - Dissertation to Award of Degree of MBA. - Kiev, 1997.

[11]   Michael G. Moore, Greg Kearsley. .Distance Education. A Systems View.. Wadsworth Publishing Company, USA, 1996, p. 290.

[12]   Osipova, Elena M., Vitrishchiak, Ilya B., Leonov, Sergei N. .The role of multimedia in distance education. //Proceedings of the International UNESCO conference on distance education, 30 June . 5 July 1997, Saint-Petersburg.

[13]   Osipova, Elena M., Kholkin, Vladimir Yu. .The concept of intellectual server - what should it include.// Proceedings of 14th Annual Conference on Distance Teaching and Learning, 5 .7 August 1998, Madison, Wisconsin.

[14]   Ovsyannikov, Eugene K., Losev, Gennady M. .Satellite-network multimedia station in Saint-Petersburg., Russian Conference for Computer Networks and Techniques, Saint- Petersburg Polytechnic University, 1997, in Russian.

[15]   Proceedings of the Conference .W.I.S.E.-Forum ., 3 December, Vienna, 1998.

[16]   Rowntree, Preparing materials for open, distance and flexible learning: an action guide for teachers and trainers, London, Kogan Page.

[17]   Stanchev, E. Niemi, N. Mileva. Teacher's Guide "How to Develop Open and Distance Learning Materials", University of Twente, Netherlands.

[18]   Tatiana Piskunova, A.I.Herzen. Using of Authorised Hypermedia Courses for Education and Training. Saint-Petersburg State Pedagogical University together with SPIIRAS, Workshop for Informatics and computer technologies, 11 Sept. 1998.

[19]   Technical Report for COPERNICUS-Project .Multiserve-1254., Brussels, 1996.

[20]   Work and Employment in the Information Society. .The Work, Information Society and Employment Forum., W.I.S.E. Forum Report, Vol.1, Sept. 1998.

MULTI-MEDIA INSTRUCTIONAL SUPPORT:
STATUS, ISSUES, AND TRENDS

Doug Vogel[1], Johanna Klassen[2], and Derrick Stone[2]
[1]*Information Systems, City University of Hong Kong, Tat Chee Avenue Kowloon, Hong Kong*

[2]*Professional Development and Quality Services, City University of Hong Kong, Tat Chee Avenue, Kowloon, Hong Kong*

Key words: learning paradigms, multi-media support, CD-ROM

Abstract:

Education is facing a broad range of challenges as we enter the new millennium with new technology, new paradigms, and new resources for learning. In response to these challenges, teachers are broadening their range of instructional methods and approaches. One area that is seeing increased attention is in the provision of instructional support for 'student-centered' and 'student-directed' learning. Many teachers are finding that 'student-centered' learning coupled with the use of technology can make the learning process more effective. This paper examines the status, issues, and trends of multimedia instructional support. Examples and experiences associated with the development of CD-ROMs and interactive websites are presented to illustrate some important considerations for the development of interactive multimedia packages. It is hoped that these considerations will result in enhanced support packages for student driven learning.

1. INTRODUCTION

Over the last 20 years educational content has changed so rapidly, especially in technologically oriented fields, that content knowledge provided to students becomes outdated almost as rapidly as they graduate. Given this, education needs to respond by shifting focus away from content based learning to a more pro-active style of learning, one that embraces process based learning. Focusing rather on providing students with creative problem solving skills, education in the new millennia will give students the abilities to adapt to changing situations, and to the speed at which these changes occur, as world circumstances dictate. Because, like never before, students are surrounded with excessive amounts of information, it becomes increasingly more important to give students the skills to identify the most relevant content and that which most appropriately serves their needs. Further, students at all levels must learn to identify outdated assumptions, and then critique, challenge, and replace them with new ones. This replacement of ideology can only occur when students are sensitive to process, to the fact of change, and to the benefits of information technology.

Systems Development Methods for Databases, Enterprise Modeling, and Workflow Management,
Edited by W. Wojtkowski, *et al.* Kluwer Academic/Plenum Publishing, New York, 1999.

The need for changing teaching methods has come about partly due to the demands of the workplace and partly because of a general re-assessment of teaching methodologies. Graduates are now expected to be versatile in the world of communications that include email, intranet, internet, conferencing systems, and the World Wide Web. These same graduates are also expected to apply higher cognitive skills, such as analysing, summarising and synthesizing information as well as thinking creatively and critically. Teaching methods that assume a single language and shared homogeneity of proficiencies, learning styles and motivational systems are increasingly inadequate and inappropriate.

Teachers who are willing to re-evaluate traditional instructional methods have begun to discover that by broadening their range of instructional methods and approaches to include the use of these new technologies, students tend to be more effective learners. This re-evaluation has brought about a shift toward 'student-centred' and 'student directed' learning. This shift has occurred because of the interactive nature of some of these new technologies. Rather than simply making technology available to students, new learning and teaching is characterized by the introduction and integration of flexible and innovative teaching/learning technology *into* teaching. What this means is that there is an integration of computer-based interactive multimedia including CD-ROMs and the World Wide Web as part of the method of content delivery.

For many, the introduction of new technology appears like the removal of the teacher from the process. But this fear is unwarranted since for university faculty this change involves altering the mode and style of teaching such that teaching/learning technologies become an integral part of their academic programs. Teachers will never be replaced by technology. What they will find, however, is that their resource base only gets larger and more varied as technology extends itself in a plurality of directions.

As part of their mandate, the universities in Hong Kong and the government of the Special Administrative Region share the goals of assuring that teaching, learning and assessment take maximum advantage of new technologies. In a practical sense, this means that student-directed learning must become a more important component in programme design. As a result of this direction, there are a variety of interactive learning environment elements, elements which extend beyond the bounds of traditional classrooms, being created and evaluated.

This paper examines the status, issues, and trends of multimedia instructional support. Examples and experiences drawn from personal experience are included to demonstrate the status of this new learning approach. The main focus is on key issues in the development of two main multimedia formats: CD-ROM and interactive web sites. Some of the issues that are considered include: degree of interactivity, performance, accessibility, dependability, and development dynamic. A brief look at present trends in multimedia development such as student driven learning, cooperative learning, and assessment is included. Conclusions about the future of technology in education and the creation of robust learning environments are drawn.

2.   BACKGROUND

Interactive multimedia learning environments have a variety of characteristics that facilitate the shift away from lecture driven towards student centered and student directed learning. In these environments, students are no longer treated as passive receptors of information; rather, they begin to actively construct, transform, and extend their knowledge. The role of faculty alters along similar lines; instead of working at making sure students have a precise set of objects of knowledge, faculty now expends its efforts at developing students' competencies and talents. This is accomplished, in part,

432

through recognizing that students encompass a wide variety of learning styles and abilities. In this way, education becomes a personal transaction among students as well as between faculty and students. Effectively, education becomes defined as working together.

Teaching in this new environment becomes a complex application of theory and research in enhancing student learning through process innovation. Technology, and its multimedia capability, becomes a crucial enabling factor for process effectiveness and efficiency. Perhaps most importantly, it aids in the removal of historical time and space constraints. Learning need no longer occur in only one space and at one time; learning becomes barrier free. For the first time, thanks mostly to this trend toward the use of technology in teaching, students are being challenged with the responsibility and accountability in controlling their own educational discovery process. This, in turn, leads to greater motivation and interest for students to learn because they can decide the pace, direction and content of that process so as to match their real-world needs, desires and expectations.

Multi-media technology provides learners with a rich base, offering students the choice of pathways of learning rather than being restricted to a single mode of delivery. It not only makes content specific instruction more compelling, but also allows learners to experience realistic simulations for more effective decision making training. While interactive teaching is only now gaining popularity and wide implementation, as early as the 1970s research showed the effectiveness of multi-sensory stimulation (Nasser & McEwen, 1976; Strang, 1973). The effectiveness of multi-sensory stimulation lies in its ability to adapt to students' individual differences. Quality interactive multi-media materials respect the learners' capabilities of controlling the learning path by facilitating a plurality of learning styles within one package. Inherent in this mode of learning is the necessity of creating material with the flexibility to allow the students to determine the pace of learning, the level of difficulty and the style of learning.

There is substantial research which indicates that learners have distinct and preferred learning styles; audio, visual, or kinaesthetic. These same learners learn best when these individual differences are taken into consideration (Anderson, 1995; Sarasin, 1998). If a particular student is a visual learner but hears a lecture without visual cues, that students' comprehension will be negatively affected. The same results hold for all three learning styles. If any students' learning style is not factored into the learning process, that student suffers significant consequences.

In a typical interactive multimedia package, multi-sensory stimulation does not discriminate against any one learning style, but, actually, emphasises all three. More recent learning theories support the conclusion that a student will have an increased comprehension and retrieval when they have access to audio, video, and text (Kemp & Smelle, 1989). Hoffstetter, 1995, believes that students will retain only 20% of the material if they are only stimulated visually. The figure rises slightly to 30% when those students are exposed to auditory stimulation. The combination of visual and auditory will facilitate a retention of almost 50%. But, the combination of all three, auditory, visual, and kinaesthetic, in other words working interactively, will raise the percentage of retention to an amazing 80%. It is then incumbent upon educators to use these facts to their advantage. In preparing materials and modes of delivery, interactive multi-media packages seem to be, by their very nature, of particular merit. By combining all three of these learning styles into one package, these may be the most appropriate tools for the enhancement of education.

A further advantage of technology delivery is the ability for the students to receive immediate individual feedback. The effect of this capacity is that it makes the learning experience less threatening. For students, uncertain of their abilities, it is less threatening to receive feedback from a computer than it is from a person. An additional advantage of computer feedback is that it can be used to complement that from a professor.

There are many studies that demonstrate that when using interactive materials, students not only learn more – more quickly and more enjoyably – they learn the much needed life skill of learning how to learn; that is, they begin to take responsibility for their own learning. Taylor, Kemp and Burgess (1993) claim substantial reduction in time for training programs when using interactive materials. Fletcher (1990, quoted in Semrau and Boyer, 1994) concludes, after reviewing 47 studies, that interactive video instruction is more effective than regular instruction. Fletcher concludes that use of interactive instruction is also more cost effective. A recent study by Morrison and Vogel (1998) demonstrates that student comprehension and retention can be significantly increased through effective use of presentation visuals, especially those that use color and some degree of animation.

There are several other positive effects that have been discovered owing to the use of interactive materials. The use of technology has a positive impact on students' perceptions of teachers. Instructors who use presentation visuals are judged to be better prepared, more concise, more professional, clearer, more credible, and more interesting. Technology can also enable teachers to support each other and work together to share material and complement each other's expertise in adding value to education (Alavi, Yoo, and Vogel, 1997). Additionally, technology can enable teachers to link students and classes together as well as actively encourage participation within and without the classroom by government and business experts.

With the new requirements of business and society coupled with this trend toward more independent learning, educators need a more proactive approach that will help students to learn autonomously and to learn how to learn. With the rise in usage of technology the role of faculty is not eliminated; rather, that role simply changes. There will always be, of course, the need for content expertise for the delineation of material, but these same content experts can find additional employ in the development of instructional interactive multimedia packages. Faculty now finds itself with an additional responsibility, that of developing a host of other skills associated with teaching *with* technology. The range of new skills that can be developed range from those that are technological in nature to some that are more artistic. What will be an additional benefit is that faculty can devote time to the incorporation of a plurality of materials, ranging over a variety of mediums into a unified content.

By integrating innovative learning technology with lectures, the role of faculty shifts from that of a delivery medium to a mentor or facilitator of learning. As such, faculty is responsible for creating a learning environment that accentuates the impact of the content and accelerates student learning. Technology accentuates the faster and more efficient involvement of more people to tackle bigger problems with more degree of freedom, higher interest level, and, ultimately, higher levels of "buy-in."

3. STATUS OF MULTIMEDIA DEVELOPMENT

In this section we look at the status of current multi-media CD-ROM development as well as the development of interactive web sites to enable comparison and contrast.

3.1 CD-ROM Development

The development of interactive multimedia CD-ROMs has a much longer history than that of interactive multimedia web development. Accordingly, the maturity of this development, and the support offered, has resulted in their expansion into substantial commercial developments. There are two major considerations that need to be addressed with regards to multimedia development: scripting and programming (including graphic design).
There is now strong software tool support for multimedia programming, including packages that are designed specifically for the incorporation of animation, video, audio,

and textual elements onto a CD-ROM. Two of the most useable programs available are Macromedia Director and Authorware.

With respect to scripting of content, there is an almost unpardonable lack of guidance. While there are ample content experts and ample people familiar with interactive environments, there are deplorably few content experts available who have knowledge and experience in designing materials for interactive environments, that is, are able to write a script which can be interpreted by programmers. Because of such a lack of direction, inevitably it is incumbent upon developers to create their own scripting guidelines.

3.2   The World Wide Web

The World Wide Web has, over the past several years, burst on the scene as a real contender as a learning environment vehicle.   Up to this time in the academic environment, the web has been used primarily as an information resource for students or for the storage of instruction materials including presentation visuals, class notes, and pointers to content-linked web sites. In a few cases, the web has also been used in a more interactive fashion. For example, FaBWeb developed at the City University of Hong Kong, combines three individual packages in one facility; *Learning Resource, Meeting Space* and *Play to Learn* (Vogel, Wagner, and Ma, 1999).

The *Learner's Resource* offers a variety of teaching materials to the learner at any time and at any place. Teaching materials for lectures and tutorials are kept there, in their native format, whether that be a presentation (e.g., Powerpoint), a written document, or a spreadsheet. Furthermore, video recordings of past classes are kept there, in highly compressed, web browser accessible format (RealVideo). Also in this area are links to other WWW based resources which complement the primary information. With these resources, the student can watch a lecture video on-line (via a normal 28.8 modem), and, at the same time, review the corresponding lecture materials.

The second application, *Meeting Space,* is the social center of the FaBWeb. In this space, students can informally meet with other students and instructors. But this is not an "accidental" meeting place. Instead, it is (at least in part) an organized society, with scheduled programs and events.

Play to Learn, the third component of FaBWeb, is an interactive game playing environment where students can measure their skills against others through the playing of business games and simulations.

The problem with these extended web applications is that they tend to be very fragmented. In general, most web-based applications are actually the assemblage of a number of non-integrated pieces, each of which was designed to support a particular form of interaction, and was not designed for "piggybacking" with other packages. Those products that have some degree of integrated support such as Lotus Learning Spaces or WebCT tend to not fully cover the domain of support desired by faculty seeking to more fully exploit web capabilities. Products seem to not have an open architecture or require extensive customization to be effective (Glasson and Jarmon, 1998). Further, there exists weak tool support to create interactive web sites. In fact, most web sites tend to be developed by instructors or students without specialized knowledge or training in multi-media development. Problematically, most web sites are developed to deal with the low bandwidth constraints typically imposed by internet limitations. Yet to be seen is any wide availability of broad-band internet -- although there are a few places, such as Hong Kong, which demonstrate such capabilities.

4.   ISSUES

There are a number of issues that provide a solid foundation for comparison and contrast for CD ROM and WWW multi-media development approaches.

4.1   Degree of interactivity

The amount and degree of interactivity varies in character when comparing CD-ROMs and the WWW. CD-ROMs allow for extensive user interaction that can include, but is not limited to, assessment, pathways of learning, randomly or selectively accessed tests, scores and reports. The WWW has not historically been seen as a highly interactive environment. Interactive tasks such as self assessment is not usual within the domain of the web. Rather, the WWW has primarily been used as an information resource. Within this resource, users have the ability to navigate a broad space in search of information or to have that information stored in a pre-structured format within a particular site.

4.2   Performance

One of the more easily recognized difference between CD and web applications is the differences in their individual performance. The speed of accessibility to information on the CD-ROM allows for the inclusion of a multitude of visual elements including the playing of normal video as well as graphics and animation. Web applications, on the other hand, tend to suffer from unfortunate delays, especially when downloading large files or graphic images over slow links. Videos, graphics and animation, while accessible, often mitigate their role to individual applications rather than being incorporated into a unified package.

4.3   Development support

The amount and availability of developmental support for CD-ROM and Web technology tends to align with the maturity of the medium. The development of CD-ROMs is well supported in terms of tools designed to assist in multi-media development. Two of the more successful scripting tools are Macromedia Authorware and Director. Development for the web has been primitive by comparison. Because of the relative recentness of the Web, little exists in the form of tool support. The most that developers can hope for is the extended capabilities of word processors and some multi-media tools adapted from support for Local Area Networks (LANs).

4.4   Accessibility

Accessibility, in this case, refers to the ease with which these technologies can be used or accessed by students. The limitations of CD-ROMS is that they require a disc drive that may or may not be available on student computers. WWW applications, however, only require access to the internet. But this access, as is only too well known, may be problematic, especially at peak use times. Modem connectivity often provides users with a frustratingly slow experience.

4.5   Dependability

The dependability of CD-ROM and WWW applications varies considerably. CDs, once pressed, hold up well over time and do not degrade with use. In fact,

436

CDs are often used as backup for other, more volatile, technologies. WWW applications tend to be only as stable as the degree to which external links are included and/or effort is put into checking and sustaining such links.

4.6   Maintainability

Like most issues so far discussed, maintainability varies considerably between the Web and CD applications. Web sites can be easily modified to include substantial material changes or there can be additions or links to other web sites that reflect or comment upon current events. CDs, on the other hand, cannot be changed, once pressed, unless a new version is implemented, which carries with it added expense as well as delay.

4.7   Development Dynamic

Overall, CD-ROM and WWW applications have differing development dynamics. WWW applications can be brought up quickly, albeit with limited functionality, while CD-ROMs tend to be designed and implemented over longer periods of time but generally result in more delivered capability. A prototyping and evolutionary development approach is typically used for WWW applications while a more traditional approach is used for CD-ROM development.

4.8   Costs

Costs vary considerably within as well as between CD-ROM and WWW applications. CD-ROM expense can vary widely as a function of the amount of animation present. WWW applications can be brought up quickly (and inexpensively) but can get extremely expensive if large amounts of animation and graphic design are incorporated.

5.   TRENDS

The educational environment is undergoing rapid change, perhaps educators will see more change in the next five years than in the past fifty. Virtual universities are cropping up around the world and challenging traditional university geographic dominance. Within this rapidly changing environment, a number of trends are emerging.

Student driven learning is a perspective on education that focuses attention on the ability of students to better select the mode of delivery and timing of course material. As such, students can elect to attend class sessions and participate in a face-to-face mode in a traditional fashion or they can be connected remotely to the class (participating at the same time but at a different location) or they can access class session material and recorded video of class lectures and interactions after the class has actually occurred in an on-demand mode. In this fashion, the student selects the type of learning environment characteristics that best fit with his/her needs and constraints. This flexibility in availability serves a variety of student needs. Students increasingly have work related situations that make it difficult for them to always attend class. Students also exhibit a variety of learning styles and experiences that may require multiple exposures to material that can be accomplished off-line without absorbing class time.

Cooperative learning represents a paradigm shift in thinking that encourages students to learn from each other, not only from the instructor. Cooperative learning tenets include simultaneous interaction, equal participation, positive interdependence, individual accountability, group skills, and reflection (e.g., Johnson and Johnson, 1975). Cooperative learning teaches group and communication skills and is felt to increase job retention, academic achievement, critical thinking and problem solving skills (Slavin, 1987). It crosses traditional disciplinary and age boundaries in providing an opportunity to add technology in a synergistic fashion building on an established foundation. A variety of structures can be used within classrooms (or increasingly when students are distributed at multiple sites) to enhance team building, class building, information exchange, communication skills, mastery, and thinking skills (e.g., Kagan, 1994).

Partnering is emerging as institutions recognize that they can not do everything and realize the value inherent in sharing resources and complementing distinctive competencies. These partnerships can involve external stakeholders from the business community who may provide resources such as broad-band internet access to homes. The partners may also be other institutions who see value in sharing faculty and material as well as physical resources to offer special programs. The extent to which partnering is balanced by individual institutional desire to operate autonomously remains to be seen. Numerous examples exist of individual faculty linking students together and sharing content and teaching responsibilities (e.g., Alavi, Yoo, and Vogel, 1997). These partnerships can enable the creation of a rich learning space from which students can develop multi-cultural appreciation of a broad range of topics.

Technology is and will continue to change dramatically both in capability and availability. For example, students are being offered programs to enable them to buy laptops and procure software at attractive prices. The ability to create high quality student applications in a short time is enabling students and faculty to explore a wide range of interface and application issues. Broad-band internet is becoming increasingly available to provide a platform for sophisticated multi-media interactive applications that formerly were only achievable on CD-ROMs . Software designed for use on the internet is enabling collaborative learning using groupware and providing access to servers and locations running programs that would be cost prohibitive to most institutions. Internet-based audio and videoconferencing is widely available to complement high quality ISDN and other forms of high speed multi-media transmission. It is now possible and increasingly achievable to hold panel discussions where panel members are located on different continents. Teams of students from multiple countries can work together across many time zones and experience the reality of multi-cultural global team interaction.

Assessment is emerging as a critical issue as institutions seek to provide the flexibility demanded by students in a form that sustains traditional credibility and accountability. Administering exams becomes more complicated as students are geographically dispersed and progressing at different rates. The usefulness of traditional assessment approaches comes into question as learners are more in control of characteristics of the learning environment. Often a portfolio of assessment approaches including self-assessment as well as more traditional examinations are being implemented to serve the needs of a broadened student base. Assessment challenges not only exist with students but also with faculty and institutions as new methods of teaching are emerging and institutions take on new roles and develop new programs outside the scope of traditional classroom instruction. Overall, formative assessment plays a stronger role in creating, evaluating and evolving educational environments to meet future needs.

6. CONCLUSION

Educational environments are changing rapidly as a result of a combination of lack of effectiveness of traditional approaches and the increasingly cost-effective availability of technology to enable alternatives to the traditional approach. Pollio (1984) claims that students are not listening attentively for 40% of the lectures. Many others would claim that figure is even lower, especially after the first half hour. As increasing percentages of students attend institutions on a part time basis and the range of ages increases, traditional educational approaches and limited attention to learning style variations become ever less tolerable. Technology is coming to the fore to provide additional degrees of freedom and enable the exploration of alternatives to traditional education. Technology is becoming more robust and capable of supporting a wider-variety of needs. For example, that which has been learned and illustrated on CD-ROMs is increasingly applicable to the WWW. Student-driven and on-demand education programs are becoming available to meet the needs and desires of an ever-broadening student population. Faculties are shifting from focus on delivery of material to design of learning spaces. Institutions are recognizing that partnering may be critical to providing comprehensive educational experiences in a cost-effective fashion. Overall, there is a heightened appreciation of the interaction between technology and educational processes in creation of a robust learning environment.

7.   REFERENCES

Alavi, M.; Yoo, Y.; and Vogel, D. 1997. "Using information technology to add value to management education," Academy of Management Journal, 40(6), 1310-1333.

Anderson, J.A. 1995. "Toward a framework for matching teaching and learning styles for diverse population," in The importance of learning styles: Understanding the implications for learning, course design, and education, Ronald R. Sims & S.J. Sims. (ed) Westport CT. Greenwood Press.

Glasson, B. & Jarmon, R. 1998. "Collaborative technologies: enabling the third wave of electronic commerce," Proceedings of the IFIP World Congress, Vienna, Austria.

Hoffstetter, F.T. 1995. Multimedia Literacy. New York. Harper and Row Publishers.

Johnson, D.W. & Johnson, R.T. 1975. Learning Together and Alone: Cooperation, Competition and Individualisation. Prentice Hall, Englewood Cliffs, NJ.

Kagan, S. 1994. Cooperative Learning, Resources for Teachers, Inc.

Kemp, J.E. & Smelle, D.C. 1989. Planning, producing, and using instructional media. New York. Harper and Row Publishers.

Morrison, J. & Vogel, D. 1998. "The Impacts of presentation visuals on persuasion," Information & Management. 33, 25-135.

Nasser, L.D. & McEwen, J.W. 1976. "The impact of alternative media channels: Recall and involvement with messages," AV Communication Review, 24(3), 263-272.

Pollio, H.R. 1984. "What Students Think About and Do in College Lecture Classes", Teaching-Learning Issues no. 53. Knoxville. Learning Research Center, University of Tennessee.

Sarasin, L.C. 1998. Learning Style Perspectives. Madison. Atwood Publishing.

Semrau, P. & Boyer, B. 1994. Using Interactive Video in education. Boston. Allyn and Bacon.

Slavin, R.E. 1987. Cooperative Learning, Student Teams. Washington, DC. National Educational Association.

Strang, H.R. 1973. "Pictorial and verbal media in self-instruction procedural skills," AV Communication Review, 22(2), 225-232.

Taylor, U., Kemp, J. & Burgess, J. 1993. Mixed-Mode Approaches to Industry and Cost Effectiveness. University of Southern Queensland.

Vogel, D., Wagner, C. and Ma, L. 1999. "Student-directed learning: Hong Kong experiences," HICSS Proceedings, Maui, Hawaii.

ELECTRONIC DELIVERY OF CURRICULUM:
PREPRODUCTION OF CYBERSCRIPTS

Josie Arnold and Kitty Vigo
Media Studies and Interactive Multimedia, Swinburne University of Technology, Lilydale Campus, Melba Avenue, Lilydale, Victoria 3140, Australia

Key words: curriculum design; virtuality; interactivity; CD ROM; websites; multimedia; semiotic.

Abstract: New systems for delivering curriculum are creating new challenges for academics. The continuing development of new electronic concepts and approaches produce new literacies. These relate to new ways of 'writing' curriculum, new relationships between the learner and the teacher and new paradigms of discourse. This paper looks at how academics might go about transforming their print-based materials so as to explore the opportunities offered by the new writing technologies. It proposes that electronic textuality and discourse, like all writing, has a structure and form. Even the fluid and singular writing for the new multilayered virtual spaces provided by the emergent electronic culture needs concept planning and preproduction scripts. It argues that the fluidity provided by the new electronic textuality itself still needs to be approached through a process of planning, trying out, imagining, conceiving and communicating to oneself and to others. It investigates how this involves writing a multi-layered script and explores the possibilities offered by electronic texts such as: the provision of immediate information; an interplay of the seen, the heard and the read; the introduction of virtuality, interactivity, IMMediacy and self-authority. In doing so, it establishes a process which will enable academics to construct their curricula fully by exploiting the differences offered by new writing technologies.

1.   INTRODUCTION

Print is one of the most the deeply foundational aspects of the construction of the powerful Western culture which is the dominating force in the current push for globalisation. Prose delivered through multiple printings is itself an ideological force which since Gutenberg and Caxton has represented certain modes of discourse as 'natural' or 'normal'. In doing so it has dominated all aspects of our culture from education and science to business and relaxation.

It is also implicated in the divisive representation of some aspects of culture as more important (canonical or epistemological) than others

Systems Development Methods for Databases, Enterprise Modeling, and Workflow Management,
Edited by W. Wojtkowski, *et al.* Kluwer Academic/Plenum Publishing, New York, 1999.

441

(popular or junk). Indeed, the idea of binary opposites themselves is heavily represented in the ways in which print performs itself. Such binary oppositions have supported, or have lead to, a particular view of social structures which empower some over others, particularly linearity/objective over laterality/personal. In the late 20th century, entering the computer age, we are more aware of the possibilities of the latter. Catherine Hayles, a professor in both physics and literature says:

> 'Many scientists working on chaos speak of the need to "develop intuition". They point to the fact that most textbooks treat linear systems as if they were the norm in nature. Students consequently emerge from their training intuitively expecting that nature will follow linear paradigms. When it does not, they tend to see non-linearity as scientifically aberrant and aesthetically ugly. But non-linearity is everywhere in nature and consequently in mathematical models. Despite its prevalence, it has been ignored for good reason: except in a few special cases, nonlinear differential equations do not have explicit solutions.' (Hayles 1991:163)

Print began a deeming cultural moment for contemporary Western people. It is heavily implicated in the dominant capitalist mode of social fabrication quite obviously through constructing needs in advertising and through the popular media. It is also less evident but no less influential in constructing a style of knowledge and ideas which are culturally acceptable and which are very influential. It underpins the sortive, taxonomic, reproducible, linear style of knowing that is the academic episteme and that has led to the powerful binary oppositions of 'qualitative' and 'quantitative' and the valuing of the latter over the former.

The feminist scientist Laurel Richardson describes print as being made up of the '...truth-constituting, legitimising and deeply hidden validifying of the genre, prose' (Richardson, 1993:103). By this she means that we are attuned to prose as conveying reality when it in fact is a constitutive part of the constructed culture in which we live. It is powerful and persuasive and establishes benchmarks for knowledge, information and communication. At the same time, it enacts itself as one side of a binary opposition. It ignores the holistic side of humanity and subjugates the personal and the story to the replicable, analysed and proven templates of scientific methodology.

Richardson, indeed, is writing in 'not-prose' in scientific areas so as to critique the myth of emotion-free, person-free scientific body of knowledge. She describes this as enacting 'the sociology of emotions' instead of locking these out of the knowledge process. So she is interested in 'not-prose', or writing which transgresses the hidden powers of prose, as a way of exploring

> '...the potential for relating, merging, and being a primary presence to ourselves and each other which makes possible the validation of transgressive writing, not for the sake of sinning or thumbing one's nose at authority,...but for the sake of knowing about lived experiences which are unspeakable in the 'father's voice, the voice of objectivity.' (Richardson 1993:705.)

The questioning of the metanarratives supported by prose presents us with new opportunities when we enter the digital production of textuality and discourse. Gregory Ulmer speaks of this process as both a mystery and a story: a 'mystory'. It is singular and involves the personal and the imagined as well as the intellectual.

'To approach knowledge from the side of not knowing what it is, from the side of one who is learning, not that of the one who already knows is a mystory. What is the experience of knowing, of coming to or arriving at an understanding, characterized as following a path or criss-crossing a field, if not a narrative experience, the experience of following a narrative?' (Ulmer, 1989:106)

For Ulmer the term 'Mystory' has a specific meaning which uses the personal and imaginative as a research tool. This not only joins together the personal and imaginative; it also opens up the potential of using different forms of inscription, particularly electronic ones. It is implicated in what Ulmer also calls 'the popcycle' (Ulmer 1989) which brings together expert knowledge, explanatory knowledge and everyday discourse and common sense.

This sense of a mystory opens up for academics the conceptual space of allowing students a singular journey through a cybertext. We have done this in our Media courses at Swinburne University of Technology, Lilydale Campus, via a dynamic website peripheral to a stable CD ROM. Over 2,00 students have experienced this form of online education over the past three years (1997/8/9) and it has been evaluated both externally and internally. From this experience, we can look back on the creative challenge of scripting for cybertexts in the academic domain.

2. MAKING THE CONCEPTUAL TRANSITION FROM PRINT-BASED TO ONLINE TEXUALITY AND DISCOURSE

Making the conceptual change from print texts to cybertexts is exciting and challenging. For academics who wish to digitise their students' learning opportunities, being enabled to script for cyberconstructions is an important step in this transition.

The need for immediate information has led to the recognition that there is a global electronic virtual reality available through electronic interactions and deliveries. This space has been called cyberspace, and it is currently being colonised. According to the magazine 'Wired', the cyber is '...the terminally over-used prefix for all things digital'. (Hale. 1996:66). While the novelist William Gibson who coined the term in his novel 'Neuromancer' described it as 'consensual hallucination'.

In the colonisation of cyberspace, there is much involvement in getting online, both as deliverers and users. Much of the material on the web, however, reminds us that nothing in our culture comes from nowhere. Almost all of the material displayed there comes from print: it is text-based or visual 'readings'.

So, of course, do our ways to apply and understand critical, literary and cultural theories. The work of Roland Barthes introduces us to the empowerment of the person who brings the text to life by participating in it. This work has been an important contribution to our understanding that we live in what Barthes calls 'An Empire of Signs' and others call 'social semiotics'. In thinking about the construction of the multilayered script for cybertexts we can revisit with interest Roland Barthes's ideas about the 'death of the author' and the concomitant empowerment of the reader, and fruitfully utilise the concept he developed from this of the 'writerly-reader'.

Such a readerly-writer can easily be seen as participating in a cybertext that is:

> '...made up of multiple writings, drawn from many cultures and entering into multiple relations of dialogue, parody, contestation, but there is one place where this multiplicity is focused, and that place is the reader, not, as was hitherto said, the author.'. (Barthes 1997:148)

From such cultural critics we can see how cybertexts have a theoretical siting. For Barthes, the text is never a finished and authoritative 'work' it is:

> '...not a line of words releasing a single 'theological' meaning (the message of the Author-god) but a multi-dimensional space in which a variety of writing, none of them original, blend and clash. The text is a tissue of quotations drawn from the innumerable centres of culture.'(Barthes 1997:142-3)

Indeed, French intellectuals of the late 20th century have dominated our understandings of textuality and discourse. Their work, particularly that of Jacques Derrida, provides us with fruitful ways of considering the construction of cybertexts and the possibilities of thus deconstructing them. Derrida's work on textuality and discourse opens up the opportunity to understand that all discourse is fabricated or made. It provides us with the opportunity of understanding and practising a totally new method of discourse in cybertexts: singular journeys constructed by the reader:

> 'The end of linear writing is indeed the end of the book...one begins also to read past writing, according to a different organization of space.' (Derrida 1991:50)

We come to print writing after many years of learning to read, utilise and write print material. We have many models. For us as academics placing our work online, becoming familiar with the new writing spaces and possibilities is an important aspect of writing for cybertexts.

The first step in this process, and one that is very fruitful for web connections, is to do a search to find out what is already on the WWW in your area, or what is already available in a cybertext form such as a CD ROM. Even looking at Encarta, the Encyclopaedia Britannica, offers good ideas about electronic textuality and its possibilities.

Currently, the web is like an endless electronic book, or even a slightly more biographic telephone directory (with pictures) self-published by individuals, groups and institutions largely as information. To critique texts like this, we need go no further than revisit the criteria which we have applied so long and so fruitfully to print textuality. Yet there are other electronic textual opportunities.

> 'Those ...who take delight in the intricacies of hypertext, the twisting web rather than the clear-cut trail, are perhaps seeing it as an emblem of the inexhaustibility of the human mind: an endless proliferation of thought looping through vast humming networks whether of neurons or electrons.' (Murray 1999:91)

Computer textuality and discourse gives us the opportunity to:

(i) interact with textuality in a very personal way which makes us co-authors or readerly-writers of the text through such interactivity.

The computer is providing us with a new stage for the creation of participatory theatre. We are gradually learning to do what actors do, to enact emotionally authentic experiences that we know are not 'real'. The more persuasive the sensory representation of the digital space, the more we feel that we are present in the virtual world and the wider range of actions we will seek to perform there.' (Murray 1997:125.)

(ii) explore and exploit the potential for lateral deliveries so that linear logic is disrupted.

The French philosopher Lyotard says that we must:

'...re-work the place of information and knowledge in time and space so that it is not enclosed information between the pages of a book nor knowledge dominated by a certain mindset or episteme The complicity between political phallocracy and philosophical metalanguage is made here: the activity men reserve for themselves arbitarily as fact is posited legally as the right to decide meaning...If 'reality' lies, it follows that men in all their claims to construct meaning, to speak the Truth, are themselves only a minority in a patchwork where it becomes impossible to establish and validly determine any major order...we Westerners must re-work our space-time and all our logic on the basis of non-centralism, non-finality, non-truth.' (Lyotard, 1989:120.)

(iii) recognise the importance of the qualitative over the quantitative in delivering singular cultural stories.

The feminist Rosemary Tong asks whether we can find a place for the local within the global and hence act to explode the dominance of binary oppositions in our cultural constructions:

'Whether women can, by breaking silence, by speaking and writing history overcome binary opposition, phallocentrism and logocentrism I do not know. All I do know is that we humans could do with a new conceptual start. In our desire to achieve unity, we have excluded, ostracised and alienated so-called abnormal, deviant and marginal people...as I see it, attention to difference is precisely what will help women achieve unity.' (Tong, 1989, 223/237)

(iv) recognise the ways in which culture is constructed locally and nationally and act to display and understand the essentially manufactured nature of our societies.'

The Cultural critic Gayatari Spivak says that:

'... human textuality can be seen not only as world and self, as the representation of a world in terms of a self at play with other selves and generating this representation, but also in the world and self, all implicated in 'intertextuality'...such a concept of textuality does not mean a reduction of the world to linguistic texts, books, or a tradition composed of books, criticism in the narrow sense, and teaching.' (Spivak, 1988:78)

Thus, we found that to see the power of a multilayered/hyper/cyber text, it was only necessary to tour the web. Starting with one website, hotlinks can take you on an individual navigation that you choose from a menu of possibilities that the links on each site to other sites present. In this way, each user of your text takes a journey over which you have no AUTHORitative control. They may begin with your text on a website and may go from there via many interconnected texts to who knows where? This is the power of the internet as an endless text with multiple possibilities and countless nodes of navigation that can be visited by the touring reader/writer. For even as the player tours the net she or he is constructing an individual journey which is singular and "written" by the individual themselves.

Such a tour of the net not only provides models and ideas for the scripting of your cybertext, it also gives relevant sites which can become to some extent a directed navigation for readers of your text.

People who look for information on the web are said to be 'surfing' the web. This is a powerful image of being at the forefront of a huge wave. You are on the edge of an ocean of information as you surf the web. Your aim is to get to shore with the information you want. Once again, we can see the word 'surfing' as carrying a masculinist Western semiotics, in this case of sporting achievements or leisure activities.

So even when we talk about the web we are forced to use metaphors to try to understand it within our own terms. We use something we are familiar with...a net, surfing, a web, nodes of navigation, search engines... to bring our real world and the cyberworld together. These words are not 'innocent'. They come from the culturally embedded and developed language.

> 'Most of what is delivered in hypertext format over the World Wide Web, both fiction and non-fiction, is merely linear writing with table-of-contents links in it.' (Murray 1997:87)

Some WWW texts also utilise pictures, films and audio. However, these usually appear to be more like 'add-ons' than intrinsic parts of a multi-layered, multi-possibilities electronic text or cybertext. We have already developed critical abilities in each of these areas, just as we have for books. However, although most of what is currently on the web is in print, it is not enough to presume that the print modality still evidenced on the web will continue to dominate electronic deliveries.

The times are transitional. They call for an understanding of interactive multimedia (IMM) which will lead to an exploration of what it delivers which differs from print and which offers more excitement and opportunity than the rather restricted ways of utilising and understanding IMM and the WWW itself that we have already seen.

3.   THE TRANSITION FROM PRINT-BASED TO ONLINE LEARNING

When scripting for our cybertext 'Oz 21 Australia's Cultural Dreaming' (Arnold, Vigo, Green), we began from where we were. We constructed a concept map of this and of how we might explore it electronically as well as verbally.

We learnt that the ideas that we wanted to convey in our electronic curriculum deliveries of CD ROM and associated websites first needed to be conceptualised in a mindmap, blueprint or concept map. One way to do this is to think about your starting point - **what I know** - and the final point **what I have conveyed**. Such a concept map is about the possibilities of presenting your information from its main ideas through to its conclusion and including feedback.

We have learnt that planning and writing for interactive multimedia is much more like script-writing than any other form of presentation such as prose, poetry, an essay or a novel. Re-visiting the criteria we apply for scriptwriting in film and television gives us an understanding of, and a starting point for, the ways in which we might capture the ideas, information and communications techniques which we want to implement in composing a cybertext.

In writing a preproduction/planning script for a cybertext, we are preparing a guide which involves:

1. the provision of information;
2. an interplay of the seen, the heard and the read;
3. the introduction of virtuality, interactivity, IMMediacy and self-authority.

Clearly, it is the third activity or series of activities which makes a cybertext a very different production from anything that has gone before. Nevertheless, as we see from activities one and two, we bring to this unfamiliar situation ideas about scriptwriting which have been long practised.

A famous story about the powerful media moguls of Hollywood in the 1950's, when they put writers at the lowest level of contributors to a film, tells of the writer who threw a script on to the producer's desk and said: "If you're so good at being an artist, make a film from this script". When the producer opened it there were only blank pages to be seen. What, then, might an IMM script be?

In scripting for Interactive MultiMedia, the person who has information which they want to make into a cybertext should not begin by wondering if they have the technical expertise to do this.

3.1 Step 1: Text

Academics wanting to go online should begin by clearly identifying what they already have in print form as text which they want to work with. This is the creative germ of the final cybertext.

The transition that has to be made from a print text to a multimedia text has multiple steps, the first of which is to see how this informative material can be arranged on the screen as text.

When we think about the book as a method of transmitting knowledge we think of information captured within the covers and revealed step-by-step from an authoritative source:

- there is a title page which defines the content in the broadest terms;
- there is the name of the author which gives the ownership of the intellectual property as well as the imprimatur of knowledge gained, captured and ready to be transferred;
- we find out who is the publisher and this gives us further clues as to the status, authority and style of the book;
- there is often a summary which gives clues as to the scope of the text;
- there is often an introduction which defines the area under consideration between the covers;
- then there is a table of contents to guide you through the contents in a logical and linear way;
- and a list of diagrams, maps or illustrations;
- each chapter might have a small summary

- within each chapter there may be sub-heads to direct the line of thought and to increase the receptivity of the reader within a clear and defined path
- a conclusion may draw the information to its logical end
- often there is an alphabetical index at the back which helps you find specific points in the text
- there is usually a bibliography in information-style books which sites the work in a certain area of critical thought and provides further reading in the area.

You may take some or all of these aspects into an electronic form, but in a very different way. A word of warning here: If your script can only do what is already done in print or in book form, then leave it in the old and proved technologies. Unless you are ready to script with the new technologies as your form of delivery, then you are unnecessarily replicating print electronically.

Text on the screen is very different from text on the page. Instead of being arranged in long paragraphs containing complex thoughts, the first step should be to chunk it on the page into different screen interfaces. This might initially be quite simply done by going though it and putting in sub-heads, then by colouring with see-through text as the various 'levels' of information to show what you will put together at different levels of the electronic text or what you might bring up in frames or boxes or through hotword connections.

In a PowerPoint presentation, the sub-heads could each provide a main frame. On a CD ROM or website the hotlinks could be live and could bring up further levels of information as the user required it.

Your script begins as a print presentation that has been broken into frames and chunks and indications have been given as to what might be hotworded and what might appear as boxes, on a split screen or as mobile text.

You might now choose to rearrange the traditional columned text into different text elements such as:
- frames;
- hotlinked words and what they display when clicked on;
- split screen.

You may also wish to include elements such as:
- sound;
- video.

In doing this, you may begin to see that there is too much informative text. You are relying on finite linear print to make an infinite multilayered text that will go on to include interactive multimedia, virtuality and three-dimensionality.

Thus, even within this preliminary step, you may choose to prune your information. There is a hard choice here, because the mind trained to linear print is also trained to logical step by step thinking that goes from point to point and has supporting argument and information. Thus you may need to think differently about how you deal with your information to cover what you believe to be the necessary ground without losing sight of the new textuality.

3.2   Step 2: Sound and Sight

We are already quite familiar with using sound, still photographs and diagrams as part of a book, with the addition of videos and films as part of a lecture or presentation. The sense we have of such visual and aural applications is that they are 'add-ons' to the central verbal and prose text itself. The non-verbal information of the visuals, whether stills or filmic, is important in itself, but seldom stands alone in an informational text. Similarly, sound is usually there as a voice-over addition to the main text or as music to emphasise an area of thought or as an example of an idea. That is, it is seldom explored as an art-form with its own creative capacities or as a text in itself. It is most often used as a print adjunct in some way.

Electronic textuality which includes multimedia emphasises sound and sight as an intrinsic part of the text itself. You will need to look at the columned script you are developing. Go through it and see what you might colour with see-through text as for each medium. Ask yourself how you could present the prose text, the chunks of verbal information, visually and/or aurally. The questions that should be uppermost is 'can this be done better visually and/or aurally?' and 'what will be value added to the way the information is conveyed in this text through the addition of sight and sound, movement and action?

Now you will need to add columns to your script for video, sound and stills. We learnt at this stage to re-arrange our print script so that we had a different typeface for each of the different columns. This horizontality was easier to handle than vertical columns as the script becomes more complex. It also presents a more appropriate look for the idea of the multi-layered text.

Some questions which you need to consider here for the aural aspects of your cybertext are:
- why are you using sound?
- is it because you want your 'real' voice to be present?
- are you only putting a voice-over on to the print information?
- is it adding something to the presentation which makes it a more lively and interesting text such as asides, case studies, reminiscences or examples?
- how will people using this electronic text access the sound?
- will it have a special button?
- what is the sound quality?
- what part might music play?

Some questions to consider for the stills visuals are:
- what is the purpose of this picture?
- where does it sit in relation to the print?
- how will this information be used by navigators of this text?
- what is the quality of this picture or diagram in terms of visual reception, spatial environment, design?

Some questions which might be appropriate for audio-visual film or video clips are:
- how long should it be?
- is it well designed in its own terms?
- what is its relationship to the other aspects of this now quite complex multi-layered and non-linear text?

Your script now has many clues as to how it can be presented as a cybertext. You have not yet, however, come to grips with the most important and exciting elements offered by electronic writing: Interactivity, virtuality, three-dimensionality, animation, singular journey.

These add many very new layers to the already complex text which is coping with aspects of writing, media and multi-media with which we are already quite familiar. These aspects are the entirely new in the conceptual space of writing for cybertexts.

3.3 Step 3: Interactivity, virtuality, immediacy and self-authority

Because these three aspects offer us an entirely new way of thinking about how a text will be 'written', 'read', 'used', or 'navigated', we need to re-arrange our conceptual spaces to take them into account. This is quite a challenge, as we have been formed by and are so familiar with the era of literacy in which we have been brought up and which is the basis for the advanced dominant Western culture which is driving the emergent electronic culture.

How will your text use the potentials for interactivity, virtuality, immediacy and self-authority? These elements offer the navigator:

- the capacity to interact with the text in a way which empowers them to move as they desire between pieces of information;
- self-testing and instant feedback;
- the ability to manipulate images and information;
- spaces to import information, ideas or opinions;
- chatrooms for synchronous interaction;
- discussion threads for asynchronous interaction;
- bulletin boards.

3.3.1 Virtuality

Cybertexts take place (are written and read, performed and navigated) in cyberspace. This means that they are not in real time or in real places. They are in a 'virtual reality'. At first glance these two words appear to be in conflict and this image is a deliberate semiotic challenge to the normalcy of our expectations even of language and meaning itself. It shocks us so as to enable a re-visiting of the know and an entering into the unknown.

Virtual reality is in its textual infancy. Of course, such genres as poetry and fiction and such technologies as television and film have paved the way for us to begin to understand elements of this construction of a reality in an electronic text. What might virtuality offer our multilayered script for an information cybertext? Is there a place for serious information to intersect with storytelling and the constructions of new and individual realities?

If your script is, for example, about an engineering site, then virtuality offers you the opportunity to build a virtual site which can be navigated by the readers so that specific information can be given at relevant points. A virtual space offers information the envisualisation of what is being described. This envisualisation obviously allows interactivity. In this way, the reader becomes an active player in the text in a way which has not been possible in print.

A layer in your script for the construction of a virtual reality will add a further dimension to an already complex document. Revisiting it at each stage will probably reveal to you that some of the print information can now be presented in non-print form.

We bring to the flat screen of the computer our experiences with film and television. The filmic screen tries to get human-vision depths and dimensionalities through sets and point-of-view camera shots. The television video camera is not so ambitious, and the screen views have a curious flatness. Nevertheless, the main way in which humans view the world is three dimensional and all representations of reality have sought to depict this.

New computer programs offer us the capacity to create three dimensional representations and to manipulate them for better understandings of their real complexities. For example, one website has a three dimensional manipulable representation of triple helix which shows how DNA is constructed in ways that verbal descriptions of print diagrams could never achieve.

There is much that we seek to describe in information that can best be seen. How will you add the aspect of three dimensional figures that can be manipulated (built-up, moved, viewed from varying perspectives, seen in various parts) to your script? Are there particular parts of your information which could best be conveyed in this way?

3.4 Step 4: Programs

Each cybertext allows for an individuality which is the journey selected by the interactor or, in Barthes's terms, 'the readerly-writer'.

A major layer in your script will be the nomination of programs or parts of programs which might be used to achieve the fullness of a cybertext. The person who is coming to this from a print background may now seek computer-expert or team to help to achieve this. It is important that the technical aspects of making a cybertext do not come first because this moves the emphasis from the content of the information to the presentation of the information. As we are involved in transforming information into knowledge, the technology support group should be there to take care of the programs.

3.5 Step 5: Storyboard

How will each screen look? What will happen on it? In what ways can the pre-production be conveyed to the production team?

A storyboard for a film is a series of still pictures or sketches which show the order of activities and scenes in the film. Their purpose is to bring the written and visual script closer together. They look like a comic strip with subtitles and speech bubbles along with sound effects.

The purpose of a storyboard for films is to provide instruction for:
- camera angles;
- lighting;
- actor's positions and movements;
- how the props are sited;
- sound effects;
- music;
- narrative developments;
- dialogue;
- common understandings between all involved in the film;
- overcoming any language differences;
- showing the director's vision of the flow of the story;
- keeping correct continuity;
- any adjustments that need to be made;

Interactive visual storyboards are different because of the complexity of what has to be shown on a single screen and of the applications behind it. In designing storyboards for multimedia one might have a single large sheet of paper for each action as it occurs on each screen as a variation on what I have called the multilayered horizontal script. Storyboarding for interactive screens helps you to establish the texts, pictures, multimedia and links that will be on the screen.

> 'A storyboard is a communication tool for the people involved and should ease the process, it would be quite easy to produce wonderful storyboards which are accurate but still lose money on the application.'(http://www.alpeda.shef.ac.uk/fr)

Different tools for the creation of storyboards include:
- creating a separate storyboard through a script on a wordprocessor;
- using a specific graphics application which make simple screen shapes where pictures and words can be placed;
- object databases can be constructed to store detailed information about the construction of the screens;
- multimedia authoring packages (Director; Authorware) and multimedia planning packages (Designers Edge) which structure storyboards.

Any script or storyboard will help to establish and sustain the creative energy of the project. It will help to ensure that the deliverables are known to all and that the project comes in on time and on budget.

4. CONCLUSION

Scripting for cybertexts is an exciting challenge as we come to the possibilities offered by the new technologies almost by accident. In the earliest stages of the Age of IMMediacy, we are still searching out the ways in which the new technologies will lead us to new practices of textuality and discourse, to new understandings of research and publication. There is much to learn, for:

> 'All the major representational formats of the previous five thousand years of human history have now been translated into digital form...And the digital domain is assimilating greater powers of representation all the time, as researchers try to build within it a virtual reality that is as deep and rich as reality itself.' (Murray 1997:28).

This paper has surveyed ways in which academics might begin to script their curriculum utilising the opportunities presented by websites and CD ROMS. It puts the curriculum development for flexible online course delivery back into the hands of the academics and reminds us that the technology is there to expand our capacities, and we may call upon technology experts to help us to bring our curriculum scripts to fruition.

5. REFERENCES

Arnold, J., Vigo, K., Green, D. (1997/8/9) *Oz 21: Australia's Cultural Dreaming*. CD ROM Rock View Press: Australia.

Barthes R. (1997) *Image-music-text*. Fontana/Collins:London.

Derrida J. in Kamuf P.(ed) (1992) *A Derrida Reader Between the Blinds*. Harvester Wheatsheaf: New York.

Gibson .W.Neuromancer.(1986) Ace Books: New York.

Haraway. D. J. (1991) *Simians, Cyborgs, and women: The reinvention of nature*. Routledge: New York.

Hale. C. (1996) *Wired Style: Principles of English use in the digital age, from the editors of Wired*. Hardwired: California.

http://www.alpeda.shef.ac.uk/fr

Lyotard J.F. (1989 (ed) Benjamin A , The Lyotard Reader, Basil Blackwell: Oxford.

Murray, J. H. (1997) *Hamlet on the Holodeck: The future of Narrative in Cyberspace* The MIT Press: Cambridge, Mass.

Richardson, l. (1993) 'Poetics, dramatics and transgressive validity: the case of the skipped line'. *The Sociological Quarterly*, Vol 34, No. 4, pp695-710.

Ulmer, G. (1989) *Teletheory: Grammatology in the age of video*, Routledge: New York.

Spivak, G. C. (1988) *In other worlds. Essays in cultural politics*, Routledge: London.

A NEW PARADIGM FOR SUCCESSFUL ACQUISITION OF INFORMATION SYSTEMS

Michael Gorman
White Marsh Information Systems
2008 Althea Lane
Bowie, MD 20716, USA

Keywords: Information systems development, database, prototyping, design iteration, meta data repositories, code generation, CASE.

Abstract: Most information systems projects are over budget, under specified, delivered late, and fail to meet organizational expectations. A key reason for these characterizations is that during the information systems development, requirements both change and are being introduced. This paper presents a strategy whereby full development is started only after a series of realistic code-generator produced prototypes are created through mission-based database design techniques. The quickly accomplished prototypes are demonstrated to key user groups, and are iterated until they truly reflect a quiesced set of valid and accurate requirements. Production class systems created through CASE, metadata repositories and code generators will then be on-budget, correctly specified, timely, and will meet organization expectations.

1. OVERVIEW

Many, if not most, information technology (IT) projects exhibit these characteristics: over budget, under specified, delivered late, and fail to meet organizational expectations. (Standish 1999, Matson 1996, Strassmann, 1997)

The United States Government's General Accounting Office (GAO) has been studying IT projects for a number of years, and a review of 10 of the GAO studies clearly shows that the main reason s why IT systems fail has nothing to do with IT. (Gorman 1999), USGAO 1997).

Systems Development Methods for Databases, Enterprise Modeling. and Workflow Management.
Edited by W. Wojtkowski, *et al.* Kluwer Academic/Plenum Publishing, New York, 1999.

455

This paper presents a nine-step process that addresses many of the GAO IT findings. These nine steps presume, however, the existence of the following IT supports:

- An installed knowledge worker framework,
- Enterprise data architectures and effective data standardization,
- A data-driven oriented systems development methodology,
- The availability of high-quality code-generation tools,
- Metadata repository environment, and
- CASE tools

All these supports exist today, and when employed, their cost is negative because of the time and money savings in the first several projects in which they are employed. The nine steps are:

- Mission Development,
- Database Design,
- Prototype Generation,
- Specification Evolution through Prototyping,
- RFP Creation,
- Vendor Responses Evaluation,
- Contract Award,
- Contractor Management, and
- Conformance Testing

This paper presumes that many public-sector agencies do not contain significant IT development organizations and have no interest in creating them. Rather, this paper presumes that public-sector agencies prefer to specify IT needs through some sort of central design authority, and then through the for-profit IT vendors, procure, install, employ IT solutions that are both functionally acceptable and conformance tested to ensure common functionality across public sector agencies. (Strassmann 1998)

Whenever public sector agencies take on the full development, operation, evolution and maintenance of IT systems they commonly fail to achieve optimum results. While there can be many reasons for this, the GAO studies show the most common to be:

- Failure to meet initial end-user expectations,
- Inability to continuously infuse advances in technology in the deployed IT environments, and
- Inability to successfully accommodate the long-standing tradition of individual autonomy.

This last reason, autonomy, prevents the effective deployment of IT systems based on a unified system's design and implementation strategy across a group of public sector agencies because:

- One size does not fit all,
- Public sector agencies often require slightly different functionality, and
- The cost of evolving IT systems through old technologies and approaches is neither cost effective nor viable.

A solution that does work is one that capitalizes on the strengths of both the public- and private-sectors, while avoiding their weaknesses. The proposed approach, modeled on the one employed for the internationally recognized and accepted data management language SQL, consists of a three-part paradigm:

- A vendor community with the profit motive to develop, sell, install, evolve, and maintain products,
- A central design authority to define, validate through prototyping, and maintain functional requirements, and
- Conformance tests and testing by the central design authority to ensure that all systems sold by the for-profit vendors conform to the minimal essential functionality.

The remainder of this paper describes the nine-step approach. Even though this nine-step approach is optimum for large, heterogeneous hardware environments across multi-site public sector agencies, this approach can be simplified to accommodate homogenous and/or single-site public sector agency environments. The first four steps are the responsibility of the public sector, the next four steps are the responsibility of the for-profit private sector, and the last step is the responsibility of the public sector.

2. MISSION DEVELOPMENT

A common lament from IT professionals is that whenever new or changed requirements surface during the IT systems implementation phases, slippages, cost overruns and significant rework almost always result. Users counter that new or changed requirements arise because they didn't fully understand the "problem" that was being solved.

The impact of requirements' rework is based on the following axiom: If the costs associated with requirements and design costs $1 then the activities associated with detailed design through initial system implementation costs another $5. Figure 1 illustrates the main phases associated with the traditional first implementation life cycle of an enterprise-wide IT system such as human resources, or accounting and finance. The scale on the bottom represents the quantity of months spent in each phase. The curve clearly shows that the bulk of the effort (commonly about 70%) is expended before any demonstration is possible.

Beyond the first-implementation cost, the total life cycle expenditure for system revision cycles (not shown in the diagram) commonly costs 4 times more. The total cost is thus, 30 times the design cost. The problem, however is not that requirements change, it's that the effects of changes are too costly to reflect in the implemented system. Because $1 in requirements' changes cause $29 in additional life cycle costs, the exhortation is simple: Get the requirements right the first time because the cost of change is prohibitive.

A review of the GAO studies of IT systems' failure show that new requirements during systems development are such a common occurrence that they must be considered intrinsic. The problems that arise from new requirements fall into two categories: Database design changes and software changes.

Significant database design changes are often a result from an inadequate data driven methodology through which the database is initially designed. Experience has shown that very high quality database designs commonly return many times their design cost in reduced software development and evolution costs.

Software changes result from database design caused changes and also from intrinsic process logic changes. Database design changes can be largely eliminated through the use of a quality database design methodology. The onerous effect of intrinsic process logic changes, can be dramatically affected through the use of object-oriented analysis, design and programming techniques employed within the environment of code-generators. The axiom of 1:30 may be reduced to 1:10. (Matson, 1996)

Given that requirements "naturally" change and are also significantly affected by accelerating technologies, the ability to posit an accurate and long-term set of requirements is close to impossible, and any IT system developed on the basis of unstable requirements is doomed from the outset.

An alternative to attempting to build an IT system on an unstable platform of requirements is to build from a platform of stable enterprise missions. Mission descriptions are characterizations of the end result independent of: technology, "who" and "how." Well done, mission description documents are timeless, technology free, and apolitical.

3. DATABASE DESIGN

Database design is the natural next step as *Data is executed policy*. A database's design is thus the overall schematic of the policy domain of the problem space. A quality database design properly reflects the enterprise policy domain rather than ad hoc reporting needs.

The GAO studies show that most IT projects fail not because organizations chose the wrong technology or are not fast or flashy enough, but because they do not engage in the activities that result in proper mission descriptions and policy determinations. These are essential to form the foundations for quality database design.

Public sector agency policy experts, not data processing experts, must accomplish database designs, the schematics of policy domains. Abrogating database design responsibilities to IT professionals is a major cause of project failures. There are quality database design methodologies that are fast,

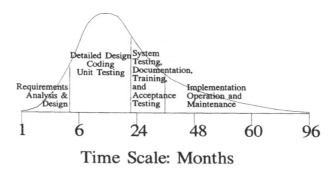

Time Scale: Months

Figure 1. Main phases of information systems life cycle.

efficient and above all engineered to properly reflect the policy domains of the enterprises.

4. PROTOTYPE GENERATION

Code generators are a class of computer software that take in database design specifications and then produce working computer software systems. Generators have been growing in sophistication and capability over the past 15 years. The code generator environments, such as Clarion for Windows (www.topspeed.com) produce first-cut working systems from database designs in one-hour or less. In contrast, hand-coded systems take upwards to 2 staff weeks per module. Thus, for a 50-table database that requires 150 modules, a hand-coded first-cut working system requires about 300 staff weeks, or 6 staff years. As a prototype, that's unacceptable. But, the equivalent system in Clarion for Windows takes less than 2 staff weeks. As a prototype that's acceptable.

5. SPECIFICATION EVOLUTION THROUGH PROTO-TYPING

The value of the first-cut working version of a system is that it provides a first "look" at what was inferred from the "requirements." If each subsequent "look" is only a few days away, then, prior to committing to the first real version of a system, a prototype can proceed through 5 to 15 iterations. The value is dramatic. First, because there has been a real attack on the cost of initial system production, and second, because there has been an elimination of many causes of the major system revision cycles. Figure 2 depicts the process of specification evolution through prototyping. The scale, also in time, but weeks, is greatly reduced because of code generation. If the resultant prototype is considered acceptable, it could be turned over to production.

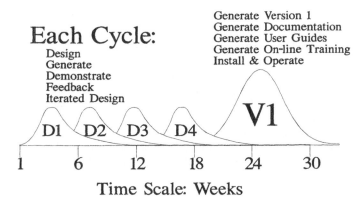

Figure 2. Specification evolution through prototyping.

Once a prototype is created, it can be taken on a "road-show" wherein critical audiences can see what the system actually does. If the "road show" appearances are a day or so apart then minor revisions to the system can be put in place and tested with the next audience.

The goal of the prototyping activities is the finalization of IT system specifications. The result of these activities is not only a valid system specification, it is also a complete set of metadata that can be employed during implementation and then subsequently evolved during the system life cycle. Prototyping also enables the creation of test data that be used for training, documentation, implemented system test cases, and the all important conformance tests.

There is a great temptation to turn the final iteration of a prototype into a production class system and be done with it. There is, however, much more to a production class system than just a working set of computer programs. For example, there are likely functional differences among the different levels and types of public sector agencies. There is the need for multiple types and sizes of hardware, operating systems, database management systems, and so on from different vendors. There are also different approaches to system implementation depending on whether the systems are to be implemented on a network of micros and servers, or mainframes with multiple LANS, servers and clients. Finally there is also the need for formal and informal training, technical support, documentation, and user group meetings.

6. RFP CREATION

The role of an RFP (request for proposal) under this approach of central specification development and for-profit vendor community implementation, delivery, support and evolution is non-traditional.

Traditionally, a procuring agency creates a requirement's specification, issues the specification to a number of bidders, receives proposals from vendors, selects a winner and then provides funding to the winning contractor as it creates the information system. Subsequent to the delivery, the contractor performs warranty maintenance for a period of time and thereafter makes functional changes on a fee for service basis. Under the traditional approach there are three main types of development contract alternatives:

- Firm fixed price
- Time and materials
- Cost plus fixed fee

Under the firm-fixed-price approach, whenever there is a requirement's change the contract's funding is increased and the original delivery date is slipped. While, under the time-and-materials approach, whenever there is a requirement's change the contract's funding is increased and the original delivery date is slipped. And, finally, under the cost-plus-fixed-fee approach, whenever there is a requirement's change the contract's funding is increased and the original delivery date is slipped. Different? Not much.

While the costing and fee structure may be different, the ultimate result is the same. That is, under any of the three RFP procurement alternatives all the risk for proper specification is with the contracting agency. Whenever there are new or changed requirements the contractor immediately requests a contract modification which almost always means a schedule slip coupled with an increase in funding. All the risk is with the public-service agency, and the relationship with the contractor is almost always adversarial.

Under the new approach proposed in this paper the procurement paradigm is fundamentally changed. Rather than adversarial, the role of the public sector agency is to promote maximum involvement by for-profit implementation vendors by providing:

- Technical support for the proven technical specification,
- A completely working prototype of the system,
- Access to the public service agency's metadata repository,
- Immediate release of any specification updates,
- A complete set of initial and iterated functional conformance tests, and
- Test and certification environments of vendor developed systems.

The implementing vendor-community has the following freedoms under this approach:

- They can choose the implementation platforms, hardware, operating system, telecommunications and interoperability environments, and database management system,
- They are free to enhance and evolve the functional specification, and
- They are free to sell to whomever they want providing of course that they have passed state mandated conformance testing
- The benefits to the public sector agency under this approach are:

- It's financial obligation is limited to the development of a completely functional prototype,
- It does not have to deploy system development organizations
- It does not incur technical support and maintenance organization costs.

The benefits to the for-profit vendor-community are:

- Vendors are relieved of having to absorb 100% of the cost of requirement's determination, system design and prototype generation
- Vendors do not have to absorb the cost of design iterations and possible loss of reputation resulting from prototype iterations through beta site testers who are seldom happy.
- Vendors do not have to find the market to which to sell its new product.
- Vendors do not have to prove their products acceptable beyond mandated functional conformance tests

All in all this strategy is WIN-WIN as contrasted with the adversarial relationship between the public-sector agency and the for-profit vendors. The RFP must clearly set out this new strategy of "procurement" and invite the for-profit vendor community to participate.

7. VENDOR RESPONSES EVALUATION

Vendor responses, which would actually be requests for acceptance into the product development effort, should contain:

- Descriptions of the system's design and development strategies,

- Methodologies and quality controls that the vendor is willing to undertake,
- The management and controls that would be placed on systems development to ensure quality product development,
- Proposed schedules,
- The expected quantity and duration of meetings with the public sector agency staff, and
- References of past efforts of a similar nature that would ensure the public sector agency that the vendor would likely complete the effort.

These types of proposal materials are critical because unlike a traditional contract-for-delivery effort, the vendors are largely on their own to produce products. The arrangement is really a partnership that is governed by voluntary participation and expectation of the mutual rewards listed in the RFP Creation section.

8. CONTRACT AWARD

Contract is certainly the wrong word to express the result of this type of cooperative process between public sector agencies and for-profit development vendors. The proper phrase should either be "cooperative agreement" or "memorandum of understanding."

Once a number of vendors are selected, memorandums of understanding can be scripted that enable vendor access to materials, key members of public sector agencies and special interest groups who would air and then resolve ongoing detailed design and implementation problems, progress status, and the testing of early releases of products.

These group meetings would enable the intended public sector agency customers to be aware of progress and to factor the impending availability of products into their budget cycles.

9. CONTRACTOR MANAGEMENT

Contractor management within this type of cooperative arrangement is unique as participation is not really mandatory. The best type of arrangement would be a series of meetings over a period of months wherein vendors could work with public sector agency committees addressing in-progress work, demonstrations, public sector agency development of test data and functional test scenarios, vendor-surfaced anomaly resolutions, central public sector agency development and promulgation of process handbooks, fundamental-process user guides, common process training materials, and the detailed formats for data exchange through the export and import functions.

10. CONFORMANCE TESTING

The final step is conformance testing. It simply answers the question: How will the public sector agencies know if they received a properly working system? Traditionally, the answer has always been, *caveat emptor.* To resolve that healthy skepticism, the procuring agencies must develop a set of functional acceptance tests. To be useful, the tests must be valid and comprehensive.

The benefits from a comprehensive and valid set of conformance tests are compelling, and include:

- All parties agree-to a minimum essential and testable set of functionality,
- The public sector agency knows that all sold products perform the minimum essential set of functionality,
- The procuring agency procurement costs are reduced because test development and testing is centralized, and
- The for-profit vendor selling costs are reduced because they only have to pass one set of centrally administered tests.

11. CONCLUSION

The solution proposed in this paper works because it capitalizes on the strengths of the government sector and the private sector, that is,

- A vendor community that has the profit motive to develop, sell, install, and maintain IT systems,
- A public sector central design group that defines, validates through prototyping, and maintains functional requirements, and
- A cooperative conformance testing activity that ensures that all systems sold by the for-profit vendor community conforms to the minimal essential functionality.

REFERENCES

Gorman, Michael M. Knowledge Worker Framework: 1999. Web: www.wiscorp.com

Matson, Eric, 1996. Speed Kills (the Competition). Fast Company (August): Page 3. Web: www.fastcompany.com/online/04/speed.html.

Standish Group. *CHAOS: 1998: A Summary Review:* 1999. The Standish Group International, Dennis, MA 02638.

Strassmann, Paul A. *40 Years of IT History:* 1997. Page 6. Web: www.strassmann.com/pubs/datamation1097/index.html

Strassman, Paul A. 1998. *Outsourcing IT: Miracle cure or emetic.* Web: www.strassmann.com/pubs/outsourcing.shtml

USGAO 1997. United States Government Accounting Office. Managing Technology: Best Practices Can Improve Performance and Produce Results,: (GAO/T-AIMD-97-38). Washington, D.C. (web: www.gao.gov)

TOTAL QUALITY MEASUREMENT:
ISSUES AND METHODS FOR MEASURING PRODUCTIVITY FOR INFORMATION SYSTEMS

Richard McCarthy
NOVA Southeastern University
171 Streamside Lane
New Britain, CT 06052-1564, USA

Executive Summary

Information Technology (I.T.) comprises an increasingly significant amount of an organizations budget each year. However, very few organizations measure how productive their programming staff is. A brief survey of seven claim information systems organizations within the property casualty insurance field uncovered only one organization that applied formal measurement techniques to their software application development. The Travelers Property Casualty Corporation Claim Information Systems unit developed a technique they refer to as complexity point calculation to measure the size (complexity) and quality (as determined by defect ratio) of each of their software versions. Complexity points are calculated by software version, providing an individual measurement scorecard of the quality of each software version. Complexity points are a streamlined version of Function Point Analysis, which was first developed by A.J. Albrecht of IBM's Data Processing Services Division and made public in 1979. This paper will examine both function point analysis and complexity point calculation, contrasting the difference between the two techniques. In addition, the importance of software quality metrics will also be discussed.

Complexity point calculation provides a simple, quick and effective means to calculate the quality and complexity of a software version. The technique was first developed to measure changes made to the Travelers Property Casualty Claim System. The system being measured consisted of over 12,000 programs, modules, screens, copybooks, jcl procedures and tables that comprised over 1.2 million lines of source code. Most of the program code is written in the COBOL and C programming languages. The system is comprised of an object-oriented workstation that is interfaced with a mainframe operational transaction system.

Background

In it simplest measure, productivity is defined as output / input. Productivity measures have not been universally embraced within information systems teams, which has led some senior executives to question the value of the supports delivered by their information technology organizations. Information technology can benefit an organization by adding new business functionality that may be needed to gain a competitive edge. Failure to be able to deliver this functionality quickly, and with quality may result in lost opportunities. Establishing a measurement system allows an organization to evaluate over time the change in the quality of the software that is delivered. An effective measurement system can be a valuable management tool to look at whether there have been process improvements in the software development process. To be effective though, the measurement system must be interrelated to the supports that

Systems Development Methods for Databases, Enterprise Modeling, and Workflow Management,
Edited by W. Wojtkowski, *et al.* Kluwer Academic/Plenum Publishing, New York, 1999.

are developed. The metrics used to measure quality should remove as much of the subjective decision making as possible. Measures should be based upon data that is easy to gather and can be analyzed quickly and efficiently. An organizational culture that recognizes the value of measures, and promotes their usage should be developed in order to maximize the value of the information that can be derived from the measures.

Software quality differs from product quality in that each piece of software is unique. The software quality definition therefore usually centers on compliance with requirements and fitness of purpose (Daily 1992). This assumes that accurate specifications are developed. If this is the case, then software quality can be evaluated using the following characteristics:

- Usability
- Correctness
- Reliability
- Efficiency
- Portability
- Integrity

These characteristics can be used to address quality issues reactively, investigating the causes of non-compliance and performing corrective actions, as well as preemptively, analyzing information to eliminate potential causes of non-compliance.

Changes in programming languages from third generation sequential languages such as COBOL or PL1 to event driven and object oriented languages has meant that information systems can be developed and changed more quickly. Few attempts have thus far been made to establish any baseline measures to understand and quantify how effective an information systems organization is. In addition, baseline measures of the quality of software version are most meaningful when they are kept up to date, so that evaluations spanning versions can be measured.

Productivity in information systems can be measured in terms of business benefits delivered as well as an individual measure of programmer effectiveness. Measuring the business benefits derived from IT will assess efficiency and effectiveness. Information technology benefits to an organization can be measured by the delivered business services divided by the information technology expense. At the enterprise level this presents a measure of the efficiency of the information systems organization. This focus on financials means that the benefits of IT must be expressed in quantifiable terms. *Intangible benefits* don't fit within this macro model. Using this simple measure can help align the IT organization with the customers they serve as they will have to work together to quantify and substantiate the benefit of systems services. A measure that is complimentary to efficiency is the effectiveness of the IT organization. An organization can be efficient without being effective, they may deliver things quickly that have little value to the organization. Effective IT organizations deliver real business value. This value may be expressed as time to market, improvements in customer service and response rate, or expansion of the customer base. To improve the effectiveness of the information systems organization, application developers have to understand how the underlying business operates. Improving the effectiveness and efficiency of the IT organization will result in a measurable productivity gain.

The need for measurement of IT results has not been universally accepted or understood. While solutions have been sought to the axiom "what gets measured gets managed", (Willcocks and Lester 1996), a single set of guiding measures has never been developed. Some of the difficulties in measuring IT contributions have not been adequately addressed. One of the difficulties has been that measurement systems have been too difficult and costly to maintain. In order for measurement to become an accepted part of the IT culture it must be viewed as a means to link individual behavior to the rewards system of the organization. To often it is viewed as an additive process that isn't focused on the concern for demonstrating the value that IT can add to an organization. Another downside to measures is the tendency to focus in only that which is being measured. *What gets measured gets managed* does not imply that the only thing that is managed are the measures. The 'IT productivity paradox' (Willcocks and Lester 1996) is that over the years, despite the large investments that organizations have made in IT, they continue to question its value because of a lack of empirical evidence that demonstrates that the IT organization operates effectively and efficiently. It has been difficult for IT management to evaluate if their organizations have been making progress in the improvement of the systems delivery process. Quality improvements add value to the IT service because they enable the organization to remain focused on growing the systems beyond their current boundaries. Otherwise, too much resource is wasted performing tasks that do not add value to the organization. Productivity measurements that have been utilized thus far fail to recognize the value of opportunity cost and cost avoidance. A lack of quality costs the organization both in terms of time spent completing the rework necessary to correct the problem, as well as the time not spent creating new business value. The latter cost may significantly impact an organizations ability to remain ahead of its competition.

Two early studies, which evaluated the productivity impact of IT, were by Roach (1986) and Loveman (1988). They reviewed deficiencies in assessment methods and measurement of productivity. These point to the lack of management focus by information systems in the implementation of productivity improvements. Productivity improvements, like the benefits that they support can take years to evaluate.

Effective measurement begins with benchmarking, establishing the quality and productivity standards by which future measures will be compared. The benchmark is the indicator of the effectiveness of improvements. It establishes the basis for evaluation of the measurement system. Comprehensive measurement should address three fundamental areas (Brown, Waldman Light 1995), productivity, technical quality and functional quality. In this definition, productivity is the efficiency rate for delivering function to the user. Technical quality is the correctness and efficiency of the production process. Functional quality is the degree to which the application meets the needs of the user. All three aspects need to be measured in order to provide the true measure of gain within the IT organization. The next step should be to look for opportunities to improve productivity or quality. Process improvement can be accomplished in many different ways, ranging from relatively minor steps such as education and training of the staff to the more comprehensive reengineering of the organization. Once improvements are defined an implementation plan should be developed and implemented. Finally, it is important to remeasure. It is in this final step that the indications of improvement will be validated.

One question that information technology organizations face is, "Why bother measuring software quality". One reason lies in one of the most commonly referenced measure, the cost per defect. The cost per defect depends upon where in the development cycle the defect is uncovered. The systems development life cycle typically identifies five major phases in the software development process; requirements definition, detailed analysis, construction, testing and implementation. Software defects found during the construction phase cost an organization less in terms of lost time because it does not carry the additional burden of having to retest the software. By extension, defects found during testing cost an organization less than those found after implementation. Defect repair costs consist of the time spent identifying the problem, the time needed to code the correction, the time needed to retest the software, and the time needed to implement the correction. In his book, *Quality Is Free*, Phil Crosby identified three cost of quality categories; Prevention costs, Appraisal costs and Failure costs. Prevention in software development consists of simplifying complexity and reducing human error. Simplifying complexity can be achieved through enforcement of structured coding techniques. The time required reviewing and enforcing structured coding. Appraisal costs consist of the costs associated with testing and inspections. Formalized code reviews are an example of the use of appraisal costs that can prove beneficial in the implementation of quality software. The failure cost is the defect repair cost. As new tools and means for developing software have been implemented over the years, measuring the total quality costs is an essential measure to understand the effectiveness of the tool. For example, Case Tools were introduced in the 1980's for the purpose of expediting the development of quality software. Without an effective baseline measurement, it is difficult to assess the effectiveness of the tools. Lowering defect rates will serve to increase user satisfaction in the long run, because it enables the organization to focus on increasing the functionality of the system.

One of the pioneers in the field of software measures was Halstead. He proposed a codified system of equations that were based upon simple counts derived from program source code, which included the number of unique operands (constants and variables) and operators (function calls, decision statements). He claimed this could be used to predict a wide range of software quality characteristics. Though flawed it proved influential in two respects. It was one of the first efforts to show software as a product that could be measured and analyzed. Second, it reinforced the focus on complexity measurement (Sheppard 1995).

One of the first attempts to measure the productivity of the software development process was to utilize source lines of code as a metric. Source lines of code refer to a physical count of the lines of code in each program in a system. This produces flawed results for a number of reasons. First of all, it fails to recognize the importance of components other than source code in the successful implementation of a quality software application. Object-oriented components provide the additional benefit of reusability, improving the time to deliver new business value. Using this measure would fail to take into account the use of tables, maintained by users to reduce the level of program maintenance needed. It also fails to recognize the mix of various program languages that can be used in the development of a software application. Programs written in COBOL for instance can be substantially longer than a program written in Visual Basic, therefore it is only comparable when used within the same programming language. When measures are linked to the rewards system in an organization, source lines of code incents the wrong behavior. One of the goals of a well-structured quality program is to minimize the future maintenance effort that will be needed. If programmers are measured based upon the number of lines of code they write per day, it drives behavior towards writing inefficient and poorly structured programs. It is possible, for instance, in the COBOL language to write a program using a single COBOL word per line. This would yield a very long, barely readable program. Fortunately, however the focus on structured programming caused significant improvement in the quality of software design. The quality improvements resulted in productivity improvements. Structured programming used walk-throughs to formalize the measurement process. Total quality management concepts suggest that quality should be built in, not inspected, therefore as a technique it lacked efficiency. Structured programming however, demonstrated that high quality programs are easier to maintain or enhance. They are operationally more efficient and incur fewer reruns. Software quality measurement encourages programmers to focus on quality software development, and as such was one of the first major developments in the recognition of the value of measurement.

Perhaps the most formal and universally accepted measurement is Function Point Analysis. Function Point Analysis is described herein, in order to provide the foundation by which complexity point measurement was established.

Function Points

Function points are a weighted sum calculation that focuses on measuring end user benefits. Function points measure five different factors of software projects:
- External Inputs
- External Outputs
- Logical Internal Files
- External Interface Files
- External Inquiries

Though there have been several modifications to the function point calculation, they center on the following concepts. First, the number of external inputs and their attributes (data fields) are counted. Inputs can consist of many sources including screens, files, mouse-based inputs, light pen based inputs, or scanned input. The number of attributes in conjunction with the number of inputs creates a complexity index (see Table 1).

The number of external outputs is then counted to determine its complexity index. External outputs consist of reports, invoices, checks, files, EDI transmissions, screens, etc. The complexity index is based upon the number of outputs and their data elements (see Table 2).

Table 1. [External Inputs]

Number of File Types Referenced	Data Elements		
	1-4	5-15	≥ 16
0-1	Low	Low	Average
2	Low	Average	High
≥ 3	Average	High	High

Table 2. [External Outputs]

Number of File Types Referenced	Data Elements		
	1-5	6-19	≥ 20
0-1	Low	Low	Average
2-3	Low	Average	High
≥ 4	Average	High	High

Table 3. [Logical Internal Files]

Number of File Types Referenced	Data Elements		
	1-19	20-50	≥51
0-1	Low	Low	Average
2-5	Low	Average	High
≥6	Average	High	High

Table 4. [External Interfaces]

Number of File Types Referenced	Data Elements		
	1-19	20-50	≥51
0-1	Low	Low	Average
2-5	Low	Average	High
≥6	Average	High	High

Logical internal files and their attributes are then counted to determine their complexity index. While these are not typically apparent to a user, they are essential in creating the end user benefit and can add significantly to the development effort needed. Examples of logical internal files include, disk files, tables within relational databases, or tape files. This index is also based upon the number of files and their attributes (see Table 3).

The complexity index for an external interface is then calculated (see Table 4). External interfaces include shared databases or files that are accessed by other systems.

The number of inquires is calculated based upon its number of input and output components. Examples of inquiries include SQL queries, help messages, or screen inquiries (see Tables 5 and 6 for the index calculation). Due to the variability of the input component, an average is frequently used to determine its complexity.

The complexity indexes are used to derive the weighting factor that will be applied to each of the five counts (see Table 7).

The weighted numbers are then summed to arrive at the initial function point value. This value is then adjusted based upon fourteen influential adjustment factors. These factors include:
- Data Communications
- Distributed Functions
- Performance Objectives
- Heavily Used Configuration
- Transaction Rate
- Online Data Entry
- End User Efficiency
- Online Update
- Complex Processing
- Reusability
- Installation Ease
- Operational Ease
- Multiple Sites
- Facilitate Change

These factors are scored individually and then summed to create a complexity multiplier. The complexity multiplier is derived by multiplying the sum by .01 and adding 0.65 (a complexity constant) to the result. The complexity multiplier is then applied to the sum of the function points to derive the final adjusted function point count. The purpose of the factors is to adjust the measure based upon the technical complexity of the application. Function point measurement can provide a useful baseline measurement to determine the effectiveness of software development and enhancements.

Function points measure systems from the view of functionality to the user, as a result they are independent of the technology utilized. The major benefits of function points (Due) include:
1. Function points are an easily understood measure, which helps communicate them to users.
2. Function points can be used to compare the relative productivity of various information technology tools and techniques.
3. Function points are used to determine the relative size of a software application; this is key to their usage as a productivity measure.
4. Different people can count function points at different times. They are independent of the developer's skill level.

Function points have limitations as well. There are several key measures which function points either don't address or inadequately address. These include, measuring the profitability of the IT investment, measuring the customer satisfaction level, measuring the value that IT adds to the organization and measuring the rate at which rework is completed. Function points have been criticized because of the effort required to count them. There is currently no automated method for counting them. The function point metric is designed for systems that have large amounts of input and output. Systems that

Table 5. [External Inquiries – Input Component]

Number of File Types Referenced	Data 5-15	Elements ≥ 16
0–1	Low	Average
2	Low	High
≥ 3	Average	High

Table 6. [External Inquiries - Output Component]

Number of File Types Referenced	Data 6-19	Elements ≥ 20
0–1	Low	Average
2-3	Low	High
≥ 4	Average	High

Table 7. [Complexity Index]

Parameter	Low	Average	High
External Inputs	X 3	X 4	X 6
External Outputs	X 4	X 5	X 7
Logical Internal File	X 7	X 10	X 15
External Interface	X 5	X 7	X 10
External Inquiry	X 3	X 4	X 6

have intensive logic built to handle edits and processing do not see the value of the logic built into their function point calculation. This can give the appearance that a system is undervalued in terms of its benefits (i.e. function points) per person hour of construction.

Function points are administered by the International Function Point Users Group (IFPUG), a nonprofit association which is comprised of several standing committees. The counting practices committee is responsible for publishing the standards for interpretation of how function points should be defined and counted. The management reporting committee is responsible for defining how function point data should be used for productivity reporting. The new environments committee is responsible for incorporating function point definition into emerging technologies, such as object oriented design. Since its inception, function points have had to undergo continual changes to remain current with the change in development tools and techniques. Function points are the de facto standard for software productivity measurement. There have been numerous other measurement techniques developed, but none have been as generally accepted by the information technology community. Complexity Point Analysis emerged in an attempt to streamline and overcome the difficulty in measuring function points.

Complexity Points

Three years ago the Travelers Property Casualty Corporation Claim Information Systems identified a need to measure the quality of the changes made to their systems. At the time, the belief was that there were problems with the quality of the systems changes and that they were primarily the result of the size of the version implementations. Changes made to the operational claim systems are implemented four times per year in versions. The versions are usually scheduled for March, June, September and November.

In order to establish a metric that would serve as a baseline measure for each systems version and also be easy enough to collect the data so that it would be kept up to date on an ongoing basis, they established the complexity point system. Complexity points are a value that is assigned to each category that is subject to change during a systems version. These categories include program modules, screens, copybooks, tables, databases, JCL procedures, and SQL Queries. Points are assigned to each category based upon whether the item is a new addition or a change to an existing item. New items are assigned substantially higher points due to the extra effort involved in both the coding and testing needed. Category points are then totaled based upon the number of new or changed items times the complexity points. The actual complexity point value was assigned by a team assessment of the relative complexity of each item. The value of the complexity points has remained as a baseline for the evaluation of versions over time. One of the benefits to utilizing this approach is that with *complexity there is simplicity*. In order for the measurement to be effective it must first be easy to administer. The complexity point model takes little time to accumulate the data for the software version and does not require subjective responses that would vary over time. Once the software version has been implemented, defect statistics are maintained. Defects consist of any production problem that occurs between versions. These include both operational errors and user reported discrepancies. It is important to consider both operational and user reported discrepancies. Operational errors may not affect a users perception of quality, particularly in instances where the error is corrected without the user having any knowledge that a error has occurred. Operational errors can affect an information systems organizations overall productivity because of the lost opportunity cost associated with having to respond to operational error. User perception of the quality of a system tends to ignore operational problems that occur in the preparation of information, unless it causes a delay. Quality metrics can therefore be as important an indicator of user perception as quality surveys. The complexity point analysis metrics are easier to administer than surveys.

Complexity points can be accumulated by individual programmer and summed for each version. In addition, a historical baseline by individual can be easily maintained. There is an additional step needed to calculate defect ratio by individual, which is to attribute a defect to an individual support(s). If the defect was caused by more than one change then there won't necessarily a one for one relationship between the individual counts and the version summary. One potential error in this count is that it is not always possible to determine the cause of a defect, particularly operational errors.

One of the shortcomings of the complexity point analysis is in dealing with contract programming staff. In many organizations which employs contract staff, they do so for a single project or version release of software. Complexity point analysis is effective after the initial measurement that serves as the baseline. For contract staff, which only participates in a single version, their individual scores do not go beyond the establishment of the baseline measure.

Complexity point analysis is better adapted to the production support environment than it is to development projects. In a software development project, the longer times associated with requirements definition and detailed design need to be accommodated by establishing a higher value for development complexity points. Once the development version is implemented, however these may no longer be the true measure of productivity for subsequent changes that require new programs to be developed.

The complexity point process takes into account both the staff productivity production support cost and the staff productivity development cost. Staff productivity production support is the rate that enhancements or removal of functionality can be applied to the existing software. Staff productivity development is the rate that new functionality is delivered. The experience level of the staff influences these measures.

Results

Total Quality Management concepts can be applied to the software development process within an information technology organization. Total quality management programs include a customer focused approach, that stresses building quality in rather than achieving it through inspection. In the development of software, inspection is achieved through code reviews or through post-production problem resolution. Effective measurement systems can help to pinpoint the causes of software defects and can be used to measure the effectiveness of software quality control. Measurement systems should point out inefficiencies in development and maintenance. They should be able to identify weaknesses in coding, testing and production.

There are four measures that are commonly derived from the function point calculation in the evaluation of the quality and productivity of results delivered by an information technology organization. Delivery productivity measures the new function points per unit of work effort. It is a measure of the effectiveness of the application development process. Maintenance productivity is the number of function points supported per unit of work effort. This differs from development productivity in that maintenance productivity does not add new functionality for the user. Functional quality measures the functional defects per function point. Technical quality can be measured as the number of technical defects per function point. It is often difficult to distinguish between technical and functional defects. For example, if a system is down because an enhancement caused a disk pack to run out of space, is that a technical defect or a functional defect? There is inconsistency in how this would be counted. However, it should be noted that it points to one of the limitations in using function points, the effort required to properly count them. These measures provide a macro view of the IT organization, but it is difficult to utilize these measures at an individual level, other than through the use of means. As was noted by Randy Mott, Walmart CIO, "Average information yields average decisions". The use of metrics that only point to an average of the staff as a whole ignores several key areas that require further analysis, such as the relationship between staff experience level and defect ratio, and the relationship between programmer type (i.e. systems analyst or technical programmer) and defect ratio. Further study of this area is needed.

Defects per 100 Complexity Points
(past 10 versions)

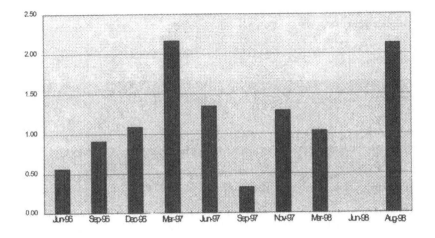

	Jun-96	Sep-96	Dec-96	Mar-97	Jun-97	Sep-97	Nov-97	Mar-98	Jun-98	Aug-98
Complexity	2300	2521	3572	2126	4221	5803	1626	3465	612	12419
Defects	13	23	39	46	57	19	21	36	0	270
Defects/100 Complexity Pts	0.57	0.91	1.09	2.16	1.35	0.33	1.29	1.04	0.00	2.13

Relative Version Complexity
(past 10 versions)

Complexity
Points

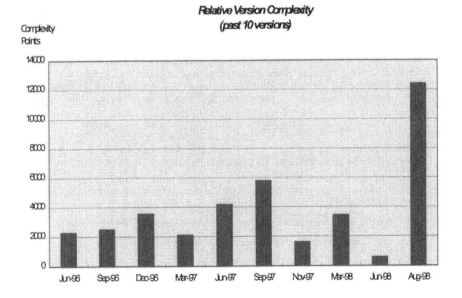

When the complexity point system was established, it was felt that it would show a strong correlation between the size of a software version and the defect ratio. The supposition being that the larger the version, the more likely that mistakes will be made due to the increased integration complexity. Even when the recent, significant Year 2000 remediation version is considered, the correlation coefficient was only 0.54. This suggests that further investigation is needed.

The next step in the effective use of complexity points as a management tool is to implement the use of the tool by programmer within a software version. Over time, this will point out or even influence patterns of behavior that affect the quality of software versions. *What gets measured get managed* can be applied as an influence technique to reinforce the value of ensuring that work is performed correctly the first time.

One of the advantages of the use of complexity points is the ease of data collection. The information that is utilized is readily available by the programmer who was responsible for the support. They are not being asked to perform a new task in order to provide complexity point data for their individual support. The additional benefit of complexity points is that it is easy to collect them by individual and to build a historical database by individual. Historical trending by individual is an excellent means to evaluate the effectiveness of the impact of training and supervisory coaching as well as a gauging how productivity improvements have affected the organization. On April 2, 1996, Travelers Insurance Group purchased the property casualty division of Aetna Life & Casualty. Over the past two years, complexity points have helped to measure the effectiveness of the integration of the two claim systems staffs into one organization. Merges and acquisitions present the additional difficulty of integrating organizations whose cultures and workflow methods may be dissimilar. Measuring productivity can be utilized as a means to integrate the *best of breed* within an organization.

Complexity point analysis does account for the increased difficulty in developing software, as opposed to changing existing software. The scale that is used can vary by organization and therefore be modified to reflect the individual time differences that exist across units. Many information technology organizations undergo continuous change. Change should not be a reason for not measuring effectiveness.

Complexity points do not measure customer satisfaction or IT profitability. However, they can be used as a measure of rework. Complexity points are collected by individual programmer, and as such can be easily documented to identify which support is a new enhancement or a correction to a previous defect. The corrections of defects can further be used to analyze the direct cost of rework to an organization. Cost measures need to be included with any measurement of productivity. Complexity points would be enhanced as a management tool if the cost associated with each support were included as part of the worksheet. This could be easily accomplished by including the person day effort required to complete the support. In order to ensure consistency this should include the time associated with each phase of the system development life cycle.

REFERENCES

Arthur, L. (1985). Measuring Programmer Productivity and Software Quality (1st ed.). New York: John Wiley & Sons.

Brown D., Waldman B., Light M., Revision: The Value of Applications Development Benchmarking, Gartner Group, September 25, 1995.

Dailey, K. (1992). Quality Management for Software (1st ed.). Oxford, England: NCC Blackwell Limited.

Due, R. (1996). The New Productivity Metrics. Information Economics. Fall 1996 60-62.

Guptill, B. Function Points vs. Lines Code: Size Does Matter. Applications Development & Management Strategies, Gartner Group, July 29, 1998.

Jones, C. (1996). Applied Software Measurement (2nd ed.). New York: McGraw-Hill. (Original work published 1991).

Krajewski, L. & Ritzman, L. (1996). Operations Management (4th ed.). New York: Addison Wesley Publishing.

Laboda, D. & Ross, J. Travelers Property Casualty Corporation: Building Object Environment for Greater Competitiveness. Center for Information Systems Research, Sloan School of Management, September 1997, pp.1-20.

Loveman, G. (1988). An Assessment of the Productivity Impact of Information Technologies. MIT Management in the Nineties Working Paper 88-054. Cambridge: Massachusetts Institute of Technology.

Mohan, S. (April 20, 1998). Measuring IT's Productivity. InfoWorld.

Moschella, D. (October 6, 1997). What does Productivity Mean? Computerworld .

Pinsonneault, A & Rivard, S. (1998, September). Information Technology and the Nature of Managerial Work: From the Productivity Paradox to the Icarus Paradox? MIS Quarterly, 22(3), 287 – 308..

Redman, B. How Can Application Developers Become More Productive? Strategic Planning, Gartner Group, September 26, 1997.

Roach, S. (1986). Macrorealities of the Information Economy. National Academy of Sciences, New York.

Ross, J. (1995). Total Quality Management Text, Case and Reading (2nd ed.). Delray Beach, Florida: St. Lucie Press.

Rosser, W. Tunick, D. Rebuilding the IS Organization. Strategic Analysis Report, Gartner Group, August 17, 1994.

Sager, I. (May 5, 1997). US Programmers are Falling off the Pace. Business Week.

Sheppard, M. (1995). Foundations of Software Measurement (1st ed.). Hertfordshire, England: Prentice Hall International.

Wateridge, J. (1998). How can IS/IT projects be measured for success? International Journal of Project Management, 16(1), 59-63.

Willcocks, L. & Lester, S. (1996, June 3). Beyond the IT Productivity Paradox. European Management Journal, 14(3), 279 - 287.

Willcocks, L. & Lester, S. (1997). In search of information technology productivity: Assessment issues. Journal of Operatinal Research Society, 48, 1082 - 1094.

MUSING ON WIENER'S CYBERNETICS AND ITS APPLICATIONS IN THE INFORMATION PROCESSES ANALYSIS

Jacek Unold

Faculty of Management and Computer Science, University of Economics
Wroclaw, ul. Komandorska 118/20, 53-345 Wroclaw, Poland
E-mail: Unold@han.ae.wroc.pl

1. INTRODUCTION

In early 1995, the members of the Group of Seven (G7) leading industrial countries met in Brussels in an attempt to respond to the growth of the information society and the information superhighway on which it would depend [3]. The information superhighway is the name given to the hardware, software and telecomunications infrastructure that allows universities, businesses, government, communities and individuals to share information accros a global network.

The exact nature of its development is very hard to predict, but the aim is outlined very precisely: *„We are commited to the goal of connecting every classroom, every library, every hospital, and every clinic to the national and global infrastructures by the end of this decade"*, as stated by A.Gore, US Vice-President [11]. While there is no agreed view on where the information superhighway will lead, its development has already started, and according to The World Bank: *„a global and highly competitive tele-economy will be born within a generation"* [11].

The information revolution of the 90s, together with the enviromental changes not seen before, make the old, standard ways of management insufficient. The main symptoms of these disadvantageous phenomena are obsolete organizational structures and ineffective ways of gaining information. These problems are especially noticeable on higher levels of management, with the predominance of semistructured and unstructured problems, encountered in indeterministic decision situations.

At the same time we should not forget, that information technology alone, no matter how well developed, is not enough for a business system to work effectively. Analyzing any Information System one must remember about all four levels constituting this area: methodological, organizational, technical and technological. The last two levels, composed of hardware, software and media, are only the top of an iceberg. What still really matters are sound information processes.

So it seems necessary, in the time of such an incredible acceleration in technological revolution, to take at least a quick look at the roots of modern theory of information, and the most important aspects of information processes encountered in management.

Systems Development Methods for Databases, Enterprise Modeling, and Workflow Management,
Edited by W. Wojtkowski, *et al.* Kluwer Academic/Plenum Publishing, New York, 1999.

479

2. THE NOTION OF INFORMATION

The fundamental role in management is played by information. The theoretical formulation of the place and role of information in the environment surrounding us is strictly connected with a human being – his/her existence and perception. This fact prejudges the directions of studies on this reality, the existence of entities, and mutual influence of various processes. The development of the recognition of phenomena surrounding us can be illustrated by significant discoveries and theories, usually analyzed separately. Here, these landmarks are presented in a certain logical sequence, leading explicitly to particular conclusions:

1683 – **matter** and gravitation – Isaac Newton,
1842 – heat and **energy** – Julius R. Mayer,
1877 – probability and **entropy** – Ludwik E. Boltzmann,
1905 – **matter** and **energy** – Albert Einstein,
1949 – **entropy** and **information** – Claude Shannon,
1951 – **matter, energy** and **information** – Norbert Wiener.

This shows that in a certain moment of the historic development of science, information emerges among other physical quantities and stays there for good [1]. Trying to interpret this course of events, one should start with the concept of two XIXth century physicists: Ludwik E. Boltzman, who was the author of basic works in kinetic theory of gases, and Julius R. Mayer, who was the first to formulate the mechanical equivalent for heat. These scientists have radically changed the conception of the world, obligatory since the times of sir Issac Newton. The Newtonian world, described by the classical mechanics, is well determined and strictly organized, where all the future is a result of all the past. The world of Boltzmann and Mayer is a dynamic one, characterized by the statistical distribution, built on the rules of probability. What is more, such a system has a natural tendency to pass from the state of the least to the highest probability, from the state of organization and differentiation to the state of chaos and similarity. The measure of this probability is called *entropy*, and its major feature is constant growth.

The beginning of the modern information theory is connected with Claude Shannon [7]. The conception of *entropy* is equivalent with that of *information*. As entropy is the measure of disorder, information is connected with organization. And, as Einstein identifies *matter* with *energy* ($E=mc^2$), Shannon subjects information to entropy. Since the message before its emission from the source is unknown to the receiver, it is treated as a random quantity. The measure of ignorance (H) is called information entropy (information rate), and it equals the mean quantity of information for a single message:

$$H = - \sum p_k log p_k$$

where: $k=1,...,N$
N – number of potential messages,
p_k – probability of message k emission.

The quantity of information contained in message k equals:

$$I_k = -log p_k$$

This way Shannon developed the basics of the Theory of Information described scientifically by R.W.L.Hartley in 1928.

3. INFORMATION, MANAGEMENT AND WIENER'S CYBERNETICS

Norbert Wiener, who is reputed to have been the father of cybernetics, finds *information* equally important as *matter* and *energy*. During World War II

Wiener made studies on trajectories of ballistic missiles, laying the foundations of the theory of controlling. He considered cybernetics as *"such a field of theory and communication, in which we deal with machines connected with the environment not only by the flow of **energy** or conversion of **matter**, but also by the exchange of **information**"* [12].

Therefore, any given system under control (U) can be described by three vector quantities: Information (I), Matter (M) and Energy (E), as shown in Figure 1.

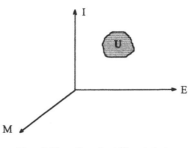

Figure 1. Three-dimensional Wiener's System

When, in Wiener's theory of controlling, a human being is taken into account, we can speak about management. In general, it is a process of making decisions on the ground of received information. This is a necesarry simplification, of course. There are many different definitions of management in literature. In productive units management is understood as controling the material subsystem by the information one, through the set of information and decision fluxes, connecting both subsystems. We encounter here a distinct separation of an information area of a system from a material-energetistic one, according to Wiener's cybernetic model.

However, since there are many units on the market where it is difficult to define a separate material unit (consulting firms, brokerage companies etc.), it seems safe to assume the approach which specifies four basic functions of management:

- planning
- organizing
- motivating
- control [8].

As it will be proven it this article, this classification complies with Wiener's presentation.

Considering all the premises described above we can construct an identification model of a general management system. A model will allow for the hiding of detail and for the concentration on general characteristics of the reality which is being modeled [5].

The model, presented in Figure 3, is a relatively isolated system (U) - working in certain environment (G). It can be a company against a background of the whole sector, a sector against a background of the whole industry, and so on. There are two basic units in the system:

- information (D) and
- material-energetistic (W).

The task of the information unit is to control the flow of material fluxes in the material-energetistic unit along the line of feedback through the set of information and decision fluxes connecting both units. So, the fundamental idea in the presented model is a distinct separation of the information area from the material-energetistic one of the analized system.

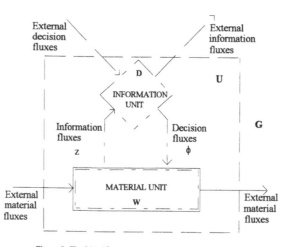

Figure 2. The Identification Model of a Management System

More detailed description of this model allows to define particular blocks composing both basic units and all the oriented fluxes connecting them, as presented in [10]. This is the first possible application of the Wiener's system, taken directly from cybernetics.

4. „THREE-DIMENSIONAL" MODEL OF A DECISION PROCESS

Another interesting conclusion results from a certain analogy of analyzed quantities. Let us notice that information, although very important, is not the only factor in management, and, what is interesting, not the only factor in Information Systems. It seems to have been proven sufficiently by Wiener. Let us take a look at the remaining quantities of the three-dimensional system.

If we limit the conception of *energy* to its mental level, where there is a certain form of transformation of information done by a given person, and which is the result of knowledge, exprerience and initiative, we will get a specific form of *energy*, called **motivation**.

If, speaking about *matter*, we take only the human factor, no matter - individual or collective - which is the basis of any managerial activity, we will obtain a specific shape of *matter*, called **competence**.

Thus, any given state of a management system or, to be more specific, any given state in an information-decision process can be described by three vectors: *I- information, M- motivation, K- competence*, as presented in Figure 3.

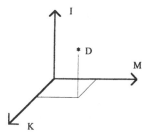

Figure 3. State of a Management System

Each consecutive decision D_t made by a manager is "produced" in a given state of the system, the state, which can be described by (I_t, M_t, K_t), and the index t shows the moment of making the decision. If any of these three vectors in the parentheses lacks its optimum value, one should continue with the analysis of the system. Only later, after completing all the necessary elements, the manager can decide properly and effectively.

This method of controlling the decision process, which means - according to real data, not expected ones, leads us to the idea of a "decision loop". It is the smallest link in the information chain, and it is based on a feedback between two objects [2]. This idea presents a situation, when two blocks are connected with two streams, one of which is an information stream and another - a decision one. This model seems to reverse the normal direction of entropy, and overcomes its natural tendency to grow.

This time the model of the loop is three-dimensional, because not only *information*, but also *competence* and *motivation* is subject to the rule of a feedback.

The higher the set of competence, either individual or collective, the greater importance we attach to the rule of supplement of competence, the fuller and more reliable the information, the longer the list of motivation incentives - the higher probability of making the proper decision.

Let us look in detail at the specific parts of the suggested model.

5. INFORMATION AND COMMUNICATION SKILLS

The first component of the three-dimensional model of a decision process is information. Its main task is to reduce uncertainty in deciding. Within each decision process one may identify this specific circulation of information:

- gathering information considering a given situation,
- information processing,
- making a decision,
- transferring information about the decision to executive units.

In the real world we encounter such decision situations, where information is incomplete and the process of deciding runs under graet uncertainty. The difference between the desired information and the information we actually have is called an information gap. Every bit of information must comply with some specific requirements, to be able to reduce uncertainty of any decision process:

- *adequacy;* information should not only accurately reflect states and events, but also these aspects of the states and events, which are essential for a given situation, and should be properly detailed,

- *authenticity;* one of the reasons of producing false information is acting purposefuly, especially in the system of central management. False information may arise without any bad intention, as well. It may result from erroneous decoding or from errors arising during data processing,
- *topicality;* it is determined by the access rate. It appears, however, that even full information is not enough. According to Iacocca: *"Too many managers admit excessive information charge, especially those overeducated.(...) You have got 95% of necessary information, but it is still not enough for you. Gaining the remaining 5% is going to take you 6 months more. The moment you get it, the facts are out of date.(...) The most important is finding the right moment"* [6].

So, it is not enough to have the information. What is crucial is the skill of making full and quick advantage of it. One should not omit the social aspects in this analysis. We approach the meaning of team work and sound interpersonal contacts. Iacocca puts it explicitly: *"the key to success is people"* [6].

Of course, very important is the area of motivation and competence. What is stressed here, however, is the problem of information flow and its clearness in the team. *"What decides is the skill of comunication with people.(...) It is very important to speak to your listeners in their own language"* [6].

So, the first condition of an effective process of decision making in management is the adequate flow of information and comunicating skills.

6. SET OF INDIVIDUAL AND COLLECTIVE COMPETENCES

The notion of *competence* refers to an individual worker and to a team as well. In the individual case, by this conception we understand education, experience or responsibility. Very often we take into consideration a set of mental attributes affording possibilities for holding a specific position or taking up a given task. Or, quite opposite, eliminating from the game. Another time, this will be a set of physical qualities, especially as far as a "blue collar" is concerned. Sometimes the features are inborn, e.g., good-fellowship, frankness, friendliness, and they are very useful in marketing, to mention but one. In the same marketing one needs some features, that have to be learned: negotiating, the art of instant evaluation of a partner or a situation, etc. Sometimes one may need features that in private life are perceived as negative: obstinacy, inflexibility, secretiveness or extreme individualism. Another time we may seek people who love the team work or even kind of *"a flash in the pan"*, but with thousands of new ideas. They are especially useful in different kinds of *"storm brains"*.

Such a wide understanding of competence leads us to the basic conclusion about the necessity of constant self-improvement and education. By the way, it is compatible with one of the Drucker's famous theses, saying that *"an enterprise is knowledge"* [4].

The set of competence in the range of the whole organization is connected with the problem of getting rid of two basic cases:
- when the competences are recurent,
- when the competences constitute a disjoint set, we have *"empty fields"*, there are some activities that are not fulfilled.

Both cases are very dangerous because, according to Bernard Tapie, an erstwhile well known French businessman and millionaire, *"one can be effective only when is responsible"* [9].

In the first case many duties are not fulfilled because at least two people are responsible for them. One often hears: *"it is not my business, it is his"*. This case also describes a situation, when each of the responsible people makes a different decision. The final result is always poor. The second case is connected with an *"idle run"*, which often paralyses the whole system.

In the optimal system one function results from the other, and the sets of competences of the employees are complementary. The organization chart of the working team should base on this rule. Of course, one should remember about a univocal flow of information, what is stressed here is the set of competences.

It is advantageous when a team consists of pesimists and optimists, young and old, technocrats and financiers. According to great Lee Iacocca: *"Each company needs two sides, constituting a kind of a mathematical equation, as the natural stress between them creates the proper system of mutual control and balance"* [6]. According to the Greiner's theory, there is a close relationship between the set of competences, the stage of growth of the company and the environment in which the company is operating.

So, the next condition of the effective decision process in management is the proper set of competences, individual and collective.

7. MOTIVATION AS A FORM OF WIENER'S ENERGY

The last, but not least, component of the three-dimensional model of an information-decision process is motivation. There are many different definitions of motivation models in psychology and management. Each model has to allow for basic values important for an emploee, values which are attainable in a proffessional career. The most visible and desirable are material goods. But equivalent are:

- accomplishment of proffessional aspirations,
- safety of work,
- esteem of an emploee,
- ethical values.

The more values one can get out of his work, the more concentrated one is going to be on the job.

To regulate the process of motivation in a business one uses the set of rules, methods, ways and forms of procedure, which are called „tools of motivation". There are many of them e.g., the work code, the mode and time of work, work conditions and so on. The motivation area is perhaps the most mysterious one. Not everything can be framed in rules and paragraphs. The real possibilities of a man can be hidden even before himself. Cocteau used to say: *"he did not know it was impossible, so he did it"* [9].

The main task of a manager is reaching those hidden potentials and taking full advantage of them. Quite often independently of regulations and paragraphs. According to Tapie, the question is always to unite people, who differ in social and proffessional background, standard of living, religion, around the main set of values; to find the common denominator, that will identify the affiliation of each one of them to the team. One should create such conditions, as to make co-workers identify with the winners, who symbolize the idea of success. The feeling of affiliation to a strong team creates the high standard and culture of the enterprise. *„Without developing the feeling of pride it is impossible to create strong motivation"* [9].

The last condition of the effective decision process in management is an adequate set of motivation incentives, not necessarily material.

8. CONCLUSIONS

The model of an effective information-decision process in management, which was proposed in this article, constitutes a link between Wiener's three-dimensional cybernetic system *(information, matter* and *energy)*, and Stoner's model of basic functions of management *(planning, organizing, motivating* and *control).*

Planning and *control* are strictly connected with the flow of *information. Organizing* business activity is based on the set of *competences*, individual and collective, here understood as a certain form of Wiener's *matter.* Finally - *motivation*, being a specific form of Wiener's *energy*, referring to mental potential, starting and sustaining the whole business process.

In the time of information revolution there is a severe danger of forgetting about the importance of sound information processes in a business activity. The information technology itself is not enough. As Business Processes Reengineering has proven, new technology imposed upon old, obsolete processes will only harm management in a company. The idea of a three-dimensional „decision loop", derived directly from the cybernetic model of control, may greatly simplify the perception of management problems in a firm, bringing back the adequate proportions of those problems. This process will facilitate decision making, especially on the highest, strategic level of management, with the predominance of indeterministic decision situations.

The presented model was succesfully applied for seven years, during the most difficult time of the system transition in Poland (1989-96), when the author was a CEO of a big housing company.

REFERENCES

[1] Bazewicz M.: *Introduction to Information Systems and Knowldege Representation* (in Polish). Technical University, Wrocław 1993.

[2] Beer S.: *Decision and Control. The Meaning of Operational Research and Management Cybernetics*. J.Wiley and Sons, Chichester 1994.

[3] Black G.: *Complications delay birth*. Financial Times, 3 July 1996.

[4] Drucker P.F.: *The coming of the new organization*. Harvard Business Review, January-February 1998.

[5] Hall V.J., Mosevich J.W.: *Information System Analysis*. Prentice Hall Canada Inc. Scarborough 1988.

[6] Iacocca L.: *Iacocca. An Autobiography* (in Polish). Książka i Wiedza, Warszawa 1991.

[7] Shannon C., Weaver W.: *The Mathematical Theory of Communication*. University of Illinois Press, Urbana 1949.

[8] Stoner J., Freeman R., Gilbert D.: *Management*. Prentice Hall Inc., New York 1995

[9] Tapie B.: *To Win* (in Polish). Tenten, Warszawa 1991.

[10] Unold J.: *On Application of a Matrix Idea in Management Systems Analysis* [in:] G.Wojtkowski et al (ed), *Systems Development Methods for the Next Century*. Plenum Press, New York 1997.

[11] Whitehorn A.: *Multimedia: The Complete Guide*. Dorling Kinderlsley, London 1996.

[12] Wiener N.: *Cybernetics* (in Polish). PWN, Warszawa 1971.

"MAN IS A CREATURE MADE AT THE END OF THE WEEK...WHEN GOD WAS TIRED":
SOME REFLECTIONS ON THE IMPACT OF HUMAN ERROR UPON INFORMATION SYSTEMS

George J. Bakehouse
School of Information Systems
The University of the West of England
Coldharbour Lane, Frenchay
Bristol Bs16 1QY England

ABSTRACT

The notion of Human error has different meanings for many disciplines. Cognitive theorists see them as an important clues to the covert processes underlying routine human action. To applied practitioners they are the main threat to the safe operation of high risk systems. The theoreticians like to collect, cultivate and categorise errors, practitioners are more interested in their elimination and where total elimination is not possible in containing their adverse effects as much as possible. This paper will concentrate on the theoretical and applied practical approaches putting theory into practice. Identification and classification of Human error is essential in the design of information systems whether computer based or manual if the elimination of error is to be achieved. As Mach (1905) so aptly stated " Knowledge and error flow from the same mental sources, only success can tell the one from the other."

Keywords: Human error, action research, modelling, information mismanagement.

INTRODUCTION

This paper will begin with a discussion of the need for modelling techniques in order to give academic disciplines a method for describing the world in relationship to their own particular paradigm. Information systems is one such discipline, in its own brief history it has developed numerous models and techniques in a quest to understand and explain phenomena within its ever widening scope of activity. The second section of the paper will describe and discuss the underlying principles and theories which have assisted the author in developing a model centred upon the notion of human error in helping to explain the occurrence of information problems detected in field research. This model has been developed as a consequence of an ongoing action research program, which spans several sectors including: health, transport, manufacturing, construction and finance. The theories, tools, techniques and methods adopted for the research program were selected on the basis of their relevance to the solution of real problems discovered in everyday working environments.

Systems Development Methods for Databases, Enterprise Modeling, and Workflow Management,
Edited by W. Wojtkowski, *et al.* Kluwer Academic/Plenum Publishing, New York, 1999.

THE NEED TO MODEL

In order to think about, understand and explain the world about us, it is necessary to develop models or abstractions of the world and ways of using them to think about it. These abstractions and approaches then become the epistemological constructs which form the basis of our reasoning, communication and discussion about the world. As new disciplines have emerged, each has developed its own ways of modelling and reasoning about the world. Many providing different descriptions of what is essentially the same set of phenomena. Religion and Philosophy, Politics and Economics, Sociology and Anthropology, all have their own brand of models and approaches, rules and evidence. The differences between them often lead us to the mistaken belief that the phenomena which they study are not the same.

The problem of fragmentation exists because of the differences in the epistemological constructs used to describe the world and its phenomena (observable or otherwise). It is true that to move from one paradigm to another constitutes a fundamental shift and that a lack of isomorphy means that the models of each may not be easily mapped or transformed into an equivalent form in the others. Through the medium of action research, different models may, be considered to be complementary because despite their differences in emphasis, focus and use, they have a common point of contact; the real world of social action. Action research also provides the means by which theory may be developed and tested in the real world.

The notion of improving theory through practice is anything but new. The action research approach, exemplified by Checkland's Soft Systems Methodology, is well established. (Checkland, 1981; Checkland and Scholes, 1990). Conventional research approaches following the paradigm of science (reductionism, experimentation, refutation and repeatability), seek to form and then verify theories. Action research promotes the emergence of theories through practice and the speculation over how such practice might be bettered in some way. In this respect action research is both descriptive/ interpretative and subjective / argumentative. (Galliers et al, 1987).

A clear point of contact for theory and practice and for their different tools and techniques, frameworks and approaches, is the world of information systems. Information systems, due to its inherent practical universal application, is an ideal area to consider this unification, as information system development and use is a natural point of contact for the 'natural' and the 'artificial', the 'hard' and the 'soft', the 'concrete' and the 'abstract' the 'physical' and the 'social'. This undertaking cannot come from abstract academic thought alone. Such a framework needs to be hammered out in the real world of human activity, through the medium of action research (Doyle, 1995).

"The modern specialist field-worker soon recognises that in order to see the facts of savage life, it is necessary to understand the nature of the cultural process. Description cannot be separated from explanation, since in the words of a great physicist, 'explanation is nothing but condensed description.' Every observer should ruthlessly banish from his work conjecture, preconceived assumptions and hypothetical schemes, but not theory". (Malinowski, 1936)

The ongoing action research project currently being undertaken by the author and a team of academics based at the University of the West of England (UWE, Bristol) spanning nearly a decade, has seen the emergence of an approach to embedding the tools and techniques of systems engineering in an action research framework (Bakehouse et al, 1995, 1997, Doyle, 1994, Waters et al, 1994). These projects have involved strategic, tactical and operational systems in transport (Lex Transfleet), health care (Frenchay Health Trust, Neuroscieces) construction (Trafalgar House, WPE Homes), banking (Citibank), manufacturing (Rolls-Royce)and other areas of the private sector. The research team has worked with a wide range of organisations, at a number of different levels of involvement in an attempt to define a general purpose framework of open utility.

INFORMATION ITS DIMENSIONS AND QUALITY

The Industrial Revolution gave us the '4 M's' of men, money, machines and materials, the Green Revolution gave us the environment (the natural world that we have inherited, that we briefly inhabit and that we must conserve for our future generations) and the Computer Revolution gave us information; today, we would re-phrase 'men' as 'people' and 'machines' as technology'. Information is widely regarded as the intangible resource. Seminal works (by Kent 1978, Stamper 1973 and others) warn us of the dangers of dealing with this abstraction whilst in the practical world of commerce, industry and administration the President of the Confederation of British Industry maintains that 'managing information is the greatest challenge facing all organisations today'. The UWE action research program attempts to address the question of how can we interpret this theoretical abstraction of the information resource into the practical reality of 'helping people to get better with information'.

The Impact of Human Error

For several decades the information systems discipline has attempted to model elements of a system, indeed it could be stated that modelling is the core tool of the analyst. Modelling techniques cover, processing, data analysis, entities, entity life histories, business processes, hardware configurations etc. The human beings involved in the systems have to a certain degree been ignored, some categorisation of users i.e. classic casual user through to dedicated user spectrum has been considered. Another approach to involve the human aspect has been in the user involvement / participation throughout the development life cycle.(SSADM 1990, Checkland et al 1990, Mumford 1995)

Somewhat surprisingly many methods of investigation / analysis pay little attention to modelling the most resourceful, flexible, multitasking, multi-media element of the systems under investigation: the Human being (Bakehouse,1996). Unfortunately human beings are prone to making errors, identifying, classifying and modelling these errors may give information systems practitioners another tool/technique that will aid them in the development of better systems. As Mark Twain aptly stated, "Man is a creature made at the end of the week ...... when God was tired."

The odds against error free performance seem overwhelmingly high, as in the majority of situations it would appear that there is a limited number of ways in which a task or process can be successfully carried out. Each process in a planned sequence must have a multitude of possible errors attributable to it. Reason (1991) states "Human error is neither as abundant nor as varied as its vast potential might suggest. Not only are errors much rarer than correct actions, they also tend to take a surprisingly limited number of forms, surprisingly, that is when set against their possible variety". He continued "Although it may be possible to accept that errors are neither as numerous nor as varied as they might first appear, the idea of a predictable error is much harder to swallow. If errors were predictable, we would surely take steps to avoid them". Evidence from the UWE research program suggests that in a complex system involving many people and numerous integrated processes that it was possible to predict the number of errors but the problem is to predict who and at what time an individual error would be made. If it were possible to predict the time and location of an individual error then it would be possible to prevent the error from occurring.

ERRORS: MISTAKES, LAPSES AND SLIPS

Even when the intended actions proceed as planned, they can still be judged as erroneous if they fail to achieve their intended outcome. In this case the problem resides in the adequacy of the plan rather than in the conformity of its constituent actions to some prior intention. Errors of this kind are often referred to as mistakes (Norman 1983, Reason and Mycielska 1982). Norman (1983) " if the intention is not appropriate, this is a mistake. If the action is not what was intended this is a slip. Mistakes involve a mismatch between the prior intention and the intended consequences. For slips and lapses however the discrepancy is between the intended actions and those that were actually executed". For the purposes of this paper lapses, mistakes and slips are all categorised under the generic term error. An error is defined as an action which was taken and consequently lead to a situation where an actor was presented with an information problem. Errors were thus categorised by their own formal characteristics, omission, repetition etc. Initially the UWE research program was not an "error spotting activity" it was essentially a method of identifying information problems, in a number of cases (probably the majority) the error lay dormant for a period of time, that is the error or cause of the information problem was undetected until information was required by the actor. For example in the health sector if a specific attribute of a patient record an x-ray had been omitted or not-replaced in the file this would not be detected as an information problem until the x-ray was required.

A distinction can be made between active errors and latent errors, the former usually associated with the performance of front-line operators, these are likely to have an immediate impact upon the system. The latter most often generated by those at the blunt end of the system may lie dormant for a long time only making there presence felt in the future.

In the study akin to the majority of information systems each and every process could be viewed as front-line, as they all effect another process. In reality complex interactive information systems often have numerous integrated subsystems, an error in one of the subsystems may lay dormant and undetected for a period of time that is until a process "up the line" requires specific data or information.

Related studies and theories

The majority of the published research into human error on the practical end of the continuum has focused on major catastrophes, Three Mile Island 1979, Chernobyl 1986, Herald of Free Enterprise 1987 and numerous other world news events, cited by Reason 1991, to a certain extent quite rightly so. These involved large in-depth studies being carried out "after a specific disaster" in an attempt to establish the cause and prevent any repetitions.

Further studies of identifying human error in a life threatening scenarios are: the US army aviation authority, the recognition of human error as the major contributing factor in the fight against accidents is made explicit. Human Error Accident Reduction Training , HEART (1998), " Human error continues to be the largest factor to plague a commanders Safety Program", The best structured safety programs, such as AR 385-95, do not focus on the factor of human error. The study continued ... there is no magic formula for avoiding accidents and even the most successful safety program can have accidents. However a commander who establishes a safety program should understand the primary cause of accidents are human error therefore the focus of any effective safety program will focus on this human angle. Likewise as information systems are developed more focus should be placed on the human component of the overall system.

Similarly the US army is developing a field known as Behaviour Based Safety BBS (1998), the major principle behind this incentive is focusing intervention on observable behaviour, "always target specific behaviour in order to produce constructive changes" that is identify the personnel that are acting in a certain way and analyse why these actions are taking place, is it because they do not know that their actions are incorrect? do they lack sufficient training? or are they just ignoring specific procedures? Following analysis some advice given to commanders is to, motivate with knowledge of consequences, focus on positive activators and apply continuous interest and evaluation. They sum up by stating " the role of human behaviour in the accident sequence is one that needs more analysis and study".

"Human error is frequently judged to be a primary contributor to high-consequence accidents in complex systems" Rouse (1990) he continued "..... total elimination of human error is a futile pursuit, Instead, systems should be designed so that they are error tolerant in the sense that errors can occur without leading to unacceptable consequences."

Perrow (1984) argues that in complex systems the inevitable human errors and equipment failures have serious results that cannot always be foreseen, especially if the system is tightly coupled, that is, errors can produce results elsewhere in the system with little room for recovery. He concludes that major accidents are so likely that they must be considered normal (as in the expression SNAFU - System Normal, All Fouled Up). This approach is rather cynical and if adopted by information system developers would not encourage the notion of "getting better with information" but on the contrary encourages common expressions such as "what do you expect", "its the systems fault", "it always happens that way" i.e. the not made here syndrome.

A major difference in the above studies and the one being adopted in the UWE research program is the "news catching" consequences. When failure occurs in the former the results are easily and immediately obvious, when errors or failures are detected in information systems they are often overcome, may cause an annoyance, are inconvenient, can be expensive to correct and often cause human misery and suffering.

CLASSIFYING HUMAN ERROR

The author has extended the research method from initially classifying information problems to incorporate an investigation of the root causes (backtracking) and quantifying these in terms of phenotypes within problem classification. The phenotype classification is used as it provides a method of describing erroneous actions as they appear in overt action, how they can be observed, hence the empirical basis for their classification, as opposed to genotypes of erroneous action which takes into account characteristics of the human cognitive system that are assumed to be a contributing cause of the action. Hollnagel (1991), Norman (1983), Reason (1979). A central theme of the UWE research program is the development of a modelling technique that identifies and classifies in terms of human errors, the causes of information problems encountered by actors in a variety of economic sectors. Bakehouse (1998).

Hollnagel presents a Human Computer Interface (HCI) perspective on "human error", in which he refers to erroneous actions; from this theory he derives a modelling framework known as 'The Phenotype of Erroneous Actions'. Hollnagel's theory is founded on the notion of a simple model which consists of a set of sequential actions leading to a goal. Purposeful action is guided by goals and carried out according to a plan. A plan can be defined as "a representation of both goal (together with its intermediate sub-goals) and the possible actions required to achieve it where both are related to some time period which may be stated either very precisely or only in the vaguest terms but which must be specified in some degree " Reason (1979).

The Phenotype of Erroneous actions in its basic form can be summarised by considering a single sequence of actions leading to a goal. Hollnagel classifies the following types: Intrusion, where an additional action is introduced into the sequence. Replacement, an action is replaced by another unplanned action. Omission, an action is omitted. Repetition, an action is repeated unnecessarily. Reversal, actions are not performed in the correct sequence often instead of A then B, B is performed before A. These five simple types do not consider the dimension of Time which in many situations is an important ingredient. The introduction of time into a model means that each step in the sequence can be described both in terms of what and when that action should occur. This then allows two further types to be identified: absence (or delay) of action and unexpected action. The authors research findings highlighted the need for further additional types to be identified: Input demand, Non-replacement, partial omission and non standard. Input demand refers to both manual and especially computer systems that require information to be given before the system allows progression,(in certain cases the information was not known but the users were forced to enter something i.e. a null response was not permitted). Non-replacement refers to an action or event occurring in the correct sequence but the result is removed from the process and not replaced. Partial omission refers to an occurrence of an action or event in the correct sequence which is incomplete, but not identified as incomplete until later in the sequence. Non standard defined data or information that entered the system not in the normal or accepted format. The final classification of phenotypes of erroneous action developed for the study was: Omission, partial omission, non-replacement, non standard, input demand, repetition, sequence and delay. Somewhat surprisingly overall in excess of seventy per cent of the information problems identified and classified during the UWE research program were found to be a direct consequence of human error, which could be modelled in terms of the above phenotypes of erroneous action.

Rasmussen (1982) " it is essential to understand the error characteristics of all the components of a system which of course includes the human beings involved... moreover the phenomenon of human error is not exclusively accounted for by cognitive mechanisms, but is determined by or related to other proximal and distal factors which may be affective, motivational or embedded in organisational and social conditions.... it is extremely important to co-ordinate the resources available to research on human error mechanisms if indeed, it is to stand a chance of influencing the design of systems still to come". The results of the project so far have influenced the processes and procedures of some of the companies which have been involved in the action research program for example: Lex Transfleet changed the format of several "input screens" which were causing problems, confusing output reports were also changed. A tracing system for x-rays and essential documents is being considered by the Health trust, wider education programmes are being initiated focusing on the consequences of errors, error trapping at source has been highlighted etc.

But hindsight, as Fischoff (1975) has demonstrated, does not equal foresight. Simply knowing how past disasters happened does not, of itself, prevent future ones he continued ...however, by combining the knowledge obtained from case studies with a more accurate theory (or theories) of error production, we not only extend our knowledge of cognitive function, we can also begin to assemble a body of principles that, when applied to the design and operation of high-risk technological systems, could reasonably be expected to reduce either the occurrence of errors or their damaging consequences." This notion of prevention can justifiably include information systems, further knowledge and understanding of the causes of information problems may enable the information systems community to deliver better systems.

DISCUSSION

A major objective of IS is to deliver the right information to the right person to support the right activities at the right time in the right place at the right cost with the right quality in the right presentation and with the right availability (in the same sense of Drucker's definition of improving organisational effectiveness and efficiency as 'doing the right thing right'). During the ethnographic period of the action research when members of the team shadowed actors throughout their working day when the actors were using information if any of the rights criteria were not met this was treated as an identification of an information problem, which doubled as a trigger to investigate the cause of the problem. In practice, if people understand and improve upon these dimensions of information then they will 'get better with information'.

The research program has identified, classified and quantified information problems across and between sectors. Investigation and analysis of the root causes of information problems has highlighted the impact of human error upon information systems i.e. they are a major cause of information problems.

SUMMARY

Table 1 presented below is a summary of the analysis of the causes of information problems that can be attributed to human error identified during several extended periods of ethnographic research which exceeds two hundred days in total.

Summary analysis of the causes of information problems in terms of phenotypes of erroneous action , these figures account for approximately seventy per cent of information problems   identified in the UWE research program, the figures quoted are a combined overview from all the various sectors studied.

492

TABLE 1

Percentages of error types as identified in fieldwork

Omission / partial omission	29
Non-replacement	23
Non standard	5
Input demand	2
Repetition	25
Sequence	11
Delay	6

The error modelling technique developed in the research program has value in classifying the initial cause of the information problem, the study suggested that the number and types of information problems detected were to a certain degree predictable, but it was not possible to determine when or who would make a specific error. Similar to many social phenomena   e.g. road accidents and the more surprising suicide (Durkhiem 1951).

Finally, our observational field research forces us to wade through mud, blood, grease and boardrooms wearing hard-hats, surgical greens, blue boiler-suits and city slickers so that we may try to understand the practical realities of information mismanagement.

REFERENCES

Bakehouse G, Davis C, Doyle K, Waters SJ. 1995. Putting Systems Theory into Practice: The Role of Observation in Analysing the Real World. Proceedings UKSS 4th International Conference, Systems Theory and Practice. Pages 631 - 635. Plenum Press, New York.

Bakehouse G. 1996. Developing a Framework for Modelling Information Mismanagement: A Cross Sectoral Study, Proceedings Third European Systems Science Congress, Rome. Pages 1043 - 1047. Ediziona Kappa.

Bakehouse G, Davis C, Doyle KG, Waters SJ. 1997. Anthropological Reflections on Systems Engineering: Seeing is Believing. In Philosophical

Aspects of Information Systems, Winder RL, Probert

SK, Beeson IA (Eds.) Pages 181 - 200.Taylor & Francis.

Bakehouse G. 1997. The Role of Quality Information in Developing Sustainable Systems. Proceedings 14th Annual Conference WACRA, Madrid Spain. Vol 1, Pages 163 - 172.

Bakehouse G. 1998. Research and Reality: Combining Theory and Practise. The Eighth Annual Conference of the International Systems Development Society. Bled, Slovenia.

Behaviour Based Safety BBS 1998. CW4 Rice A. Army Aviation and Behaviour Based Study. www.angelfire.com/al/airsafety/aabbs.

Checkland PB. 1981. Systems Thinking Systems Practice. John Wiley. Chichester.

Checkland P, Scholes J. 1990. Soft Systems Methodology in Action. John Wiley. Chichester.

Doyle KG. 1995. Uniting Systems Theory With Practice. Proceedings UKSS 4th International Conference, Systems Theory and Practice. Pages 297 - 302. Plenum Press, New York.

Durkheim E. 1951. Suicide. Reprinted The Free Press, Glencoe Routledge and Kegan. London.

Fischhoff B. 1975. Hindsight does not equal foresight: The effect of outcome knowledge on judgement under uncertainty. Journal of Experimental Psychology: Human Performance and Perception. 1, 288-299.

Galliers RD, Land FF. 1987, Choosing appropriate information systems research methodologies. Communications of the ACM. Vol 30. No 11. Pages 900 - 902.

Human Error Accident Reduction Training (HEART). 1998. Taking Human Error accidents To Heart. Other Safety Links. www.angelfire.com/al/airsafety/toheart.

Hollnagel E. 1991. The Phenotype of Erroneous Actions: Implications for HCI Design. Weir GRS, Alty JL. (eds).Human Computer Interaction and Complex Systems. AcademicPress. London.

Kent WA. 1978. Data and Reality. North Holland. Amsterdam.

Mach E. 1905. Knowledge and Error. Dordrecht: Reidel Publishing Company

Malinowski B. 1936. Anthropology. The Encyclopaedia Britannica. (first supplementary vol).

Mumford E. 1995. Effective Requirements Analysis and Systems Design: The ETHICS Method. Macmillan. Basingstoke.

Norman DA. 1983. Position Paper on Human Error. NATO Advanced Research Workshop on Human Error. Bellagio, Italy

Perrow C. 1984, Normal Accidents, Basic Books, New York

Rasmussen J. 1982. Human Errors: A Taxonomy for Describing Human Malfunction in Industrial Installations.

Journal of Occupational Accidents.Research Workshop on Human Error. Bellagio. Italy.

Reason J. 1979. Actions not as planned: The price of automation. Underwood G. Stevens S. (eds). Aspects of Consciousness. Psychological Issues. Academic Press London.

Reason JT. 1987 The Chernobyl Errors. Bulletin of the British Psychological Society.

Reason JT, Mycielska K. 1982. Absent-Minded? The Psychology of Mental Lapses and Everyday Errors. Englewood Cliffs. N.J. Prentice-Hall.

Reason JT. 1991. Human Error. Cambridge University Press.

Rouse WB. 1990. Designing for Human Error: Concepts for Error Tolerant Systems. An Approach to Systems Integration. Booher HR (ed). Van Nostrand Reinold. New York.

Stamper R. 1973. Information in Business and Administative Systems. John Wiley.

Structured Systems and Design Methods (SSADM). 1990. Version 4. Reference Manuals. National Computing Centre (NCC). Blackwell.

Waters SJ, Bakehouse G, Davis C, Doyle KG. 1994. Integrated Clinical Workstation: User requirements for a Neurosciences Directorate. National Health Service (NHS) Executive. Information Management Group (IMG).

AUTHOR INDEX

Arnold, Josie, 441-453

Aus, H. M., 357-365

Babka, Otakar, 77-86

Baik, Doo-Kwon, 31-49, 95-105

Bakehouse, George J., 487-494

Barrios, Judith,61-75

Benediktsson, Oddur., 257-262

Burstein, Frada, 279-298

Chester, Myrvin, 375-385

Chae, Jin-Seok, 95-105

Chiou, Yin-Wah, 87-94

Drbohlav, Milan, 199-208

Ezeife, C. I., 51-60

Ferguson, John D., 183-197

Garanito, Laurinda A. G., 77-86

Georges, Grosz, 61-75

Gnaho, Christophe, 61-75

Goldkuhl, Goran, 233-244

Gorman, Michael, 455-463

Gray, Edwin M., 257-262

Harindranath, G., 327-336

Haucke, M., 357-365

Jere, Uroš, 313-325

Klassen, Johanna, 431-440

Král, Jaroslav, 387-398

Krawczyk, Piotr, 405-416

Lee, Sun-jung, 31-49

Lei, Celestino, 77-86

Linger, Henry, 279-298

Lings, Brian, 169-181

Lundell, Björn, 169-181

Luntovskiy, Andrey Oleg, 417-430

Macdonald, Fraser, 183-197

McCarthy, Richard, 465-477

Melin, Ulf, 233-244

Miller, James, 183-197

Moe, Carl Erik, 157-168

Moon, Chang-Joo, 31-49

Moreton, Robert, 349-356, 375-385

Na, Hong-Seok, 95-105

Nilsen, Hallgeir, 157-168

Nowicki, Adam, 367-374

Olszak, Celina M., 399-404

Ørvik, Tore U., 157-168

Osipova, Elena M., 417-430

Ovsyannikov, Eugene K., 417-430

Page, Steve, 263-278

Park, Soo-Hyun, 107-136

Polovina, Rubina, 1-29

Probert, Stephen K., 245-255

Rafaeli, Neto Silvio Luis, 139-155

Repa, Vaclav, 337-348

Rodrigues, Marcos, 139-155

Rogers, Athena, 245-255

Salam, Al F., 299-311

Schumann, Christian, 417-430

Sillince, John A. A., 327-336

Smith, Warren., 257-262

Steimer, A., 357-365

Stone, Derrick, 431-440

Torres, Martinez Diego Ricardo, 209-231

Unold, Jacek, 479-486

Verber, Borut, 313-325

Vigo, Kitty, 441-453

Vogel, Doug, 431-440

Wagner, Karsten, 417-430

Zheng, Jian, 51-60

Zupančič, Jože, 313-325

Zurada, Jozef, 299-311

CPSIA information can be obtained at www.ICGtesting.com
Printed in the USA
LVOW020123241212

313028LV00007B/34/A

9 780306 462993